The Plant Cell Wall

Annual Plant Reviews

A series for researchers and postgraduates in the plant sciences. Each volume in this series will focus on a theme of topical importance and emphasis will be placed on rapid publication.

Titles in the series:

1. Arabidopsis
Edited by M. Anderson and J. Roberts

2. Biochemistry of Plant Secondary Metabolism
Edited by M. Wink

3. Functions of Plant Secondary Metabolites and their Exploitation in Biotechnology
Edited by M. Wink

4. Molecular Plant Pathology
Edited by M. Dickinson and J. Beynon

5. Vacuolar Compartments
Edited by D. G. Robinson and J. C. Rogers

6. Plant Reproduction
Edited by S. D. O'Neill and J. A. Roberts

7. Protein–Protein Interactions in Plant Biology
Edited by M. T. McManus, W. A. Laing and A. C. Allan

8. The Plant Cell Wall
Edited by J. Rose

9. The Golgi Apparatus and the Plant Secretory Pathway
Edited by D. G. Robinson

The Plant Cell Wall

Edited by

JOCELYN K. C. ROSE
Department of Plant Biology
Cornell University
Ithaca, New York
USA

Blackwell
Publishing

CRC Press

© 2003 by Blackwell Publishing Ltd

Editorial offices:
Blackwell Publishing Ltd, 9600 Garsington
Road, Oxford OX4 2DQ, UK
    Tel: +44 (0)1865 776868
Blackwell Publishing Asia Pty Ltd, 550
Swanston Street, Carlton, Victoria 3053,
Australia
    Tel: +61 (0)3 8359 1011

ISBN 1-84127-328-7
ISSN 1460-1494
Originated as Sheffield Academic Press

Published in the USA and Canada (only) by
CRC Press LLC, 2000 Corporate Blvd., N.W.
Boca Raton, FL 33431, USA
Orders from the USA and Canada (only) to
CRC Press LLC

USA and Canada only:
ISBN 0-8493-2811-X
ISSN 1097-7570

First published 2003

Library of Congress Cataloging-in-
Publication Data:
A catalog record for this title is available from
the Library of Congress

British Library Cataloguing-in-Publication
Data:
A catalogue record for this title is available
from the British Library

Set in 10/12 pt Times
by Sparks Computer Solutions Ltd, Oxford
http://www.sparks.co.uk
Printed and bound in Great Britain using
acid-free paper
by MPG Books Ltd, Bodmin, Cornwall

For further information on Blackwell
Publishing, visit our website:
www.blackwellpublishing.com

Contents

5 Towards an understanding of the supramolecular organization of the lignified wall   155
ALAIN-M. BOUDET

6 Plant cell wall biosynthesis: making the bricks   183
MONIKA S. DOBLIN, CLAUDIA E. VERGARA,
STEVE READ, ED NEWBIGIN and ANTONY BACIC

7 WAKs: cell wall associated kinases   223
JEFF RIESE, JOSH NEY and BRUCE D. KOHORN

8 Expansion of the plant cell wall   237
DANIEL J. COSGROVE

9 Cell wall disassembly 264
JOCELYN K. C. ROSE, CARMEN CATALÁ, ZINNIA H.
GONZALEZ-CARRANZA and JEREMY A. ROBERTS

10 Plant cell walls in the post-genomic era 325
WOLF-RÜDIGER SCHEIBLE, SAJID BASHIR
and JOCELYN K. C. ROSE

List of Contributors

Professor Antony Bacic

Plant Cell Biology Research Centre,
School of Botany, University of
Melbourne, VIC 3010, Australia

Dr Sajid Bashir

Department of Plant Biology,
347 Emerson Hall, Cornell University,
Ithaca, New York NY 14853, USA

Professor Alain-M. Boudet

UMR CNRS-UPS 5446,
Pôle de Biotechnologie Végétale,
BP 17 Auzeville, F-341326 Castanet
Tolosan, France

Dr Carmen Catalá

Department of Plant Biology,
Cornell University, Ithaca, NY 14853,
USA

Dr Daniel J. Cosgrove

Department of Biology, 208 Mueller Lab,
Penn State University, University Park,
PA 16802, USA

Dr Monika S. Doblin

Plant Cell Biology Research Centre,
School of Botany, University of
Melbourne, VIC 3010, Australia

Dr Zinnia H. Gonzalez-Carranza

Plant Science Division, School
of Biosciences, University of
Nottingham, Sutton Bonington Campus,
Loughborough, Leicester LE12 5RD, UK

Ms Kim L. Johnson

Plant Cell Biology Research Centre,
School of Botany, University of
Melbourne, VIC 3010, Australia

Dr Brian J. Jones   Plant Cell Biology Research Centre,
                                     School of Botany, University of
                                     Melbourne, VIC 3010, Australia

Dr J. Paul Knox   Centre for Plant Sciences, University of
                                     Leeds, Leeds LS2 9JT, UK

Dr Bruce D. Kohorn   Department of Biology, Bowdoin
                                     College, Brunswick, ME 04011, USA

Dr A. J. MacDougall   Institute of Food Research, Norwich
                                     Research Park, Colney, Norwich
                                     NR4 7UA, UK

Dr V. J. Morris   Institute of Food Research, Norwich
                                     Research Park, Colney, Norwich
                                     NR4 7UA, UK

Dr Ed Newbigin   Plant Cell Biology Research Centre,
                                     School of Botany, University of
                                     Melbourne, VIC 3010, Australia

Mr Josh Ney   Department of Biology, Bowdoin
                                     College, Brunswick, ME 04011, USA

Dr Malcolm A. O'Neill   Complex Carbohydrate Research Center
                                     and Department of Biochemistry and
                                     Molecular Biology, The University of
                                     Georgia, 22 Riverbend Road, Athens,
                                     GA 30602-4712, USA

Dr Steve Read   School of Resource Management and
                                     Forest Science Centre, University of
                                     Melbourne, Creswick, VIC 3363,
                                     Australia

Mr Jeff Riese   Department of Biology, Bowdoin
                                     College, Brunswick, ME 04011, USA

Dr S. G. Ring   Institute of Food Research, Norwich
                                     Research Park, Colney, Norwich
                                     NR4 7UA, UK

Professor Jeremy A. Roberts

Plant Science Division, School
of Biosciences, University of
Nottingham, Sutton Bonington Campus,
Loughborough, Leicester LE12 5RD, UK

Dr Jocelyn K. C. Rose

Department of Plant Biology,
331 Emerson Hall, Cornell University,
Ithaca, New York, NY 14853, USA

Dr Wolf-Rüdiger Scheible

Max-Planck Institute of Molecular Plant
Physiology, Am Mühlenberg 1,
14476 Golm, Germany

Dr Carolyn J. Schultz

Department of Plant Science,
The University of Adelaide, Waite
Campus, Glen Osmond, SA 5064,
Australia

Dr Claudia E. Vergara

Plant Cell Biology Research Centre,
School of Botany, University of
Melbourne, VIC 3010, Australia

Dr William G. T. Willats

Centre for Plant Sciences, University of
Leeds, Leeds LS2 9JT, UK

Dr R. H. Wilson

Institute of Food Research, Norwich
Research Park, Colney, Norwich
NR4 7UA, UK

Dr William S. York

Complex Carbohydrate Research Center
and Department of Biochemistry and
Molecular Biology, The University of
Georgia, 22 Riverbend Road, Athens,
GA 30602-4712, USA

Preface

Plant cell wall research has advanced dramatically on numerous fronts in the last few years, in parallel with many related technical innovations. Analytical tools associated with molecular biology, biochemistry, spectroscopy and microscopy, immunology, genomics and proteomics, have all been brought to bear on elucidating plant cell wall structure and function, providing a degree of resolution that has never been possible before. Furthermore, as an appreciation develops of the critical role of cell walls in a broad range of plant developmental events, so does the strength and diversity of cell wall-related scientific research.

This book, written at professional and reference level, provides the growing number of scientists interested in plant cell walls with an overview of some of the key research areas, and provides a conceptual bridge between the wealth of biochemistry-oriented cell wall literature that has accumulated over the last fifty years, and the technology-driven approaches that have emerged more recently. The timing is especially appropriate, given the recent completion of the first plant genome sequencing projects and our entry into the 'post-genomic' era. Such breakthroughs have given an exciting glimpse into the substantial size and diversity of the families of genes encoding cell wall-related proteins and, as with most areas of biological complexity, the greater the apparent resolution, the greater the number of questions that are subsequently raised. A common approach of the chapters is therefore to provide suggestions and predictions about where each of the fields of wall research is heading and which milestones are likely to be reached.

Due to size limitations, it has not been possible to cover all the areas of cell wall research, and there are several topics that are not addressed here, such as the role of the wall in plant-pathogen interactions and the significance of apoplastic signaling and metabolism. However, this volume illustrates many of the molecular mechanisms underlying wall structure and function.

The first chapter provides an overview of primary cell wall polysaccharide composition and structure – a long-established field but one that remains extraordinarily challenging and open to debate. Developing clearer visions of secondary walls and wall structural proteins, covered in Chapters 4 and 5, respectively, are also formidable goals, and Chapters 2 and 3 describe analytical approaches that promise to help address these challenges. The dynamic multifunctional nature of plant walls, including mechanisms of information exchange with the protoplast, and the exquisite regulation of wall synthesis, restructuring and disassembly, are discussed in subsequent chapters. The volume concludes with a summary of some of the genome-scale approaches that are providing remarkable new opportunities and perspectives on wall biology.

I would like to dedicate this book to Peter Albersheim, whose remarkable insights have continued to drive the field forward and who has mentored and inspired not only this editor but a remarkable number of 'cell-wallers' worldwide.

Jocelyn K.C. Rose

1 The composition and structure of plant primary cell walls

Malcolm A. O'Neill and William S. York

1.1 Introduction

The diversity in shape and size of flowering plants results from the different morphologies of the various cell types that make up the vegetative and reproductive organs of the plant body (Raven *et al.*, 1999; Martin *et al.*, 2001). These cell types may vary in form and often have specialized functions. Nevertheless, they are all derived from undifferentiated cells that are formed in regions known as meristems. Meristematic cells are typically isodiametric and are surrounded by a semi-rigid, polysaccharide-rich matrix (0.1–1 μm thick) that is referred to as a primary wall. This wall is sufficiently strong to resist the internal turgor generated within the cell yet must accommodate controlled, irreversible extension to allow turgor-driven growth (Cosgrove, 1999; see Chapter 8).

Most plant scientists agree that the changes in tissue and organ morphology that occur during plant growth and development result in large part from controlled cell division together with the structural modification and reorganization of wall components, and the synthesis and insertion of new material into the existing wall (Cosgrove, 1999; Rose and Bennett, 1999; Martin *et al.*, 2001; Meijer and Murray, 2001; Smith, 2001). Nevertheless, the biochemical and physical factors that regulate wall modification and expansion are not fully understood (Cosgrove, 1999; see Chapter 8).

Primary walls are the major textural component of many plant-derived foods. The ripening and 'shelf-life' of fruits and vegetables is associated with changes in the structure and organization of primary wall polymers. Fermented fruit products, including wine, contain quantitatively significant amounts of primary wall polysaccharides (Doco *et al.*, 1997). Primary wall polysaccharides are used commercially as gums, gels and stabilizers (Morris and Wilde, 1997). The results of several studies have suggested that primary wall polysaccharides are beneficial to human health as they have the ability to bind heavy metals (Tahiri *et al.*, 2000, 2002), regulate serum cholesterol levels (Terpstra *et al.*, 2002), and stimulate the immune system (Yu *et al.*, 2001a). Thus, the structure and organization of primary wall polysaccharides is of interest to the food processing industry and the nutritionist as well as the plant scientist.

Cell walls have been studied for many years by specialist research groups who often worked in isolation from one another. However, diverse researchers including chemists, biophysicists, biochemists and molecular biologists have begun to join forces even though they may not as yet have a entirely 'common language'. Such multidisciplinary approaches are essential if primary wall structure and function is to be understood in the context of plant growth and development.

In this chapter we briefly review the major structural features of the components of the primary cell walls of dicotyledonous plants. The effects on plant growth and development that result from altering primary wall polysaccharide structures will be discussed. Finally, some of the current models of the organization and architecture of dicotyledon primary walls will be considered in relation to plant cell expansion and differentiation, with particular emphasis on the cellulose–hemicellulose and borate cross-linked pectic networks. This chapter is intended to highlight emerging ideas and concepts rather than provide a simple overview of primary wall structure, as this has been reviewed extensively elsewhere (Albersheim, 1976; Selvendran and O'Neill, 1985; Carpita and Gibeaut, 1993; Ridley *et al.*, 2001).

1.2   Definition of the wall

Both plant and animal cells are composed of a cytoplasm that is bounded by a plasma membrane, but only plant cells are surrounded by a 'wall' (Raven *et al.*, 1999). This wall, which is exterior to the plasma membrane, is itself part of the apoplast. The apoplast, which is largely self-contiguous, contains *everything* that is located between the plasma membrane and the cuticle. Thus, the apoplast includes the primary wall, the middle lamella (a polysaccharide-rich region between primary walls of adjacent cells), intercellular air spaces, water, and solutes. The symplast is another major feature of plant tissues that distinguishes them from their animal counterparts. This self-contiguous phase exists because of the tube-like structures known as plasmodesmata that connect the cytoplasm of adjacent plant cells (Fisher, 2000).

In growing plant tissues the primary wall and middle lamella account for most of the apoplast. Thus, in the broadest sense the wall corresponds to the contents of the apoplast. However, for the purposes of the analytical chemist the wall is the insoluble material that remains after plant tissue or cells have been lysed and then treated with aqueous buffers, organic solvents and enzymes. This isolated wall contains much of the apoplastic content of the tissue but may also contain some cytoplasmic and vacuolar material. Some of the apoplastic material is inevitably lost during the isolation of walls even though it may be a component of the wall *in vivo*.

Several investigators have proposed that the terms 'extracellular matrix' (Roberts, 1989) or 'exocellular matrix' (Wyatt and Carpita, 1993) are more appropriate than 'cell wall' because they suggest a dynamic organelle rather than an inert rigid box. These new terms were not met with universal approval partly because plant scientists have yet to agree on the relationship between a plant cell and its 'wall' (Staehelin, 1991). Nevertheless, this debate did serve to draw the attention of a much wider audience to the biological significance of the 'wall'. Most, if not all, plant scientists now agree that '… walls do not a prison make…' (Roberts, 1994) even though they still '*call a wall a wall*'.

1.3   The composition of the primary cell wall

Primary walls isolated from higher plant tissues and cells are composed predominantly of polysaccharides (up to 90% of the dry weight) together with lesser amounts of structural glycoproteins (2–10%), phenolic esters (<2%), ionically and covalently bound minerals (1–5%), and enzymes. Lignin is a characteristic component of secondary walls and is discussed in Chapter 5 of this book. In living tissue water may account for up to 70% of the volume of a primary wall (Monro *et al.*, 1976).

Twelve different glycosyl residues (Figure 1.1) have been shown to be constituents of all primary walls, albeit in different amounts. These glycosyl residues include

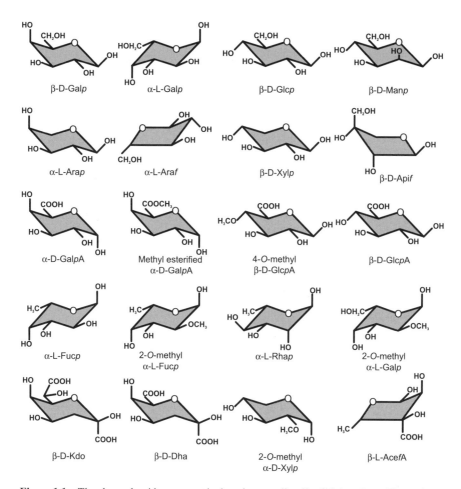

Figure 1.1   The glycosyl residues present in the primary cell walls of higher plants. These glycosyl residues are present, albeit in different amounts, in the primary walls of all higher plants. 2-*O*-Me L-Gal has only been detected in the walls of the fucose-deficient *Arabidopsis* mutant *mur1*.

the hexoses (D-Glc, D-Man, D-Gal and L-Gal), the pentoses (D-Xyl and L-Ara), the 6-deoxy hexoses (L-Rha and L-Fuc), and the hexuronic acids (D-GalA and D-GlcA). D-GalA is present both as the acid and as its C6 methyl esterified derivative. Primary walls contain a branched pentosyl residue (D-Api) and a branched acidic glycosyl residue (3-C-carboxy-5-deoxy-L-xylose; referred to as aceric acid, AceA). Two keto sugars (2-keto-3-deoxy-D-manno-octulosonic acid (Kdo) and 2-keto-3-D-lyxo-heptulosaric acid (Dha)) are also present in primary walls, as are the mono-O-methyl glycosyl residues 2-O-Me L-Fuc, 2-O-Me D-Xyl, and 4-O-Me D-GlcA.

The primary walls of lower plants (hornworts, liverworts, mosses, lycophytes, horsetails, and ferns) have not been studied in detail. Nevertheless, the available data suggests that lower and higher plants have walls with similar glycosyl residue compositions (Popper *et al.*, 2001). Interestingly, the walls of lycophytes, including *Lycopodium pinifolium* and *Selaginella apoda*, have been shown to contain 3-O-Me D-Gal (Popper *et al.*, 2001). This glycosyl residue was not detected in the walls of other lower plants or the walls of gymnosperms and angiosperms, which led the authors to suggest that the presence of 3-O-Me D-Gal is one of the characteristics that uniquely defines the lycophytes.

Hydroxyproline (Hyp) may account for up to 10% of the amino acid content of purified primary walls and is derived from the Hyp-rich glycoproteins that are present in most if not all primary walls (Kieliszewski and Shpak, 2001). In contrast, phenolic residues including ferulate and coumarate are, with the exception of the Caryophyllidae (e.g. spinach and sugar beet), rarely present in the walls of dicotyledons (Ishii, 1997a).

Primary cell walls may contain hydrophobic molecules such as waxes. In addition ions and other inorganic molecules such as silicates may also be present (Epstein, 1999). These quantitatively minor components are often more abundant in specific plants or cell types. For example, silicates are abundant in grasses and seedless vascular plants such as horsetails (*Equisetum*) (Epstein, 1999).

1.4   The macromolecular components of primary walls

Some general features of the polysaccharide composition of primary walls have emerged from the cumulative results of studies over the last 40 years. The walls of angiosperms and gymnosperms are composed of cellulose, hemicelluloses (xyloglucan, glucomannan, or arabinoxylan), and pectic polysaccharides (homogalacturonan, rhamnogalacturonans, and substituted galacturonans) albeit in different amounts (see Table 1.1). There are two general types of wall based on the relative amounts of pectic polysaccharides and the structure and amounts of hemicellulosic polysaccharides. Type I walls (Carpita and Gibeaut, 1993), which typically contain xyloglucan and/or glucomannan and 20–35% pectin, are found in all dicotyledons, the non-graminaceous monocotyledons (e.g. Liliidae) and gymnosperms (e.g.

Douglas fir). Type II walls are present in the Poaceae (e.g. rice and barley) and are rich in arabinoxylan, but contain <10% pectin (Carpita, 1996).

For the purposes of chemical analyses, a primary wall is operationally defined as the insoluble material remaining after a growing plant tissue has been extracted with buffers and organic solvents (Selvendran and O'Neill, 1985; York *et al.*, 1985). An additional treatment with α-amylase to remove starch, which is not a component of the apoplast, may also be required. Pectic polysaccharides are components of the wall solubilized by treatment with aqueous buffers, dilute mineral acids, and calcium chelators. Hemicellulosic wall polysaccharides are often defined as those that are solubilized with strong alkali (Selvendran and O'Neill, 1985). Such chemical treatments may cause partial depolymerization or degradation of the polysaccharides and often result in the solubilization of complex mixtures of different polysaccharides. The problems associated with solubilizing wall components with chemical extractants can be overcome to a large extent by using homogenous glycanases that cleave specific glycosidic bonds and thereby selectively solubilize specific polysaccharide classes (York *et al.*, 1985). For example, treating walls with endopolygalacturonase solubilizes material rich in pectic polysaccharides whereas oligosaccharide fragments of hemicellulosic polysaccharides are solubilized by treating walls with glycanases that include endoglucanase, endomannanase, and endoxylanase. A combination of glycanase treatments and chemical extractants are used in many cell wall studies. Nevertheless, pectic and hemicellulosic polysaccharides may not be completely solubilized by these treatments, which has led to the suggestion that some of these polymers are covalently linked to or entrapped within cellulose fibres.

A primary wall can be analysed in situ or after it has been isolated and purified using solid state NMR spectroscopy, Fourier transform infrared spectroscopy, atomic force microscopy (see Chapter 2), and immunocytochemistry (see Chapter 3). These techniques have begun to yield new information on the physical properties of wall polymers, the organization of polymers within a wall, and the distribution of polysaccharides and glycoproteins in the walls of different cells and tissues. Such techniques when combined with improvements in conventional wall analysis now provide the investigator with a powerful battery of experimental approaches to probe primary wall composition, organization, and function.

1.5 Determination of the structures of primary wall polysaccharides

The ultimate goal for the characterization of primary wall polysaccharides is to relate the primary structures of these molecules to their three-dimensional conformations, their physical and dynamic properties, their interactions with themselves and other polymers in the wall, and their biological functions.

Table 1.1 The principal polymeric components of primary cell walls.

Polymer	% of wall (dry wt)[1]		Backbone	Side chain glycose(s)	Non-carbohydrates
	Type I[2]	Type II[3]			
Cellulose	20–30		1,4 β-D-Glcp	none detected	none detected
Hemicellulose					
Xyloglucan	10–20	1–5	1,4 β-D-Glcp	α-D-Xylp, β-D-Galp, α-L-Fucp, β-L-Araf	acetyl esters
Glucomannan	5–10	nd	1,4-β-D-Manp 1,4-β-D-Glcp	α-D-Galp	
Xylan	~5	20–40	1,4-β-D-Xylp	α-L-Araf, β-L-GlcpA, 4-O-Me β-D-GlcpA	acetyl, feruloyl and coumaroyl esters
Mixed-linkage Glucan	absent	10–30	1,4- and 1,3-β-D-Glcp	none detected	none detected
Callose[4]	variable if present		1,3 β-D-Glcp	none detected	none detected
Pectin					
Homogalacturonan	15–20	1–5	1,4 α-D-GalpA	none detected	methyl and acetyl esters
Substituted galacturonans Rhamnogalacturonan II	2–5	1	1,4 α-D-GalpA	α-L-Rhap, β-L-Rhap, β-L-Araf, α-L-Arap, β-D-Galp, α-L-AcefA, β-D-Apif, α-L-Fucp, α-D-GalpA, β-D-GalpA, β-D-GlcpA, 2-O-Me α-L-Fucp, β-D-KdopA, 2-O-Me α-D-Xylp, β-D-DhapA	acetyl esters borate diester

Xylogalacturonan	5	nd	1,4 α-D-GalpA	β-D-Xylp
Apiogalacturonan[5]	nd	20	1,4 α-D-GalpA	β-D-Apif
Rhamnogalacturonans				
Rhamnogalacturonan I	10–15	2	1,4 α-D-GalpA and 1,2 α-L-Rhap	α-L-Araf, β-D-Galp, α-L-Fucp, β-D-GlcpA, 4-O-Me β-D-GlcpA, acetyl esters
Structural glycoproteins[6]				
Hydroxyproline-rich protein	1–10	1	Hyp and Ser-rich protein	α-L-Araf, β-L-Araf, α-D-Galp
Glycine-rich protein	1	1	Gly-rich protein	α-L-Araf, β-L-Araf, α-D-Galp
Threonine-rich protein	?	?	Thr-rich protein	L-Araf, β-L-Araf, α-D-Galp
Proteoglycans				
Arabinogalactan proteins	+	+	Hyp-rich and Hyp-poor	β-D-Galp, α-L-Araf, β-D-GlcpA, α-L-Fucp
Phenolic esters	0–3	0–5	feruloyl and coumaroyl esters linked to xylan and pectins	

[1] The amounts of a particular wall component may vary depending on the plant and tissue type.
[2] Type I walls are found in dicots, non-graminaceous monocots, and gymnosperms.
[3] Type II walls are found in the *poaceae*.
[4] Callose is usually absent from the primary wall but is typically present in the developing cell plate and is often formed after wounding of plant cells.
[5] Limited to *Lemnacea* and *Zosteracea*.
[6] Primary walls also contain quantitatively small amounts of non-structural protein including enzymes and 'expansin'.

The primary sequence of a polysaccharide is known when the following have been determined:

1. The quantitative glycosyl residue composition;
2. The absolute configuration (D or L) of each glycosyl residue;
3. The ring form (furanose or pyranose) of each glycosyl residue;
4. The linkages (1→3, 1→4, etc.) of the glycosyl linkages;
5. The anomeric configuration (α or β) of each glycosyl residue;
6. The sequence of glycosyl residues;
7. The location of non-carbohydrate substituents (e.g. O-acetyl esters)

Numerous detailed methods have been described for the determination of points 1–7 and have been described elsewhere (Aspinall, 1982; McNeil et al., 1982a; van Halbeek, 1994).

Determining the primary sequence of a polysaccharide, unlike nucleic acids and proteins, often requires a considerable amount of time and effort together with the use of sophisticated and expensive equipment. No single universally applicable method has been developed to date for glycosyl sequencing of a polysaccharide. Moreover, only an average structure can be deduced for some primary wall polymers because they are highly branched and not composed of discrete oligosaccharide repeating units (Stephen, 1982). Nevertheless, a wealth of structural information can now be obtained when mass spectrometry and nuclear magnetic resonance spectroscopy are used together with chemical and enzymic fragmentation of a polysaccharide.

1.5.1 Mass spectrometry

Mass spectrometry, by virtue of its ability to measure the mass of a molecule or of well-defined fragments of the molecule, provides information on the composition and glycosyl sequence of oligosaccharides. Each different glycosyl residue (e.g. hexose, pentose, and uronic acid) contributes a characteristic mass to the glycan in which it resides. However, mass spectral data can rarely be interpreted in the absence of glycosyl residue composition data because structurally distinct glycosyl residues often have the same mass. For example, all hexoses (Glc, Gal, Man, etc.) contribute a mass of 162 Da to the glycan.

Glycosyl sequence information is obtained by analysing fragment ions generated in the MS source itself or by tandem MS. The advantage of tandem MS is that parent ions with a specific mass to charge (m/z) ratio can be selected for fragmentation (either spontaneous or induced by collision with gas molecules or atoms) to produce a daughter ion spectrum. The major fragmentation pathways are well characterized, allowing the glycosyl sequence to be derived from the daughter ion spectrum (Domon and Costello, 1988). Nevertheless, unambiguous determination of the glycosyl sequence is not always possible, due to the mass degeneracy of isomeric glycosyl residues (e.g. Gal and Glc), molecular rearrangements, or the generation of 'inner fragments' by multiple cleavage processes (Reinhold et al., 1995). The likelihood that specific, well-characterized fragmentation reactions will occur is often facilitated by converting the oligosaccharide to its per-*O*-acetylated or per-*O*-methylated derivative, which can also reduce the complexity of the daughter ion spectra. In general, the structural information provided by mass spectral analysis depends on the ionization technique and the physical properties of the glycan being analysed.

1.5.1.1 Matrix-assisted laser-desorption ionization (MALDI) with time-of-flight (TOF) mass analysis

Matrix-assisted laser-desorption ionization with time-of-flight mass spectrometry (MALDI-TOF-MS) can provide both molecular weight and sequence information (Harvey, 1999). MALDI-TOF has been successfully used to analyse neutral oligosaccharides but rarely give high quality spectra of anionic oligosaccharides (e.g. pectic fragments) (Jacobs and Dahlman, 2001). A MALDI-TOF spectrum is obtained by applying a solution containing the analyte and a UV-absorbing matrix (such as dihydroxy benzoic acid, DHB), onto a metal target and then concentrating it to dryness. The target is introduced into the spectrometer and irradiated with brief (nanosecond) pulses of ultraviolet laser light. The matrix efficiently absorbs the laser's energy and heats up rapidly, thereby vaporizing itself and the analyte within a small area on the target. The vaporized analyte molecules are ionized in this process. The target is held at a high positive voltage, so that the positively charged ions that are generated are accelerated away from the target by electrostatic forces.

The TOF mass analyser consists of an evacuated tube with a detector at the end (Mamyrin, 2001). The m/z ratio for an ion is determined by measuring the time between the laser pulse and the arrival of the ion at the detector. More massive ions travel more slowly and take more time to reach the detector. A reflectron (ion mirror) is incorporated into more sophisticated TOF instruments and compensates for slight differences in the kinetic energy of the ions and improves the resolution of the spectrometer (Mamyrin, 2001). The reflectron is also used to separate fragment ions that are formed after a parent ion has exited from the ion source. This makes it possible to obtain sequence-specific data using a technique called MALDI-TOF with post-source decay (PSD) (Harvey, 1999). Daughter ions formed by PSD have the same velocity as the parent ion, but different momenta, and are separated from the parent ion and from each other by the reflectron. PSD analysis is thus a type of tandem MS that selects and analyses a set of daughter ions originating from a parent ion having a specific m/z ratio. Sequence information can often be obtained by PSD analysis, as multiple fragmentation and molecular rearrangement processes can be minimized, due to the relatively short residence time of ions in the analyser.

1.5.1.2 Electrospray ionization (ESI)

Electrospray ionization mass spectrometry (ESI-MS) (Griffiths *et al.*, 2001) is often the method of choice when analysing anionic oligosaccharides such as pectic fragments. Unlike most other ionization techniques, ESI occurs at atmospheric pressure and is often referred to as atmospheric pressure ionization (API). An ESI mass spectrum is obtained by introducing a solution (usually aqueous) containing the oligosaccharide into the ion-source through a capillary tube, which itself is held at high voltage. Small positively charged droplets are ejected from the tip of the capillary tube. A drying gas (usually warm N_2) is passed over the droplets, evaporating the solvent. Analyte molecules in the droplet are progressively desolvated, until electrostatic forces within the droplet eject ionized analyte molecules. A small orifice allows the ions to enter the spectrometer's mass analyser, which is kept under high vacuum and at a relatively low electric potential. The ions are guided through the orifice and accelerated by electrostatic forces. ESI, in contrast to MALDI, is a continuous rather than a pulsed-ion generation method, so it is not convenient to use a TOF mass analyser. Rather, a scanning mass analyser (e.g. a quadrupole or field sector) is used in conjunction with ESI to determine the mass of the analyte. A scanning mass analyser filters out all ions except those with an m/z ratio that lie within a very narrow range. This mass window is moved over time (scanned), so ions with different m/z ratios will be detected at different times during the scan. Scanning mass analysers produce high quality mass spectra, but are less sensitive than TOF analysers, as only a small portion of the ions being generated reach the detector. TOF mass analysers allow the detector to 'see' virtually all of the ions that make it out of the ionization source.

Tandem MS techniques are also used with ESI. Daughter ions are usually generated in a collision cell, where the selected parent ion collides with gas molecules and breaks into fragments. However, unambiguous glycosyl sequence information can be difficult to obtain by tandem ESI-MS, because the ions have a relatively long residence time in the analyser, providing more opportunity for molecular rearrangement or multiple fragmentation processes.

1.5.1.3   Fast-atom bombardment mass spectrometry (FAB-MS)

Fast-atom bombardment mass spectrometry (FAB-MS) involves dissolving the glycan in a liquid matrix (e.g. glycerol), which is then introduced into the ion source and bombarded with atoms that have been accelerated by an atom gun. Kinetic energy is transferred to the liquid matrix, and some of the analyte at the surface of the matrix is vaporized/ionized (Dell, 1987; Dell and Morris, 2001). Typically, the resulting ions are singly charged and continuously generated. Mass analysis is usually performed using scanning techniques, which are not generally well suited for the analysis of ions with high m/z values. Therefore, FAB-MS usually provides molecular weight and sequence information only for oligosaccharides with molecular weights less than 3 kDa. Either in-source fragmentation (for pure compounds) or tandem MS can be used to obtain sequence information.

1.5.2   Nuclear magnetic resonance spectroscopy (NMR)

Nuclear magnetic resonance spectroscopy (NMR) is a non-destructive technique that, in principle, allows the complete structural characterization of an oligosaccharide (Duus et al., 2000). The structural assignment is based on analysis of several spectroscopic parameters for each magnetically active nucleus (e.g. ^{1}H or ^{13}C) in the glycan. A nucleus can be identified by its resonance frequency (chemical shift), which depends on its molecular environment. For example, protons attached to the anomeric carbon (C1), which itself is directly attached to two electronegative oxygen atoms, are readily distinguished from protons attached to other sugar-ring carbons, which have only one directly attached oxygen atom. Magnetic nuclei in a glycan interact with each other, and are thereby 'magnetically coupled'. This coupling arises by different mechanisms. Direct (dipolar) coupling provides information regarding distances between nuclei (e.g. by analysis of the nuclear Overhauser effect) (Neuhaus and Williamson, 1989) and molecular geometry (e.g. by measurement of 'residual' dipolar coupling in partially aligned molecules) (Prestegard and Kishore, 2001). Indirect, electron-mediated (scalar) coupling gives rise to the familiar splitting of resonances in ^{1}H-NMR spectra, and provides information regarding the geometry of the molecular bond networks connecting the coupled nuclei (Bush et al., 1999). Analysis of these magnetic phenomena should allow the complete structure of the glycan to be determined. However, the complete, unambiguous structural analysis of a complex glycan by NMR is not always possible, due to factors such as signal overlap, higher order coupling effects, and the effects of conformational dynamics, which can lead to line broadening and increased spectral complexity. Furthermore, a complete determination of a glycan's primary structure typically requires a highly purified sample, although NMR analysis of mixtures can provide a significant amount of structural information.

The one- and two-dimensional NMR techniques commonly used for determining a complete primary structure require approximately 1 micromole of *pure* oligosaccharide. This criterion is often difficult to meet, especially with wall polysaccharides isolated from small amounts of a specific tissue or cell type, or when analysing a large number of different oligosaccharides generated by chemical or enzymic fragmentation of a complex polysaccharide. The sensitivity problem becomes more acute for commonly used heteronuclear NMR experiments including HSQC (Bodenhausen and Ruben, 1980) and HMBC (Bax and Summers, 1986) that involve 'dilute' nuclei such as ^{13}C, which has a natural abundance of only 1.1%.

The sensitivity of an NMR experiment can be increased by isotopic enrichment. For a *fixed* sampling time, the NMR signal (S) increases linearly with the concentration of magnetically active nuclei. Thus, ^{13}C-enrichment may decrease the minimum sample requirement by almost 100 fold. Isotopic enrichment also reduces the spectrometer time required to analyse a sample. For a heteronuclear (^{1}H-^{13}C) NMR experiment, doubling the number of ^{13}C atoms produces the same S in half the time (t). But decreasing the sampling time also decreases the noise (N), which is proportional to $\sqrt{t}$. Taking this noise reduction into account, a doubling of the concentration of ^{13}C atoms makes it possible to obtain the same signal to noise (S/N) in one-fourth of the time. Extending this logic further, it would require 8264 times as long (i.e. (100% ÷ 1.1%)^2) to obtain a given S/N for a natural abundance sample than it would for the same sample that was 100% ^{13}C-enriched. Thus, an experiment that requires 2 hours of instrument time for a 100% ^{13}C-enriched sample would take 1.88 years for the natural abundance sample.

Plants are photosynthetic organisms, and cell walls that are enriched in ^{13}C content can be obtained from plants grown in an ^{13}C-enriched atmosphere. One of the authors of this chapter (W.S. York) has constructed a growth chamber that is routinely used to produce ^{13}C-enriched plant cell walls and cell wall polysaccharides.

1.5.2.1 The structural reporter approach and spectral databases

The insensitivity of NMR and difficulties in separating complex mixtures of oligosaccharides can be overcome to some extent by using the 'structural reporter' approach (Vliegenthart *et al.*, 1983) that was originally developed for the ^{1}H-NMR spectroscopic analysis of N-linked glycans. This technique only requires material in amounts sufficient to record a one-dimensional ^{1}H-NMR spectrum. The oligosaccharides are identified by virtue of the correlation of structural features to specific, well-characterized resonances in their ^{1}H-NMR spectrum. Using this approach, one can obtain, for example, quantitative information regarding the identity and linkage patterns of the oligosaccharide components of a mixture (i.e. a non-destructive 'glycosyl linkage analysis'). This type of linkage analysis can be more quantitatively accurate than chemical glycosyl linkage analysis as it depends only on the correct identification and integration of NMR resonances and does not depend on the completeness of chemical reactions. The structural reporter method can provide a complete determination of the primary structure of a pure oligosaccharide, even if the oligosaccharide has not been previously characterized. However, care must be exercised when assigning structures to a new oligosaccharide by this method, as the inference of structural information is based solely on correlations between structural features and chemical shifts, which may vary significantly in different overall molecular environments.

The characterization of oligosaccharides using the structural reporter approach requires a database containing NMR chemical shift data for many (usually more than 20) rigorously characterized oligosaccharides. For example, a database for the endoglucanase-generated oligosaccharide subunits of xyloglucans is available at the Complex Carbohydrate Research Center (http://www.ccrc.uga.edu/web/specdb/nmr/xg/xgnmr.html). The ^{1}H-NMR spectra of these oligosaccharides are simplified by chemically (sodium borohydride reduction) converting the glucose residues at the reducing termini into glucitol. The anomeric proton resonances of the resulting oligoglycosyl alditols are resolved from the other resonances, making them especially useful for rapid structural determination by NMR. To a first approximation, the chemical shifts of anomeric resonances in the NMR spectra of xyloglucan oligoglycosyl alditols depend on a few, well-defined parameters (York *et al.*, 1989, 1993, 1994, 1996; Hisamatsu *et al.*, 1992; Hantus *et al.*, 1997; Vierhuis *et al.*, 2001).

1. The identity, anomeric configuration, and linkage of the sugar residue (e.g. a 4,6-linked β-D-Glc*p*) in which the anomeric proton is located.

2. The identity of the substructure containing the sugar residue. (In this context, a substructure comprises a backbone Glcp residue and its pendant side chain(s), as represented by X, L, F, G, S; see Figure 1.2.)
3. The environment of the substructure containing the sugar residue, including end effects arising from the proximity of the residue to the non-reducing or alditol end of the oligomer and the presence of other side chains in the immediate vicinity.

The xyloglucan NMR database at the CCRC was developed expressly so that it could be searched by specifying these and other structural parameters that are characteristic of xyloglucan oligosaccharides. Other carbohydrate NMR databases, including 'sugabase' (http://www.boc.chem.uu.nl/sugabase/sugabase.html), are organized somewhat differently.

1.6   Oligosaccharide profiling of cell wall polysaccharides

Cell wall polysaccharides that are composed of a limited number of discrete oligosaccharide subunits can in principle be characterized by determining the identity and relative proportion of each subunit. A polysaccharide isolated from a new source can be rapidly characterized by this procedure providing that:

1. it is fragmented into subunits by an endolytic enzyme;
2. chromatographic methods to separate and identify each subunit have been developed; and
3. the structures of the most abundant subunits are known.

This procedure has been successfully used to characterize xyloglucans, where methods to separate the native oligosaccharides (by high-performance anion-exchange chromatography) and their UV-absorbing derivatives (by reversed-phase chromatography) have been developed (Pauly et al., 1999a, 2001a, b). Chromatographic analysis requires much less material than NMR spectroscopic analysis and provides a quantitative estimation of the relative amount of each oligosaccharide. In addition, chromatographic profiling can, depending on the derivatization and/or chromatographic methods used, provide information regarding the relative amounts of xyloglucan oligosaccharides that differ only in the number or position of O-acetyl substituents (Pauly et al., 2001a, b).

Oligosaccharide profiling analysis has made it possible to characterize structural differences in xyloglucans isolated from different tissues of the same plant or different 'domains' of the xyloglucan polymer within the cell wall (Pauly et al., 1999a). Such an approach could also be used to determine the relative amounts of structural subunits of any complex polysaccharide including methyl esterified pectins, rhamnogalacturonans, and substituted galacturonans. Indeed, oligosaccharide profiling in combination with MALDI-TOF and ESI-MS has been used to examine the distribution of methyl esters in commercial and cell wall-derived pectins (Daas *et al.*, 1998; Limberg *et al.*, 2000). Nevertheless, oligosaccharide profiling of pectic polysaccharides has not been exploited to its fullest extent because of the lack of homogeneous endoglycanases that efficiently fragment the backbone of the naturally occurring polysaccharides. For example, the rhamnogalacturonan backbone is not fragmented by the currently available hydrolases and lyases unless many of the oligosaccharide side chains have been enzymically or chemically removed (Azadi *et al.*, 1995). No homogeneous endoglycanases are available that fragment the RG-II backbone.

1.7   The structures of the polysaccharide components of primary walls

1.7.1   The hemicellulosic polysaccharides

Hemicelluloses are operationally defined as those plant cell wall polysaccharides that are not solubilized by hot water or chelating agents, but are solubilized by aqueous alkali. According to this definition, the hemicelluloses include xyloglucan, xylans (including glucuronoxylan, arabinoxylan, glucuronoarabinoxylan), mannans (including glucomannan, galactomannan, galactoglucomannan), and arabinogalactan. Hemicelluloses may also be defined chemically as plant cell wall polysaccharides (usually branched) that are structurally homologous to cellulose, in that they have a backbone composed of 1,4-linked β-D-pyranosyl residues such as glucose, mannose, and xylose, in which O4 is in the equatorial orientation. Xyloglucan, xylans, and mannans but not arabinogalactan are included under this chemical definition of hemicelluloses. The structural similarity between hemicellulose and cellulose most likely gives rise to a conformational homology that can lead to a strong, noncovalent association of the hemicellulose with cellulose microfibrils.

1.7.2   Xyloglucan

Xyloglucan is the most abundant hemicellulosic polysaccharide in the primary cell walls of non-graminaceous plants, often comprising 20% of the dry mass of the wall. Xyloglucan has a 'cellulosic' backbone consisting of 1,4-linked β-D-Glc*p* residues. Up to 75% of the backbone residues are branched, bearing α-D-Xyl*p* residues at O6. Many of the Xyl*p* residues bear glycosyl substituents at O2, thereby extending the side chain (Figure 1.2). The cellulosic backbone itself does not vary among xyloglucans from different plant species and tissues and only a limited number of xyloglucan side chain structures have been described. Therefore the structure of a xyloglucan molecule can be completely and unambiguously described by listing, in order, the pattern of side chain substitution for each Glc*p* residue in the backbone (see Figure 1.2; Fry *et al.*, 1993). For example, an uppercase G designates an unbranched Glc*p* residue and a Glc*p* residue bearing a single α-D-Xyl*p* residue at O6 is designated by an uppercase X. A Glc*p* residue bearing the trisaccharide α-L-Fuc*p*-(1,2)-β-D-Gal*p*-

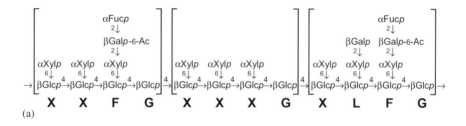

(a)

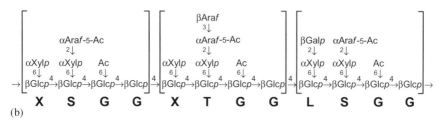

(b)

Figure 1.2 Primary structures of xyloglucans. (a) A representative structure of xyloglucan that is present in the primary cell walls of most higher plants (other than the Poaceae, Solanaceae, and Lamiaceae). (b) A representative structure of the xyloglucan that is present in the primary cell walls of plants in the family Solanaceae. The oligosaccharide fragments indicated by brackets [ ] are generated by endoglucanase treatment of the xyloglucan. This enzyme hydrolyses the glycosidic bond of those 4-linked β-D-glucosyl residues that are not substituted at O6.

$(1,2)$-α-D-Xyl$p$ at O6 is designated by an uppercase F. Thus, the most commonly occurring, fucose-containing xyloglucan sequence is XXFG (Figure 1.2).

Xyloglucans are classified as 'XXXG-type' or 'XXGG-type' based on the number of backbone glucosyl residues that are branched (Vincken *et al.*, 1997). XXXG-type xyloglucans have three consecutive backbone residues bearing an α-D-Xyl$p$ substituent at O6 and a fourth, unbranched backbone residue. In XXGG-type xyloglucans, two consecutive backbone residues bear an α-D-Xyl$p$ substituent at O6, and the third and fourth backbone residues are not branched. The glycosidic bond of the unbranched Glc$p$ residue in XXXG-type xyloglucans is cleaved by many endo-β-1,4-glucanases. Thus, endoglucanase-treatment of XXXG-type xyloglucans typically generates a well-defined set of oligosaccharide fragments that have a tetraglucosyl backbone (Figure 1.2). In contrast, the glycosidic bonds of both unbranched Glc$p$ residues of XXGG-type xyloglucans can be hydrolysed by endoglucanases. However, the type and the amount of the oligosaccharide fragments that are generated depends on the substrate specificity of the endoglucanase and on the presence or absence of acetyl substituents at O6 of some of the unbranched Glc$p$ residues.

1.7.3 Variation of xyloglucan structure in dicotyledons and monocotyledons

The major structural features of primary wall polymers are generally conserved among higher plants, although some structural variation is observed in different plant species, tissues, cell-types, and perhaps even in different parts of the wall surrounding an individual cell (Freshour *et al.*, 1996). Xyloglucans are the most thoroughly characterized cell wall polysaccharides, other than cellulose, and their general structure is conserved among most higher plants (Figure 1.2). The data available to date indicate that fucosylated xyloglucans with XXXG-type structure, in which four subunits (XXXG, XXFG, XLFG, and XXLG) constitute the majority of the polymer, are present in the primary walls of gymnosperms, a wide range of dicotyledonous plants and all monocotyledonous plants with the exception of the Poaceae (Figure 1.3, taxonomy). The xyloglucans synthesized by the Poaceae contain little or no fucose and are less branched than dicotyledon xyloglucans. These xyloglucans have not been as thoroughly characterized as dicotyledon xyloglucans, although the available evidence suggest that they have an XXGG-type structure.

Species-specific variation of xyloglucan structure is evident in the Asteridae, a dicotyledon subclass that includes the families Solanaceae and Oleaceae. Many of the Asteridae produce xyloglucans that contain little, if any, fucose. In the species examined to date, xyloglucans produced by the Oleaceae have an XXXG-type structure (Vierhuis *et al.*, 2001) and those produced by the Solanaceae have an XXGG-type structure (York *et al.*, 1996). Typically, one of the two unbranched Glc*p* residues in each solanaceous xyloglucan subunit has an acetyl substituent at O6 (Figure 1.2) (Sims *et al.*, 1996). The 6-*O*-acetyl glucosyl residues of solanaceous xyloglucans are resistant to hydrolysis by most endo-1,4–β-glucanases, so well-defined XXGG-type oligosaccharide fragments are generated. Both the Solanaceae and Oleaceae produce xyloglucans with a distinctive α-L-Ara*f*-(1,2)-α-D-Xyl*p* side chain (designated as S), which may functionally replace the α-L-Fuc*p*-(1,2)-β-D-Gal*p*-(1,2)-α-D-Xyl*p* side chain that is present in most other dicotyledon xyloglucans (Figure 1.3).

The existence of variations in xyloglucan structure indicates that conservation of fucose-containing side chains in the xyloglucans of taxonomically diverse plants is not due to an absolute requirement for this structure. Indeed, *A. thaliana* plants in which the gene (*AtFUT1*) encoding the transferase responsible for adding fucose to xyloglucan has been inactivated by mutation (Vanzin *et al.*, 2002) or by a T-DNA insertion (Perrin *et al.*, 2003) synthesize xyloglucan that contains little or no detectable fucose. Nevertheless, these plants grow normally under laboratory conditions. The *AtFUT1* gene product only uses xyloglucan as an acceptor substrate for fucosyl transfer (Perrin *et al.*, 1999). Thus, the fucose residues of xyloglucans are not just 'along for the ride' and there must be some selective pressure to maintain a viable copy of the *AtFUT1* gene in wild-type populations. It is possible that fucosyl residues are conserved in the xyloglucans of taxonomically diverse plants because they confer some advantage for growth in the natural environment. An analysis of xyloglucans from a large number of individual plants in populations of wild-type and *fut1* plants that are exposed to a broad range of environmental and biological challenges may provide insight into why many diverse plant species synthesize fucosylated xyloglucan.

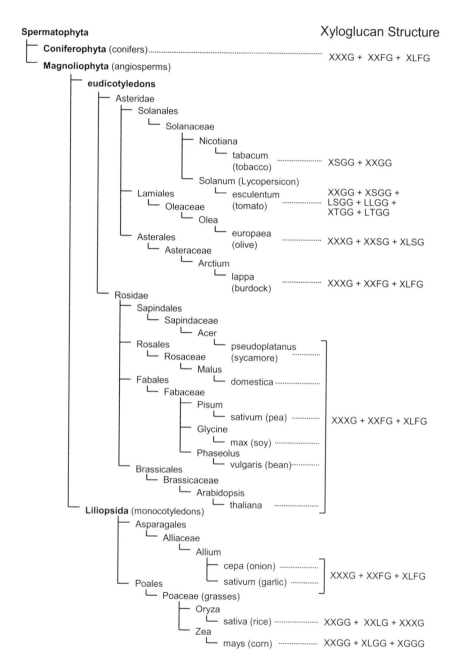

Figure 1.3 Phylogenetic relationships of xyloglucan oligosaccharide subunit structures. Each oligosaccharide structure is represented using specific code letters (Fry *et al.*, 1993) for each segment. See Figure 1.2 for xyloglucan nomenclature. Phylogenetic relationships are derived from the National Center for Biotechnology Information (NCBI) Taxonomy Browser (http://www.ncbi.nlm.nih.gov/Taxonomy/taxonomyhome.html/).

The structure of xyloglucan has been shown to differ in a tissue-specific manner in individual plants. For example, fucosyl residues are typically absent in seed xyloglucans, which are generally considered to be a fixed-carbon source for the germinating embryo, while the xyloglucan in other tissues of the same plant usually contain fucose. This suggests that the fucosylation of xyloglucans is important only in the context of the growing cell wall. More subtle structural changes are observed when xyloglucans from primary cell walls of different tissues of the same plant are compared. For example, subunits in which the central side chain is terminated by a β-D-Galp residue (e.g. XLFG) are more abundant in pea leaf xyloglucan than in pea stem xyloglucan (Pauly et al., 2001a).

The immunocytochemical analysis of primary cell walls (described in Chapter 3) suggests that xyloglucan structure may vary from cell to cell or even within different regions of the wall surrounding a single cell. For example, different cells and even different parts of the same wall in the developing root of A. thaliana plants are differentially labelled with the CCRC M1 antibody that recognizes fucosylated xyloglucans (Freshour et al., 1996). Furthermore, cell walls in the developing roots of A. thaliana plants carrying the mur1 mutation are differentially labelled by the CCRC-M1 antibody (Freshour et al., 2003). Only a subset of the root cells of mur1 plants are competent to produce GDP-fucose, the glycosyl donor required for fucosylation of xyloglucan. These observations are consistent with the idea that the extent to which xyloglucan is fucosylated in a specific tissue or cell is, at least in part, metabolically controlled. However, differential labelling with CCRC-M1, or any other xyloglucan-specific antibody, may reflect differences in the total amount of xyloglucan or the accessibility of the antibody's epitope, as well as differences in xyloglucan structure.

Structurally distinct xyloglucan 'domains' (Pauly et al., 1999a) have been isolated by sequentially treating depectinated pea-stem cell walls with a xyloglucan-specific endoglucanase (XEG) (Pauly et al., 1999b), with 4N KOH, and finally with a non-specific cellulase. Each extract is composed of a slightly different collection of xyloglucan oligosaccharide subunits. For example, oligosaccharides that appear to result from endogenous enzymatic processing are found in quantitatively greater amounts in the XEG-extracted xyloglucan domain than in the KOH-extracted and cellulase-released xyloglucan domains. These enzymically modified subunits include GXXG, which lacks the xylosyl residue normally found at the non-reducing end of the main chain, and XXG, which may be generated from a GXXG subunit at the non-reducing end of the polysaccharide by hydrolysis of the β-Glcp residue. The amount of XXG present in the XEG-extract increases as the tissue matures but similar amounts of GXXG are present irrespective of the tissue's developmental stage. These observations are consistent with the idea that the modified subunits are generated by enzymatic processing during cell wall development, and that GXXG is a non-accumulating intermediate in a metabolic pathway leading to XXG (Pauly et al., 2001a). The low abundance of these subunits in the KOH-extracted and cellulase-released domains is consistent with the enzyme-inaccessibility of these domains in muro, presumably due to their close association with cellulose microfibrils (see below).

1.7.4   Xylans

Xylans, including arabinoxylans, glucuronoxylans, and glucuronoarabinoxylans, are quantitatively minor components of the primary cell walls of dicotyledons and non-graminaceous monocotyledons (Darvill *et al.*, 1980), and are abundant in the primary cell walls of the Gramineae and in the secondary cell walls of woody plants (Ebringerova and Heinze, 2000). Xylans have a backbone composed of 1,4-linked β-D-Xyl$p$ residues, many of which are branched, bearing α-L-Ara$f$ residues at O2 or O3 (Gruppen *et al.*, 1992), and β-D-Glc$p$A or 4-O-methyl-β-D-Glc$p$A residues at O2 (Ebringerova and Heinze, 2000). Other side chains, including β-D-Xyl$p$-(1,3)-β-D-Xyl$p$-(1,2)-α-L-Ara$f$, β-D-Xyl$p$-(1,2)-α-L-Ara$f$ (Wende and Fry, 1997), and α-L-Ara$f$-(1,2)-α-L-Ara$f$ (Verbruggen *et al.*, 1998), have also been reported. The α-L-Ara$f$ residues often bear a feruloyl ester at O5 in the side chains of arabinoxylans produced by the Gramineae (Wende and Fry, 1997), which may lead to the oxidative cross-linking of xylan chains (Ishii, 1997a). The backbone Xyl$p$ residues of some xylans bear O-acetyl substituents at O2 and or O3.

1.7.5   Mannose-containing hemicelluloses

Mannose-containing polysaccharides include mannans, galactomannans, and galactoglucomannans. Homopolymers of 1,4-linked β-D-Man$p$ are found in the endosperm of several plant species including, for example, ivory nut (Stephen, 1982). Galactomannans, which are abundant in the seeds of many legume species, have a 1,4-linked β-D-Man$p$ backbone that is substituted to varying degrees at O6 with α-D-Gal$p$ residues (Stephen, 1982). Glucomannans, which are abundant in secondary cell walls of woody species, have a backbone that contains both 1,4-linked β-D-Man$p$ and 1,4-linked β-D-Glc$p$ residues (Stephen, 1982). Galactoglucomannans, which are found in both primary and secondary cell walls, have a similar backbone but some of the β-D-Man$p$ residues bear α-D-Gal$p$ and β-D-Gal$p$ (1→2)-α-D-Gal$p$ side chains at O6. Galactoglucomannans have been isolated from the walls of tobacco leaf midribs, (Eda *et al.*, 1984), suspension-cultured tobacco cells (Eda *et al.*, 1985), and from the culture filtrate of suspension-cultured *Rubrus fruticosus* (Cartier *et al.*, 1988), tobacco (Sims *et al.*, 1997) and tomato cells (Z. Jia and W.S. York, unpublished results). Galactoglucomannans are especially abundant in primary cell walls of solanaceous species, which also contain non-fucosylated XXGG-type xyloglucans.

1.8   The pectic polysaccharides

Three pectic polysaccharides have been isolated from primary cell walls and structurally characterized. These are homogalacturonan, substituted galacturonans, and rhamnogalacturonans.

1.8.1 Homogalacturonan

Homogalacturonan (HG) is a linear chain of 1,4-linked α-D-galactopyranosyluronic acid (GalpA) residues in which some of the carboxyl groups are methyl esterified (Figure 1.4). HG polymers with a high degree of methyl esterification are referred to as 'pectin' whereas HG with low or no methyl esterification is termed 'pectic acid'. HGs may, depending on the plant source, also be partially O-acetylated (Ishii 1997b;

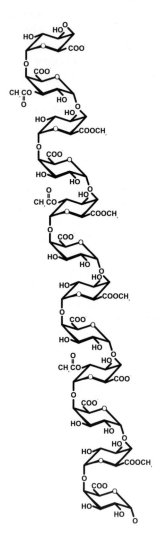

Figure 1.4 The primary structure of homogalacturonan. Homogalacturonan is a linear polymer composed of 1,4-linked α-D-GalpA residues. Some of the GalpA residues are methyl-esterified at C6. The GalpA residues may also be O-acetylated.

Perrone *et al.*, 2002). There are also reports that HGs contain other, as yet, unidentified esters (Kim and Carpita, 1992; Brown and Fry, 1993).

Homogalacturonan may account for up to 60% of the pectin in the primary walls of dicotyledons and non-graminaceous monocotyledons and thus is the predominant anionic polymer. Many of the properties and biological functions of HG are believed to be determined by ionic interactions (Ridley *et al.*, 2001; Willats *et al.*, 2001a). The degree of methyl esterification of HG has a major influence on its ability to form gels (Goldberg *et al.*,1996; Willats *et al.*, 2001b). HGs with a high degree of methyl esterification do not gel in the presence of Ca^{2+}, although they do gel at low pH in the presence of high concentrations of sucrose. A decrease in the degree of methyl esterification of HG is often observed as cells mature and this is believed to result in a increase in Ca^{2+} cross-linking of HG together with a increase in wall strength (Goldberg *et al.*, 1996; Willats *et al.*, 2001b). The degree of methyl esterification of HG in the middle lamella has also been implicated in cell separation as has its degree of *O*-acetylation (Liners *et al.*, 1994; Bush and McCann, 1999) and the extent of branching of the rhamnogalacturonan backbone with arabinosyl and galactosyl-containing side chains (Redgwell *et al.*, 1997). HG with low and high degrees of methyl esterification have been reported to be present in the junction regions that form between cells of an *Arabidopsis* mutant that exhibits postgenital organ fusions. Some of the tissues of this mutant lack an intact cuticle, and it is believed that the walls of closely appressed epidermal cells fuse by copolymerization of HG (Sieber *et al.*, 2000).

Approximately 52 genes encoding putative polygalacturonases (PG) have been identified in *Arabidopsis* (The Arabidopsis Genome Initiative, 2000). Little is known about the function or specificities of these pectic-degrading enzymes. Nevertheless, there is increasing evidence that PGs are expressed in a wide range of plant tissues and at various stages during plant development (Hadfield and Bennett, 1998). These PGs are likely to be involved in modifying the structure and properties of wall-bound pectin during normal plant growth and development.

The function of homogalacturonan in plant development has been examined using transformed plants that either over-express a specific polygalacturonase (PG) or that have had the level of endogenous PG reduced. For example, the suppression of endogenous PG using antisense mRNA reduced tissue breakdown during fruit senescence but did not alter the ripening process (Tieman and Handa, 1994). Over-expression of the tomato fruit-specific EPG in the ripening-inhibited (*rin*) mutant resulted in increased pectin depolymerization but caused no apparent increase in fruit softening (Giovannoni *et al.*, 1989). Over-expression of pTOM6 (a tomato fruit-specific PG gene) in tobacco had no discernible affect on wall pectin nor was there any visible effect on the plant phenotype (Oosteryoung *et al.*, 1990). In contrast, transgenic apple trees that contained one or two additional copies of a fruit-specific apple PG exhibited several phenotypes including premature leaf shedding, reduced cell adhesion, and brittle leaves (Atkinson *et al.*, 2002). Somewhat unexpectedly, the transgenic apples also produced stomata that were frequently malformed and did not function normally. Such effects may result from the separation of guard cells from their adjacent epidermal cells which reduced the ability of stomata to open and close. The authors concluded that all the observed phenotypes are likely to be a consequence of reduced cell adhesion resulting from changes in wall pectin structure.

A low-esterified HG-enriched polymer together with a 9 kDa cysteine-rich basic protein (SCA) have been shown to have a role in the adhesion of lily pollen tubes to the stylar matrix (Mollet *et al.*, 2000). The pectic material and SCA alone were not effective at promoting adhesion. The most active pectic material was composed predominantly of Gal*p*A residues (73 mol%) but also contained quantitatively significant amounts of Ara, Gal, Rha, and GlcA and thus is likely to be composed of both HG and rhamnogalacturonan regions, although it is not known which of the components is required for pollen adhesion (Mollet *et al.*, 2000).

Some progress has been made in characterizing the enzymes involved in the biosynthesis of HG (Ridley *et al.*, 2001), and this is discussed in more detail in Chapter 6.

1.8.2   *Rhamnogalacturonans*

Rhamnogalacturonans (RGs) are a group of closely related cell wall pectic polysaccharides that contain a backbone of the repeating disaccharide 4)-α-D-Gal*p*A-(1,2)-α-L-Rha*p* (Lau *et al.*, 1985). Between 20 and 80% of the Rha*p* residues are, depending on the plant source and the method of isolation, substituted at C-4 with neutral and acidic oligosaccharides (McNeil *et al.*, 1982b; Lau *et al.*, 1987; Ishii *et al.*, 1989; see Figure 1.5). These oligosaccharides predominantly contain linear and branched α-L-Ara*f*, and β-D-Gal*p* residues (McNeil *et al.*, 1980; Schols and Voragen, 1994), although their relative proportions may differ depending on the plant source. α-L-Fuc*p*, β-D-Glc*p*A, and 4-*O*-Me β-D-Glc*p*A residues may also be present (An *et al.*, 1994). The number of glycosyl residues in the side chains is variable and may range from a single glycosyl residue to more than twenty (Lerouge *et al.*, 1993). The oligosaccharide side chains in RGs from some plants (e.g. sugar beet) may be esterified with phenolic acids (e.g. ferulic acid) (Ishii, 1997a). In many RGs the backbone Gal*p*A residues are *O*-acetylated on C-2 and/or C-3 (Perrone *et al.*, 2002) but there is no evidence that the Gal*p*A residues are methyl esterified (Komalavilas and Mort, 1989; Perrone *et al.*, 2002). The backbone Gal*p*A residues are not usually substituted with other glycosyl residues although there has been one report (Renard *et al.*, 1999) showing that a single Glc*p*A residue is attached to Gal*p*A in sugar beet RG.

Little is known about the biological function of rhamnogalacturonans (Willats *et al.*, 2001a); nevertheless, immunocytochemical studies have provided evidence that changes in the structures of the arabinan and galactan side chains are correlated with cell and tissue development (Willats *et al.*, 1999; Orfila and Knox, 2000; Willats *et al.*, 2001a; see Chapter 3). The function of rhamnogalacturonans has been investigated using plants transformed with endoglycanases that fragment pectin. For example, potato plants transformed with a fungal rhamnogalacturonan lyase produce small tubers that exhibit abnormal cell development (Oomen *et al.*, 2002). The tuber walls contain somewhat less RG-I than normal walls and also have an

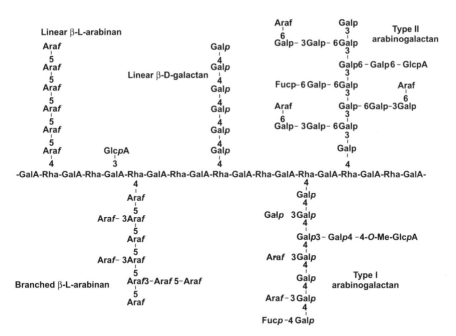

Figure 1.5 A schematic representation of the primary structure of rhamnogalacturonan I. The backbone repeat unit [→4)-α-D-GalpA-(1→2)-α-L-Rhap-(→] is predominantly substituted with arabinosyl and galactosyl-containing side chains. Some of the side chains may also contain quantitatively small amounts of α-L-Fucp, β-D-GlcpA and 4-O-Me β-D-GlcpA residues. The distribution of the side chains along the backbone is not known. Moreover, it is not known whether individual RG-I molecules contain either arabinosyl- or galactosyl-containing side chains or mixtures of both these side chains.

altered pattern of pectin deposition. Potato plants transformed with an apoplastically targeted fungal endo-1,5-α-L-arabinanase resulted in plants lacking flowers, stolons and tubers (Skjot *et al.*, 2002). Such a severe phenotype may be a stress response that is induced by the presence in the apoplast of the fungal arabinanase and/or the arabinosyl-containing oligosaccharides. Indeed, potato plants transformed with a Golgi membrane-anchored endo-1,5-α-L-arabinanase had a normal phenotype even though the arabinosyl content of their walls was reduced by 70% (Skjot *et al.*, 2002). Potato plants transformed with a fungal endo-β-1,4-galactanase also produce tubers with no visible phenotype even though the RG present in the tuber walls contain much less galactose than the RG of wild-type plants (Sorensen *et al.*, 2000). Transforming plants with endo- or exoglycanases that fragment wall polysaccharides has considerable potential for investigating the role of these polymers in plant growth and development. However, the results of such studies need to be interpreted with caution because pectin-derived oligosaccharides are known to elicit defence responses in plant cells and tissues (Ridley *et al.*, 2001).

Some progress has been made in studying the enzymes involved in the biosynthesis of RGs (Geshi *et al.*, 2000; Ridley *et al.*, 2001), as discussed in more detail in Chapter 6 of this book.

1.8.3   Substituted galacturonans

Substituted galacturonans are a group of polysaccharides that contain a backbone of linear 1,4-linked α-D-Gal*p*A residues.

1.8.3.1   Apiogalacturonans and xylogalacturonans

Xylogalacturonans contain β-D-Xyl*p* residues attached to C-3 of the backbone (Figure 1.6a). Such polysaccharides have only been detected in the walls of specific plant tissues, such as soybean and pea seeds, apple fruit, carrot callus, and pine pollen (Bouveng, 1965; Schols *et al.*, 1995; Kikuchi *et al.*, 1996; Yu and Mort, 1996; Huisman *et al.*, 2001). Apiogalacturonans, which are present in the walls of some aquatic monocotyledonous plants, including *Lemna* and *Zostera* (Cheng and Kindel, 1997; Golovchenko *et al.*, 2002), contain β-D-Api*f* residues attached to C-2 of the backbone Gal*p*A residues either as a single Api*f* residue or as the disaccharide β-D-Api*f*-(1,3′)-β-D-Api*f*-(1, (Figure 1.6b). Oligosaccharides composed of α-L-Ara*f*, β-D-Gal*p*, and

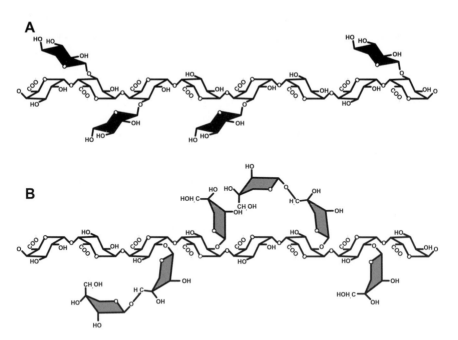

Figure 1.6 A schematic representation of the primary structure of substituted galacturonans. A. Xylogalacturonan. Xylosyl residues are linked to C3 of the 1,4-linked α-D-galacturonan backbone. B. Apiogalacturonan. Apiosyl and apiobiosyl residues are linked to C2 of the 1,4-linked α-D-galacturonan backbone.

β-D-Xyl*p* residues have also been reported to be linked to the galacturonan backbone of an apiogalacturonan (lemnan) isolated from *L. minor* (Golovchenko *et al.*, 2002).

1.8.3.2 Rhamnogalacturonan II

A third substituted galacturonan, which is referred to as rhamnogalacturonan II (RG-II), is found in the walls of all higher plants (Stevenson *et al.*, 1988; O'Neill *et al.*, 1990). The glycosyl sequence of RG-II is to a large extent conserved in gymnosperms (Thomas *et al.*, 1987; Shimokawa *et al.*, 1999), monocotyledons (Thomas *et al.*, 1989; Kaneko *et al.*, 1997), and dicotyledons (Stevenson *et al.*, 1988; Pellerin *et al.*, 1996; Ishii and Kaneko, 1998; Shin *et al.*, 1998; Strasser and Amado, 2002). RG-II has also been reported to be present in the sporophyte cell walls of the fern *Adiantum* (Matoh and Kobayashi, 1998). However, additional studies are required to determine if RG-II is indeed present in the walls of lower vascular plants, including the Pteridopsida (ferns, e.g. *Pteridium* and *Ophioglossum*), the Psilotidae (whisk ferns, e.g. *Psilotum*), the Equisetopsida (horsetails, e.g. *Equisetum*), and the Lycophyta (club mosses, e.g. *Huperzia* and *Selaginella*). Little, if anything, is known about the occurrence of RG-II in the walls of non-vascular land plants such as the Bryophyta (mosses, e.g. *Sphagnum*), the Hepatiphyta (liverworts, e.g. *Marchantia*), and the Anthocerophyta (hornworts, e.g. *Anthoceros*).

RG-II is a low molecular mass (~5–10 kDa) pectic polysaccharide that can be solubilized from the cell wall by treatment with *endo*polygalacturonase. RG-II contains eleven different glycosyl residues including the unusual sugars Api*f*, Ace*f*A, 2-*O*-Me Fuc*p*, 2-*O*-Me Xyl*p*, Dha, and Kdo (see Figure 1.7). The backbone of RG-II contains at least eight 1,4-linked α-D-Gal*p*A residues (Whitcombe *et al.*, 1995). Thus, RG-II is *not* structurally related to the RGs that have a backbone composed of the repeating disaccharide 4)-α-D-Gal*p*A-(1,2)-α-L-Rha*p*. Two structurally distinct disaccharides (C and D in Figure 1.7) are attached to C-3 of the backbone and two structurally distinct oligosaccharides (A and B in Figure 1.7) are attached to C-2 of the backbone. The locations of the side chains on the backbone with respect to each other has not been established. Nevertheless, some evidence for their locations (see Figure 1.7) has been obtained by NMR spectroscopic analysis of RG-II (du Penhoat *et al.*, 1999) and a enzymically-generated oligoglycosyl fragment of RG-II (Vidal *et al.*, 2000).

The RG-IIs solubilized by treating walls with endopolygalacturonase all contain four oligoglycosyl side chains (A–D) linked to a backbone that contains between 7 and 15 1,4-linked α-D-Gal*p*A residues (see Figure 1.7a). However, there is increasing evidence that side chain B may not be structurally identical in all plants. This side chain has been reported to exists as a hepta-, octa-, and nonasaccharide in the walls of suspension-cultured sycamore (*Acer pseudoplatanus*) RG-II (Whitcombe *et al.*, 1995) and red wine RG-II (Vidal *et al.*, 2000; Glushka *et al.*, 2003), and as a nonasaccharide in ginseng (*Panax ginseng*) leaf (Shin *et al.*, 1998). Side chain B has been reported to be a heptasaccharide in *Arabidopsis* RG-II (B. Reuhs, J. Glenn, S. Stephens, J. Kim, D. Christie, J. Glushka, M. O'Neill, S. Eberhard, P. Albersheim, and A. Darvill, manuscript in preparation), an octasaccharide in bamboo

Side chain A

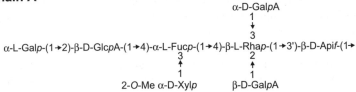

α-D-GalpA
1
↓
3
α-L-Galp-(1→2)-β-D-GlcpA-(1→4)-α-L-Fucp-(1→4)-β-L-Rhap-(1→3')-β-D-Apif-(1→
3   2
↑   ↑
1   1
2-O-Me α-D-Xylp   β-D-GalpA

Side chain B

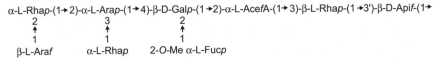

α-L-Rhap-(1→2)-α-L-Arap-(1→4)-β-D-Galp-(1→2)-α-L-AcefA-(1→3)-β-L-Rhap-(1→3')-β-D-Apif-(1→
2   3   2
↑   ↑   ↑
1   1   1
β-L-Araf   α-L-Rhap   2-O-Me α-L-Fucp

Side chain C

α-L-Rhap-(1→5)-β-D-Kdo-(1→

Side chain D

β-L-Araf-(1→5)-β-D-Dha-(1→

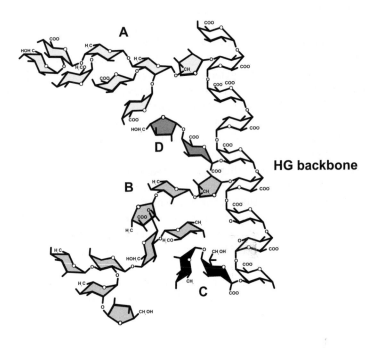

Figure 1.7   The primary structure of rhamnogalacturonan II. **Top**. The four side chains (A – D) that are attached to the 1,4-linked α-D-galacturonan backbone of RG-II. In a RG-II molecule side chains A and B are linked to C2 of different backbone GalpA residues whereas side chains C and D are linked to C3 of different backbone GalpA residues. **Bottom**. A schematic representation of the locations of the side chains along the 1,4-linked α-D-galacturonan backbone. The backbone of endopolygalacturonase-released RG-II contains, on average, between 7 and 15 1,4-linked α-D-GalpA residues. The hydroxyl groups are shown as lines extending from the glycosyl ring.

(*Phyllostachys edulis*) shoot RG-II (Kaneko *et al,.* 1997), and a hexasaccharide in the RG-II isolated from sugar beet (*Beta vulgaris*) pulp (Ishii and Kaneko, 1998), akamutsu (*Pinus densiflora*) hypocotyls (Shimokawa *et al.*, 1999), and red beet (*Beta vulgaris* L var *conditiva*) tubers (Strasser and Amado, 2002). Thus, the number of glycosyl residues in this side chain may vary depending on the plant source, although the possibility cannot be discounted that the structural variations result from differences in the procedures used to isolate and characterize the RG-II. The limited data available suggest that the structural variations result from the presence or absence of substituents linked to C2 and/or C3 of the Ara*p* residue. This residue is 2,3-linked in red wine and ginseng RG-II but is 2-linked in *Arabidopsis* RG-II but has been reported to be present as a terminal non-reducing residue in beet RG-II (Ishii and Kaneko, 1998; Strasser and Amado, 2002). Thus, some plants may lack the glycosyl transferases required to add the Ara*f* and Rha*p* residues to the Ara*p* residue. Alternatively, all plants may synthesize a nonasaccharide but only some of them may produce exoglycanases that 'trim' the nonasaccharide to smaller oligosaccharides.

The biosynthesis of RG-II has not been investigated. Nevertheless, it is likely to require at least 22 glycosyltransferases (Ridley *et al.*, 2001). These glycosyl transferases alone account for almost half of the transferases required for the synthesis of primary wall HG and RGs (Ridley *et al.*, 2001). Each glycosyltransferase transfers a monosaccharide from its activated donor (a nucleotide diphosphate-sugar) in a linkage- and anomer-specific manner to an acceptor molecule. The activated forms of ten of the glycosyl residues (UDP-D-Api*f,* UDP-L-Ara*p,* UDP-L-Ara*f,* GDP-L-Fuc*p,* UDP-D-Gal*p,* GDP-L-Gal*p,* UDP-D-Gal*p*A, UDP-D-Glc*p*A, UDP-L-Rha*p,* and UDP-D-Xyl*p*) present in RG-II are enzymatically formed from pre-existing nucleotide sugars including UDP-D-Glc*p,* GDP-D-Man*p,* and UDP-D-GlcA (Reiter and Vanzin, 2001; Tanner, 2001). The biosynthetic origin of L-Ace*f*A is not known, nor has the activated form of this glycosyl residue been identified. RG-II contains an Ara*p* residue and at least two Ara*f* residues (see Figure 1.7a). UDP-L-Ara*f* is formed from UDP-L-Ara*p* although the mechanism of formation in plants is not understood (Rodgers and Bolwell, 1992). Interestingly, a bacterial UDP-D-galactopyranose mutase that inter-converts UDP-D-Gal*p* and UDP-D-Gal*f* has been shown to inter-convert UDP-L-Ara*f* and UDP-L-Ara*p* which has led to the suggestion that plants contain a UDP-L-Ara*p* mutase (Zhang and Liu, 2001). RG-II is, as far as we are aware, the only naturally occurring pectic polysaccharide that contains a D-Gal and a L-Gal residue (B. Reuhs, J. Glenn, S. Stephens, J. Kim, D. Christie, J. Glushka, M. O'Neill, S. Eberhard, P. Albersheim and A. Darvill, manuscript in preparation). These glycosyl residues have different biosynthetic origins because UDP-D-Gal is formed from UDP-D-Glc whereas GDP-L-Gal is formed from GDP-D-Man (Tanner, 2001). CMP-Kdo, the activated form of Kdo, is formed from CTP and Kdo-8-phosphate (Brabetz *et al.*, 2000). The Kdo-8-P itself is generated by an aldol-type condensation of phosphoenol pyruvate (PEP) and D-Ara-5-phosphate, a reaction that is catalysed by Kdo-8-P synthase (Brabetz *et al.*, 2000). The synthesis of Dha has not been investigated but it may be formed in an analogous manner to Kdo, except that the initial reactants would be PEP and D-threose-4-P. Two of the glycosyl residues

(Fuc and Xyl) are present in RG-II as their 2-O-methyl ether derivatives and two glycosyl residues (2-O-Me Fuc and Ace$f$A) are O-acetylated. Thus, the biosynthesis of RG-II requires two O-methyl transferases and two O-acetyl transferases, in addition to the glycosyl transferases.

It is apparent that a large number of enzymes are required for RG-II biosynthesis and that the synthesis of RG-II must come with a significant entropic cost to the plant. Moreover, polysaccharides are secondary gene products and their sequences are not encoded by DNA, thus, the expression of the genes encoding the various transferases and the activities of the transferases must be tightly coordinated to ensure that the RG-II side chains are correctly assembled. Plant polysaccharides are believed to be synthesized by the step-wise addition of glycosyl residues to the growing polymer (Ridley *et al.*, 2001). However, the possibility cannot be discounted that the side chains of RG-II are assembled on a lipid intermediate, as are the repeat units of bacterial polysaccharides and the N-linked oligosaccharides of glycoproteins (Kornfeld and Kornfeld, 1985), and then transferred to a pre-existing HG chain. The isolation and characterization of all the enzymes involved in RG-II synthesis, together with the determination of the factors that regulate RG-II biosynthesis, is a major challenge for cell wall researchers.

The demonstration that RG-II is cross-linked by a tetravalent 1:2 borate-diol ester (Matoh *et al.*, 1993; Kobayashi *et al.*, 1995, 1996) was a major advance in our understanding of the structure and function of this pectic polysaccharide. Subsequent studies have confirmed that B cross-links two chains of RG-II to form a dimer in the primary walls of numerous plants (Ishii and Matsunaga, 1996; O'Neill *et al.*, 1996; Pellerin *et al*, 1996; Kaneko *et al.*, 1997).

The location of the borate ester in RG-II has been investigated using selective acid hydrolysis of the per-O-methylated RG-II dimer (Ishii *et al.*, 1999). The results of these studies suggest that the apiosyl residue of side chain A (see Figure 1.7), but not the apiosyl residue of side chain B, in each RG-II monomer is cross-linked by borate. The cross-link is a diester in which borate is covalently linked to four oxygen atoms (O2 and O3) of two D-apiosyl residues (Figure 1.8). The B atom in this cross-link is chiral and thus two diasteroisomers can form (see Figure 1.8). Indeed, two 1:2 borate-diol esters are formed when methyl β-D-apiofuranoside reacts with borate at pH 8 (Ishii and Ono, 1999). It is not known whether the naturally occurring RG-II dimer contains one or both diasteroisomers.

Some of the factors involved in the inter-conversion of the RG-II dimer and monomer have been investigated using *in vitro* studies. For example, the dimer is rapidly converted to the monomer and boric acid at pH 1 because borate-diol esters are hydrolysed at low pH (Kobayashi *et al.*, 1996; O'Neill *et al.*, 1996). The RG-II dimer is formed *in vitro* between pH 3 and 4 by treating the monomer with boric acid (O'Neill *et al.*, 1996). The anionic nature of RG-II is likely to be a factor that allows the dimer to form rapidly at low pH because virtually no diester is formed when apiose itself is reacted with borate at pH <5 (Ishii and Ono, 1999). Dimer formation is much more rapid in the presence of divalent cations (Sr^{2+}, Ba^{2+} and Pb^{2+}) with ionic

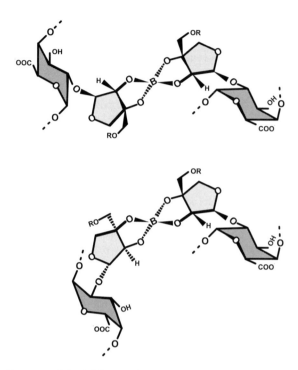

Figure 1.8 Structure of the 1:2 borate-diol ester that cross-links two RG-II molecules. The cross-link is a diester in which borate is covalently linked to four oxygen atoms (O2 and O3) of two β-D-apiosyl residues. The B atom in this cross-link is chiral and thus two diastereoisomers can form. It is not known whether one or both of these forms exist in naturally occurring RG-II.

radii >1.1A (O'Neill *et al.*, 1996; Ishii *et al.*, 1999). Somewhat unexpectedly, Ca^{2+} ions are somewhat less effective at promoting dimer formation *in vitro* than cations with a larger ionic radius (O'Neill *et al.*, 1996; Ishii *et al.*, 1999). Nevertheless, higher concentrations (>10 mM) of Ca^{2+} do promote dimer formation and calcium ions are likely to be important in stabilizing the RG-II dimer *in muro* (Matoh and Kobayashi, 1998; Fleischer *et al.*, 1999; Kobayashi *et al.*, 1999).

Little is known about how the structural complexity of RG-II contributes to its biological function. However, some clues have begun to emerge. The side chains of RG-II are believed to have a role in promoting dimer formation and stabilizing the dimer once it has formed. For example, the *Arabidopsis mur1* mutant synthesizes RG-II that contains L-Gal rather than L-Fuc residues (B. Reuhs, J. Glenn, S. Stephens, J. Kim, D. Christie, J. Glushka, M. O'Neill, S. Eberhard, P. Albersheim, and A. Darvill, manuscript in preparation). The *mur1* RG-II forms a dimer less rapidly and is less stable than the normal dimer (O'Neill *et al.*, 2001). This result also suggests that hydrophobic interactions have a role in dimer formation since L-Gal differs from L-Fuc by having a hydroxymethyl rather than a methyl group at C-6.

The ability of RG-II to form a dimer but not a trimer or a larger complex *in muro* (O'Neill *et al.*, 1996) and *in vitro* (O'Neill *et al.*, 1996; Ishii *et al.*, 1999) suggests that the chemical structure and conformation of RG-II are major factors that regulate its interaction with borate. RG-II may only be able to adopt a limited number of conformations because of the steric crowding that results from the presence of four oligosaccharides attached to an octagalacturonide backbone (see Figure 1.7b). Such conformations may allow the borate ester to form between the apiosyl residue in each side chain A yet prevent the random cross-linking of the apiosyl residues in side chains A and B. To confirm such a hypothesis requires a detailed knowledge of the solution conformation and dynamic properties of RG-II but such information is lacking. Several possible three-dimensional structures for the side chains of RG-II have been suggested on the basis of computer modelling procedures (Perez *et al.*, 2000). These studies have predicted that side chain A is somewhat rigid whereas side chain B is flexible. The conformational flexibility of oligosaccharides results in large part from rotations around their glycosidic bonds, although other factors including the flexibility of the pyranose and furanose rings may also contribute to the molecular motions. For example, the 2,3-linked α-L-Ara$p$ residue present in side chain B generated from wine RG-II has been shown to exhibit conformational flexibility and may exist in an equilibrium between a $^1C_4$ and a boat-like chair configuration (Glushka *et al.*, 2003). These data are consistent with the notion that side chain B is flexible, although the effect of this motion on the properties of RG-II is not understood.

The homogalacturonan backbone of RG-II is not fragmented by endopolygalaturonases suggesting that the side chains sterically prevent the hydrolysis of the 1,4-linked Gal$p$A residues. Indeed, computer modelling procedures have predicted that side chains C and D extend along the longitudinal axis of the backbone and that hydrogen bonds between the side chain and backbone stabilize these conformations (Perez *et al.*, 2000). The RG-II side chains themselves may also be resistant to glycanases, including those secreted by plant cells and microorganisms, because they contain glycosyl residues whose anomeric configurations and glycosidic linkages (see Figure 1.7) are not typically present in other wall polysaccharides. A borate cross-link whose stability is controlled by oligosaccharide side chains that are resistant to fragmentation by endogenous glycanases may be essential for maintaining the integrity of the pectic network, whilst at the same time providing a framework that allows the enzymic restructuring of this network during plant growth and development.

RG-II is one of the most unusual polysaccharides yet identified in nature. Understanding the structure, dynamics, and function of this molecule is challenging and progress in this area has until recently been slow. There is now considerable interest in the polysaccharide because of its ability to specifically interact with borate, its role in plant growth, and its ability to selectively bind heavy metals.

1.9 Other primary wall components

1.9.1 Structural glycoproteins

Primary cell walls typically contain O-glycosylated proteins, as reviewed in detail in Chapter 4. These include the hydroxyproline-rich glycoproteins (HRGPs, often referred to as 'extensin', even though their role in plant growth remains unclear) which are glycosylated with arabinose, arabinobiose, arabinotriose, and arabinotetraose, and with galactose (Kieliszewksi and Shpak, 2001). Some primary walls (e.g. maize) contain threonine-rich HRGPs or proline-rich glycoproteins (e.g. soybean) that are also glycosylated with Ara and Gal but to a lower extent than the HRGPs. Glycine-rich proteins have also been detected in various primary walls (e.g. petunia). Numerous roles have been ascribed to these (glyco)proteins but their actual functions remain to be fully elucidated (Jose-Estanyol and Puigdomenech, 2000; Ringli *et al.*, 2001; see Chapter 4).

1.9.2 Arabinogalactan proteins (AGPs)

Arabinogalactan proteins are a family of structurally complex proteoglycans (Gaspar *et al.*, 2001). The polysaccharide portions of AGPs typically account for more than 90% of the molecule and is rich in galactose and arabinose, while the protein moieties have diverse amino acid sequences, although they are often enriched in Hyp, Ala, and Ser. Recent data have shown that quantitatively small amounts of AGP are linked to the plasma membrane by a glycosylphosphatidylinositol (GPI) membrane anchor (Youl *et al.*, 1998). Those AGPs that are not bound to the plasma membrane are present in the apoplast. However, the apoplastic AGPs are readily solubilized by aqueous buffers and thus may not be structural components of the wall. Nevertheless, the wall-associated AGPs may have a role in cell expansion and cell differentiation (see Chapter 4).

1.9.3 Enzymes

Primary cell walls contain numerous enzymes (Fry, 1995), including those involved in wall metabolism (endo and exoglycanases, methyl and acetyl esterases, and trans-glycosylases) and enzymes that may generate cross-links between wall components (e.g. peroxidases). Walls also contain proteins referred to as expansins that have been proposed to break hydrogen bonds between XG and cellulose and thus are believed to regulate wall expansion (Cosgrove, 1999; and reviewed in Chapter 8).

1.9.4 Minerals

The mineral content of a primary wall is to a large extent dependent on the conditions of plant growth and the methods used to prepare the walls. Nevertheless, minerals such as Ca, K, Na, Fe, Mg, Si, Zn and B together can account for up to 5% of the dry weight of dicotyledon walls (Welch, 1995; Epstein, 1999). Calcium is typically present ionically linked to the anionic pectic polysaccharides and is believed to have a major influence on the rheological properties of the primary wall. Boron is also intimately involved with the organization of primary wall pectin since *in muro* it cross-links two chains of RG-II (Hu and Brown, 1994; Matoh *et al.*, 1996). The function of silicon in plants and their cell walls remains controversial (Epstein, 1999).

1.10 General features of wall ultrastructural models

Many different models have been proposed for the ultrastructure of primary cell walls. Several of the early models predicted that many if not all of the wall matrix polymers (xyloglucan, pectin, and glycoprotein) are covalently linked to one another, and that the spontaneous binding of xyloglucan to the surface of cellulose microfibrils could lead to the cross-linking of these rigid structural elements and impart tensile strength to the wall (Keegstra *et al.*, 1973; Albersheim, 1976). However, the lack of compelling evidence for the existence of covalent linkages between the noncellulosic wall components led some workers to question whether the primary wall was a covalently cross-linked macromolecular complex (Monro *et al.*, 1976). Currently, the most popular models (McCann and Roberts, 1991; Talbot and Ray, 1992; Carpita and Gibeaut, 1993; McCann and Roberts, 1994; Ha *et al.*, 1997) emphasize non-covalent interactions between wall polymers and stipulate two independent but interacting networks (see Figure 1.9). One network is composed of pectic polysaccharides (HG, RGs, and RG-II) and the second consists of cellulose and xyloglucan.

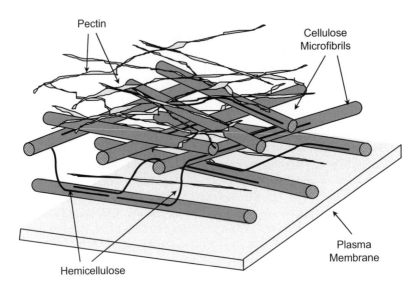

Figure 1.9 Model of the primary cell wall in most higher plants (adapted from McCann and Roberts, 1991; McCann and Roberts, 1994). See text for details.

An additional network composed of wall glycoprotein (extensin) may also be present in the primary wall.

The cellulose/xyloglucan network is believed to be the major load-bearing structure in the cell wall. Load-bearing functions are less frequently attributed to the pectin network, which may rather function as a 'scaffolding' that controls wall porosity and electrostatically binds to positively charged molecules, such as enzymes in the cell wall. Thus, the pectin network may compartmentalize the apoplastic space, preventing enzymes and other macromolecules from diffusing to inappropriate sites. Such a network may also confer orientational order on other cell wall components, thereby directing, for example, the effects of enzymes that catalyse the macromolecular assembly or reorganization of the wall. Another possibility is that the pectic network may function as a sensor of mechanical stress or elastic strain in the cell wall, thereby allowing the cell to respond by modulating the activities of cell wall modifying enzymes or by controlling the rate at which cell wall polysaccharides are synthesized.

1.10.1 The xyloglucan/cellulose network

Rigid cellulose microfibrils interact with soluble xyloglucan at the cell surface to form the xyloglucan/cellulose network. The xyloglucan is synthesized in the Golgi and exported to the apoplast for incorporation into this network (Levy and Staehelin, 1992). Xyloglucan export is presumably facilitated by its high solubility, as xyloglucan purified from primary cell walls is soluble in water. However, xyloglucan spontaneously and avidly binds to the surface of cellulose *in vitro* (Valent and Albersheim, 1974). Thus, a fundamental step in the assembly of cellulose/xyloglucan network is likely to occur when cellulose is extruded from rosettes in the plasma membrane into a matrix that contains high concentrations of xyloglucan. The xyloglucan is believed to 'coat' the surfaces of nascent microfibrils, limiting their aggregation and connecting them via tethers that directly or indirectly regulate the mechanical properties of the cell wall (Whitney *et al.*, 1995).

Molecular tethers connecting cellulose microfibrils have been visualized by electron microscopy (EM) of rotary shadowed replicas of rapidly frozen, deep etched (RFDE) cell walls (McCann *et al.*, 1990; Satiat-Jeunemaitre *et al.*, 1992; Itoh and Ogawa, 1993). These tethers persist in images of depectinated cell walls, but are much less abundant in images of cell walls that have been subjected to conditions that remove xyloglucan. Furthermore, removal of xyloglucan destroys the ordered spacing of the cellulose microfibrils in the wall, suggesting that xyloglucans have a role in defining the spatial organization of the microfibrils (McCann *et al.*, 1990; Itoh and Ogawa, 1993). A fully extended xyloglucan molecule with a MW of 300 kDa would have a length of ~500 nm. In contrast, the tethers observed in the RFDE cell walls have a well-defined length of between 16 and 40 nm (McCann *et al.*, 1990; Satiat-Jeunemaitre *et al.*, 1992). Tethers of approximately the same length are also observed in the composite material produced when cellulose is synthesized by *Acetobacter xylinum* in the presence of xyloglucan (Whitney *et al.*, 1995). The consistency of tether length in these systems has been attributed (Whitney et al., 1995) to a balance between enthalpic factors (favouring the binding of xyloglucan segments to the microfibril) and entropic factors (favouring the dissociation of xyloglucan segments from the microfibril). It was suggested (Whitney et al., 1995) that, due to the rigidity of the xyloglucan backbone over short distances, there would be little entropic advantage for the formation of a tether less than 20 nm in length (i.e. the tether's entropy would not be significantly greater than that of the same segment bound directly to the microfibril surface). Formation of tethers greater than 100 nm in length would be entropically favoured, but presumably disfavoured by enthalpic factors promoting the binding of xyloglucan segments to exposed microfibril surfaces. The observed tether length can be explained by this argument if, as the tether length increases, the entropic advantage to tether formation increases more slowly than the enthalpic disadvantage.

Unexpectedly, the contour lengths (i.e. the distance from end to end of a fully extended molecule) of alkali-extracted cell-wall xyloglucans, as measured by EM (McCann et al., 1992), have a clear 30 nm periodicity, which corresponds to the average length of the tethers in RFDE walls (McCann et al., 1990; Itoh and Ogawa, 1993). Although the reason for this periodicity is not understood, one possibility is that xyloglucan in the cell wall is assembled from pre-formed, 30 nm long xyloglucan blocks (McCann et al., 1992).

Albersheim recognized that cleavage of cross-links in the cell wall would allow the wall to expand and grow under osmotic stress, but would also weaken it. Therefore, he proposed the existence of enzymes that break and reform covalent (e.g. glycosidic) bonds in the cross-linking molecules (Albersheim, 1976). The incorporation of soluble xyloglucan oligosaccharides into polysaccharides that had been previously deposited into the growing cell wall led to the proposal that enzyme-catalysed transglycosylation reactions are responsible for this phenomenon (Baydoun and Fry, 1989). A transglycosylase was also suggested to be responsible for the increase in the molecular weight of pea-stem xyloglucan soon after its synthesis (Talbot and Ray, 1992). Similar conclusions were reached by Nishitani and coworkers, based on changes in the molecular weight of xyloglucan in azuki bean (Vigna angularis) walls (Nishitani and Matsuda, 1993). Fry and coworkers detected an enzyme in suspension-cultured cells that catalyses the glycosyl transfer reaction responsible for these effects and named it 'xyloglucan endotransglycosylase' (XET) (Fry et al., 1992). Following the development of a confusing and conflicting nomenclature for this class of proteins and their corresponding genes, XETs were recently renamed xyloglucan endotransglucosylase/hydrolases (XTHs) (Rose et al., 2002). XTHs constitute a large family of related but distinct enzymes (Fry, 1995; Nishitani, 1997) that can catalyse transglycosylation or hydrolysis of xyloglucan. When XTH functions as a transglycosylase, it breaks a glycosidic bond in the backbone of its donor substrate (a xyloglucan polysaccharide), forming two fragments. This process is likely to proceed by a mechanism similar to that used by configuration-retaining endoglucanases (Jakeman and Withers, 2002), in which the fragment containing the reducing end of the original polysaccharide is released and a covalent bond is formed between the enzyme and the other fragment. The acceptor substrate (a xyloglucan polysaccharide or oligosaccharide) then binds to the XTH-xyloglucan complex, displacing the covalently bound xyloglucan fragment, which is transferred to the non-reducing end of the acceptor substrate. XTH thereby has the capacity to catalyse the assembly of high-molecular-weight xyloglucan and its incorporation into the xyloglucan/cellulose network.

At least two xyloglucan domains are implicit in current models describing the cross-linked xyloglucan/cellulose network. One domain, comprising regions of xyloglucan that are not in direct contact with the microfibril, would include cross-links between microfibrils. A second domain comprises xyloglucan that is bound directly to the surface of cellulose microfibrils. A third possible domain comprises xyloglucan that is trapped within microfibrils (e.g. between 'crystalline units' of the microfibril). The greater mobility of the first domain should allow it to be distinguished from the latter two. Indeed, xyloglucan domains with different mobilities in hydrated onion cell walls have been distinguished by their proton relaxation times (T_2 and $T_{1\rho}$) measured by solid state NMR techniques (Ha et al., 1997). The more mobile xyloglucan component was attributed to xyloglucan chains present in the 'hydrated matrix between the microfibrils,' and presumably includes tethers in the xyloglucan/cellulose network. The more rigid xyloglucan component was attributed to xyloglucan chains that are 'spatially associated (within about 2 nm) with the cellulose chains.' The rigid xyloglucan is likely to be bound to the microfibril surface and/or trapped within the microfibril.

The presence of distinct xyloglucan domains in primary cell walls is supported by data obtained using sequential enzymic and chemical methods to release xyloglucan fractions from the wall (Pauly et al., 1999a). Treating depectinated pea-stem cell walls with a xyloglucan-specific endoglucanase (XEG) solubilizes the 'enzyme accessible' xyloglucan domain, which appears to undergo metabolic processing during tissue maturation. Extraction of the insoluble residue with 4M KOH solubilizes the xyloglucan that is presumably associated with the microfibril surface, and which undergoes minimal metabolic processing. Cellulase treatment of the insoluble residue remaining after KOH extraction degrades cellulose microfibrils and releases xyloglucan that is presumably trapped within the cellulose. The total amount of xyloglucan (18.4% of the cell wall) released by sequential XEG and KOH treatment was virtually the same as the amount of xyloglucan (18.3% of the cell wall) released by KOH treatment without prior XEG treatment. Thus, there do not appear to be numerous covalent attachments linking the trapped xyloglucan domain to the XEG-accessible xyloglucan domain. Otherwise, XEG treatment followed by KOH treatment would release more xyloglucan than KOH treatment alone because XEG treatment would cleave molecules that include a trapped domain, increasing the amount of xyloglucan that is susceptible to subsequent extraction by KOH. Xyloglucan solubilized by KOH treatment after XEG treatment had a molecular weight of approximately 30 kDa, and xyloglucan solubilized by KOH without prior XEG treatment had a molecular weight of approximately 100 kDa. These data suggest that the XEG-accessible xyloglucan is covalently linked to the xyloglucan that is extractable by KOH treatment.

Cell growth is accompanied by expansion of the xyloglucan/cellulose network. It is likely that this expansion depends on the XTH-catalysed cleavage and reformation of glycosidic bonds in the xyloglucan tethers, which would also facilitate the incorporation of new polysaccharides into the network. However, some researchers have questioned the role of XTH in regulating wall expansion (Kutschera, 2001). Fry and coworkers (Thompson *et al.*, 1997; Thompson and Fry, 2001) have presented evidence supporting the role of XTH in the integration of newly synthesized xyloglucan into the xyloglucan/cellulose network and the restructuring of the network during cell growth. It is likely that XTH acts on the 'enzyme accessible' (tether) domain rather than the domain that is closely associated with the microfibril surface.

The binding of xyloglucan to cellulose is likely to be a complex topological process. Based on conformational energy calculations, it has been proposed that binding requires the xyloglucan backbone to adopt a 'flat ribbon' conformation (Figure 1.10), whose surface is complementary to that of the cellulose microfibril (Levy *et al.*, 1991). These energy calculations also suggested that in solution xyloglucan molecules adopt a 'twisted' conformation that is not complementary to cellulose. All of the xyloglucan side chains must fold onto the same face of the xyloglucan backbone upon binding to cellulose, so as not to interfere with the interaction of the complementary xyloglucan and cellulose surfaces. According to this model (Figures 1.10 and 1.11), initiation of the binding process requires local flattening of the xyloglucan backbone. This flattened region would spread out as the binding interface is extended to adjacent segments of the xyloglucan, resulting in the rotation of the xyloglucan that extends away from the microfibril. If the distal portion of the xyloglucan mol-

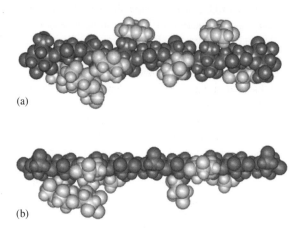

(a)

(b)

Figure 1.10 Proposed conformations of the xyloglucan backbone. (a) In aqueous solution, xyloglucan is likely to adopt a 'twisted ribbon' conformation. (b) Upon binding to the cellulose microfibril, xyloglucan is likely to adopt a 'flat ribbon conformation' where one surface is complementary to the microfibril surface.

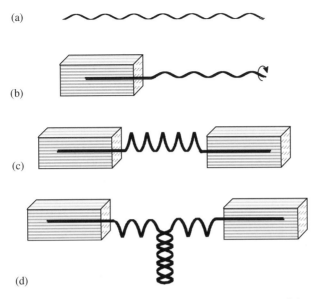

Figure 1.11   Illustration of the possible topological effects of the binding of the xyloglucan to cellulose. (a) Xyloglucan in solution, with a twisted conformation. (b) Binding of the xyloglucan 'untwists' the backbone, which can lead to rotation of the free end of the xyloglucan. (c) Coiled structures may evolve in xyloglucan molecules that do not have a free end. (d) Duplex structures could form if the energy barrier to rotation around glycosidic bonds in the xyloglucan backbone is sufficiently high and the anti-parallel coil is energetically stable.

ecule is also attached to a microfibril, this rotation could result in 'twining' to form coiled structures (Figure 1.11), such as those observed when xyloglucan molecules are dried down on mica surfaces (McCann *et al.*, 1990). However, the putative torque that is generated upon binding could be accommodated by rotation around glycosidic bonds in the xyloglucan backbone. Therefore, the extent of coiling would depend on kinetic considerations (e.g. the energetic barriers to rotations around glycosidic bonds) and thermodynamic considerations (e.g. the enthalpic and entropic contributions to the free energy of coil formation).

The XTH-catalysed cleavage and religation of glycosidic bonds in the xyloglucan backbone may lead to an increase or decrease in the topological complexity of the cell wall. Indeed, the reactions catalysed by XTH are related, in a topological sense, to the cleavage and religation reactions catalysed by DNA topoisomerases. This suggests that XTH functions as a polysaccharide topoisomerase during cell wall development, acting to generate or relax topologically constrained structures that arise during the assembly and restructuring of the xyloglucan/cellulose network.

Duplex structures such as that illustrated in Figure 1.11 are likely to occur in the cell wall only if the interaction of two segments of the xyloglucan backbone is thermodynamically favourable. Several observations are consistent with the idea that molecular-weight dependent interactions between segments of the xyloglucan backbone are thermodynamically accessible in solution. For example, the transient presence of molecular 'hyper-entanglements' (Gidley *et al.*, 1991) have been invoked to explain the rheological properties of tamarind xyloglucan. The intensity of xyloglucan staining with iodine increases as the molecular weight of the xyloglucan increases: xyloglucans with a molecular weight of less than 10 kDa are not stained (Hayashi, 1989) and deep staining is not observed for xyloglucans with a molecular weight of less than 20 kDa (W.S. York, unpublished results). It is possible that the linear array of iodine atoms required for the development of colour during iodine staining (Bluhm and Zugenmaier, 1981) depends on the presence of a binding pocket defined by the coils of a xyloglucan duplex (or multiplex) structure, which could be stabilized in solution by cooperative effects that require a minimum chain length. The affinity of monoclonal antibody CCRC-M1 to fucosylated xyloglucan is strongly influenced by the MW of the polysaccharide. This antibody binds to xyloglucans with a molecular weight greater than 20 kDa much more avidly than xyloglucan oligosaccharides (Puhlmann *et al.*, 1994). A multiple-valence effect has been ruled out for this behaviour, suggesting that the epitope is partially defined by conformational or topological states of xyloglucan, which might arise due to the cooperative (MW-dependent) interaction of xyloglucan segments. Interestingly, the molecular weight at which xyloglucan molecules appear to change their behaviour in solution (approximately 20 kDa) corresponds to a contour length of approximately 33 nm, which also corresponds to the length of xyloglucan tethers observed in xyloglucan/cellulose networks (McCann *et al.*, 1990; Satiat-Jeunemaitre *et al.*, 1992; Itoh and Ogawa, 1993; Whitney *et al.*, 1995) and to the contour length periodicity of xyloglucan (McCann *et al.*, 1992).

1.10.2 The pectic network of dicotyledon primary walls

Most researchers now agree that dicotyledon primary wall pectin is comprised of HGA, RG-I, and RG-II, albeit in different proportions. In some reproductive tissues some, or maybe all, of the HGA may be replaced by XGA (Huisman *et al.*, 2001). Current models of the organization of these polysaccharides envision a macromolecular complex in which HGA, RG-I and RG-II are covalently linked to one another, although direct evidence for these linkages is still lacking (Willats *et al.*, 2001a). The models for the structural organization of pectin have been inferred from the observations that (1) HGA, RG-I, and RG-II are all solubilized by treating walls with aqueous buffers and chelators, but are not separated by size-exclusion chromatography (SEC) and (2) RG-I and RG-II together with oligogalacturonides are also solubilized by treating walls with endopolygalacturonase, a enzyme that specifically fragments 1,4-linked α-D-Gal*p*A residues.

Homogalacturonan and RG-II both have 1,4-linked α-D-galacturonan backbones (see Figures 1.4 and 1.7), suggesting that RG-II is a highly branched domain within HG. Additional evidence that HG and RG-II are linked together has been obtained by treating high-molecular-weight pectin with endopolygalacturonase (Ishii and Matsunaga, 2001; B. Reuhs, J. Glenn, S. Stephens, J. Kim, D. Christie, J. Glushka,

M. O'Neill, S. Eberhard, P. Albersheim and A. Darvill, manuscript in preparation). The products generated by enzymic fragmentation consist of RG-I (100 kDa), RG-II (5–10 kDa) and oligogalacturonides with degrees of polymerization between 1 and 5. The high-molecular-weight pectin that has not been treated with endopolygalacturonase elutes at the column void volume. The molecular mass of a high-molecular-weight RG-II-containing pectin is also decreased by low pH, a treatment known to cleave borate esters (Ishii and Matsunaga, 2001). More importantly, reacting the low pH-treated pectin under conditions that favour the formation of borate cross-links generates a product whose molecular mass is comparable to the untreated material. Such results provide additional support for the notion that *in muro* B cross-linking of RG-II generates a macromolecular pectic network (Fleischer *et al.*, 1999).

The linkage of HGA (or XGA) to RG-I remains controversial since no oligosaccharide fragment has been isolated and shown to contain portions of HGA and RG-I.

A covalent linkage between pectin and xyloglucan was a major feature of the first model of the primary wall proposed by Albersheim and colleagues (Keegstra *et al.*, 1973). However, no oligosaccharide fragments containing portions of pectin and xyloglucan were isolated and structurally characterized. Subsequently, Thompson and Fry (2000) have reported that up to 12% of the xyloglucan in the walls of suspension-cultured rose cells is covalently linked to pectin. The authors suggested that the anionic xyloglucan was most likely attached, via arabinogalactan side chains, to HG but the nature of the covalent bond remains to be determined. Several investigators have suggested that pectin is covalently linked to hemicellulose, glycoprotein, and/or cellulose even though the oligosaccharides containing the putative cross-links have not yet been identified (reviewed in Ridley *et al.*, 2001). Similarly, additional data are required to substantiate the claims for the existence in primary walls of molecules including galacturonoyl-L-lysine amides that may covalently cross-link pectin and wall proteins (Perrone *et al.*, 1998).

The arabinan and galactan side chains of RG-I are esterified with feruloyl and coumaroyl residues in the primary walls of some dicotyledonous plants, such as beet and spinach (Ishii, 1997a). The oxidative coupling of pectin-bound phenolic residues *in vitro* has been shown to result in the formation of dehydrodiferuloyl cross-linked pectin (Saulnier and Thibault, 1999), although there is no evidence that such cross-links are present in the primary wall. The existence of, as yet unidentified, non-methyl esters on HG remains a subject of debate (Needs *et al.*, 1998). Such esters, if they exist, may be formed between the carboxyl group of a Gal*p*A residue on one HG chain and a hydroxyl group on either a Gal*p*A residue in a separate chain or a hydroxyl group on a neutral glycosyl residue. In addition, internal esters (lactones) may form between adjacent or non-adjacent Gal*p*A residues in a HG chain. AFM images of tomato pectin have revealed branched structures which has provided additional evidence that some HG chains may be covalently linked to one another (Round *et al.*, 1997, 2001; see Chapter 2), although further studies are required to establish the nature of the cross-link.

1.10.3 Borate cross-linking of RG-II and the pectic network of primary walls

A relationship between boron and the primary wall pectic polysaccharides was first suggested by Schmucker in 1933 and subsequently supported by the observations of Smith in 1944 and Yamanouchi in 1971 (reviewed in Loomis and Durst, 1992; Matoh and Kobayashi, 1998). Subsequently, studies have confirmed that the B requirement and wall pectin content of plants are correlated (Hu and Brown, 1994; Hu *et al.*, 1996; Matoh *et al.*, 1996). Such studies are consistent with the fact that the Poaceae, whose primary walls contain quantitatively small amounts of pectin, have a much lower B requirement than the dicotyledons and non-graminaceous monocotyledons (Hu *et al.*, 1996; Matoh *et al.*, 1996).

There is a considerable amount of evidence showing that RG-II exists in the primary wall as a dimer that is cross-linked by a borate-diol ester. However, the function of this cross-link has only recently begun to emerge. Brown and Hu (1997) hypothesized that B has only a structural role in plants and that its function is to cross-link primary wall pectin and thereby control the organization and physical and biochemical properties of the pectic network. The symptoms of B deficiency in plants would be a direct result of changes in the physical properties of the wall that result from abnormal pectic network formation. Evidence supporting this notion was obtained when suspension-cultured *Chenopodium album* cells were shown to grow and divide in the absence of added B as long as they are maintained in the logarithmic phase of growth (Fleischer *et al.*, 1998). The walls of the B deficient cells ruptured when they entered the stationary phase. The authors concluded that B had a role in controlling the physical properties of the wall because the walls of B deficient cells have a larger size-exclusion limit (~6 nm) than the walls of cells grown with normal amounts of B (~3.5 nm). Subsequently, the walls of B deficient cells were shown to contain monomeric RG-II but no borate cross-linked RG-II dimer (Fleischer *et al.*, 1999). Adding physiological amounts of boric acid to the B-deficient cells resulted, within 10 minutes, in a decrease of wall pore size to near wild-type values, together with the formation of the RG-II dimer. The B-treated cells remain viable in the stationary phase. B-deficient suspension-cultured tobacco cells have swollen walls and only 40% of the RG-II in the wall is cross-linked by borate (Matoh *et al.*, 2000). Boron deficiency in pumpkin plants results in a substantial reduction in growth and is accompanied by cell wall thickening and a decrease in borate cross-linking of RG-II (Ishii *et al.*, 2001). Normal growth is restored, wall thickening is reduced, and the amount of RG-II cross-linking is increased to normal levels by supplying borate to the B-deficient plants. These results, when taken together, add support to earlier studies that had suggested that the pectin network controlled the pore size and physical properties of the primary wall, and that altering the structure of this network has a major influence on plant growth (O'Neill *et al.*, 1996; Fleischer *et al.*, 1999; O'Neill *et al.*, 2001).

Germanium (Ge) has been reported to delay the symptoms of B deficiency in plants and to substitute for B in suspension-cultured carrot cells (Loomis and Durst, 1992). These authors speculated that Ge may form diester cross-links, albeit with different dimensions and geometry than borate diesters. Subsequent studies have demonstrated that reacting monomeric RG-II *in vitro* with germanic acid resulted in the formation of a product that had the chromatographic characteristics of the RG-II dimer (Kobayashi *et al.*, 1997). However, no evidence was presented to show that the RG-II-Ge dimer contained a 1:2 germante-diol diester. The putative RG-II-Ge dimer is less stable than the borate cross-linked RG-II dimer, consistent with the notion that a germanate ester cross-link is less stable than a borate cross-link. Monomeric RG-II also forms a dimer when reacted with germanium dioxide *in vitro* (T. Ishii, personal communication); however, germanium dioxide treatment did not rescue the growth of B-deficient pumpkin plants nor did it result in the formation of quantitatively significant amounts of dimeric RG-II-Ge even though germanate accumulated in the leaf cell walls (T. Ishii, personal communication). Thus, it would appear that Ge does not substitute for B in the cross-linking of RG-II at least in pumpkin plants.

Many investigations of the function of B in plants have used plants grown with sub-optimal amounts of borate (Dell and Huang, 1997). The results of such studies are often ambiguous because the primary and secondary effects of B deficiency are difficult to distinguish (Dell and Huang, 1997). Thus, the generation of *Arabidopsis* mutants that require increased amounts of B for their normal growth has provided a unique opportunity to investigate the biological role(s) of this essential micro-element.

The aerial portions of the *Arabidopsis thaliana mur1-1* and *mur1-2* mutants, which are dwarfed and have brittle stems (Reiter *et al.*, 1993), contain less than 2% of the amount of L-fucose (6-deoxy-L-galactose) present in wild-type plants (Reiter *et al.*, 1997). The altered gene in *mur1* plants has been shown to encode an isoform of GDP-mannose 4,6-dehydratase, a enzyme required for the biosynthesis of L-fucose (Bonin *et al.*, 1997). L-fucosyl residues are present in several *Arabidopsis* wall polysaccharides including RG-I, RG-II, and XG (Zablackis *et al.*, 1995), the oligosaccharide side chains of AGPs and the N-linked oligosaccharides of glyco-proteins (Rayon *et al.*, 1999). Thus, the absence of L-fucosyl residues in one or more of these complex glycans may result in the dwarf phenotype of *mur1* plants. The absence of fucosyl residues in xyloglucan is not likely to be responsible for the dwarf phenotype of *mur1* plants since *Arabidopsis mur2* plants, which are deficient in a xyloglucan-specific fucosyl transferase, synthesize xyloglucan that contains less than 2% of the normal amounts of L-Fuc but grow normally under laboratory conditions (Vanzin *et al.*, 2002). Similarly, the absence of L-Fuc residues in the N-linked complex glycan side chains of glycoproteins synthesized by *mur1* plants (Rayon *et al.*, 1999) is also unlikely to result in the dwarf phenotype, since an *Arabidopsis* mutant (*cgl1*), which synthesizes glycoproteins that lack L-Fuc, grows normally (von Schaewen *et al.*, 1993).

In plants carrying the *mur1* mutation the L-fucosyl (L-Fuc) and 2-*O*-methyl L-fucosyl (2-*O*-Me L-Fuc) residues of RG-II are replaced by L-galactosyl (L-Gal) and 2-*O*-methyl L-galactosyl (2-*O*-Me L-Gal) residues, respectively (O'Neill *et al.*, 2001; B. Reuhs, J. Glenn, S. Stephens, J. Kim, D. Christie, J. Glushka, M. O'Neill, S. Eberhard, P. Albersheim, and A. Darvill, manuscript in preparation). Somewhat unexpectedly, only 50% of the RG-II in the rosette leaves of *mur1* plants is cross-linked by borate, whereas at least 95% of the RG-II is cross-linked in wild-type plants (O'Neill *et al.*, 2001). The altered structure of *mur1* RG-II was shown using *in vitro* studies to result in a reduction of the rate of dimer formation and to decrease the stability of the borate cross-link (O'Neill *et al.*, 2001). Such a result suggested that the dwarf phenotype of *mur1* plants was a consequence of reduced cross-linking of RG-II. This hypothesis was confirmed by the demonstration that spraying *mur1* plants with aqueous borate rescues their growth and also results in an increase in the extent of borate cross-linking of RG-II. The growth of *mur1* plants is also rescued by exogenous L-Fuc treatment (Reiter *et al.*, 1993; O'Neill *et al.*, 2001). The L-Fuc-treated *mur1* plants synthesize RG-II that contains L-Fuc and 2-*O*-Me L-Fuc residues (O'Neill *et al.*, 2001). Thus, the rate of dimer formation, the stability of the cross-link dimer, and the B requirement of the L-Fuc-treated *mur1* plants are comparable to wild-type plants (O'Neill *et al.*, 2001).

The primary walls of *mur4* plants contain reduced amounts of arabinose, although the plants themselves have no visible phenotype (Reiter *et al.*, 1997). The *MUR4* gene is believed to encode a UPD-xylose 4-epimerase, an enzyme involved in the conversion of UDP-D-Xyl to UDP-L-arabinose (Burget and Reiter, 1999). An *Arabidopsis* mutant with an extreme dwarf phenotype has been generated by crossing *mur1* and *mur4* plants (Reiter *et al.*, 1997) and preliminary studies (M.A. O'Neill and W.-D. Reiter, unpublished results) indicate that borate treatment rescues the growth of the double mutant, again suggesting that reduced RG-II cross-linking is in part responsible for the dwarf phenotype.

Arabidopsis plants carrying the *bor1* mutation are extremely dwarfed (Noguchi *et al.*, 1997). The altered gene in these plants has not been identified but is believed to encode a protein that has a role in the uptake and/or transport of B to the leaves, since this mutant requires higher concentrations of B than wild-type plants for normal growth (Takano *et al.*, 2001). Wild-type and *bor1* RG-II have comparable glycosyl residue compositions; however, ~60% of the RG-II in the rosette leaf cell walls of *bor1* plants is present as the monomer (O'Neill *et al.*, 2003). Borate treatment rescues the growth of *bor1* plants and the extent of RG-II cross-linking in their walls is comparable to that of wild-type plants (O'Neill *et al.*, 2003). Thus, reduced cross-linking of RG-II is likely to be responsible for the dwarf phenotype of *bor1* plants. Nevertheless, the possibility cannot be discounted that in *bor1* plants B also functions in an as yet unidentified manner.

The cumulative results of several studies suggest that boron's primary role is in turgor-driven wall expansion rather than in cell division and that covalent cross-linking of pectin has a role in determining the physical and biochemical properties of the primary wall (O'Neill *et al.*, 2001, 2003). Nevertheless, the mechanism(s) by which reduced cross-linking of RG-II affects plant growth is not understood. Borate cross-links are unlikely to be load-bearing since decreasing the number of such bonds would be expected to result in 'wall loosening' together with uncontrolled wall expansion. This is in direct contrast to the available experimental data which show that reducing borate cross-linking results in a decrease in cell expansion and plant growth. The covalent cross-linking of pectin may generate a structurally well-defined three-dimensional network that facilitates restructuring of the cellulose-hemicellulose network that itself is required for wall expansion. Alternatively, the borate cross-link may act as a mechanical sensor that provides the cell with information concerning the physical state of the wall (O'Neill *et al.*, 2003). In the absence of this cross-link, it is possible that the cell cannot 'perceive' the physical status of the wall and therefore does not synthesize additional xyloglucan and cellulose or restructure the existing wall architecture to allow cell expansion.

The ability of RG-II to self assemble *in vitro* into a dimer in the presence of borate suggests that a pectin network may also self assemble *in muro* (Fleischer *et al.*, 1999). The organisation of this network is likely to be controlled in large part by the number and distribution of RG-II molecules among HG chains. A typical dicotyledon primary wall will contain, on average, one RG-II molecule per 50 Gal*p*A residues in a HG chain. Some of these HG chains must contain at least two RG-II molecules that cross-link with RG-II molecules on other HG chains if the pectic network is covalently cross-linked only by borate esters. A network that is also interconnected by ionic cross-links would only require a single RG-II molecule in each HG chain to be linked by borate esters. A borate ester may provide increased structural order to a pectic network because this cross-link forms only between specific apiosyl residues. In contrast, a Ca^{2+} cross-link may form between any two appropriately positioned Gal*p*A residues. Nevertheless, Ca^{2+} ions are believed to be required to maintain the stability of the borate cross-link *in muro* (Fleischer *et al.*, 1999; Kobayashi *et al.*, 1999), although it is not known if the Ca^{2+} forms a co-ordination complex with the borate ester or ionically cross-links nearby Gal*p*A residues. Additional studies are required to determine the relationship between B, Ca^{2+}, and the pectic network of primary cell walls (Koyama *et al.*, 2001). Similarly, the reported relationship between B deficiency and an increase in the levels of cytoskeletal proteins in *Arabidopsis* root cells (Yu *et al.*, 2001b) needs further study as this may shed light on the biological function of the putative primary wall–plasma membrane–cytoskeleton continuum (see Chapter 5).

Little is known about the cellular location of borate ester cross-linking of RG-II, although the available evidence suggests that this reaction is most likely to occur in the apoplast. For example, the walls of B-deficient *C. album* cells (Fleischer *et al.*, 1999) and the leaves of B-deficient pumpkin plants (T. Ishii, personal communication) contain virtually no dimeric RG-II. The addition of borate to these tissues results within 10–30 min in the conversion of most of the monomeric RG-II to the dimer. The total amounts of RG-II in the walls was not increased by borate treatment, suggesting that the treatment did not induce the synthesis of significant amounts of RG-II, which itself could have been cross-linked prior to its insertion into the wall. Rather, the data suggests that previously deposited RG-II was cross-linked upon addition of borate. Further experiments are required to determine whether RG-II must be deposited into the cell wall before the borate cross-link can form.

1.11 Conclusions

Models of the primary cell wall are based in large part on incomplete structural information obtained from the analysis of polysaccharide mixtures. Nevertheless, such analyses have provided much useful information regarding the overall structures of the polysaccharides. For example, the backbones of all the quantitatively major polysaccharides have been fully characterized. Somewhat less is known about the distribution of side chains along the backbones of these polysaccharides. In particular, little is known about the distribution of the side chains in rhamnogalacturonans such as RG-I. Furthermore, it has become apparent that the structures of cell wall polysaccharides vary in different tissues and even within regions of the same wall. A comprehensive understanding of the structure, assembly, and organization of the primary wall during the different phases of plant growth will require the development of more sophisticated and sensitive methods for determining polysaccharide structure and the integration of these techniques with recent advances in molecular biology, immunocytochemistry, enzymology, spectroscopy, and computational chemistry.

Acknowledgements

We gratefully acknowledge funding from the United States Department of Energy (Grant number DE-FG02–96ER20220) and the United States National Science Foundation (Grant number MCB-9974673).

References

Albersheim, P. (1976) The primary cell wall. In *Plant Biochemistry*, 3rd edn (eds J. Bonner and J. Varner), Academic Press, New York, pp. 225–274.

An, J., O'Neill, M.A., Albersheim, P. and Darvill, A.G. (1994) Isolation and structural characterization of β-D-glucosyluronic acid and 4-O-methyl β-D-glucosyluronic acid-containing oligosaccharides from the cell-wall pectic polysaccharide rhamnogalacturonan I. *Carbohydr. Res.*, 252, 235–243.

Aspinall, G.O. (1982) Chemical characterization and structure determination of polysaccharides, in *The Polysaccharides*, *Vol 1* (ed. G.O. Aspinall), Academic Press, New York, pp 35–131.

Atkinson, R.G., Schroder, R., Hallett, I.C., Cohen, D. and MacRae, E.A. (2002) Over expression of polygalacturonase in transgenic apple trees leads to a range of novel phenotypes involving changes in cell adhesion. *Plant Physiol.*, 129, 122–133.

Azadi, P., O'Neill, M.A., Bergmann, C., Darvill, A.G. and Albersheim, P. (1995) The backbone of the pectic polysaccharide rhamnogalaturonan I is cleaved by an *endo*hydrolase and an *endo*lyase. *Glycobiology*, 5, 783–789.

Bax, A. and Summers, M.F. (1986) ^{1}H and ^{13}C assignments from sensitivity-enhanced detection of heteronuclear multiple-bond connectivity by 2D multiple quantum NMR. *J. Am. Chem. Soc.*, 108, 2093–2094.

Baydoun, E.A.-H. and Fry, S.C. (1989) *In vivo* degradation and extracellular polymer-binding of xyloglucan nonasaccharide, a naturally-occurring anti-auxin. *J. Plant Physiol.*, 134, 453–459.

Bluhm, T.L. and Zugenmaier, P. (1981) Detailed structure of the V_h-amylose-iodine complex: a linear polyiodide chain. *Carbohydr. Res.*, 89, 1–10.

Bodenhausen, G. and Ruben, D.J. (1980) Natural abundance nitrogen-15 NMR by enhanced heteronuclear spectroscopy. *Chem. Phys. Lett.* 69, 185–189.

Bonin, C.P., Potter, I., Vanzin, G.F. and Reiter, W.-D. (1997) The *MUR1* gene of *Arabidopsis thaliana* encodes an isoform of GDP-D-mannose-4,6-dehydratase, catalyzing the first step in the *de novo* synthesis of GDP-L-fucose. *Proc Natl. Acad. Sci. USA*, 94, 2085–2090.

Bouveng, H.O. (1965) Polysaccharides in pollen. 2. Xyloglacturonan from mountain pine (*Pinus mugoturra*) pollen. *Acta Chem. Scand.* 19, 953–963.

Brabetz, W., Wolter, F.P. and Brade, H. (2000) A cDNA encoding 3-deoxy-D-*manno*-oct-2-ulosonate-8-phosphate synthase of *Pisum sativum* L. (pea) functionally complements a *kdsA* mutant of the gram-negative bacterium *Salmonella enterica*. *Planta*, 212, 136–143.

Brown J.A. and Fry, S.C. (1993) Novel *O*-galacturonosyl esters in the pectic polysaccharides of suspension-cultured plant cells. *Plant Physiol.*, 103, 993–999.

Brown, P.H. and Hu, H. (1997) Does boron play only a structural role in the growing tissues of higher plants? *Plant Soil*, 196, 211–215.

Burget, E.G. and Reiter, W.-D. (1999) The *mur4* mutant of arabidopsis is partially defective in the de novo synthesis of uridine diphospho L-arabinose. *Plant Physiol.*, 121, 383–389.

Bush, M.S. and McCann, M.C. (1999) Pectic epitopes are differentially distributed in the cell walls of potato (*Solanum tuberosum*) tubers. *Physiol Plant.*, 107, 201–213.

Bush, C.A., Martin-Pastor, M. and Imberty, A. (1999) Structure and conformation of complex carbohydrates, glycoproteins, glycolipids, and bacterial polysaccharides. *Annu. Rev. Biophys. Biomol. Struct.*, 28, 269–293.

Carpita, N.C. (1996) Structure and biogenesis of the cell walls of grasses. *Annu. Rev. Plant Physiol. Mol. Biol.*, 47, 445–471.

Carpita, N.C. and Gibeaut, D.M. (1993) Structural models of primary-cell walls in flowering plants – consistency of molecular structures with the physical properties of the wall during growth. *Plant J.*, 3, 1–30.

Cartier, N., Chambat, G. and Joseleau, J.-P. (1988) Cell wall and extracellular galactoglucomannans from suspension-cultured *Rubus fructicocus* cells. *Phytochemistry*, 27, 1361–1364.

Cheng, L. and Kindell, P. (1997) Detection and homogeneity of cell wall pectic polysaccharides of *Lemna minor. Carbohydr. Res.*, 301, 205–212.

Cosgrove, D.J. (1999) Enzymes and other agents that enhance cell wall extensibility. *Annu. Rev. Plant Physiol. Mol. Biol.*, 50, 391–417.

Daas, P.J.H., Arisz, P.W., Schols, H.A., DeRuiter, G. and Voragen, A.G.J. (1998) Analysis of partially methyl-esterified galacturonic acid oligomers by high-performance anion-exchange chromatography and matrix-assisted laser desorption/ionization time-of-flight mass spectrometry. *Anal. Biochem.*, 257, 195–202.

Darvill, J.E., McNeil, M., Darvill, A.G. and Albersheim, P. (1980) Structure of plant cell walls. XI. Glucuronoarabinoxylan, a second hemicellulose in the primary cell walls of suspension-cultured sycamore cells. *Plant Physiol.*, 66, 1135–1139.

Dell, A. (1987) F.A.B.-Mass spectrometry of carbohydrates. *Adv. Carbohydr. Chem. Biochem.*, 45, 19–72.

Dell, B. and Huang, L (1997) Physiological responses of plants to low boron. *Plant Soil*, 193, 103–120.

Dell, A. and Morris, H.R. (2001) Glycoprotein structure determination by mass spectrometry. *Science*, 291, 2351–2356.

Doco, T., Williams, P., Vidal, S. and Pellerin, P. (1997) Rhamnogalacturonan II, a dominant polysaccharide in juices produced by enzymic liquefaction of fruits and vegetables. *Carbohydr. Res.*, 297, 181–186.

Domon, B. and Costello, C.E. (1988) A systematic nomenclature for carbohydrate fragmentations in FAB-MS/MS spectra of glycoconjugates. *Glycoconjugate J.*, 5, 397–409.

Duus, J.Ø., Gotfredsen, C.H. and Bock, K. (2000) Carbohydrate structural determination by NMR spectroscopy: modern methods and limitations. *Chem. Rev.*, 100, 4589–4614.

Ebringerova, A. and Heinze, T. (2000) Xylan and xylan derivatives – biopolymers with valuable properties, 1. Naturally occurring xylans structures, isolation procedures and properties. *Macromol. Rapid Commun.*, 21, 542–556.

Eda, S., Akiyama, Y., Kato, K. *et al.* (1984) Structural investigation of a galactoglucamannan from cell walls of tobacco (*Nicotiana tabacum*) midrib. *Carbohydr. Res.*, 131, 105–118.

Eda, S., Akiyama, Y., Kato, K., Ishizu, A. and Nakano J. (1985) A galactoglucamannan from cell walls of suspension-cultured tobacco (*Nicotiana tabacum*) cells. *Carbohydr. Res.*, 137, 173–181.

Epstein, E. (1999) Silicon. *Annu. Rev. Plant Physiol. Plant Mol. Biol.*, 50, 641–664.

Fisher, D.B. (2000) Long-distance transport. In *Biochemistry and Molecular Biology of Plants* (eds B. Buchanan, W. Gruisseman and R. Jones), American Society of Plant Physiologists, Rockville, MD, pp. 730–784.

Fleischer, A., Titel., C. and Ehwald, R. (1998) The boron requirement and cell wall properties of growing and stationary-phase suspension-cultured *Chenopodium album* L. cells. *Plant Physiol.*, 117, 1401–1410.

Fleischer, A., O'Neill., M.A. and Ehwald, R. (1999) The pore size of non-graminaceous plant cell walls is rapidly decreased by borate ester cross-linking of the pectic polysaccharide rhamnogalacturonan II. *Plant Physiol.*, 121, 829–838.

Freshour, G., Fuller, M.S., Albersheim, P., Darvill, A.G. and Hahn, M.G. (1996) Developmental and tissue-specific structural alterations of the cell-wall polysaccharides of *Arabidopsis thaliana* roots. *Plant Physiol.*, 110, 1413–1429.

Freshour, G. Bonin, C.P., Reiter, W.-D., Albersheim, P., Darvill, A.G. and Hahn, M.G. (2003) Distribution of fucose-containing xyloglucans in cell walls of the *mutl* mutant of *Arabidopsis*. *Plant Physiol.*, 131, 1602–1612.

Fry, S.C. (1995) Polysaccharide-modifying enzymes in the plant cell wall. *Annu. Rev. Plant. Physiol. Mol. Biol.*, 46, 497–520.

Fry, S.C., Smith, R.C., Renwick, K.F., Martin, D.J., Hodge, S.K. and Matthews K.J. (1992) Xyloglucan endotransglycosylase, a new wall-loosening enzyme activity from plants. *Biochem. J.*, 282, 821–828.

Fry, S.C., York, W.S., Albersheim, P. *et al.* (1993) An unambiguous nomenclature for xyloglucan-derived oligosaccharides. *Physiol. Plant.*, 89, 1–3.

Gaspar, Y., Johnson, K.L., McKenna, J.A., Bacic, A. and Schultz, C.J. (2001) The complex structures of arabinogalactan-proteins and the journey towards understanding function. *Plant Mol. Biol.*, 47, 161–176.

Geshi, N., Jorgensen, B., Scheller, H.V. and Ulvskov, P. (2000) In vitro biosynthesis of 1,4-β-galactan attached to rhamnogalacturonan I. *Planta*, 210, 622–629.

Gidley, M.J., Lillford, P.J., Rowlands, D.W. *et al.* (1991) Structure and properties of tamarind-seed polysaccharide. *Carbohydr. Res.*, 214, 299–314.

Giovannoni, J.J., DellaPena, D., Bennett, A.B. and Fischer, R.L. (1989) Expression of chimeric polygalacturonase gene in transgenic rin (ripening inhibitor) tomato fruit results in polyuronide degradation but not fruit softening. *Plant Cell*, 1, 53–63.

Glushka, J., Terrell, M., York, W.S. *et al.* (2003) Primary structure of the 2-O-methyl fucose-containing side chain of the pectic polysaccharide, rhamnogalacturonan II. *Carbohydr. Res.*, 338, 341–352.

Goldberg, R., Morvan, C., Jauneau, A. and Jarvis, M.C. (1996) Methyl esterification, de-esterification and gelation of pectins in the primary cell wall. In *Pectins and Pectinases* (eds J. Visser and A.G.J. Voragen), Elsevier Science, Amsterdam, pp. 151–172.

Golovchenko, V.V., Ovodova, R.G., Shaskov, A.S. and Ovodov, Y.S. (2002) Structural studies of the pectic polysaccharide from duckweed *Lemna minor* L. *Phytochemistry*, 60, 89–97.

Griffiths, W.J., Jonsson, A.P., Liu, S., Rai, D.K. and Wang, Y. (2001) Electrospray and tandem mass spectrometry in biochemistry. *Biochem. J.* 355, 545–561.

Gruppen, H., Hoffman, R.A., Kormelink, F.J.M., Voragen, A.G.J., Kamerling, J.P. and Vliegenthart, J.F.G. (1992) Characterization by ^{1}H NMR spectroscopy of enzymically derived oligosaccharides from alkali-extractable wheat-flour arabinoxylan. *Carbohydr. Res.*, 233, 45–64.

Ha, M.A., Apperley, D.C. and Jarvis, M. (1997) Molecular rigidity in dry and hydrated onion cell walls. *Plant Physiol.*, 115, 593–598.

Hadfield, K.A. and Bennet, A.B. (1998) Polygalacturonases: many genes in search of a function. *Plant Physiol.*, 117, 337–343.

van Halbeek, H. (1994) NMR developments in structural studies of carbohydrates and their complexes. *Curr Opin. Struct. Biol.*, 4, 697–709.

Hantus, S., Pauly, M., Darvill, A.G., Albersheim, P. and York, W.S. (1997) Structural characterization of novel L-galactose-containing oligosaccharide subunits of jojoba seed xyloglucans. *Carbohydr. Res.*, 304, 11–20.

Harvey, D J. (1999) Matrix-assisted laser desorption/ionization mass spectrometry of carbohydrates. *Mass Spectrom. Reviews*, 18, 349–451.

Hayashi, T. (1989) Measuring β-glucan deposition in plant cell walls. In *Modern Methods of Plant Analysis, New Series*, Vol. 10 (eds H.F. Linskens and J.F. Jackson), Springer Verlag, Berlin, pp. 138–160.

Hisamatsu, M., York, W.S., Darvill, A.G. and Albersheim, P. (1992) Structure of plant cell walls XXXIV. Characterization of seven xyloglucan oligosaccharides containing from seventeen to twenty glycosyl residues. *Carbohydr. Res.*, 227, 45–71.

Hu, H. and Brown, P.H. (1994) Localization of boron in the cell walls of squash and tobacco and its association with pectin. *Plant Physiol.*, 105, 681–689.

Hu, H., Brown, P.H. and Labavitch, J.M. (1996) Species variability in boron requirement is correlated with cell wall pectin. *J. Exp. Bot.*, 47, 227–232.

Huisman, M.M.H., Fransen, C.T.M., Kamerling, J.P., Vliegenthart, J.F.G., Schols, H.A. and Voragen, A.G.J. (2001) The CDTA-soluble pectic substances from soybean meal are composed of rhamnogalacturonan and xylogalacturonan but not homogalacturonan. *Biopolymers*, 58, 279–294.

Ishii, T. (1997a) Structure and function of feruloylated polysaccharides. *Plant Sci.*, 127, 111–127.

Ishii, T. (1997b) *O*-acetylated oligosaccharides from pectins of potato tuber cell walls. *Plant Physiol.*, 113, 1265–1272.

Ishii, T. and Kaneko, S. (1998) Oligosaccharides generated by partial hydrolysis of the borate-rhamnogalacturonan II complex from sugar beet. *Phytochemistry*, 49, 1195–1202.

Ishii, T. and Matsunaga, T. (1996) Isolation and characterization of a boron-rhamnogalacturonan II complex from cell walls of sugar beet pulp. *Carbohydr. Res.*, 284, 1–9.

Ishii, T. and Matsunaga, T. (2001) Pectic polysaccharide rhamnogalacturonan II is covalently linked to homogalacturonan. *Phytochemistry*, 57, 969–974.

Ishii, T. and Ono, H. (1999) NMR spectroscopic analysis of the borate diol esters of methyl apiofuranoside. *Carbohydr. Res.*, 321, 257–260.

Ishii, T., Thomas, J.R., Darvill, A.G. and Albersheim, P. (1989) Structure of plant cell walls. XXVI. The walls of suspension-cultured sycamore cells contain a family of rhamnogalacturonan-I-like pectic polysaccharides. *Plant Physiol.*, 89, 421–428.

Ishii, T., Matsunaga, T., Pellerin, P., O'Neill, M.A., Darvill, A.G. and Albersheim, P. (1999) The plant cell wall polysaccharide rhamnogalacturonan II self-assembles into a covalently cross-linked dimer. *J. Biol. Chem.*, 274, 13098–13109.

Ishii, T., Matsunaga, T. and Hayashi, N. (2001) Formation of rhamnogalacturonan II-borate dimer in pectin determines cell wall thickness of pumpkin tissue. *Plant Physiol.*, 126, 1698–1705.

Itoh, T. and Ogawa, T. (1993) Molecular architecture of the cell wall of poplar cells in suspension culture, as revealed by rapid-freezing and deep-etching techniques. *Plant Cell Physiol.*, 34, 1187–1196.

Jacobs, A. and Dahlman, O. (2001) Enhancement of the quality of MALDI mass spectra of highly acidic oligosaccharides by using a nafion-coated probe. *Anal. Chem.*, 73, 405–410.

Jakeman, D.L. and Withers, S.G. (2002) Engineering glycosidases for constructive purposes. In *Carbohydrate Bioengineering – Interdisciplinary Approaches* (eds T.T. Teeri, B. Svensson, H.J. Gilbert and T. Feizi). The Royal Society of Chemistry, Cambridge, UK, pp. 5–8.

Jose-Estanyol, M. and Puigdomenech, P. (2000) Plant cell wall glycoproteins and their genes. *Plant Physiol. Biochem.*, 38, 97–108.

Kaneko, S., Ishii, T. and Matsunaga, T. (1997) A boron-rhamnogalacturonan-II complex from bamboo shoot cell walls. *Phytochemistry*, 44, 243–248.

Keegstra, K., Talmadge, K.W., Bauer, W.D. and Albersheim, P. (1973) Structure of plant cell walls. III. A model of the walls of suspension-cultured sycamore cells based on the interconnections of the macromolecular components. *Plant Physiol.*, 51, 188–197.

Kieliszewksi, M.J. and Shpak, E. (2001) Synthetic genes for the elucidation of glycosylation codes for arabinogalactan-proteins and other hydroxyproline-rich glycoproteins. *Cell. Mol. Life Sci.*, 58, 1386–1398.

Kikuchi, A., Edashige, Y., Ishii, T. and Satoh, S. (1996) A xyloglacturonan whose level is dependent on the size of cell clusters is present in the pectin from cultured carrot cells. *Planta*, 200, 369–372.

Kim, J.-B. and Carpita, N.C. (1992) Changes in esterification of the uronic acid groups of cell wall polysaccharides during elongation of maize. *Plant Physiol.*, 98, 646–653.

Kobayashi, M., Matoh, T. and Azuma, J. (1995) Structure and glycosyl composition of the boron-polysaccharide complex of radish roots. *Plant Cell Physiol.*, 36S, 139.

Kobayashi, M., Matoh, T. and Azuma, J. (1996) Two chains of rhamnogalacturonan II are cross-linked by borate-diol ester bonds in higher plant cell walls. *Plant Physiol.*, 110, 1017–1020.

Kobayashi, M., Ohno, K. and Matoh, T. (1997) Boron nutrition of cultured tobacco BY-2 cells. Characterization of the boron-polysaccharide complex. *Plant Cell Physiol.*, 38, 676–683.

Kobayashi, M., Nakagawa, H., Asaka, T. and Matoh, T. (1999) Borate-rhamnogalacturonan II bonding reinforced by Ca2+ retains pectic polysaccharides in higher plant cell walls. *Plant Physiol.*, 119, 199–203.

Komalavilas, P. and Mort, A.J. (1989) The acetylation of O-3 of galacturonic acid in the rhamnose-rich regions of pectins. *Carbohydr. Res.*, 189, 261–272.

Kornfeld, R. and Kornfeld, S. (1985) Assembly of asparagine-linked oligosaccharides. *Annu. Rev. Biochem.*, 54, 631–664.

Koyama, H., Toda, T. and Hara, T. (2001) Brief exposure to low-pH causes irreversible damage to the growing root in *Arabidopsis thaliana*: pectin-Ca interaction may play an important role in proton rhizotoxicity. *J. Exp. Bot.*, 52, 361–368.

Kutschera, U. (2001) Stem elongation and cell wall proteins in flowering plants. *Plant Biol.*, 3, 466–480.

Lau, J.M., McNeil, M., Darvill, A.G. and Albersheim, P. (1985) Structure of the backbone of rhamnogalacturonan I, a pectic polysaccharide in the cell walls of plants. *Carbohydr. Res.*, 137, 111–125.

Lau, J.M., McNeil, M., Darvill, A.G. and Albersheim, P. (1987) Treatment of rhamnogalacturonan I with lithium in ethylenediamine. *Carbohydr. Res.*, 168, 245–274.

Lerouge, P., O'Neill, M.A., Darvill, A.G. and Albersheim, P. (1993) Structural characterization of endo-glycanase-generated oligoglycosyl side chains of rhamnogalacturonan I. *Carbohydr. Res.*, 243, 359–371.

Levy, S., and Staehelin, L.A. (1992) Synthesis, assembly and function of plant cell wall macromolecules. *Curr. Opin. Cell Biol.*, 4, 856–862.

Levy, S., York, W.S., Struike-Prill, R., Meyer, B. and Staehelin, L.A. (1991) Simulations of the static and dynamic molecular conformations of xyloglucan. The role of the fucosylated sidechain in surface-specific sidechain folding. *Plant J.*, 1, 195–215.

Limberg, G., Korner, R., Bucholt, H.C., Christensen, T.M.I.E., Roepstorff, P. and Mikklesen, J.D. (2000) Quantitation of the amount of galacturonic acid residues in block sequences in pectin homogalacturonan by enzymatic fingerprinting with *exo*- and *endo*-polygalacturonase II from *Aspergillus niger. Carbohydr. Res.*, 327, 321–332.

Liners, F., Gaspar, T. and van Cutsem, P. (1994) Acetyl and methyl-esterification of pectins of friable and compact sugar-beet calli: consequences for intercellular adhesion. *Planta*, 192, 545–556.

Loomis, W.D. and Durst, R.W. (1992) Chemistry and biology of boron. *BioFactors,* 3, 229–239.

Mamyrin, B.A. (2001) Time-of-flight mass spectrometry (concepts, achievements, and prospects). *Int. J. Mass Spectrom.*, 206, 251–266.

Martin, C., Bhatt, K. and Baumann, K. (2001) Shaping in plant cells. *Curr. Opin. Plant Biol.*, 4, 540–546.

Matoh, T. and Kobayashi, M. (1998) Boron and calcium, essential inorganic constituents of pectic polysaccharides in higher plant cell walls. *J. Plant Res.*, 111, 179–190.

Matoh, T., Ishigaki, K., Ohno, K. and Azuma, J. (1993) Isolation and characterization of a boron-polysaccharide complex from radish roots. *Plant Cell Physiol.*, 34, 639–642.

Matoh, T., Kawaguchi, S. and Kobayashi, M. (1996) Ubiquity of a borate rhamnogalacturonan II complex in the cell walls of higher plants. *Plant Cell Physiol.*, 37, 636–640.

Matoh, T., Takasaki, M., Kobayashi, M. and Takabe, K. (2000) Boron nutrition of cultured tobacco BY-2 cells. III. Characterization of the boron rhamnogalacturonan II complex in cells acclimated to low levels of boron. *Plant Cell Physiol.*, 41, 363–366.

McCann, M.C. and Roberts, K. (1991) Architecture of the primary cell wall. In *The Cytoskeletal Basis of Plant Growth and Form* (ed. C.W. Lloyd), Academic Press, London, pp. 109–129.

McCann, M.C. and Roberts, K. (1994) Changes in cell wall architecture during cell elongation. *J. Exp. Bot.*, 45, 1683–1691.

McCann, M.C., Wells, B. and Roberts, K. (1990) Direct visualization of cross-links in the primary plant cell wall. *J. Cell Sci.*, 96, 323–334.

McCann, M.C., Wells, B. and Roberts, K. (1992) Complexity in the spatial localization and length distribution of plant cell-wall matrix polysaccharides. *J. Microscopy*, 166, 123–136.

McNeil, M., Darvill, A.G. and Albersheim, P. (1980) Structure of plant cell walls. X. Rhamnogalacturonan I, a structurally complex pectic polysaccharide in the walls of suspension-cultured sycamore cells. *Plant Physiol.*, 66, 1128–1134.

McNeil, M., Darvill, A.G., Aman, P., Franzen, L.-E. and Albersheim, P. (1982a) Structural analysis of complex carbohydrates using high-performance liquid chromatography, gas chromatography, and mass spectrometry. *Methods Enzymol.*, 83, 3–45.

McNeil, M., Darvill, A.G. and Albersheim, P. (1982b) Structure of plant cell walls. XII. Identification of seven differently linked glycosyl residues attached to O-4 of the 2,4-linked L-rhamnosyl residues of rhamnogalacturonan I. *Plant Physiol.*, 70, 1586–1591.

Meijer, M. and Murray, J.A.H. (2001) Cell cycle controls and the development of plant form. *Curr. Opin. Plant Biol.*, 4, 44–49.

Mollet, J.-C., Park, S.-Y., Nothnagel, E.A. and Lord, E.M. (2000) A lily stylar pectin is necessary for pollen tube adhesion to an in vitro stylar matrix. *Plant Cell*, 12, 1737–1749.

Monro, J.A., Penny, D., and Bailey, R.W. (1976) The organization and growth of primary cell walls of lupin hypocotyls. *Phytochemistry*, 15, 1193–1198.

Morris, V.J. and Wilde, P.J. (1997) Interactions of food biopolymers. *Curr. Opin. Colloid Interface Sci.*, 2, 567–572.

Needs, P.W., Rigby, N.M., Colquhoun, I.J. and Ring, S.G. (1998) Conflicting evidence for non-methyl galacturonosyl esters in *Daucus carota. Phytochemistry*, 48, 71–77.

Neuhaus, D. and Williamson, M. (1989) *The Nuclear Overhauser Effect in Structural and Conformational Analysis*, VCH Publishers, New York.

Nishitani, K. (1997) The role of endoxyloglucan transferase in the organization of plant cell walls. *Int. Rev. Cytol.*, 173, 157–206.

Nishitani, K. and Matsuda, Y. (1993) Acid pH-induced structural changes in cell wall xyloglucans in *Vigna angularis* epicotyl segments. *Plant Sci. Let.*, 28, 87–94.

Noguchi, K., Yasumori, M., Imai, T. *et al.* (1997) borl–1, an *Arabidopsis thaliana* mutant that requires a high level of boron. *Plant Physiol.*, 115, 901–906.

O'Neill, M.A., Albersheim, P. and Darvill, A. (1990) The pectic polysaccharides of primary cell walls. In *Methods in Plant Biochemistry*, Vol. 2 (ed. P.M. Dey), Academic Press, London, pp. 415–441.

O'Neill, M.A., Warrenfeltz, D.W., Kates, K. *et al.* (1996) Rhamnogalacturonan II, a pectic polysaccharide in the walls of growing plant cells, forms a dimer that is covalently cross-linked by a borate ester – *in vitro* conditions for the formation and hydrolysis of the dimer. *J. Biol Chem.*, 271, 22923–22930.

O'Neill, M.A., Eberhard, S., Darvill, A.G. and Albersheim, P. (2001) Requirement of borate cross-linking of cell wall rhamnogalacturonan II for *Arabidopsis* growth. *Science,* 294, 846–849.

O'Neill, M.A., Eberhard, S., Reuhs, B. *et al.* (2003) Covalent cross-linking of primary cell wall polysaccharides is required for normal plant growth. In *Advances in Pectin and Pectinase Research* (eds F. Voragen, H. Schols and R. Visser), Kluwer, Dordrecht, pp. 61–73.

Oomen, R.J.F.J., Doeswijk-Voragen, C.H.L., Bush, M.S. *et al.* (2002) *In muro* fragmentation of the rhamnogalacturonan I backbone in potato (*Solanum tuberosum*) results in a reduction and re-location of the galactan and arabinan side-chains and abnormal periderm development. *Plant J.*, 30, 403–413.

Oosteryoung, K.W., Toenjes, K., Hall, B., Winkler, V. and Bennett, A.B. (1990) Analysis of tomato polygalacturonase expression in transgenic tobacco. *Plant Cell*, 2, 1239–1248.

Orfila, C. and Knox, J.P. (2000) Spatial regulation of pectic polysaccharides in relation to pit fields in cell walls of tomato fruit pericarp. *Plant Physiol.*, 122, 775–781.

Pauly, M., Albersheim, P., Darvill, A. and York, W.S. (1999a) Molecular domains of the cellulose/xyloglucan network in the cell walls of higher plants. *Plant J.*, 20, 629–639.

Pauly, M., Andersen, L.N., Kaupinen, S. *et al.* (1999b) A xyloglucan-specific *endo*-β-1,4-glucanase from *Aspergillus aculeatus*: expression cloning in yeast, purification, and characterization of the recombinant enzyme. *Glycobiology*, 9, 93–100.

Pauly, M., Qin, Q., Greene, H., Albersheim, P., Darvill, A. and York, W.S. (2001a) Changes in the structure of xyloglucan during cell elongation. *Planta*, 212, 842–850.

Pauly, M., Eberhard, S., Albersheim, P., Darvill, A. and York, W.S. (2001b) Effects of the *murl* mutation on xyloglucans produced by suspension-cultured *Arabidopsis thaliana* cells. *Planta*, 214, 67–74.

Pellerin, P., Doco, T., Vidal, S., Williams, P., Brillouet, J.-M. and O'Neill, M.A. (1996) Structural characterization of red wine rhamnogalacturonan II. *Carbohydr. Res.*, 290, 183–197.

du Penhoat, C.H., Gey, C., Pellerin, P. and Perez, S. (1999) An NMR solution study of the mega-oligosaccharide rhamnogalacturonan II. *J. Biomol. NMR.*, 14, 253–271.

Perez, S., Mazeau, K. and du Penhoat, C. (2000) The three-dimensional structures of the pectic polysaccharides. *Plant Physiol. Biochem.*, 38, 37–55.

Perrin, R.M., DeRocher, A.E., Bar-Peled, M. *et al.* (1999) Xyloglucan fucosyltransferase, an enzyme involved in plant cell wall biosynthesis. *Science,* 284, 1976–1979.

Perrin, R.M., Jia, Z., Wagner, T.A., O'Neill, M.A., Sarria, R., York, W.S., Raikhel, N.V. and Keegstra, K. (2003) Analysis of xyloglucan fucosylation in *Arabidopsis. Plant Physiol.* 132 (in press).

Perrone, P., Hewage, C.M., Sadler, I.H. and Fry, S.C. (1998) N alpha- and N epsilon-D-galacturonoyl-L-lysine amides: properties and possible occurrence in plant cell walls. *Phytochemistry*, 49, 1879–1890.

Perrone, P., Hewage, C.M., Thomson, A.R., Bailey, K., Sadler, I.H. and Fry, S.C. (2002) Patterns of methyl and O-acetyl esterification in spinach pectins: new complexity. *Phytochemistry*, 60, 67–77.

Popper, Z.A., Sadler, I.H. and Fry S.C. (2001) 3-O-methyl-D-galactose residues in lycophyte primary cell walls. *Phytochemistry*, 57, 711–719.

Prestegard, J.H. and Kishore, A.I. (2001) Partial alignment of biomolecules: an aid to NMR characterization. *Curr. Opin. Chem. Biol.*, 5, 584–590.

Puhlmann, J., Bucheli, E., Swain, M.J. *et al.* (1994) Generation of monoclonal antibodies against plant cell-wall polysaccharides. I. Characterization of a monoclonal antibody to a terminal α-(1→2)-linked fucosyl-containing epitope. *Plant Physiol.*, 104, 699–710.

Raven, P.H., Evert, R.F. and Eichorn, S.E. (1999) *The Biology of Plants*, 6th edn, W.H. Freeman Worth Publishers, New York.

Rayon, C., Cabanes-Macheteau, M., Loutelier-Bourhis, C., Salliot-Maire, I., Lemoine, J., Reiter, W.-D., Lerouge, P. and Faye, L. (1999) Characterization of N-glycans from *Arabidopsis*. Application to a fucose deficient mutant. *Plant Physiol.*, 119, 725–734.

Redgwell, R.J., Melton, L.D. and Brasch, D.J. (1991) Cell-wall polysaccharides of kiwifruit (*Actinidia deliciosa*). Effect of ripening on the structural features of cell-wall materials. *Carbohydr. Res.*, 209, 191–202.

Redgwell, R.J., Fischer, M., Kendal, E. and MacRae, E.A. (1997) Galactose loss and fruit ripening: high-molecular weight arabinogalactans in the pectic polysaccharides of fruit cell walls. *Planta*, 203, 174–181.

Reinhold, V.N., Reinhold, B.B. and Costello, C.E. (1995) Carbohydrate molecular weight profiling, sequence, linkage and branching data: ES-MS and CID. *Anal. Chem.*, 67, 1772–1784.

Reiter, W.-D. and Vanzin, G.F. (2001) Molecular genetics of nucleotide sugar interconversion pathways in plants. *Plant Mol. Biol.*, 47, 95–113.

Reiter, W.-D., Chapple, C.C.S. and Somerville, C.R. (1993) Altered growth and cell walls in a fucose-deficient mutant of *Arabidopsis*. *Science*, 261, 1032–1035.

Reiter, W.-D., Chapple, C.C.S. and Somerville, C.R. (1997) Mutants of *Arabidopsis thaliana* with altered cell wall polysaccharide composition. *Plant J.*, 12, 335–345.

Renard, C.M.C.G., Crepeau, M.J. and Thibault, J.-F. (1999) Glucuronic acid directly linked to the galacturonic acid in the rhamnogalacturonan backbone of beet pectins. *Eur. J. Biochem.*, 266, 566–574.

Ridley, B.L., O'Neill, M.A. and Mohnen, D. (2001) Pectins: structure, biosynthesis, and oligogalacturonide-related signaling. *Phytochemistry*, 57, 929–967.

Ringli, C., Keller, B. and Ryser, U. (2001) Glycine-rich proteins as structural components of plant cell walls. *Cell. Mol. Life Sci.*, 58, 1430–1441.

Roberts, K. (1989) The plant extracellular matrix. *Curr. Opin. Cell Biol.*, 1, 1020–1027.

Roberts, K. (1994) The plant extracellular matrix: in a new expansive mood. *Curr. Opin. Cell Biol.*, 6, 688–694.

Rodgers, M.W. and Bolwell, G.P. (1992) Partial purification of Golgi-bound arabinosyltransferase and 2 isoforms of xylosyltransferase from French bean (*Phaseolus vulgaris* L.). *Biochem. J.*, 288, 817–822.

Rose, J.K.C. and Bennett, A.B. (1999) Cooperative disassembly of the cellulose-xyloglucan network of plant cell walls: parallels between cell expansion and fruit ripening. *Trends Plant Sci.*, 4, 176–183.

Rose, J.K.C., Braam, J., Fry, S.C. and Nishitani, K. (2002) The XTH family of enzymes involved in xyloglucan endotransglucosylation and endohydrolysis: current perspectives and a new unifying nomenclature. *Plant Cell Physiol.*, 43, 1421–1435.

Round, A.N., MacDougall, A.J., Ring, S.G. and Morris, V.J. (1997) Unexpected branching in pectin observed by atomic force microscopy. *Carbohydr. Res.*, 303, 251–253.

Round, A.N., Rigby, N.M., MacDougall, A.J., Ring, S.G. and Morris, V.J. (2001) Investigating the nature of branching in pectin by atomic force microscopy and carbohydrate analysis. *Carbohydr. Res.*, 331, 337–342.

Satiat-Jeunemaitre, B., Martin, B. and Hawes, C. (1992) Plant cell wall architecture is revealed by rapid-freezing and deep-etching. *Protoplasma*, 167, 33–42.

Saulnier, L. and Thibault, J.-F. (1999) Ferulic acid and diferulic acids as components of sugar-beet pectins and maize bran heteroxylans. *J. Sci. Food Agric.*, 79, 396–402.

von Schaewen, A., Strum, A., O'Neill, J. and Chrispeels, M.J. (1993) Isolation of a mutant Arabidopsis plant that lacks N-acetyl glucosaminyl transferase I and is unable to synthesize Golgi-modified complex N-linked glycans. *Plant Physiol.*, 102, 1109–1118.

Schols, H.A. and Voragen, A.G.J. (1994) Occurrence of pectic hairy regions in various plant cell wall materials and their degradability by rhamnogalacturonase. *Carbohydr. Res.*, 256, 83–95.

Schols, H.A., Bax, E.J., Schipper, D. and Voragen, A.G.J. (1995) A xylogalacturonan subunit present in the modified hairy regions of apple pectin. *Carbohydr. Res.*, 279, 265–279.

Selvendran, R.R. and O'Neill, M.A. (1985) Isolation and analysis of cell walls from plant material. In *Methods of Biochemical Analysis*, Vol. 32 (ed. D. Glick), John Wiley & Sons, London, pp. 25–153.

Shimokawa, T., Ishii, T. and Matsunaga, T. (1999) Isolation and structural characterization of rhamnogalacturonan-borate complex from *Pinus densiflora*. *J. Wood Sci.*, 45, 435–439.

Shin, K.-S., Kiyohara, H., Matsumoto, T. and Yamada, H. (1998) Rhamnogalacturonan II dimers cross-linked by borate diesters from the leaves of *Panax ginseng* C.A. Meyer are responsible for expression of their IL-6 production enhancing activities. *Carbohydr. Res.*, 307, 97–106.

Sieber, P., Schorderet, M., Ryser, U. *et al.* (2000) Transgenic *Arabidopsis* plants expressing a fungal cutinase show alterations in the structure and properties of the cuticle and postgenital organ fusions. *Plant Cell*, 12, 721–737.

Sims, I.M., Craik, D.J. and Bacic, A. (1997) Structural characterization of galactoglucomannan secreted by suspension-cultured cells of *Nicotiana plumbaginigolia*. *Carbohydr. Res.*, 303, 79–92.

Sims, I.M., Munro, S.L.A., Currie, G., Craik, D. and Bacic, T. (1996) Structural characterization of xyloglucan secreted by suspension-cultured cells of *Nicotiana plumbaginifolia*. *Carbohydr. Res.*, 293, 147–172.

Skjot, M., Pauly, M., Bush, M.S., Borkhardt, B., McCann, M.C. and Ulvskov, P. (2002) Direct interference with rhamnogalacturonan I biosynthesis in Golgi vesicles. *Plant Physiol.*, 129, 95–102.

Smith, L.G. (2001) Plant cell division: building walls in the right places. *Nature Reviews, Mol. Cell Biol.*, 2, 33–39.

Sorensen, S.O., Pauly, M., Bush, M. *et al.* (2000) Pectin engineering: modification of potato pectin by *in vivo* expression of an endo-1,4-β-D-galactanase. *Proc. Natl. Acad. Sci. USA*, 97, 7639–7644.

Staehelin, A. (1991) What is a plant cell? A response. *Plant Cell*, 3, 553.

Stephen, A.M. (1982). Other plant polysaccharides. In *The Polysaccharides*, Vol. 2 (ed. G.O. Aspinall), Academic Press, New York, pp. 97–193.

Stevenson, T.T., Darvill, A.G. and Albersheim, P. (1988) Structure of plant cell walls. 23. Structural features of the plant cell-wall polysaccharide rhamnogalacturonan II. *Carbohydr Res.*, 182, 207–226.

Strasser, G.R. and Amado, R. (2002) Pectic substances from red beet (*Beta vulgaris* L. var. conditiva). Part II. Structural characterization of rhamnogalacturonan II. *Carbohydr. Polymers*, 48, 263–269.

Tahiri, M., Pellerin, P., Tressol, J.C. *et al.* (2000) The rhamnogalacturonan-II dimer decreases intestinal absorption and tissue accumulation of lead in rats. *J. Nutr.*, 130, 249–253.

Tahiri, M., Tressol, J.C., Doco, T., Rayssiguier, Y. and Coudray, C. (2002) Chronic oral administration of rhamnogalacturonan-II dimer, a pectic polysaccharide, failed to accelerate body lead detoxification after chronic lead exposure in rats. *Brit. J. Nutr.*, 87, 47–54.

Takano, J., Yamagami, M., Noguchi, K., Hayashi, H. and Fujiwara, T. (2001) Preferential translocation of boron to young leaves in *Arabidopsis thaliana* regulated by the BOR1 gene. *Soil Sci. Plant Nutr.*, 47, 345–357.

Talbot, L.F. and Ray, P.M. (1992) Molecular size and separability features of pea cell wall polysaccharides: implications for models of primary wall structure. *Plant Physiol.*, 98, 357–368.

Tanner, M.E. (2001) Sugar nucleotide-modifying enzymes. *Curr. Org. Chem.*, 5, 169–192.

Terpstra, A.H.M., Lapre, J.A., de Vries, H.T. and Beynen, A.C. (2002) The hypocholesterolemic effect of lemon peels, lemon pectin, and the waste stream material of lemon peels in hybrid F1B hamsters. *Eur. J. Nutr.* 19–26.

The Arabidopsis Genome Initiative (2000) Analysis of the genome sequence of the flowering plant *Arabidopsis thaliana. Nature*, 408, 796–815.

Thomas, J.R., McNeil, M., Darvill, A.G. and Albersheim, P. (1987) Structure of plant cell walls. XIX. Isolation and characterization of wall polysaccharides from suspension-cultured Douglas fir cells. *Plant Physiol.*, 83, 659–671.

Thomas, J.R., Darvill, A.G. and Albersheim, P. (1989) Isolation and structural characterization of the pectic polysaccharide rhamnogalacturonan II from walls of suspension-cultured rice cells. *Carbohydr. Res.*, 185, 261–277.

Thompson, J.E. and Fry, S.C. (2000) Evidence for covalent linkage between xyloglucan and acidic pectins in suspension-cultured rose cells. *Planta*, 211, 275–286.

Thompson, J.E. and Fry, S.C. (2001) Restructuring of wall-bound xyloglucan by transglycosylation in living plant cells. *Plant J.*, 26, 23–34.

Thompson, J.E., Smith, R.C. and Fry, S.C. (1997) Xyloglucan undergoes interpolymeric transglycosylation during binding to the plant cell wall *in vivo*: evidence from $^{13}C/^{1}H$ dual labelling and isopycnic centrifugation in caesium trifluoroacetate. *Biochem. J.*, 327, 699–708.

Tieman, D. and Handa, A. (1994) Reduction in pectin methyl-esterase activity modifies tissue integrity and cation levels in ripening tomato (*Lycopersicon esculentum*). *Plant Physiol.*, 106, 429–436.

Valent, B.S. and Albersheim, P. (1974) The structure of plant cell walls. V. On the binding of xyloglucan to cellulose fibers. *Plant Physiol.*, 54, 105–108.

Vanzin, G.F., Madson, M., Carpita, N.C., Raikhel, N.V., Keegstra, K. and Reiter, W.-D. (2002) The *mur2* mutant of *Arabidopsis thaliana* lacks fucosylated xyloglucan because of a lesion in fucosyltransferase AtFUT1. *Proc. Natl. Acad. Sci. USA.*, 99, 3340–3345.

Verbruggen, M.A., Spronk, B.A., Schols, H.A. *et al.* (1998) Structures of enzymically derived oligosaccharides from sorghum glucuronoarabinoxylan. *Carbohydr. Res.*, 306, 265–274.

Vidal, S., Doco, T., Williams, P. *et al.* (2000) Structural characterization of the pectic polysaccharide rhamnogalacturonan II: evidence for the backbone location of the aceric acid-containing oligoglycosyl side chain. *Carbohydr. Res.*, 326, 277–294.

Vierhuis, E., York, W.S., Vincken, J.-P., Schols, H.A., Van Alebeek, G.-J.W.M. and Voragen, A.G.J. (2001) Structural analyses of two arabinose containing oligosaccharides derived from olive fruit xyloglucan: XXSG and XLSG. *Carbohydr. Res.*, 332, 285–297.

Vincken, J.-P., York, W.S., Beldman, G. and Voragen, A.G.J. (1997) Two general branching patterns of xyloglucan: XXXG and XXGG. *Plant Physiol.*, 114, 9–12.

Vliegenthart, J.F.G., Dorland, L. and van Halbeek, H. (1983) High resolution ^{1}H-nuclear magnetic resonance spectroscopy as a tool in the structural analysis of carbohydrates related to glycoproteins. *Adv. Carbohydr. Chem. Biochem.*, 41, 209–374.

Welch, R.M. (1995) Micronutrient nutrition of plants. *Crit. Rev. Plant Sci.*, 14, 49–82.

Wende, G. and Fry, S.C. (1997) *O*-Feruloylated, *O*-acetylated oligosaccharides as side-chains of grass xylans. *Phytochemistry,* 44, 1011–1018.

Whitcombe, A.J., O'Neill, M.A., Steffan, W., Darvill, A.G. and Albersheim, P. (1995) Structural characterization of the pectic polysaccharide rhamnogalacturonan II. *Carbohydr Res.,* 271, 15–29.

Whitney, S.E.C., Bringham, J.E., Darke, A.H., Reid, J.S.G. and Gidley, M.J. (1995) *In vitro* assembly of cellulose/xyloglucan networks: ultrastructural and molecular aspects. *Plant J.,* 8, 491–504.

Willats, W.G.T., Steele-King, C.G., Marcus, S.E. and Knox, J.P. (1999) Side chains of pectic polysaccharides are regulated in relation to cell proliferation and cell differentiation. *Plant J.,* 20, 619–628.

Willats, W.G.T., McCartney, L., Mackie, W. and Knox, J.P. (2001a) Pectin: cell biology and prospects for functional analysis. *Plant Mol. Biol.,* 47, 9–27.

Willats, W.G.T., Orfila, C., Limberg, G. *et al.* (2001b) Modulation of the degree and pattern of methyl-esterification of pectic homogalacturonan in plant cell walls – Implications for pectin methyl esterase action, matrix properties, and cell adhesion. *J. Biol. Chem.,* 276, 19404–19413.

Wyatt, S.E. and Carpita, N.C. (1993) The plant cytoskeleton-cell-wall continuum. *Trends in Cell Biol.,* 3, 413–417.

York, W.S., Darvill, A.G., McNeil, M., Stevenson, T.T. and Albersheim, P. (1985) Isolation and characterization of plant cell walls and cell wall constituents. *Methods Enzymol.,* 118, 3–40.

York, W.S., van Halbeek, H., Darvill, A.G. and Albersheim, P. (1989) Structural analysis of xyloglucan oligosaccharides by ^{1}H-n.m.r. spectroscopy and fast atom bombardment mass spectrometry. *Carbohydr. Res.,* 200, 9–31.

York, W.S., Harvey, L.K., Guillén, R., Albersheim, P. and Darvill, A.G. (1993) Structural analysis of tamarind seed xyloglucan oligosaccharides using β-galactosidase digestion and spectroscopic methods. *Carbohydr. Res.,* 248, 285–301.

York, W.S., Impallomeni, G., Hisamatsu, M., Albersheim, P. and Darvill, A. (1994) Eleven newly characterized xyloglucan oligoglycosyl alditols: the specific effects of sidechain structure and location on ^{1}H-NMR chemical shifts. *Carbohydr. Res.,* 267, 79–104.

York, W.S., Kolli, V.S., Orlando, R., Albersheim, P. and Darvill, A.G. (1996) The structures of arabinoxyloglucans produced by solanaceous plants. *Carbohydr. Res.,* 285, 99–128.

Youl, J.J., Bacic, A. and Oxley, D. (1998) Arabinogalactan-proteins from *Nicotiana alata* and *Pyrus communis* contain glycosylphosphatidylinositol membrane anchors. *Proc. Natl. Acad. Sci. USA,* 95, 7921–7926.

Yu, L. and Mort, A.J. (1996) Partial characterization of xylogalacturonans from cell walls of ripe watermelon fruit: inhibition of endopolygalacturonase activity by xylosylation. In *Pectins and Pectinases* (eds J. Visser and A.G.J. Voragen), Elsevier, Amsterdam, pp. 79–88.

Yu, K.W., Kiyohara, H., Matsumoto, T., Yang, H.C., and Yamada, H. (2001a) Characterization of pectic polysaccharides having intestinal immune system modulating activity from rhizomes of *Atractylodes lancea* DC. *Carbohydr. Polymers,* 46, 125–134.

Yu, Q., Wingender, R., Schulz, M., Baluska, F. and Goldbach, H.E. (2001b) Short-term boron deprivation induces increased levels of cytoskeletal proteins in *Arabidopsis* roots. *Plant Biol.,* 3, 335–340.

Zablackis, E., Huang, J., Muller, B., Darvill, A. and Albersheim, P. (1995) Characterization of the cell-wall polysaccharides of *Arabidopsis thaliana* leaves. *Plant Physiol.,* 107, 1129–1138.

Zhang, Q. and Liu, H.-W. (2001) Chemical synthesis of UDP-β-L-arabinofuranose and its turnover to UDP-β-L-arabinopyranose by UDP-galactopyranose mutase. *Bioorg. Med. Chem. Let.,* 11, 145–149.

2 Biophysical characterization of plant cell walls

V.J. Morris, S.G. Ring, A.J. MacDougall and R.H. Wilson

2.1 Introduction

Biophysical techniques provide methods for probing the structure and structural changes of plant cell walls. This chapter describes new insights obtained through the use of infrared and atomic force microscopy, and the use of physical methods to probe the response of cell wall networks to changing environments and stresses. Infrared techniques yield information on the composition of the cell wall and the location and orientation of cell wall components. Atomic force microscopy permits the imaging of cell wall networks, cell wall polysaccharides and their interactions. Finally, biophysical methods permit investigation of the way that cell wall networks respond to environmental changes such as pH, ionic strength, or the presence of other molecular components.

2.2 Infrared spectroscopy of plant cell walls

Infrared spectroscopy has been used to characterize the structures of biopolymers for many years (Mathlouthi and Koenig, 1986; Kacurakova and Wilson, 2001). The technique can reveal which components are present, how much there is present, and their physical and/or chemical form. In addition it provides information on molecular alignment and molecular interactions. This information is now available from very small samples, and the relatively new technique of infrared micro-spectroscopy enables us to probe directly the structure of the plant cell wall. This section describes the recent developments in this field, and their application to the plant cell wall.

An infrared spectrum is produced when infrared light is passed through a sample. Some of the transmitted light will be absorbed at certain characteristic frequencies. These frequencies correspond to vibrational modes of bonds within the molecules of the sample. The absorption frequency depends upon the masses of the atoms forming the bond, and the strength of the linkage between them. Other groups that are directly attached to, or positioned further along the molecules (vibrational coupling) can also affect the precise value of the frequency. A full description of the origins of infrared spectra, and the conventional means of collecting them, can be found elsewhere (Griffiths and de Haseth, 1986). For the present purposes it is only necessary to know that the infrared spectrum shows what types of functional groups are present and, consequently, what types of biopolymers are present in the sample. In addition, vibrational coupling and hydrogen bonding generate complex spectral patterns that are often unique for individual compounds, and are thus often referred to as 'fingerprints'. Pure compounds can, in many instances, be identified by their infrared spectra. However, biological systems are normally mixtures of many compounds, and the result is usually a complex spectral profile from which it may be possible only to identify the presence of a few compounds, or classes of compound.

Spectral assignments for many of the key functional groups of biopolymers are shown in Table 2.1. For the spectrum of onion epidermis (Figure 2.1) the most notable features are peaks at 1740, 1650, 1450 cm^{-1} and a complex envelope in the region 1150–900 cm^{-1}. The 1740 cm^{-1} band denotes ester residues and the 1650 and 1450 cm^{-1} peaks are from carboxylic acid groups. These two groups are characteristic of pectin. The complex envelope, when examined in detail, reveals other characteristic peaks for pectin, cellulose and hemicellulose (Kacurakova and Wilson, 2001). Other compounds that can be observed include proteins, for which a very large literature exists detailing the extraction of secondary structural information (Mathlouthi and Koenig, 1986), in addition to lipids, waxes and phenolics. The precise shape of the profiles is thus the sum of the contributions from all of the infrared-absorbing molecules present. As expected, the spectra of biological materials vary considerably from sample to sample.

Table 2.1 Assignments for infrared vibrations.

Frequency (cm^{-1})	Assignment	Bond, orientation	Origin
1740	ν(C=O)		ester, P
1426	δ_sCH$_2$		C
1371	δCH$_2$w		XG
1362, 1317	δ_s CH$_2$ w	$\perp$	C
1243	ν(C-O)		P
1160	ν_{as}(C-O-C)	glycosidic link, ring, II	
1146	ν_{as}(C-O-C)	glycosidic link, ring, II	P
1130	ν_{as}(C-O-C)	glycosidic link, ring, II	XG
1115	ν(C-O), ν(C-C)	C2-O2	C
1100	ν(C-O), ν(C-C)	ring	P
1075	ν(C-O), ν(C-C)	ring	XG
1060	ν(C-O), ν(C-C)	C3-O3, II	
1042	ν(C-O), ν(C-C)	ring, II	XG
1030	ν(C-O), ν(C-C)	C6H$_2$-O6, II	
1015, 1000	ν(C-O), ν(C-C)	C6H$_2$-O6, II	C
1019	ν(C-O), ν(C-C)	C2-C3, C2-O2, C1-O1	P
960	δ(CO)	II	P
944	ring	II	XG
895	δ(C1-H)	β- anomeric link	C, XG
833	ring	II	P

Key: IR vibrations: ν, stretching; δ, bending; w, wagging; $_s$ symmetric; $_{as}$ asymmetric; C, cellulose; P, pectin; XG, xyloglucan; II parallel; $\perp$ perpendicular orientation
*IR band assignment is based on literature data.

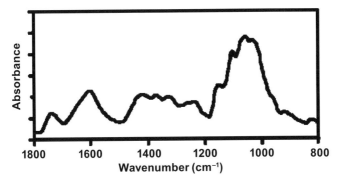

Figure 2.1   Infrared spectrum of onion epidermis in the region 1800–800 cm^{-1}. Spectral features are as described in the text.

2.2.1   Infrared micro-spectroscopy

The development of the infrared microscope (micro-spectrometer) has enabled the routine acquisition of spectra from a wide range of biological samples, and has allowed the direct study of the plant cell wall. In a micro-spectrometer light from an infrared spectrometer is condensed to a spot size of approximately 1 mm diameter at the sample position. The sample can be viewed using conventional light optics and is positioned within the centre of the beam. A region from which a spectrum is to be collected is selected using a variable aperture located in a remote image plane. In this way only a selected part of the sample is illuminated. The light, having passed through the sample, is focused onto a small area detector. The maximum sample area is typically 100 μm and the smallest area is 10 μm. One manufacturer includes a second aperture after the sample in order to reject diffracted light, and they claim this offers improved elimination of spectral contamination from outside the sample area. The microscope is also able to operate in a reflection mode with samples mounted on a reflecting surface. This mode of operation is rarely used for cell wall samples. In the standard operating mode the sample is mounted in position, suspended across a hole, or mounted on a suitable substrate, such as a barium fluoride or a potassium bromide disc. A suitable aperture size is selected (e.g. sufficient to examine a single cell or a collection of epidermal cells) and a spectrum is collected. Sample thickness is crucial and the technique is usually limited to samples of less than 25 μm thick and, more often, 10 μm in thickness. Thus individual cell walls are fine but leaves, roots and hypocotyls may also be suitable for examination.

2.2.2   Polarization

It is possible to incorporate infrared polarizers into the optical path in order to measure dichroism. Infrared dichroism reflects the mean orientation of the transition moments of the corresponding vibrational modes. Intrinsic anisotropy or anisotropy induced by an applied deformation can be characterized by the *dichroic difference*

$\Delta A = (A_{\parallel} - A_{\perp})$ and/or *dichroic ratio*, defined as $R = A_{\parallel}/A_{\perp}$, where the $A_{\parallel}$ and $A_{\perp}$ are the absorbances measured with parallel and perpendicular polarization (see Griffiths and de Haseth, 1986; Koenig, 1999). For a polymer network, the segmental orientation (F) detected by infrared spectroscopy can be expressed in terms of the dichroic ratio: $F = C[(R - 1)/(R + 2)]$, where $C = (2\cot^2\alpha + 2)/(2\cot^2\alpha - 1)$. For a given absorption band, α is the angle between the transition moment vector of the vibrational mode, and a directional vector characteristic of the chain segment. In a molecule such as cellulose, the glycosidic band at 1162 cm^{-1} is approximately co-aligned with both the molecular long axis and the stretching direction. In this case $\alpha = 0°$ and the equation reduces to: $F = (R - 1)/(R + 2)$.

Polarization has been used to study orientation in many synthetic polymers and biopolymers such as cellulose (Cael *et al.*, 1975; Mathlouthi and Koenig, 1986). Polarization, and hence molecular orientation, has also been observed in complex biological tissues such as elongating carrot cells (McCann *et al.*, 1993), onion epidermis (Chen *et al.*, 1997) and algal cell walls (Toole *et al.*, 2000).

In studies on onion epidermis, ΔA was measured before, and after, the application of controlled strains. ΔA in unstrained material was found to be close to zero (Figures 2.2 and 2.3), but ΔA increased with increasing strain (Figure 2.4), showing that molecular orientation resulted from the applied strain. Examination of the ΔA spectra revealed that both the cellulose and the pectin networks were oriented. For the algal system *Chara corallina*, molecular orientation was also observed, but this time the cellulose orientation was at 90° to the applied strain. This result has been interpreted in terms of the known structure of the cell wall of the algae (Toole *et al.*,

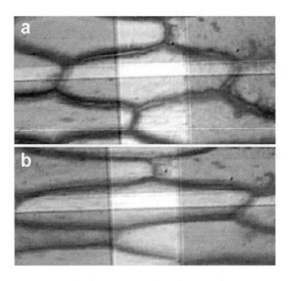

Figure 2.2 Photographs of onion epidermis cells (a) before and (b) after stretching *in situ* on an infrared microscope. Spectra collected in Figures 2.3 and 2.4 were taken from the region delineated by apertures superimposed on the image (approximate size $100 \times 20\,\mu\text{m}$).

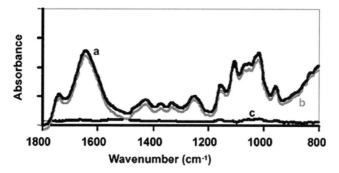

Figure 2.3 Infrared spectra of unstretched onion epidermis taken with the polarization parallel (a) and perpendicular (b) to the long axis of the cell. The lower trace (c) is the difference spectrum, and shows that there is no molecular orientation.

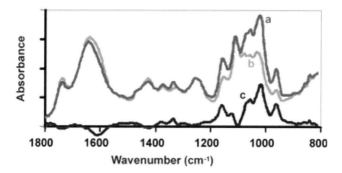

Figure 2.4 Infrared spectra of onion epidermis taken with the polarization parallel (a) and perpendicular (b) to the long axis of the cell. The lower trace (c) is the difference spectrum, and shows that there is significant orientation of the molecules in the stretched cell wall. Positive features arise from the cellulose and the pectin backbone vibrations. Negative features arise from the pectin ester and acidic side-groups.

2000). A more detailed study of molecular alignment on mixed biopolymer systems, based on composites of *Acetobacter* cellulose and pectin or hemicellulose, has been carried out recently (Kacurakova *et al.*, 2002).

In order to analyse the strain-induced spectral changes in these composites, ΔA and F values for selected bands were used to determine the degree of orientation of the cellulose within individual composites that had been subjected to uniaxial strain. One experimental problem that occurs on stretching is a change in the thickness of the polymer film, requiring normalization of the data to a corresponding unstretched absorption whose intensity is not orientation dependent. For molecular orientation analysis the ΔA used caused the bands from the vibrational dipoles parallel to the polymer chain axis to appear with a positive intensity, whereas those that are perpendicular to the chain had a negative intensity.

For all composites the initial alignment was very small, but an increased dichroism in the parallel direction was observed with increasing strain, indicating that the cellulose was aligned along the stretch direction (see Figure 2.5). In pure cellulose, bands at 1058 and 1033 cm^{-1}, assigned to vibrations of C-3H –O-3H and C-6H$_2$–O-6H groups and pyranosyl-ring vibrations, increased in intensity without a significant frequency shift. In the cellulose-pectin mixed composite the intensity at 1162 cm^{-1} was more intense relative to the two ring vibrations, which broadened, reflecting a wider distribution of the intra-molecular and inter-molecular hydrogen bond energies, compared to pure cellulose alone. For cellulose in the presence of xyloglucan, only the cellulose glycosidic 1062 cm^{-1} band showed a significantly positive ΔA. Frequency shifts in the relatively weak xyloglucan peaks were observed, attributable to molecular deformation. When the water content of the samples was reduced, less molecular orientation generally occurred, and the samples failed at lower strains.

The orientation function (F) of the cellulose glycosidic bond, a measure of the distribution of cellulose chain segments, provided similar results to that obtained from ΔA measurements. F increased in all the samples (Figure 2.6), but was generally larger at higher moisture content. The enhanced ability of cellulose to orient in a wetter environment was interpreted in terms of water lubricating the molecular motion in the fibrillar layers within the samples.

Results of this work generally agreed well with other studies on these composites, including X-ray analysis. The present technique has raised questions as to the role of pectin in determining wall strength, and is currently being applied to study wild-type and mutant *Arabidopsis* hypocotyls with altered composition and mechanical phenotype. The work is shedding new light on the relationship between composition and cell wall mechanical properties. However, this experiment does not provide any information on the nature of interactions between molecules within the cell wall. Such information can be obtained through the use of a method called two-dimensional FTIR spectroscopy.

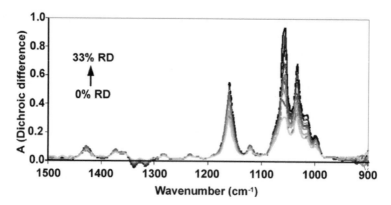

Figure 2.5 Molecular orientation in *Acetobacter* cellulose. Dichroic difference spectra show increasing intensity with increased strain (relative deformation), demonstrating enhanced molecular orientation at high strain.

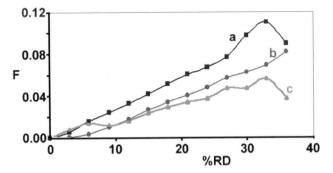

Figure 2.6 Orientation function, F, as a function of applied strain for cellulose-pectin (a), cellulose (b) and cellulose-xyloglucan (c) composites. Results are discussed in the text.

2.2.3 Mapping

The infrared microscope can be used to map the distribution of components in a sample. There are two ways of doing this, and the simplest way is to move the sample beneath a fixed aperture under computer control. This generates a matrix of spectra from which plots can be generated by selecting the intensity of various peaks (Carpita *et al.*, in press). However, this process is relatively slow and the maximum spatial resolution (10 μm) is rarely achieved. A recent alternative approach is based on a focal plane array detector that comprises (typically) 128 × 128 detector elements of 5 μm each. An image of high spatial resolution is generated in a relatively short acquisition time. Images based on peak intensities are produced in a similar way to the step-wise mapping approach.

2.2.4 Mutant screening methods

The variability of spectral profiles has been put to use in the identification of samples of unusual composition. Early work by Sene *et al.* (1995) showed that significant differences could be observed in the infrared spectra of epidermal cell walls of a range of plants. These differences could be interpreted in terms of the known composition of the cell walls. For example, graminaceous monocotyledons such as rice and maize exhibited peaks arising from phenolic compounds present in relatively high amounts. These compounds were not present in onion or carrot cell walls. Examination of such spectra led to two hypotheses: firstly, that perhaps the spectra of the cell walls could be used to identify the species, and secondly, that such spectra may be useful for taxonomic purposes. The second hypothesis has not been explored, but the first has led directly to a number of developments. The spectra of epidermal cell walls of a range of fruits and vegetables were examined using chemometric methods such as PCA and discriminant analysis (Kemsley *et al.*, 1995). This showed that spectra could indeed be assigned to a specific species, even against a natural compositional variation. This work formed part of many subsequent studies aimed at authenticating fruit-based materials based on infrared spectra of fruit pulps (Defernez and Wilson, 1995), where techniques such as partial least squares, PLS, discriminant analysis (Kemsley *et al.*, 1996; Holland *et al.*, 1998) and artificial neural networks were used to extract information. These mathematical techniques, when combined with micro-spectroscopic data collection of *Arabidopsis* leaves and hypocotyls (Figure 2.7), has led directly to a rapid method for screening for compositional mutants (Chen *et al.*, 1998), greatly speeding up the identification of new cell wall mutants. The method is based on identifying subtle compositional variations in cell wall components that are greater than the normal compositional variation seen in wild-type plants. The methods employed were also able to provide an indication of the source of the variation, e.g. cellulose or pectin expression.

2.2.5 Analysis of cell walls

Spectra of plant cell walls have been used to follow compositional changes during the sequential extraction of components from the cell wall (Figure 2.8). These spectra confirmed the nature of the biopolymer that had been extracted, but also highlighted some differences between the nature of extracted polymers, and their form in the cell wall. Differences between species were revealed (McCann *et al.*, 1992; Sene *et al.*, 1995).

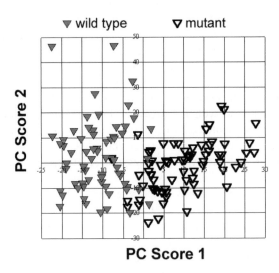

Figure 2.7 Principal component scores plot derived from infrared spectra of *Arabidopsis* leaves. This shows the ability of the technique to discriminate between wild-type and mutant *Arabidopsis* plants on the basis of compositional differences in the cell wall, detectable by infrared spectroscopy.

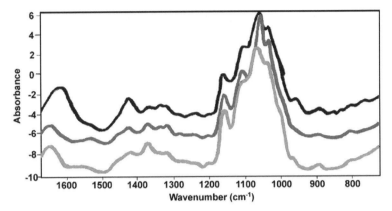

Figure 2.8   Infrared spectra of algal cell walls. The top trace is the spectrum for the native cell wall. The middle trace is the spectrum obtained after removal of the pectin components (note the loss of features at 1600 and 1450 cm⁻¹). The bottom trace is the spectrum after treatment with KOH, showing the effects of removal of hemicellulose, and change in cellulose crystallinity witnessed by the increased line width of the cellulose peaks.

2.2.6   Two-dimensional FTIR spectroscopy

In this experimental approach, dynamic spectra are collected, i.e. the samples are not stationary during the measurement. In dynamic 2D FT-IR (Figure 2.9) a small-amplitude oscillatory strain is applied to a sample and the resulting spectral changes are measured as a function of time (Noda, 1990; Wilson *et al.*, 2000). The dynamic perturbation generates directional changes in the transition moments of the functional groups whose relaxations are affected by inter- and intra-molecular couplings. The resulting dynamic absorbance spectra, which vary sinusoidally with the stretching frequency, are deconvoluted into two separate spectra: the in-phase (IP) spectrum that reveals the re-orientational motions of the electrical dipole moments occurring simultaneously with the applied strain, and the quadrature (Q) spectrum that reveals those that are $\pi/2$ out-of-phase with respect to the applied strain. From the IP and Q spectra, synchronous 2D correlation spectra can be obtained, with correlation peaks at wavenumber coordinates where the IR signal responses are in phase with each other.

This technique, which has been used predominantly to study interactions in synthetic polymer systems (Noda *et al.*, 1999), has recently been applied to biological systems such as cellulose, plant cell walls and cellulose composites (Hinterstoisser and Salmen, 2000; Wilson *et al.*, 2000; Hinterstoisser *et al.*, 2001; Åkerholm and Salmen, 2001). The technique can be used to determine the presence of, and extent of interactions between molecules in mixed systems. For the plant cell wall the methodology is potentially able to answer one of the long-standing questions in plant cell wall science: are the cellulose and pectin networks independent of each other (Wilson *et al.*, 2000).

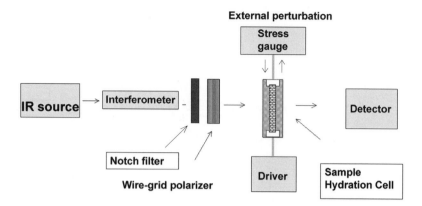

Figure 2.9 Schematic diagram of a two-dimensional infrared experiment. The sample in the hydration cell is pre-stretched and mechanically oscillated. Infrared radiation is encoded in the step-scan interferometer before passing through the sample via the polarizer. The detector monitors the signals that are in phase, and 90° out of phase, with the applied oscillation. From these two dynamic signals the synchronous correlation plot is generated.

For studies on onion epidermal tissue (Wilson *et al.*, 2000), dynamic 2D FT-IR spectral measurements were made on an epidermal strip (1 cm × 1.5 cm) placed in a polymer stretcher, which was modified to enclose the sample in a hydration cell that allowed the onion samples to maintain a water content of about 70% of the total weight during the measurements (2 to 3 hours). The onion samples were pre-stretched to about 35% relative elongation, in order to partially align the cellulose and pectin, and then subjected to a small periodic strain with a sample modulation amplitude of 200 μm. 2D spectra were obtained from the modulated step-scan spectra using digital signal processing. The phase modulation frequency was 400 Hz. The in-phase static spectra were normalized against a single beam background spectrum, and the in-phase and quadrature dynamic spectra were normalized against the in-phase static spectrum.

Dynamic in-phase and quadrature spectra showed only the responses of the individual spectral bands to the applied strain. In the onion spectra with 70% water content the 1640 cm^{-1} band of adsorbed water was not seen in the dynamic spectra, and neither the pectin ester bands (1740, 1444 cm^{-1}) nor the carboxyl vibrations of the pectate form (1610, 1415 cm^{-1}) were observed. Other modes arising from side chains, or side groups (including ester and OH) were apparently not sensitive to the applied perturbation. However, a significant response was found in the bands related to the backbones of the polysaccharides in the 1200–900 region (Figure 2.10). For example it was apparent that the cellulose glycosidic (C-O-C) stretching band at 1165 cm^{-1} had a very high dynamic intensity in parallel polarization. The other cellulose bands, e.g. 1080, 1056, 1029 cm^{-1}, and the pectin bands at 1150, 1113, 1006 cm^{-1} attributed to the backbone pyranosyl ring vibrational modes, were of medium intensity.

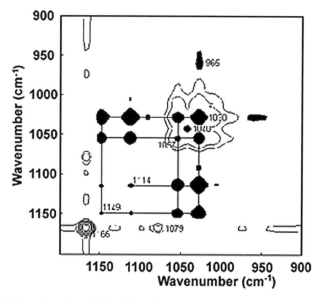

Figure 2.10 Two-dimensional infrared (synchronous correlation) plot for onion epidermis.

The cellulose band at 1165 cm^{-1} was positive in both in-phase and quadrature spectra with parallel polarized light. This indicated that the cellulose glycosidic linkage became more oriented in the direction of the applied stress. In contrast, the bands assigned to pectin were predominantly negative. Moreover, the pectin bands were stronger in the in-phase spectrum than in the quadrature spectrum, whilst the cellulose bands were much more intense in the quadrature spectrum. This suggested a difference in the response rate of the two polymers to the applied stress.

With parallel-polarized light, positive cross-peaks in the synchronous correlation spectrum showed intensity maximum for cellulose at 1168, 1132, 1098, and 975 cm^{-1}, that indicated that these modes responded together to the applied strain. The negative peaks corresponded to pectin bands at 1150, 1112, 1056, 1028, 1006 and 962 cm^{-1}. There were no cross-peaks between the pectin and the cellulose peaks. The independent re-orientation responses of the cellulose and the pectin chain functional groups suggested that the cellulose and the pectin molecules were not directly interacting.

Recently the same techniques have been applied to cellulose composite samples (Wilson *et al.*, 2000). Pure cellulose gave cross-peaks at 1162, 1112, 1062 and 1030 cm^{-1} representing similar time-dependent movements of specific cellulose groups. In cellulose-pectin mixtures the result was essentially the same, and showed no evidence of connected motion between the two networks in this composite, consistent with earlier evidence obtained for onion epidermis tissue. The dynamic spectra of the cellulose in a mixture with xyloglucan exhibited bands at 1162 and 1080 cm^{-1} in the in-phase spectrum. In the 2D correlation maps, the intense cellulose and xyloglucan cross-peaks, appearing at the off-diagonal positions at 1162 and 1080 cm^{-1}, indicated a strong synchronous correlation between the cellulose and the xyloglucan, providing direct evidence, for the first time, that the two macromolecules move collectively.

2.3 Atomic force microscopy of cell walls

Infrared microscopy gives information on the location and orientation of different polysaccharides within the cell wall. Probe microscopy provides an alternative to electron microscopy for direct molecular imaging of plant cell wall polysaccharides and cell wall structure. In particular, atomic force microscopy (AFM) offers comparable resolution to the electron microscope but with minimal sample preparation. The textbook by Morris *et al.* (1999a) discusses the principles of the operation of the AFM, the different modes of imaging, and applications to biological systems. AFMs image by detecting the changes in the force acting on a sharp tip as the sample is scanned in a raster fashion beneath this probe (Figure 2.11). The tip is attached to a flexible cantilever (Figure 2.12) and the bending of the cantilever in response to changes in force is monitored optically. In the normal mode of operation the cantilever deflection is preset to a given value, and deviations from this position are corrected for through a feedback circuit that moves the sample towards or away from the probe at each sample point, in order to keep the cantilever deflection constant. The resultant changes in position of the sample are amplified to generate a

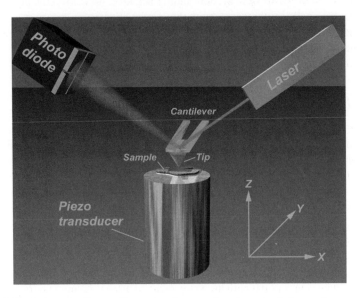

Figure 2.11   Schematic diagram illustrating the operation of an atomic force microscope.

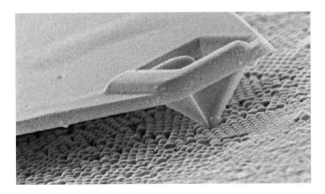

Figure 2.12   Scanning electron microscope picture of an AFM tip positioned above a sample surface. The magnification is approximately × 10,000. The probe is used to 'feel' the surface structure.

three-dimensional profile of the sample surface. If the sample surface is uniform in structure then these images represent the topography of the surface. For crystalline samples the AFM can produce atomic resolution images of biological samples. For non-crystalline materials molecular resolution is usually possible. The force between the tip and the sample surface will be sensitive to heterogeneity of charge, surface elasticity and adhesion, all of which will then contribute to the contrast in the images. Different imaging modes are available on modern microscopes that allow these different contrast mechanisms to enhance or supplement normal topographic images.

Because the AFM images by 'feeling', rather than 'looking' at samples, it can be operated at ambient temperatures in gaseous or liquid environments. Provided the processes are slow enough it is possible to generate real-time molecular movies. Thus the AFM can provide an alternative view of the molecular structure of plant cell walls.

2.3.1   Plant cells

Some of the first living systems to be examined by AFM were plant cells (Butt *et al.*, 1990). Cut sections of plant leaves were stuck to stainless steel discs and imaged under water. Cellular features were resolved in the images of the undersides of the leaves of *Lagerstroemia subcostata*, a small Indian tree, but high-resolution images failed to resolve features less than 200 nm in size, possibly due to the presence of thick cuticle layers. More detail was seen for the leaves of the water lily *Nymphaea odorata*, which are thought to have thinner cuticles. Fibrous structures were observed in addition to features resembling cells. Canet *et al.* (1996) reported studies on isolated ivy leaf cuticles that were extracted enzymatically from the leaves. Transverse sections were cut and examined after embedding in Epon, and images of the inner and outer faces of the cuticles were obtained after binding them in place with double-sided 'sellotape'. AFM images of sections showed stacked lamellae in the outer lamella zone. The inner reticulate regions were found to be largely amorphous, although some evidence was seen for fibrous inclusions at regions that may have been close to the epidermal cell wall. The outer surface of the cuticle was difficult to image and appeared featureless, mainly due to problems caused by the probe tip adhering to the sample. At low resolution the internal cuticle surface showed imprints of epidermal cells surrounded by high cell walls while at higher resolution the imprints revealed a helicoidal stacking of fibres. These fibrous structures could be removed by acid treatment, suggesting that they were polysaccharide material from the epidermal cell walls penetrating through into the wax cuticle. The fibrous structures on the inner faces observed by AFM were consistent with AFM and transmission electron microscopy images of fibres seen within the transverse sections near the cell wall (Canet *et al.*, 1996). Woody tissue is more rigid and might be expected to yield more details of molecular structure. Hanley and Gray (1994) examined the surfaces of mechanically pulped fibres and transverse and radial sections of black spruce (*Picea marianna*) wood. AFM images of sectioned Epon-embedded wood samples revealed details of the cell wall and characteristic features such as bordered pits. The cell wall region appears layered and the middle lamella and different regions of the secondary cell wall were resolved. It is believed that different orientations of microfibrils within the wall, relative to the cut direction, lead to different degrees of roughness of the cut surface that undergo different extents of deformation during scanning, generating contrast in the images. A specialized technique of imaging called phase imaging has been used to highlight lignified regions in woody tissue (Hansma *et al.*, 1997). This contrast enhancement is considered to arise because the lignified areas are more hydrophobic than the cellulose regions.

Because of the present difficulties in examining intact tissue, attention has been largely focused on studies of isolated cell wall material, and on the characterization of cell wall polysaccharides and their interactions.

2.3.2 Plant cell walls

It has been possible to image the cellulose microfibrillar network in isolated cell wall fragments (Kirby *et al.*, 1996a; Round *et al.*, 1996; van der Wel *et al.*, 1996; Morris *et al.*, 1997). Samples were imaged in air after drying down onto suitable substrates. The size and orientation of the microfibrils observed by AFM in extracts of cell wall material from root hairs of *Zea mays* and *Rhaphanus sativus* were consistent with data obtained on platinum/carbon coated specimens imaged by electron microscopy (EM) (van der Wel *et al.*, 1996). Similar results have been found for the microfibrillar structures from *Linderina pennispora* sporangia (McKeown *et al.*, 1996). Cell wall fragments from Chinese water chestnut, potato, apple and carrot (Kirby *et al.*, 1996a) were deposited onto freshly cleaved mica and imaged wet. Excess water was removed with blotting paper and the material imaged in air before dehydration occurred. The sample preparation technique presents the cell wall face previously adjacent to the plasma membrane. The roughness of the sample is apparent in the normal topographic images that show bright (high) and dark (low) regions of the samples (Figure 2.13a). Because of the large number of grey levels needed to describe the whole image, molecular detail is only perceived in certain regions of the image (Figure 2.13a). The error signal mode image of the same area shows that molecular detail is present throughout the image area (Figure 2.13b). Error signal mode images effectively repress the low frequency information characterizing the curvature of the sample and emphasize the high frequency molecular information. Although this form of imaging demonstrates that structure is present, the picture is not strictly a real image in that the contrast arises from momentary changes in force as the probe scans the sample. Real images showing detailed molecular structure can be obtained either by high pass filtering or by subtracting the low frequency background curvature from the topographic image (Figure 2.13c). The latter procedure is best (Round *et al.*, 1996) and the background function can be generated by locally smoothing the topographic image.

Unlike EM images of cell walls, the AFM images only appear to show the cellulose fibres. AFM does not provide information on the other components of the cell wall, such as pectin or hemicelluloses, that are supposed to interpenetrate or cross-link the cellulose fibrils. For algal cell walls it is possible to show that the AFM can discriminate between different molecular species (Gunning *et al.*, 1998). Semi-refined carrageenan is extracted from algal cell walls using a milder extraction procedure than normal that does not completely remove all the cellulose component of the cell wall. The AFM images show interpenetrating networks in which the individual components can be easily identified (Figure 2.14): the stiffer, thicker cellulose fibres appear brighter than the thinner carrageenan fibres. Therefore the failure to image non-cellulose components in cell wall fragments must be due to the higher mobility of these polymers, and the consequent positional averaging that occurs during the AFM scan. The sample preparative methods for EM freeze this motion, allowing visualization of these molecules.

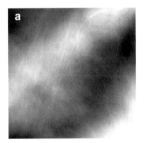

Figure 2.13   AFM images of hydrated Chinese water chestnut plant cell walls. Scan size is 2 × 2 μm. (a) Topographic image: the bright and dark bands indicate peaks and troughs in the rough surface. (b) Error signal mode image emphasising the high frequency molecular structure present in image 'a'. (c) Background subtracted image showing 'true' molecular structure: this form of processing effectively selects the molecular structure and projects it onto a flat plane. Note that the image area in 'c' is different to that in 'a' and 'b'.

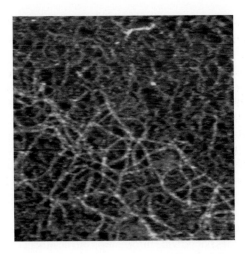

Figure 2.14   AFM image showing interpenetrating networks of carrageenan and cellulose. The stiffer and thicker cellulose fibres appear brighter than the thinner and more flexible carrageenan fibres. Image size 700 × 700 nm.

Despite these limitations it is still worth pursuing these studies. An advantage of imaging cell walls under aqueous conditions is the possibility of following enzymatic degradation. Lee *et al.* (1996) have reported low resolution imaging of the addition of cellulase to cotton fibres. Addition of cellobihydrolase I (CBH I) was found to disrupt the microfibrillar structure, whereas a control experiment involving addition of a catalytically inactivated CBH I resulted in no detectable change in structure. A long-term goal of studying aqueous cell walls would be eventually to develop methods of imaging cell wall structures within intact plant cells under physiological conditions, with the potential for investigation of biological processes such as growth or cell elongation.

Difficulties in imaging intermolecular interactions within cell walls have prompted studies on isolated cell wall polysaccharides.

2.3.3   Cellulose

Cellulose is the major structural component of the plant cell wall. A number of researchers have used AFM to investigate the isolated cellulose fibres (Hanley *et al.*, 1992, 1997; Baker *et al.*, 1997, 1998, 2000; van der Wel *et al.*, 1996; Kuutti *et al.*, 1995) and high resolution images and analysis of the surface structure have been reported (Hanley *et al.*, 1992, 1997; Baker *et al.*, 1997, 1998, 2000; Kuutti *et al.*, 1995). Provided adequate account is taken of probe broadening effects, then the sizes of the cellulose fibres are consistent with observations from electron microscopy. The highest resolution images have been obtained for *Valonia* cellulose. In order to probe the surface structure of *V. macrophysa* cellulose AFM images were Fourier processed and compared with model Connolly surfaces generated from electron diffraction data for the two expected allomorphs I_α (triclinic) and I_β (monoclinic). The surface structures observed were assigned to the monoclinic phase (Kuutti *et al.*, 1995). More recent AFM images of *V. ventricosa* cellulose have revealed the repeating cellobiose unit along the cellulose chains, through identification of the location of the bulky hydroxymethyl group, thus permitting assignment of the triclinic phase directly from the images (Baker *et al.*, 1997, 1998, 2000). This required detecting differences in the displacement of cellulose chains along their axes by 0.26 nm. This was achieved without filtering or averaging the data in the images, and is believed to be the highest achieved resolution for an AFM image of a biological specimen at the present time. The ability to image cellulose surfaces under water (Baker *et al.*, 1997, 1998) suggests that it may be possible to investigate the binding and action of cellulases.

2.3.4 Pectins

Pectins are important both as cell wall components and as industrial gelling agents. There is a considerable literature on the chemical structure and mechanisms of gelation (Voragen *et al.*, 1995; Morris, 1998). Because of the difficulties in observing the pectin networks within cell walls their modes of association are normally inferred from models of gelation. As described in Chapter 1, the pectic polysaccharides are structurally complex and heterogeneous (Schols and Voragen, 1994; Schols *et al.*, 1994). They consist of a backbone of $(1{\to}4)$ α-D-galacturonosyl residues interrupted with typically a 10% substitution of $(1{\to}2)$-α-L-rhamnopyranosyl residues. A fraction of the rhamnosyl residues are branch points for neutral sugar side-chains that contain L-arabinose and D-galactose. The rhamnosyl substitution is thought to cluster in 'hairy' regions leaving 'smooth' sequences of the galacturonan backbone (Figure 2.15). The backbone may be partially acetylated and may be further substituted with terminal xylose. A fraction of the galacturonosyl residues of extracted pectins are partially methyl esterified. The ester substitution may be random, or present as blocks of esterified or

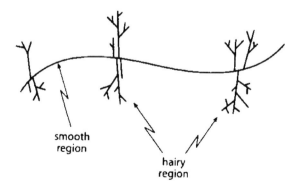

smooth
region

hairy
region

Figure 2.15 Schematic diagram illustrating the distribution of smooth and hairy regions on a pectin chain.

unesterified regions. The level and distribution of free uronic acid residues is considered to be very important in influencing the form of intra-molecular associations.

The ability to visualize individual polysaccharide chains offers a possibility for probing the heterogeneity of pectin chains and their mode of interaction. AFM studies have been made on pectin isolated from green tomato cell walls using sequential extractions with CDTA and then Na_2CO_3. The CDTA is believed to complex calcium, freeing pectin predominately from the middle lamellae, whereas the mild base is considered to cleave ester linkages, releasing pectin from the primary cell wall. The AFM images showed extended rigid molecules (Figure 2.16). The stiffness of the individual molecules seen in the images suggests that the observed pectin molecules are helical, but it is not known whether this is the structure adopted in solution, or whether deposition onto mica promotes formation of the ordered 3-fold helical structure. An unexpected finding (Figure 2.16) was that, whilst the majority of the molecules were linear, a significant fraction (about 30%) contained long branches, with a smaller fraction being multiply branched (Round *et al.*, 1997, 2001). The contour length and branching length distributions differed for the two extracts, with the CDTA extracts being longer, and thus of higher molecular weight (Figure 2.17). In order to ascertain the nature of the branches the Na_2CO_3-extracted pectin was subjected to a mild acid hydrolysis (0.1 M HCl at 80°C for 1, 4, 8, 24 and 72 hours) (Round, 1999). After each hydrolysis step the pectin was analysed chemically and imaged by AFM to determine the contour length distribution (Figure 2.18). After 8 hours of hydrolysis, which had removed the galactose and arabinose residues, the contour length distribution was essentially unchanged (Figure 2.18b) and the branches remained, demonstrating that the branches were not composed of

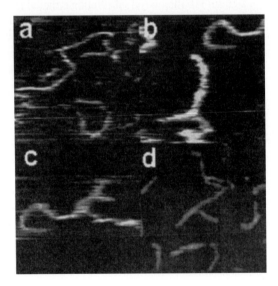

Figure 2.16   AFM images of pectin molecules showing (a) linear, (b) single branched, (c) double branched and (d) multiple branched structures. The scan size is 250 × 250 nm.

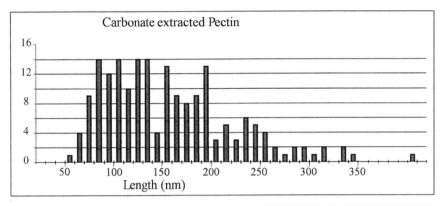

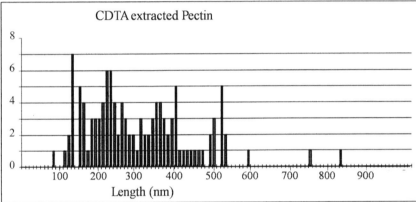

Figure 2.17   Contour length distributions for carbonate and CDTA pectin extracts. The figure is based on data from Round (1999).

neutral sugars. The rhamnose linkages were cleaved after 24 hours but no reduction in contour length (Figure 2.18c), or loss of branching was observed. This observation suggests that the rhamnose residues are not distributed along the chains as shown in Figure 2.15, but probably clustered at the ends of the chains (Figure 2.19). Only after 72 hours did the contour length (Figure 2.18d) and the branch lengths decrease, suggesting that the branches are formed from galacturonic acid residues. Thus the study of individual pectin molecules has revealed new information on their structural heterogeneity. The branched structures observed presumably reflect previously unknown modes of association within the cell wall. The approach used to investigate the pectin structure has been to hydrolyse particular sugars or linkages, and then to infer their location by examining the fragmentation caused by hydrolysis. A better approach would be to label specific sites, such as rhamnogalactan regions or uronic acid blocks, and then map their location along individual chains. The potential of this technique is discussed further in the following section on arabinoxylans.

THE PLANT CELL WALL

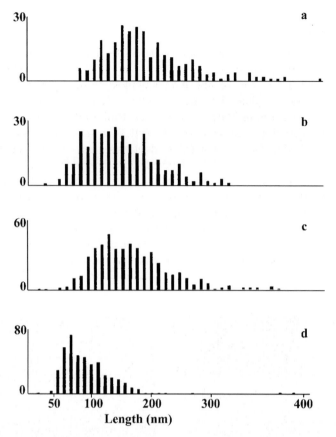

Figure 2.18 Contour length distributions for carbonate-extracted pectin showing the effects of partial acid hydrolysis (0.1 M HCl, 80°C for (a) 1 hour, (b) 8 hours, (c) 24 hours and (d) 72 hours). The figure is based on data from Round (1999).

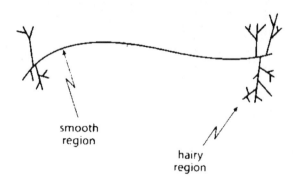

Figure 2.19 Schematic diagram showing an alternative model for pectin structure based on AFM images of partially hydrolysed pectin.

2.3.5 Arabinoxylans

Water-soluble wheat endosperm arabinoxylans can be extracted from cereal grains. They are considered to consist of a linear backbone of β (1–4) linked D-xylose residues containing O-2 and/or O-3 linked α L-arabinose residues (Izydorczyk & Biliaderis, 1993). Arabinoxylans also contain ferulic acid dimers that are considered to play the role of cross-linking the polysaccharides (Ishii, 1991; Waldron *et al.*, 1996). AFM images of arabinoxylans deposited onto mica substrates reveal stiff extended molecules suggesting that the molecules adopt an ordered helical conformation (Gunning *et al.*, 2000; Adams, 2001). Detailed light scattering studies of arabinoxylans (Chanliaud *et al.*, 1996; Pinel *et al.*, 2000, 2001) have shown that the polymers adopt a stiff coil-like structure in aqueous solution, suggesting that helix formation is induced on adsorption to the mica substrate. Adoption of this ordered helical structure on adsorption makes the molecules easier to image. The images reveal mainly linear molecules with a small fraction (15%) of branched chains (Figure 2.20). The molecules can be completely hydrolysed with xylanases, showing that both the backbone and branches are based on β(1–4) D-xylose (Adams, 2001). The nature of

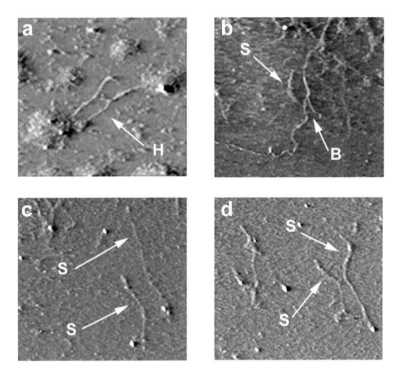

Figure 2.20 AFM images of arabinoxylan molecules showing (a) an unusual 'H' shaped structure (H), (b) linear (S) and branched (B) molecules, and (c & d) linear (S) molecules. Scan sizes 1 × 1 μm. The figure is based on data from Adams (2001).

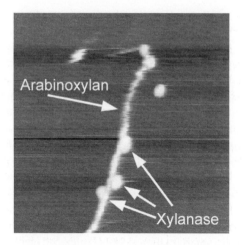

Figure 2.21   AFM image showing the binding of genetically inactivated xylanases to an arabinoxylan chain. Scan size 500 × 500 nm. The figure is based on data from Adams (2001).

the branch points remains to be determined but a likely candidate is diferulic acid cross-links. Interestingly the level of branching observed is less than the measured content of ferulic acid dimers, suggesting that these linkages may also play a role in polymerizing smaller arabinoxylan chains into the longer structures observed by the AFM (Adams, 2001). As with pectin, it would be possible to specifically degrade specific linkages chemically, or enzymatically, and observe the fragmentation patterns by AFM, in order to ascertain their roles in the structure of the arabinoxylans. A better approach would be to label and identify specific linkages on the polymer chains. One way of achieving this would be to observe the binding of inactivated enzymes to their specific binding sites on the polymer chains. The feasibility of this approach is illustrated in Figure 2.21, which shows AFM images revealing the binding of genetically inactivated xylanases to an arabinoxylan chain (Adams, 2001). It is intended to develop and use this 'molecular mapping' approach for the analysis of chemically heterogeneous populations of plant cell wall polysaccharides.

2.3.6   Carrageenans

Many cell wall polysaccharides are extracted commercially and used as industrial gelling and thickening agents. This functionality arises because the polymers naturally associate to form network structures that are considered to mimic their form of association within the plant cell wall. Therefore the mechanisms of gelation are usually used as models for describing the network structures within the plant cell walls. Carrageenans are perhaps the most studied of the gelling polysaccharides (Piculell, 1995; Morris, 1998). The junction zones, or regions of association of the carrageenan chains, have been studied extensively, and can be modelled at atomic resolution (Piculell, 1995). The polymers form thermo-setting, thermo-reversible

gels. Cooling polymer sols results in the adoption of an ordered double helical structure. Electron microscopy (Hermansson, 1989; Hermansson *et al.*, 1991) and atomic
force microscopy (Ikeda *et al.*, 2001) suggest that, at this stage, the polymers can
further 'polymerize' into longer fibril structures. These fibrils then associate to form
networks composed of thicker fibres, within which the fibrils are assembled side-
by-side due to specific binding of cations. This type of structure can be visualized
by AFM (Kirby *et al.*, 1996b; Ikeda *et al.*, 2001) for gel precursors and in aqueous
carrageenan films (Figure 2.22). The gelling mechanism appears to be generic for
this class of gelling polysaccharide, and has been studied in more detail for the bacterial polysaccharide gellan gum. Here, AFM has been used to study (Gunning *et al.*,
1996; Morris *et al.*, 1999b) the gel precursors, and to visualize the network structures
formed in hydrated films and even in bulk hydrated gels (Figure 2.23). It is likely
that these types of fibrous networks occur within the cell wall and AFM provides
a route to examining such structures, or model films formed at the higher polymer
concentrations that are expected to be present within the plant cell wall.

AFM is providing new information on the nature of cell wall polysaccharides
and the networks they form within the plant cell wall. It provides unique methods
for probing chemical heterogeneity and unexpected forms of association that may
be present at too low a concentration to be detectable by chemical or enzymatic
analysis. AFM provides complementary information to electron microscopy, but
with the potential for probing structure in hydrated cell wall extracts. In the future,
it may be possible to image cell walls within intact cells, or plant tissue, offering the
possibility of studying processes such as enzymatic degradation, or elongation and
growth under more realistic conditions.

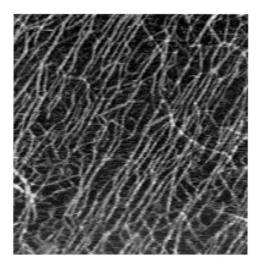

Figure 2.22 AFM image of a hydrated film of aggregated kappa carrageenan. Scan size
1 × 1 μm.

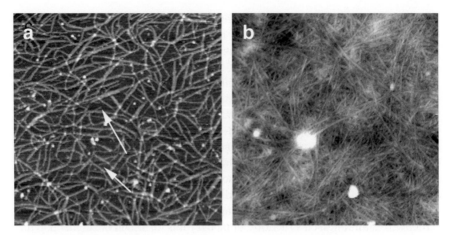

Figure 2.23   AFM images of hydrated gellan films and gels. (a) Gellan film: the arrows indicate stubs or precursors to branches, scan size 700 × 700 nm. (b) Surface of a hydrated bulk acid-set gellan gel, scan size 3 × 3 μm.

2.4   Molecular interactions of plant cell wall polymers

Biophysical analysis of the interactions between plant cell wall polysaccharides mediated by solutes and ions, and by the solvent water molecules, is providing fresh insight into the way the mechanical properties and porosity of the cell wall polymer networks come under physiological control. This approach shows how small changes in the pH and ionic composition of the apoplast, together with changes in the level of osmotic stress exerted on the cell wall by the cell contents, have the potential to contribute to rapid and reversible changes in cell wall properties. It is leading to the development of models of the cell wall that begin to predict its response to changing physiological conditions, rather than simply provide a static pictorial representation of the polymers and their interactions.

2.4.1   Plant cells and their wall polymers

The primary cell walls of higher plants, and many lower plants, are constructed on the same basic principle (Carpita and Gibeaut, 1993). Crystalline cellulose microfibrils, several nanometres in diameter, are embedded in a matrix of more highly hydrated polymers, which also form the junction zones between adjacent cells (Bacic *et al.*, 1988; Cosgrove, 1997). Fungal hyphal cell walls have a similar pattern of construction, using chitin in place of cellulose. Matrix polysaccharides show great variability between different plant groups, with pectin and xyloglucans predominating in dicotyledons, gymnosperms, and some monocotyledons (e.g. onion), and glucuronoarabinoxylans and mixed linkage glucans predominating in the Gramineae and other monocotyledons (Bacic *et al.*, 1988). In some ways, these structures can be likened to fibre-reinforced elastomers, with the fibres increasing and reinforcing the stiffness of the matrix. The common feature of all the matrix polysaccharides is their tendency to be hydrated.

The plant cell wall and middle lamella perform a range of functions. Many tissues, such as parenchyma, can be likened to a closed cell, fluid-filled cellular solid. The mechanical properties of these structures are influenced by cell adhesion, the mechanical properties of the wall (including its behaviour on bending and extension), and the rate at which fluid can be transferred between cells on deformation (Ashby and Gibson, 1997). The plant cell wall must provide resistance to osmotic swelling of the protoplast, whilst retaining the capacity to expand. In some instances the wall must expand in a highly directional way as in root hairs. The walls of neighbouring cells must also adhere to each other, and at times lose their adhesion, as in the root cap, ripening fruit, and abscission zones. The cell wall also functions as a barrier restricting pathogen invasion, and potentially restricting the access of endogenous hydrolytic, cell wall degrading enzymes, to the cell wall during processes such as fruit ripening.

The widespread use across the plant kingdom of one essential form of wall construction suggests that this arrangement is highly adaptable to the range of functions described above. Although matrix polysaccharides can be changed by turnover of the cell wall components, or modified by enzymes exported into the cell wall, one of the keys to the success of hydrated polysaccharides as major components of the cell wall appears to be the physiological control that can be exercised over their behaviour by regulation of the aqueous environment of the apoplast. Although the apoplast is often referred to as the extracellular space, it is clearly not 'exterior' to the physiology of the plant, and it is best thought of as a metabolic compartment in its own right. In addition to cell wall polysaccharides, the apoplast may contain a range of low molecular weight and macromolecular solutes including sugars, inorganic and organic cations, organic anions such as citrate, and structural cell wall proteins and enzymes. There is a range of potential ionic interactions, with the various ionic species being involved in different complex equilibria. The cell wall polymer networks can be isolated for analysis relatively free from contamination with intracellular lipids, proteins and polysaccharides, but characterisation of the soluble components of the apoplast has proved difficult (Grignon and Sentenac, 1991). The fact that the apoplast cannot be isolated as a membrane bounded entity like organelles or protoplasts, its narrowness (100 nm), and the presence of charged polymers, all present problems, so that no single method of analysis is adequate on its own. Various methods have been used to obtain fluid from the apoplast. These include centrifugation of hypocotyl sections, washes from leaves from which the epidermis has been removed, and pressure-expression of sap from tomato fruit. Whilst contamination from cell contents can be assessed using marker enzymes, and these approaches are valid for determining the distribution of enzymes, the value of the ionic analyses is less certain. Direct measurement of apoplastic concentrations of unbound ions has been attempted with microelectrodes (Harker and Venis, 1991). For calcium and magnesium, where a substantial proportion of the total is located in the apoplast a more successful approach has been to use non-aqueous fractionation to determine the total amount present in the apoplast

(MacDougall *et al.*, 1995). In this method the tissue is fast-frozen to avoid disruption of the plasma membrane by ice-crystal formation and the tissue is disintegrated while frozen and freeze-dried. A cell wall enriched fraction is obtained by sieving in a non-aqueous solvent, and after correction for cytoplasmic contamination (assessed from the presence of marker enzymes) the total ion content of the apoplast is obtained. Various methods of fixation, followed by spectroscopic analysis (e.g. X-ray emission or electron energy loss spectroscopy) have also been reported. These suffer from the potential for redistribution of the ionic species during fixation. For those plants that lack a cuticle and are openly exposed to the environment, the apoplast acts as a buffer against environmental variation. This is particularly true for intertidal algae that experience wide fluctuations in the salt content of the water they are bathed in. While most plants are not exposed in this way, some halophytes (notably mangrove) tolerate elevated levels of salt in the apoplast. Large local variations in apoplastic salt content are also associated with plant movements.

There is likely to be a high degree of complexity in the exact relationship between the structure and function of different matrix polysaccharides from different plant species and cell types. However, to illustrate some of the potential effects, we will focus on the primary cell wall of dicotyledonous plants and examine the properties of the pectin network. Although this network performs a range of functions, key generic aspects of the physical chemistry, which will have a major impact both on mechanical and barrier properties of the cell wall, include the extent of network cross-linking and how cross-linking and the affinity of the cell wall polymers for water influences cell wall hydration.

2.4.2   The pectic polysaccharide network

The swelling and hydration of the pectin network in the plant cell wall and middle lamella will depend on the extent of cross-linking and the affinity of the polymer for water. In pectin networks there is the potential for both covalent and non-covalent cross-links, and the greater the extent of cross-linking, the greater the potential restorative force resisting swelling.

The characterisation of the cross-linking of the pectic network is still a matter of research. In some plant families e.g. the Chenopodiaceae, of which the most investigated is the sugar beet (*Beta vulgaris* L.), the potential for covalent cross-linking through phenolic residues, such as diferulic acid, has been demonstrated (Saulnier and Thibault, 1999). There is also the possibility of cross-linking of the pectic network through ester cross-links, involving the D-galacturonic acid of the pectic polysaccharide backbone (Brown and Fry, 1993). As yet, the cross-link has not been isolated. The cross-link for which there is a more detailed structural characterisation is a 1:2 borate ester diol of a pectic polysaccharide fragment (rhamnogalacturonan II) that can be released from the primary cell wall of plants by treatment with *endo*-α-1,4-polygalacturonase (Ishii *et al.*, 1999), as described in Chapter 1. In addition to these covalent cross-links, there is also the potential for secondary interactions to contribute to network formation. Of these, the most studied is the interaction of pectic polysaccharides with calcium ions (Kohn, 1975; Garnier *et al.*, 1994), that can result in network formation and gelation of moderately concentrated solutions of the pectic polysaccharides (Morris *et al.*, 1982; MacDougall *et al.*, 1996). The main requirement for gelation is the presence of stretches of unsubstituted D-galacturono-syl residues in the pectic polysaccharide backbone, and a sufficient number of such regions to form an interconnected network. From studies on the solution behaviour of pectic polysaccharides, it is proposed that the conformation of pectic polysaccharides in the junction zone or cross-link is that of an 'egg box' (Morris *et al.*, 1982), although other types of association are also possible (Jarvis and Apperley, 1995).

2.4.3   Ionic cross-linking of the pectic polysaccharide network

In the cell wall literature, there is a focus on the role of calcium ions in cross-linking the pectic polysaccharide network, yet the ionic environment of the apoplast contains a range of other species, including macro-ions such as polyamines and the structural protein extensin, which could be involved in cross-linking. Why is there then the focus on Ca^{2+}? Part of the reason stems from studies on pectic polysaccharide gels. It is found that moderately concentrated solutions of pectins (1–2% w/w), with a degree of methyl esterification of less than ~65% and a suitable blockwise distribution of charged residues, form gels on addition of Ca^{2+}. As addition of other simple cations, such as potassium and magnesium, does not lead to network formation, the reason for the focus on calcium is clear. However, this view fails to take into account a number of other relevant features of the problem. The first of these is polymer concentration. In the pectin gel, polymer concentration is typically a few percent. In the plant cell wall and middle lamella, polymer concentration may be as high as 30 to 50% w/w. This huge difference in concentration must have an enormous impact on ionic equilibria. Divalent counterions such as Mg^{2+}, which have a relatively weak affinity for the pectin chain (Kohn, 1975), may well function as cross-linking agents at the very much higher concentration of pectin found in the plant cell wall. As oligo- and polygalacturonate crystallize in the presence of monovalent ions (Na^+, K^+) at room temperature (Walkinshaw and Arnott, 1981; Rigby *et al.*, 2000), it is even possible to imagine that pectins of a suitable structure could be cross-linked by these ions under appropriate conditions, with the cross-link being a microcrystallite. An additional aspect to be considered is the possible role of structural proteins in the plant cell wall (Showalter, 1993). Extensin is a basic protein and therefore could form charge complexes with pectic polysaccharides under appropriate conditions of pH and ionic strength. Support for the view that it could function as a cross-linker comes from studies on the cross-linking of pectic polysaccharide gels with basic peptides (Bystrický *et al.*, 1990; MacDougall *et al.*, 2001a). A range of ionic equilibria could therefore potentially affect the properties of the pectin network of the plant cell wall, and network cross-linking *in vivo* has the potential to be modulated in a range of ways. Further research is necessary to establish the extent to which non-Ca^{2+} mediated cross-linking is important.

A number of approaches can be suggested. Firstly, it would be helpful to have mechanical data on concentrated pectic polysaccharide films to establish the extent to which ionic species can cross-link the network in systems of comparable concentration to that found in the plant cell wall. Secondly, more information is needed on apoplast ion contents and the associated ionic equilibria that might potentially affect behaviour. To attempt to calculate speciation within the cell wall, information is required on the components present, their concentration, and the stability constants describing the various ionic equilibria. As will be discussed later in this chapter, pectin concentration can be estimated from the *in vitro* experiments on cell wall swelling. For the determination of cation content, and the content of other anions such as organic acids, which could compete with the pectic polysaccharides for counterions, two complementary approaches may be used. The solute and ionic content of apoplastic sap, expressed after application of pressure to tomato fruit (Ruan *et al.*, 1996), gives an indication of the free cation and organic acid concentration within the cell wall. For example, the main cations found in tomato fruit apoplastic sap (24–26 days after anthesis) were K^+, Na^+, NH_4^+, Mg^{2+}, and Ca^{2+}, at concentrations of ~20, 0.5, 0.5, 5.2 and 6.3 mM, respectively (Ruan *et al.*, 1996). Non-aqueous methods give the total ion content of the apoplast, which will include free and bound forms (MacDougall *et al.*, 1995). In both ripe and unripe tomato cell walls the Mg^{2+} and Ca^{2+} levels were ~18 and ~60 $\mu mol\ g^{-1}$, respectively. If an *in vivo* swelling of the cell wall of 3.0 g/g is assumed, this would lead to a total concentration of Ca^{2+} of ~36 mM. The large difference between the total and free Ca^{2+} levels (36 and 6.3 mM) presumably reflects the extent of binding of Ca^{2+} by the pectic polysaccharides in the wall. To test this proposition, information is required on the amount of anhydrogalacturonic acid in the cell wall and the affinity of the various species within the cell wall, including pectic polysaccharides and organic acids, for the different ions. The affinity of ions interacting with pectin is conventionally described by a stability constant, K, for the equilibrium

$$K = [Ca^{2+}2COO^-]/[Ca^{2+}][2COO^-]$$

for which, following convention and the observed stoichiometry of binding, one Ca^{2+} ion is considered to bind to a fragment of pectin containing two carboxyl functions. The change in stability constant with both the degree of methyl esterification and the galacturonate chain length has been determined: the stability constant increases with decreasing degree of methyl esterification. For pectin with a degree of methyl esterification of 65%, the interpolated value of log K is ~2.4 (Kohn, 1975). This constant describes the binding in solution, which could involve both intra- and intermolecular associations. For tomato pectin (degree of methyl esterification 68%), the Ca^{2+} binding behaviour in aqueous solution was comparable to that of other pectins (log K ~2.7) with a similar degree of methyl esterification (Tibbits *et al.*, 1998). The binding of Ca^{2+} in the tomato pectin network was determined by following the dissolution of a pectin gel in water, and a value of log K of ~3.9 was found (Tibbits *et al.*, 1998). This is more than an order of magnitude greater than the comparable solution case, indicating that a cross-link of sufficient permanence to create an elastic gel has a higher affinity for Ca^{2+} than the chain in solution. Having estimates of free uronic acid concentration in the cell wall, its affinity for Ca^{2+} ions, and the ionic composition of the apoplast, it is possible to predict the speciation behaviour as a function of pH. This type of calculation is relatively easy to perform, as software is readily available to help describe ionic equilibria in multicomponent systems such as wastewater. The predicted concentrations of Ca^{2+}, pectate and calcium pectate as a function of pH in the range 3–7 is shown in Figure 2.24 (based on a log stability constant for the formation of calcium pectate of 3.7).

This simple calculation reveals a number of features. Firstly, pH has an important effect on the extent of interaction and the charge on the network. Secondly, for the uronic acid and calcium contents found in the cell wall, there is still free uronic acid present that can potentially participate in other equilibria and, as discussed later, can potentially have a role in cell wall swelling through a Donnan type effect. Although there are a number of assumptions involved in these calculations, the most serious of which is that the ionic equilibrium can be described through a simple stability constant, the approach has value. For example, it is possible to assess the potential effect of other equilibria on behaviour. Organic acids are components of the apoplast and it might be speculated that they could have a role in promoting cell separation through their ability to complex Ca^{2+} ions, reducing the cross-linking of the pectin network of the middle lamella. Calculation of their potential effect shows that it is likely to be small (MacDougall *et al.*, 1995). Similar calculations on the pectin extractant CDTA (cyclohexane diamine tetraacetic acid; a Ca^{2+} chelating agent), confirm its usefulness, but suggest that it might not extract pectins with a low degree

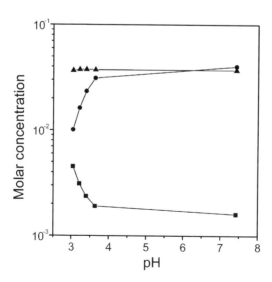

Figure 2.24 Predicted speciation of pectate at ionic concentrations estimated for the tomato cell wall. Pectate (●), Ca^{2+} (■), Ca^{2+} pectate (▲).

of methyl esterification at the commonly used pH of 6.8 – higher pHs should be more effective.

2.4.4 The significance of polymer hydration for the plant cell wall

For animal biochemists, polymer hydration and its role in determining the properties of extracellular matrices has been an active field of study since the 1970s (Grodzinsky, 1983). The hydration of proteoglycans and glycosaminoglycans in cartilage has been shown to give rise to a swelling pressure that plays a significant role in the ability of this tissue to resist compressive loads (Grodzinsky, 1983). Similarly, in the cornea, hydration forces associated with proteoglycans help maintain the correct spacing between collagen fibrils that is necessary for translucence (Elliot and Hodson, 1998). Isolated matrix components from plant cell walls find extensive use as food additives, partly on account of their hydration behaviour (Stephen, 1995), but this has had surprisingly little influence on conceptual development of the functional role of polymer hydration in plant cell walls. One reason is the narrowness of the wall that makes changes in its hydration hard to detect, except in the more dramatic examples of fruit ripening or cooking. Another reason appears to have been the predominance of the view that the role of water in plant cell walls can be adequately understood in terms of the behaviour of water in soils (Nobel, 1999). Recently evidence has been obtained suggesting that hydration of matrix polysaccharides influences cell wall porosity (Zwieniecki, 2001), and that increased hydration is a significant step on the path to loss of cell adhesion in ripening tomato fruit (MacDougall et al., 2001b). In general, a thorough examination of the hydration behaviour of the isolated polysaccharides and extension of this to the in vivo situation is justified.

2.4.5 Swelling of the pectin network

Swelling and cross-linking of networks are interlinked. For a network in a good solvent the network will swell until the swelling force is balanced by the extent of network cross-linking that produces a restorative force (Flory, 1953). The physico-chemical basis of the swelling of synthetic polymer networks has been the subject of continuing study (Skouri et al., 1995; Rubinstein et al., 1996). For neutral, hydrophilic polymer networks, hydration can be characterized in terms of a single parameter describing the affinity of polymer and solvent (Flory, 1953). At high polymer concentrations this contribution to network hydration can be large. Literature on the hydration behaviour of weakly charged synthetic polymers suggests that polyelectrolyte effects can be very important and should be considered in any physicochemical description of the role of the pectin network in the plant cell wall. For polyelectrolyte gels, the requirement for electrical neutrality leads to an excess of counterion within the gel compared to the external medium. This excess generates an osmotic pressure difference between the gel and external medium, which increases with decreasing ionic strength. The excess osmotic pressure leads to the swelling of the gel until a balance is achieved between the osmotic pressure that drives swelling and the restorative force arising from the deformation of the cross-linked network. At intermediate salt concentrations, an estimate of the contribution to osmotic pressure, π, due to a polyelectrolyte may be obtained from (Flory, 1953; Skouri *et al.*, 1995; Rubinstein *et al.*, 1996):

$$\pi \approx \frac{RTc^2}{A(c + 4Ac_s)} \tag{2.1}$$

for univalent electrolytes, where c and c_s are the molar concentrations of polymer segment and salt, and A is the number of monomers between effective charges. The greater the charge on the polymer, and the lower the ionic strength, the greater the osmotic pressure generated. However, at high charge densities, the phenomenon of counterion condensation can reduce the counterion fraction that can contribute to this Donnan effect (Manning and Ray, 1998). In the extreme case, counterion condensation on the backbone can lead to reduced swelling and a fall in the average dielectric constant of the material, leading to further counterion condensation, and a 'catastrophic' collapse of the network (Grosberg and Khokhlov, 1994). Highly charged polyelectrolytes can therefore exhibit minimal swelling in water. Polygalacturonate is an example (Ryden *et al.*, 2000) and this suggests that enzymes such as pectin methyl esterase might have a potential role in regulating cell wall behaviour. Limited random attack on pectin with a high degree of methyl esterification would lead to increased affinity for water through a Donnan-type effect. For a cross-linked pectin network this would increase swelling and porosity. Extensive pectin methyl esterase action, producing blocks of unsubstituted anhydrogalacturonic acid, could lead to collapsed structures.

Where pectin networks are cross-linked by interaction with Ca^{2+} ions, the galacturonate sequences play a dual role. They contribute to swelling when ionized, but when involved in Ca^{2+} mediated cross-linking they no longer contribute to swelling and instead play a role in resisting network expansion. Experiments on isolated pectin gels and films confirm the above expectations. For Ca^{2+} cross-linked tomato pectin gels, a fraction of the uronic acid does not participate in calcium mediated cross-linking but contributes to gel swelling (Tibbits *et al.*, 1998). Experiments on films prepared from cell wall pectic polysaccharides show that weakly charged pectins have the highest affinity for water and that this is reduced on increasing ionic strength, and on increasing charge on the counterion. Highly charged pectic polysaccharides swell to a very limited extent (Ryden *et al.*, 2000). Relationships of the form of equation 2.1 give a good generic description of the swelling of pectin networks *in vitro*. It is now necessary to consider how *in vitro* studies on isolated polysaccharides relate to *in vivo* behaviour of the cell wall network.

Extracted cell wall materials swell extensively in water or dilute salt solution. The extent of swelling may be determined through measurement of sediment volume after centrifugation, or more directly by measurement of the exclusion of a macro-molecular probe such as a high-molecular-weight dextran. Sediment from extracted cell wall material is mixed with a solution of the macromolecular probe. Through measurement of the dilution of the probe it is possible to determine the volume of sediment that is inaccessible to the probe, and hence the volume of the cell wall material. For a tomato cell wall preparation, a swelling of 8 g/g was found in 50 mM KCl (MacDougall *et al.*, 2001b). With a typical cell wall content on a dry weight basis of 1 to 2% this might indicate that the cell wall would occupy 8 to 16% of the volume of the tomato. Light microscope observations clearly demonstrate that this is not the case. What constrains cell wall swelling *in vivo*? There is increasing interest in how the osmotic stress of the cellular environment affects biomolecular assembly and the mode of enzyme action (Parsegian *et al.*, 1995). The plant cell wall is exposed to the osmotic stress of the cell contents and therefore the potential exists for this osmotic stress to regulate cell wall swelling.

To help test this proposal, the swelling of isolated tomato cell wall material in 50 mM KCl was examined as a function of osmotic stress (MacDougall *et al.*, 2001b). The cell wall swelling falls from 8 g/g to 3 g/g as the osmotic stress is increased to 0.15 MPa (Figure 2.25). The osmotic stress being exerted on the cell wall by solutes contained within plant cells can be determined from microprobe measurements of the hydrostatic (turgor) pressure in individual cells (Tomos and Leigh, 1999). Reported values for the expanding cells of higher plants are generally in the range 0.1 to 1 MPa. For mature green tomato fruit pericarp cells the turgor pressure has been estimated at 0.13 MPa. The compressive effect of this pressure on the cell wall is illustrated in Figure 2.26. On ripening, a fall in the turgor pressure of tomato cells to 0.03 MPa has been observed (Shackel *et al.*, 1991) – the decrease being associated with failure of the cell membranes to continue to act as an effective barrier to the movement of solutes. These observations suggest that during ripening the osmotic stress that the cell wall is exposed to will fall with a consequent potential increase in

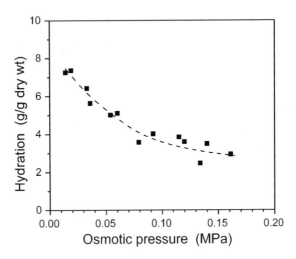

Figure 2.25 The effect of externally applied osmotic pressure on the hydration of tomato cell walls (■). Reproduced with permission from *Biomacromolecules* 2, 450–455. Copyright 2001 Am. Chem. Soc.

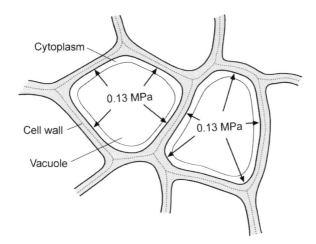

Figure 2.26 Diagrammatic representation of the compressive effect of turgor pressure on the cell walls of unripe tomato pericarp cells.

cell wall swelling and change in cell wall properties. These observations illustrate the potential importance of osmotic stress in influencing cell wall behaviour.

References

Adams, E.L. (2001) Developing molecular probes in arabinoxylan structural analysis using atomic force microscopy. PhD thesis, University of East Anglia, Norwich, UK.

Åkerholm, M. and Salmen, L. (2001) Interactions between wood polymers studied by dynamic FT-IR spectroscopy. *Polymer,* 42, 963–969.

Ashby, M.F. and Gibson, L.J. (1997) *Cellular Solids Structure and Properties,* Cambridge University Press, Cambridge.

Bacic, A., Harris, P.J. and Stone, B.A. (1988) Structure and function of plant cell walls. In *The Biochemistry of Plants,* Vol. 14, Academic Press, London, pp. 297–371.

Baker, A.A., Helbert, W., Sugiyama, J. and Miles, M.J. (1997) High resolution atomic force microscopy of native *Valonia* cellulose I microcrystals. *J. Struct. Biol.,* 119, 129–138.

Baker, A.A., Helbert, W., Sugiyama, J. and Miles, M.J. (1998) Surface structure of native cellulose microcrystals by AFM. *Appl. Phys. A-Mater. Sci. Processing,* 66, S559-S563.

Baker, A.A., Helbret, W., Sugiyama, J. and Miles, M.J. (2000) New insight into cellulose structure by atomic force microscopy shows the I$_\alpha$ crystal phase at near atomic resolution. *Biophys. J.,* 79, 1139–1145.

Brown, J.A. and Fry, S.C. (1993) The preparation and susceptibility to hydrolysis of novel O-galacturonyl derivatives of carbohydrates. *Plant Physiol.,* 103, 993–999.

Butt, H.-J., Wolff, E. K., Gould, S.A.C., Northern, B.D., Peterson, C.M. and Hansma, P.K. (1990) Imaging cells with the atomic force microscope. *J. Struct. Biol.,* 105, 54–61.

Bystrický, S., Malovíková, A. and Sticzay, T. (1990) Interaction of alginate and pectins with cationic polypeptides. *Carbohydr. Polymers,* 13, 283–294.

Cael, J.J., Gardner, K.H., Koenig, J.L. and Blackwell, J. (1975) Infrared and Raman spectroscopy of carbohydrates. Normal coordinate analysis of cellulose I. *J. Chem. Phys.,* 62, 1145–1153.

Canet, D., Rohr, R., Chamel, A. and Guillian, F. (1996) Atomic force microscopy of study of isolated ivy leaf cuticles observed directly and after embedding in Epon. *New Phytol.*, 134, 571–577.

Carpita, N.C. and Gibeaut, D.M. (1993) Structural models of primary cell walls in flowering plants: consistency of molecular structure with the physical properties of the walls during growth. *Plant J.*, 3, 1–30.

Carpita, N.C., Findlay, K., Wells, B. *et al.* (in press). Cell wall architecture of the developing maize coleoptile. *J. Cell Sci.*

Chanliaud, E., Roger, P., Saulnier, L. and Thibault, J.-F. (1996) Static and dynamic light scattering studies of heteroxylans from maize bran in aqueous solution. *Carbohydr. Polymers*, 31, 41–46.

Chen, L., Wilson, R.H. and McCann, M.C. (1997) Investigation of macromolecule orientation in dry and hydrated walls of single onion epidermal cells by FTIR microspectroscopy. *J. Mol. Struct.*, 408/9, 257–260.

Chen, L., Carpita, N.C., Reiter, W.-D., Wilson, R.H., Jeffries, J. and McCann, M.C. (1998) A rapid method to screen for cell-wall mutants using discriminant analysis of Fourier transform infrared data. *Plant J.*, 16, 385–392.

Cosgrove, D.J. (1997) Assembly and enlargement of the primary cell wall in plants. *Annu. Rev. Cell Dev. Biol.*, 13, 171–201.

Defernez, M. and Wilson, R.H. (1995) Mid-infrared spectroscopy and chemometrics for determining the type of fruit used in jam. *J. Sci. Food Agric.*, 67, 461–467.

Elliott, G.F. and Hodson, S.A. (1998) Cornea, and the swelling of polyelectrolyte gels of biological interest. *Rep. Progr. Phys.*, 61, 1325–1365.

Flory, P. J. (1953) *Principles of Polymer Chemistry*, Cornell University Press.

Garnier, C., Axelos, M.A.V. and Thibault, J.-F. (1994) Selectivity and cooperativity in the binding of calcuim ions by pectins. *Carbohydr. Res.*, 256, 71–81.

Griffiths, P.R. and de Haseth, J.A. (1986) Fourier transform infrared spectrometry. In *Chemical Analysis*, Vol. 83 (eds P.J. Elving, J.D. Winefordner and I.M. Kolthoff), Wiley, New York, pp. 437–445.

Grignon, C. and Sentenac, H. (1991) pH and ionic conditions in the apoplast. *Annu. Rev. Plant Physiol. Plant Molecular Biol.*, 42, 102–128.

Grodzinsky, A.J. (1983) Electrochemical and physicochemical properties of connective tissue. *CRC Critical Reviews in Biomed. Engng*, 9, 133–199.

Grosberg, A.Y. and Khokhlov, A.R. (1994) *Statistical Physics of Macromolecules*, American Institute of Physics, New York.

Gunning, A.P., Kirby, A.R., Ridout, M.J., Brownsey, G.J. and Morris, V.J. (1996) Investigation of gellan networks and gels by atomic force microscopy. *Macromolecules*, 29, 6791–6796 [correction, *ibid* (1997) 30, 163–164.].

Gunning, A.P., Cairns, P., Kirby, A.R., Bixler, H.J. and Morris, V.J. (1998) Characterising semi-refined iota-carrageenan networks by atomic force microscopy. *Carbohydr. Polymers*, 36, 67–72.

Gunning, A.P., Mackie, A.R., Kirby, A.R., Kroon, P., Williamson, G. and Morris, V.J. (2000) Motion of a cell wall polysaccharide observed by atomic force microscopy. *Macromolecules*, 33, 5680–5685.

Hanley, S.J. and Gray, D.G. (1994) Atomic force microscope images of black spruce wood sections and pulp fibres. *Holzforsch.*, 48(1), 29–34.

Hanley, S.J., Giasson, J., Revol, J.-F. and Gray, D.G. (1992) Atomic force microscopy of cellulose microfibrils: comparison with transmission electron microscopy. *Polymer*, 33, 4639–4642.

Hanley, S.J., Revol, J.-F., Godbout, L. and Gray, D.G. (1997) Atomic force microscopy and transmission electron microscopy of cellulose from *Micrasterias denticulata*: evidence for a chiral helical microfibril twist. *Cellulose*, 4, 209–220.

Hansma, H.G., Kim, K.L., Laney, D.E. *et al.* (1997) Properties of biomolecules measured from atomic force microscopic images: a review. *J. Struct. Biol.*, 119, 99–108.

Harker, F.R. and Venis, M.A. (1991) Measurement of intracellular and extracellular free calcium in apple fruit cells using calcium-selective microelectrodes. *Plant Cell Environ.*, 14, 525–530.

Hermansson, A.-M. (1989) Rheological and microstructural evidence for transient states during gelation of kappa-carrageenan in the presence of potassium. *Carbohydr. Polymers*, 10, 163–181.

Hermansson, A.-M., Ericksson, E. and Jordansson, E. (1991) Effects of potassium, sodium and calcium on the microstructure and rheological behaviour of kappa carrageenan gels. *Carbohydr. Polymers*, 16, 297–320.

Hinterstoisser, B. and Salmen, L. (2000) Application of dynamic 2D FTIR to cellulose. *Vibrat. Spectrosc.*, 22, 111–118.

Hinterstoisser, B., Åkerholm M. and Salmen, L. (2001) Effect of fiber orientation in dynamic FTIR study on native cellulose. *Carbohydr. Res.*, 334, 27–37.

Holland, J.K., Kemsley, E.K. and Wilson, R.H. (1998) Use of Fourier transform infrared spectroscopy and chemometrics for the detection of adulteration of strawberry purees. *J. Sci. Food Agric.*, 76, 263–269.

Ikeda, S., Morris, V.J. and Nishinari, K. (2001) Microstructure of aggregated and nonaggregated κ-carrageenan helices visualised by atomic force microscopy. *Biomacromolecules*, 2, 1331–1337.

Ishii, T. (1991) Isolation and characterisation of a diferuloyl arabinoxylan hexasaccharide from bamboo shoot cell-walls. *Carbohydr. Res.*, 210, 15–22.

Ishii, T., Matsunaga, T., Pellerin, P., O'Neill, M.A., Darvill, A. and Albersheim, P. (1999) The plant cell wall polysaccharide rhamnogalacturonan II self assembles into a covalently cross-linked dimer. *J. Biol. Chem.*, 274, 13098–13104.

Izydorczyk, M.S. and Biliaderis, C.G. (1993) Structural heterogeneity of wheat endosperm arabinoxylans. *Cereal Chem.*, 70, 641–646.

Jarvis, M.C. and Apperley, D. C. (1995) Chain conformation in concentrated pectic gels: evidence from 13C NMR. *Carbohydr. Res.*, 275, 131–145.

Kacurakova, M. and Wilson, R.H. (2001) Developments in mid-infrared FT-IR spectroscopy of selected carbohydrates. *Carbohydr. Polymers*, 44, 291–303.

Kacurakova, M., Smith, A.C., Gidley, M.J. and Wilson, R.H. (in press) Molecular interactions in bacterial cellulose composites studied by 1D FT-IR and dynamic 2D FT-IR spectroscopy. *Carbohydr. Res.,* 337(12), 1145–1153.

Kemsley, E.K., Belton, P.S., McCann, M.C., Ttofis, S., Wilson, R.H. and Delgadillo, I. (1995) The identification of vegetable matter using Fourier transform infrared spectroscopy. *Food Chem.*, 54, 437–441.

Kemsley, E.K., Holland, J.K., Defernez, M. and Wilson, R.H. (1996) Detection of adulteration of raspberry purees using infrared spectroscopy and chemometrics. *J. Agric. Food Chem.*, 44, 3864–3870.

Kirby, A.R., Gunning, A.P., Waldron, K.W., Morris, V.J. and Ng, A. (1996a) Visualisation of plant cell walls by atomic force microscopy. *Biophys. J.*, 70, 1138–1143.

Kirby, A.R., Gunning, A.P. and Morris, V.J. (1996b) Imaging polysaccharides by atomic force microscopy. *Biopolymers*, 38, 355–366.

Koenig, J.L. (1999) *Spectroscopy of polymers*, Elsevier, Amsterdam, pp. 147–206.

Kohn, R. (1975) Ion binding on polyuronates – alginate and pectin. *Pure Appl. Chem.*, 42, 371–397.

Kuutti, L., Peltonen, J., Pere, J. and Teleman, O. (1995) Identification and surface structure of crystalline cellulose studied by atomic force microscopy. *J. Microsc.*, 178, 1–6.

Lee, I., Evans, B.R., Lane, L.M. and Woodward, J. (1996) Substrate enzyme interactions in cellulase systems. *Bioresource Technol.*, 58,163–169.

MacDougall, A.J., Parker, R. and Selvendran, R.R. (1995) Nonaqueous fractionation to assess the ionic composition of the apoplast during fruit ripening. *Plant Physiol.*, 108, 1679–1689.

MacDougall, A.J., Needs, P.W., Rigby, N.M. and Ring, S.G. (1996) Calcium gelation of pectic polysaccharides isolated from unripe tomato fruit. *Carbohydr. Res.*, 923, 255–349.

MacDougall, A.J., Brett, G.M., Morris, V.J., Rigby, N.M., Ridout, M.J. and Ring, S.G. (2001a) The effect of peptide–pectin interactions on the gelation behaviour of a plant cell wall pectin. *Carbohydr. Res.*, 335, 115–126.

MacDougall, A.J., Rigby, N.M., Ryden, P., Tibbits, C.W. and Ring, S.G. (2001b) Swelling of the tomato cell wall network. *Biomacromolecules*, 2, 450–455.

Manning, G.S. and Ray, J. (1998) Counterion condensation revisited. *J. Biomolecular Struct. Dynamics,* 16, 461–476.

Mathlouthi, M. and Koenig, J.L. (1986). Vibrational spectroscopy of carbohydrates. *Adv. Carbohydr. Chem. Biochem.*, 44, 7–89.

McCann, M.C., Hammouri, M.K., Wilson, R.H., Roberts, K. and Belton, P.S. (1992) Fourier transform infrared micro-spectroscopy: a new way to look at cell walls. *Plant Physiol.*, 100, 1040–1947.

McCann, M.C., Stacey, N.J., Wilson, R.H. and Roberts, K. (1993) Orientation of macromolecules in the walls of elongating carrot cells. *J. Cell Sci.*, 106, 1347–1356.

McKeown, T.A., Moss, S.T. and Jones, E.B.G. (1996) Atomic force and electron microscopy of sporangial wall microfibrils in *Linderina pennispora*. *Mycol. Res.*, 100, 821–826.

Morris, V.J. (1998) Gelation of polysaccharides, in *Functional Properties of Food Macromolecules*, 2nd edn (eds S.E. Hill, D.A. Ledward and J.R. Mitchell), Aspen Publishers, MD, pp. 143–226.

Morris, E.R., Powell, D.A., Gidley, M.J. and Rees, D.A. (1982) Conformations and interactions of pectins: 1. Polymorphism between gel and solid states of calcium polygalacturonate. *J. Mol. Biol.*, 155, 507–516.

Morris, V.J., Gunning, A.P., Kirby, A.R., Round, A.N., Waldron, K. and Ng, A. (1997) Atomic force microscopy of plant cell wall polysaccharides and gels. *Int. J. Biol. Macromolecules*, 21, 61–66.

Morris, V.J., Kirby, A.R. and Gunning, A.P. (1999a) *Atomic Force Microscopy for Biologists*, Imperial College Press, London.

Morris, V.J., Kirby, A.R. and Gunning, A.P. (1999b) A fibrous model for gellan gels from atomic force microscopy studies. *Progr. Colloid Polymer Sci.*, 114, 102–108.

Nobel, P.S. (1999) *Physicochemical and Environmental Plant Physiology*, 2nd edn, Academic Press, London.

Noda, I. (1990) Two-dimensional infrared (2DIR) spectroscopy: theory and applications. *Appl. Spectrosc.*, 47, 1329–1336.

Noda, I., Dowrey, A.E. and Marcott, C. (1999) Two-dimensional infrared (2D IR) spectroscopy. In *Modern Polymer Spectroscopy*, (ed G. Zerbi), Wiley-VCH Weinheim, pp. 1–32.

Parsegian, V.A., Rand, R.P. and Rau, D.C. (1995) Macromolecules and water-probing with osmotic stress. *Methods Enzymol.*, 259, 43–94.

Piculell, L. (1995) Gelling carrageenans. In *Food Polysaccharides and Their Applications* (ed. A. M. Stephen), Marcel Dekker, New York, pp. 205–244.

Pinel, G.D., Saulnier, L., Roger, P. and Thibault, J.-F. (2000) Isolation of homogeneous fractions from wheat water-soluble arabinoxylans. Influence of the structure on their macromolecular characteristics. *J. Agric. Food Chem.*, 48, 270–278.

Pinel, G.D., Thibault, J.-F. and Saulnier, L. (2001) Experimental evidence for a semi-flexible conformation for arabinoxylan. *Carbohydr. Res.*, 330, 365–372.

Rigby, N.M., MacDougall, A.J., Ring, S.G., Cairns, P., Morris, V.J. and Gunning, P.A. (2000) Observations on the crystallization of oligogalacturonates. *Carbohydr. Res.*, 328, 235–239.

Round, A.N. (1999) Atomic force microscopy of plant cell wall polysaccharides. PhD thesis, University of East Anglia, Norwich, UK.

Round, A.R., Kirby, A.R. and Morris, V.J. (1996) Collection and processing of AFM images of plant cell walls. *Microsc. Anal.*, 55, 33–35.

Round, A.N., MacDougall, A.J., Ring, S.G. and Morris, V.J. (1997) Unexpected branching in pectins observed by atomic force microscopy. *Carbohydr. Res.*, 303, 251–253.

Round, A.N., Rigby, N.M., MacDougall, A.J., Ring, S.G. and Morris, V.J. (2001) Investigating the nature of branching in pectin by atomic force microscopy and carbohydrate analysis. *Carbohydr. Res.*, 331. 337–342.

Ruan, Y-L., Patrick, J.W. and Brady, C.J. (1996) The composition of apoplast fluid recovered from intact developing tomato fruit. *Austral. J. Plant Physiol.*, 23, 9–13.

Rubinstein, M., Colby, R.H., Dobrynin, A.V. and Joanny, J.-F. (1996) Elastic modulus and equilibrium swelling of polyelectrolyte gels. *Macromolecules*, 29, 398–406.

Ryden, P., MacDougall, A.J., Tibbits, C.W. and Ring, S.G. (2000) Hydration of pectic polysaccharides. *Biopolymers*, 54, 398–405.

Saulnier, L. and Thibault, J.-F. (1999) Ferulic acid and diferulic acids as components of sugar beet pectins and maize bran heteroxylans. *J. Sci. Food Agric.*, 79, 396–402.

Schols, H.A. and Voragen, A.G.J. (1994) Occurrence of pectic hairy regions in various plant cell wall materials and their degradability by rhamnogalacturonase. *Carbohydr. Res.*, 256, 83–95.

Schols, H.A., Voragen, A.G.J. and Colquhoun, I.J. (1994) Isolation and characterisation of rhamnogalacturonan oligomers, liberated during degradation of pectic hairy regions by rhamnogalacturonase. *Carbohydr. Res.*, 256, 97–111.

Sene, C.B., McCann, M.C., Wilson, R.H. and Grinter, R. (1995) FT Raman and FTIR spectroscopy: an investigation of five higher plant cell walls. *Plant Physiol.*, 106, 1623–1631.

Shackel, K.A., Greve, C., Labavitch, J.M. and Ahmadi, H. (1991) Cell turgor changes associated with ripening in tomato pericarp tissue. *Plant Physiol.*, 97, 814–816.

Showalter, A.M. (1993) Structure and function of cell wall proteins. *Plant Cell*, 5, 9–23.

Skouri, R., Schosseler, F., Munch, J.P. and Candau, S.J. (1995) Swelling and elastic properties of polyelectrolyte gels. *Macromolecules*, 28, 197–210.

Stephen, A.M. (1995) *Food Polysaccharides and Their Applications*, Marcel Dekker, New York.

Tibbits, C.W., MacDougall, A.J. and Ring, S.G. (1998) Calcium binding and swelling behaviour of a high methoxyl pectin gel. *Carbohydr. Res.*, 310, 101–107.

Tomos, A.D. and Leigh, R.A. (1999) The pressure probe is a versatile tool in plant cell physiology. *Annu. Rev. Plant Physiol. Plant Mol. Biol.*, 50, 447–480.

Toole, G.A., Smith, A.C. and Waldron, K.W. (2002) The effect of physical and chemical treatment on the mechanical properties of the cell wall of the alga *Chara corallina*. *Planta* 214(3), 468–475.

Voragen, A.G.J., Pilnik, W., Thibault, J.-F., Axelos, M.A.V. and Renard, M.G.C. (1995) Pectins. In *Food Polysaccharides and Their Applications* (ed. A. M. Stephen), Marcel Dekker, New York, pp. 287–339.

Waldron, K.W., Parr, A.J., Ng, A. and Ralph, J. (1996) Cell wall esterified phenolic dimers: identification and quantification by reverse phase high performance liquid chromatography and diode detection. *Phytochem. Anal.*, 7, 305–312.

Walkinshaw, M.D. and Arnott, S. (1981) Conformations and interactions of pectins. I. X-ray diffraction analyses of sodium pectate in neutral and acidified forms. *J. Mol. Biol.*, 153, 1055–1073.

van der Wel, N.H., Putman, C.A.J., van Noort, S.J.T., de Grooth, B.G. and Emons, A.C.M. (1996) Atomic force microscopy of pollen grains, cellulose microfibrils, and protoplasts. *Protoplasma*, 194, 29–39.

Wilson, R.H., Smith, A.C., Kacurakova, M., Saunders, P.K., Wellner, N. and Waldron, K.W. (2000) Mechanical properties and molecular dynamics of cell wall biopolymers. *Plant Physiol.*, 124, 397–405.

Zwieniecki, M.A., Melcher, P.J. and Holbrook, N.M. (2001) Hydrogel control of xylem hydraulic resistance in plants. *Sci.*, 291, 1059–1062.

3 Molecules in context: probes for cell wall analysis

William G.T. Willats and J. Paul Knox

3.1 Introduction

Cell walls are a defining feature of plants and have many fundamental roles during growth and development. They are central to determining the mechanical properties of all organs and are critical for a wide range of cell functions including cell expansion and cell adhesion. The multifunctionality of plant cell walls is reflected in the fact that they are amongst the most sophisticated and abundant of nature's biomaterials. In structural terms, cell walls are fibrous composites consisting of load-bearing components embedded in a hydrated pectic polymer matrix. The structures of cell walls are highly dynamic and throughout development undergo modulations in composition and configuration in response to functional requirements. In order to reveal the complexity of cell walls in relation to cellular processes it is necessary to be able to analyse individual components *in situ* in relation to intact cell wall architecture. To achieve this, a series of specific probes are required. Currently, the generation of antibodies is the best way to produce such probes.

Although the number of antibodies directed against cell wall epitopes has grown steadily over recent decades they still cover only a minute fraction of the molecular structures that make up cell walls and this shortfall is reflected by gaps in our understanding of cell wall processes. This chapter is concerned with recent developments in the generation and use of antibodies in relation to our understanding of plant cell wall biology. Our focus here is on the challenges and technologies involved in antibody generation and their characterization and use, rather than the cell biological insights that antibodies can provide. The generation of antibodies to the polysaccharide and phenolic components of cell walls is not a straightforward matter. In contrast, the generation of antibodies to proteins and the peptide components of glycoproteins is generally straightforward (with numerous examples in the literature) and will not be our concern here.

The reader is directed to recent reviews covering antibodies and cell walls (Knox, 1997; Willats *et al.*, 2000b) and also to a review discussing insights gained from the use of antibodies to pectic polysaccharides – the group of cell wall polysaccharides that has seen the greatest progress in antibody generation in recent years (Willats *et al.*, 2001a).

3.2   Technologies for the generation of antibodies

The generation of antisera is the easiest way to prepare antibodies. Antisera are often sufficiently specific and of high enough titre to be useful for a large number of analyses, and this is particularly the case where the immunogen is a protein. However, the generation of defined monoclonal antibodies, derived from immortal cell lines, that have the capacity to become standard reagents that can be used in a range of systems by a community of cell wall researchers is an important goal that will aid greatly the full understanding of cell wall structure and function. Table 3.1 is a list of the widely used monoclonal antibodies and antisera to cell wall components that have been of use in a range of plant systems.

    The development of monoclonal antibodies using hybridoma technologies and the advantages of having unlimited amounts of a defined antibody are well known. However, cell fusion procedures and the isolation of hybridoma cell lines carried out subsequent to immunization are time consuming and expensive and are restricted to only a few laboratories interested in the generation of antibodies to cell walls. Although the isolation of hybridomas remains the major technique for the generation of antibodies to cell walls there are increasing possibilities for the generation and manipulation of probes based on molecular technologies. In recent years, a method of producing recombinant monoclonal antibodies, termed phage display, has been developed. This technology offers the prospect not only of greatly extending our range of antibody specificities but has also created possibilities for using antibodies in novel ways for the direct functional analysis of epitopes (see section 3.4.2). Both conventional hybridoma and phage display antibody production exploit the vast diversity of the mammalian antibody repertoire. The fundamental difference is that with hybridoma antibody production this diversity is captured by the immortalization of antibody-producing B-cells, while with phage display it is the genes that encode antibody variable regions (V-genes) that are immortalized. V-genes encode the variable regions of antibody (Fab) domains that are responsible for antigen binding and define specificity (Figure 3.1).

    In practical terms, antibody phage display involves the cloning of the V-gene sequences which are then combined in many different permutations (sometimes with the addition of random coding sequences) to create diversity. Libraries are then constructed in which sequences encoding synthetic antibody fragments are fused to phage coat protein genes, thus ensuring the expression of antibody fragments at the phage surface (Figure 3.1). The use of bacteriophage as a display platform allows successive rounds of affinity selection against target molecules and subsequent amplification – a process that mimics the immune selection process in mammals. Some phage antibodies display heterodimeric antibody fragments that resemble natural antibody Fab fragments, while others display single chain fragments (scFvs) in which heavy and light chain fragments are joined by a flexible polypeptide linker. V-gene sequences may be derived from un-immunized donors (to create naïve libraries) or from animals previously immunized with the target molecule (to create antigen-specific libraries). Naïve libraries have the significant advantage that, if

Table **3.1** Antibodies to plant cell wall components.

Antibody	T	Antigen/epitope	References	Commercial availability
BG1	m	Callose/(1→3)-β-D-glucan	Meikle et al., 1991	Biosupplies
BGMC6	m	(1→3) (1→4)-β-D-glucan	Meikle et al., 1994	Biosupplies
	m	Galactomannan/(1→4)-β-D-mannan	Pettolino et al., 2001	
CCRC-M1	m	Xyloglucan/t-α-fucose-(1→2)-linked to D-galactosyl[a]	Puhlmann et al., 1994	
JIM5	r	HG/low/no methyl-esters	VandenBosch et al., 1989; Knox et al., 1990; Willats et al., 2000a	
JIM7	r	HG/methyl-esterified	Knox et al., 1990; Willats et al., 2000a	
2F4	m	HG/un-esterified, calcium cross-linked	Liners et al., 1989, 1992	
PAM1	s	HG/un-esterified	Willats et al., 1999a, 2000a	
LM7	r	HG/non-blockwise pattern of de-esterification	Willats et al., 2001b	
	as	RG-II dimer	Matoh et al., 1998	
CCRC-R1	s	RG-II monomer	Williams et al., 1996	
CCRC-M2	m	RG-I/unknown	Puhlmann et al., 1994	
CCRC-M7	m	RG-I/arabinosylated (1→6)-β-D-galactan[b]	Puhlmann et al., 1994; Steffan et al., 1995	
LM5	r	RG-I/(1→4)-β-D-galactan	Jones et al., 1997; Willats et al., 1999b	PlantProbes
LM6	r	RG-I/(1→5)-α-L-arabinan[b]	Willats et al., 1998, 1999b	PlantProbes
	as	uronosylated (1→3,6)-β-D-galactosyl	Andème-Onzighi et al., 2000	
11.D2	m	HRGP	Meyer et al., 1988	
MC-1[c]	m	HRGP	Fritz et al., 1991	
JIM11[c]	r	HRGP	Smallwood et al., 1994; Knox et al., 1995	

Antibody	Type	Epitope	Reference	Source
LM1	r	HRGP	Smallwood et al., 1995	PlantProbes
	as	AGP/(1→6)-β-D-galactan	Kikuchi et al., 1993	
PCBC3[c]	m	AGP/α-L-arabinofuranosyl	Anderson et al., 1984	
PN16.4B4	m	AGP	Norman et al., 1986	
MAC207	r	AGP	Pennell et al., 1989	
JIM4	r	AGP	Knox et al., 1989	
JIM8	r	AGP	Pennell et al., 1991, 1992	
JIM13[c]	r	AGP	Knox et al., 1991	
ZUM15[c]	m	AGP	Kreuger and van Holst, 1995	
LM2	r	AGP/β-D-glucuronosyl	Smallwood et al., 1996	PlantProbes
JIM101[c]	r	AGP (liverworts)	Basile et al., 1999	
	as	p-hydroxyphenylpropane (H) lignin	Ruel et al., 1994; Joseleau and Ruel, 1997	
	as	guaiacyl (G) lignin	Ruel et al., 1994; Joseleau and Ruel, 1997	
	as	guaiacyl-syringyl (GS) lignin	Ruel et al., 1994; Joseleau and Ruel, 1997	
	as	5-O-t-feruloyl-α-L-arabinofuranose	Migné et al., 1998	

Table codes:

HG = homogalacturonan, RG-I = rhamnogalacturonan-I; RG-II = rhamnogalacturonan-II; HRGP = hydroxyproline-rich glycoprotein; AGP = arabinogalactan-protein

T = antibody type: as = rabbit antiserum; m = mouse monoclonal antibody; r = rat monoclonal antibody; s = synthetic (recombinant) antibody

[a] Epitope also occurs in RG-I.

[b] Epitope may also occur in arabinogalactan-protein proteoglycans.

[c] Related antibodies also covered in same references.

Biosupplies Australia Pty Ltd. PO Box 835, Parkville 3052 Australia. Fax: +613 9347 1071, enquiries@biosupplies.com.au

PlantProbes. ULCL, 3 Cavendish Road, Leeds, LS2 9JT, UK. Fax: +44–113–34–33144, plantprobes@leeds.ac.uk

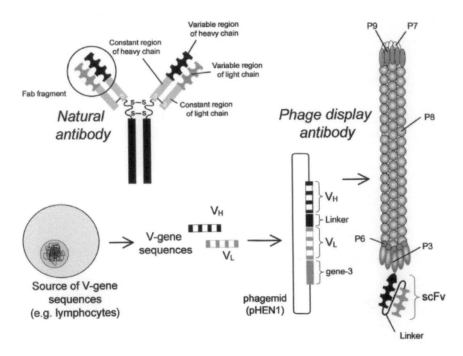

Figure 3.1 Schematic representation of natural antibody structure and the production of phage display antibodies. This scheme applies to the production of phage display libraries expressing scFv, e.g. the Human Synthetic Library #1, used to isolate PAM1 (Willats *et al.*, 1999a). Heavy (V_H) and light (V_L) V-gene sequences are combined and fused to a sequence encoding a phage coat protein. In this example the coat protein is p3 of M13. The fusion product is expressed at the surface of phage particles which act as carriers of the scFv and provide a means of replication.

sufficiently diverse, the same library can be used to obtain a wide range of antibodies with different specificities, thereby avoiding the laborious process of repeated library construction.

A major advantage of phage display antibody production is that, since the whole process occurs *in vitro*, there is no requirement for target antigens to be immunogenic and the range of feasible target antigens is therefore extended considerably. The amount of target antigen required is much less than is typically required for hybridoma antibody production (micrograms compared with milligrams) and the time required to generate monoclonal antibodies is also much reduced (a few weeks compared with several months). Phage display antibody production is relatively simple and cheap, requiring no special facilities, and because immunization is bypassed (if naïve libraries are used), the ethical and financial burdens of animal use are also avoided. The protocols for antibody generation and antibody use from our laboratory have recently been published (Willats *et al.*, 2002a).

3.3   Targets, immunogens and antigens

Which molecules in plant cell walls is it desirable to have probes for? The short answer, of course, is all of them. However, one of the major needs for the elucidation of the structure and function of both primary and secondary cell walls is antibody probes to defined structural features of the complex carbohydrates that comprise the major macromolecular networks of these walls.

A major problem with antibodies to cell wall carbohydrates is that monosaccharides such as arabinose, galactose, fucose and rhamnose are common to many polymer groups and immunization with one polymer may result in antibodies to common epitopes. The key points to consider when generating anti-glycan probes are to define as much as possible an epitope in structural terms, and to be aware of its possible occurrence in diverse polymers. A defined oligosaccharide (in the region of four to seven monosaccharides) can often confer antigen class specificity. Ideally, such an oligosaccharide is coupled to an immunogenic protein. This neoglycoprotein can then be used for antibody isolation by hybridoma or phage display technology. The specificity of an antibody is defined by analysis of the antibody binding properties to the immunogen and panels of related oligosaccharides. For most plant cell wall polysaccharides, appropriate oligosaccharides are rarely available in amounts (>20 mg) suitable for coupling chemistry. If an appropriate oligosaccharide is not available, then a polysaccharide can be used directly or coupled to a protein to prepare an immunogen. However, the more complex and ill-defined the immunogen, or target, then the more difficult can be the subsequent characterization of antibody specificity. This is particularly the case for complex branched heteropolymers, for which antibody characterization ideally requires a range of appropriately sized and defined component oligosaccharides in amounts sufficient for inhibition assays.

Strategies involving structurally simple polysaccharides or neoglycoproteins have proved successful for the generation of monoclonal antibodies to $(1\rightarrow3)$-β-D-glucan (Meikle et al., 1991), $(1\rightarrow3,1\rightarrow4)$-$\beta$-D-glucan (Meikle et al., 1994), $(1\rightarrow4)$-β-D-galactan (Jones et al., 1997), $(1\rightarrow5)$-α-L-arabinan (Willats et al., 1998) and $(1\rightarrow4)$-β-D-mannan (Pettolino et al., 2001) epitopes (Table 3.1). Immunization with mixtures of cellular components, or complex polysaccharides and proteoglycans containing a range of structural features, should be avoided wherever possible – although such strategies can turn up intriguing positional markers as discussed below.

3.3.1   Pectic polysaccharides

The pectic and the hemicellulosic polysaccharides are important targets for antibody generation. Our strategy to extend the range of available antibodies to pectic polymers involves both hybridoma and phage display technologies, in conjunction with a series of model polysaccharides and defined oligosaccharides obtained from pectin biochemists and synthetic chemist collaborators. The pectic network is complex and consists of structurally and functionally distinct domains. Three domains in particular appear to be ubiquitous and have been the focus of antibody production: homogalacturonan (HG), rhamnogalacturonan-I (RG-I) and rhamnogalacturonan-II (RG-II) (Ridley *et al.*, 2001; Willats *et al.*, 2001a).

HG is the most abundant domain of the pectic network (see Chapter 1) and appears to be subject to extensive *in muro* modification, most notably through the action of pectin methyl-esterases (PMEs) and pectin acetyl-esterases. The distribution patterns of the methyl-ester and acetyl groups along the HG backbone is variable and likely to be of functional significance. Three monoclonal antibodies have been widely used to detect the HG component of pectin (JIM5, JIM7 and 2F4). 2F4 binds to calcium cross-linked de-esterified HG (Liners *et al.*, 1992). JIM5 and JIM7 appear extensively in the literature and are claimed to bind to relatively high methyl-ester and relatively low methyl-ester pectins respectively – or often as methyl-esterified and un-esterified pectin, respectively. This latter claim is wrong. Recent work looking at the specificity of these two antibodies indicates that both appear to bind to a wide range of methyl-ester containing HG epitopes. In addition, JIM5 can bind to un-esterified oligomers of GalA – but optimal binding is achieved when some methyl-ester groups are present (Willats *et al.*, 2000a). In short, although these two antibodies are excellent and specific probes for HG – they cannot be used to derive precise information about the spatial or developmental distribution of the methyl-ester status of HG – and therefore results obtained through their use in immunolocalization studies needs to be interpreted with caution.

One of the key tools that has allowed characterization of anti-HG antibodies has been the development of a series of model pectins. Using a relatively highly methyl-esterified lime pectin as a starting sample, it has been possible to create series of pectins with varying degrees and patterns of methyl-esterification by treatment with a plant PME from orange (P series), a fungal PME from *Aspergillus niger* (F series) and by base catalysis (B series) (Limberg *et al.*, 2000). These three methods of de-esterification produce different patterns of de-esterification, with the plant PME acting in a blockwise manner and the fungal PME and base catalysis acting in different but both non-blockwise manners (Limberg *et al.*, 2000). In this way, pectin samples varying in both the degree (DE) and pattern of methyl-esterification have been produced. Using these model pectin samples we have been able to characterize the specificities of two more recently produced probes in detail – one (LM7) was produced by hybridoma methods and another (PAM1) was generated by phage display technology. PAM1 binds to long unesterified stretches or 'blocks' of HG. In the region of 30 contiguous unesterified GalA residues are required for PAM1 binding and it is likely that PAM1 binds to HG in a conformationally dependent manner (Willats *et al.*, 1999a). A conformational and length dependent epitope has also been reported for the sialic acid containing type III group B *Streptococcus* capsular polysaccharide (Zou *et al.*, 1999).

The epitope recognized by LM7 is susceptible to digestion by both pectin lyase and polygalacturonase, indicating that stretches of both methyl-esterified and un-esterified HG are included in the epitope (Willats *et al.*, 2001b). The PAM1 and LM7 epitopes, generated most readily by blockwise and non-blockwise de-esterification processes respectively, occur in all cell walls so far examined in spatially restricted patterns around developing intercellular spaces. These observations suggest that altered patterns and extents of HG methyl-esterification are key contributors to both matrix properties and cell adhesion/separation. Such observations are complementary to the detailed analysis of the action patterns of plant PME isoforms (Catoire *et al.*, 1998) and provide further support to the idea that the large multigene families of plant PMEs may encode proteins with subtly different enzyme activities.

A further important aspect of HG structure and heterogeneity is acetylation, but little is known of the patterns of acetylation along HG chains. The vast potential for different HG structures with varying levels and patterns of both acetylation and methyl-esterification make the generation of antibodies specific to particular levels of acetylation a difficult task. Series of model pectins with defined levels of acetylation are being developed as complementary series to the methyl-esterified pectins discussed above. Ideally, as with methyl-esterification, oligosaccharides with defined substitutions will be required for both generation and characterization of antibodies. In this case, collaboration with chemists will be essential, although coupling chemistry with acidic sugars is not always straightforward.

The family of pectic polymers known as RG-I (see Chapter 1) comprises a complex series of branched heteropolymers that appear to be highly variable within cell walls and between cells. The backbone of RG-I consists of GalA alternating with rhamnose residues that act as sites of substitution with side chains containing largely neutral sugars – often galactose and arabinose. We have made two antibodies that bind to epitopes that occur as components of side chains of RG-I molecules. Both antibodies were made using defined oligosaccharides coupled to an immunogenic protein carrier. These monoclonal antibodies, LM5 and LM6, recognize $(1\rightarrow4)$-β-D-galactan and $(1\rightarrow5)$-α-L-arabinan epitopes, respectively (Jones *et al.*, 1997; Willats *et al.*, 1998). The CCRC-M7 antibody binds to an arabinosylated $(1\rightarrow6)$-β-D-galactan epitope occurring on RG-I (Steffan *et al.*, 1995). A possible problem is that components of pectic side chains may also occur on glycoproteins or proteoglycans. This is the case for CCRC-M7, the epitope of which also occurs on glycoproteins (Puhlmann *et al.*, 1994), and for the LM6 arabinan epitope, which in certain systems is carried by AGPs in addition to RG-I (unpublished observations). It will be of considerable interest to generate antibodies using the neoglycoprotein approach to other structural features of RG-I, including the Rha/GalA backbone. The side chains of RG-I show great diversity and may contain uronic acids (An *et al.*, 1994) and an internal $(1\rightarrow5)$-linked arabinofuranose residue within galactan chains (Huisman *et al.*, 2001). An antiserum to an epitope of $(1\rightarrow3,6)$-β-D-galactosyl linked to uronic acid of the branched region of the *Bupleurum* 2IIc pectic polysaccharide (a polymer that has pharmacological properties) has been characterized (Sakurai *et al.*, 1998). This probe has been used to determine the occurrence of this structure in flax seedlings and has provided evidence for it developmental regulation with particular abundance in epidermal and fibre cell walls (Andème-Onzighi *et al.*, 2000).

As yet, the significance of the extensive developmental regulation of epitopes that occur in the side chains of pectins is far from clear (Freshour *et al.*, 1996; Bush *et al.*, 2001; Willats *et al.*, 1999b; Vicré *et al.* 1998; Ermel *et al.*, 2000; Bush and

McCann, 1999; Orfila *et al.*, 2001; Serpe *et al.*, 2001). One consistent aspect of the developmental regulation of the LM5 $(1\rightarrow4)$-β-D-galactan epitope appears to be its post-meristematic appearance in a range of species (Vicré *et al.*, 1998; Willats *et al.*, 1999b; Bush *et al.*, 2001; Serpe *et al.*, 2001). Studies of pea cotyledons indicate that the appearance of this epitope may relate to altered mechanical properties (Mc-Cartney *et al.*, 2000). Moreover, these RG-I side chain epitopes also have spatial restrictions within cell walls in relation to cell wall architecture (Freshour *et al.*, 1996; Jones *et al.*, 1997; Orfila and Knox, 2000). The basis of these epitope dynamics is not clear – whether they reflect the turnover of new pectic polymers or the masking or *in muro* modification of polymers in the cell wall is not known. This is an important point to address in future work.

Like RG-I, RG-II is highly complex, but in contrast to RG-I, RG-II is a highly conserved structure (see Chapter 1). The function of the RG-II pectic domain is not clear but its borate-mediated dimerization appears to be important for growth (O'Neill *et al.*, 2001). Antisera to the borate-RG-II dimer and a recombinant antibody to the RG-II monomer suggest that RG-II occurs throughout primary cell walls of angiosperms (Williams *et al.*, 1996; Matoh *et al.*, 1998). Antibodies to defined regions of the RG-II domain would be useful to further study the functions of this intriguing molecule.

3.3.2 Hemicellulosic polysaccharides

Complex branched heteropolymers such as xyloglucans and glucuronoarabinoxylans are still largely uncharted in terms of antibody generation. The monoclonal antibody CCRC-M1 binds to a terminal α-1,2 linked fucosyl-containing epitope (Puhlmann *et al.*, 1994), but the epitope also occurs to a lesser extent in RG-I from many dicotyledons and therefore observations require careful interpretation. Antisera can be readily made to neutral polysaccharides such as xyloglucans (Suzuki *et al.*, 1998), but there is a critical need for probes to defined structural features of these and related polymers. The difficulties and strategies for overcoming these are similar to those encountered in the generation of anti-pectin probes, i.e. antibody generation requires large amounts of defined oligosaccharides for immunogen preparation and epitope characterization. A key step forward has been made with the generation of antibodies to highly substituted glucuronoarabinoxylan (hsGAX) and unbranched xylan. These were generated by the conjugation of an hsGAX (isolated from maize shoots) and a synthetic xylopentaose to protein for use as immunogens (Suzuki *et al.*, 2000). These antisera showed good specificity and their use indicated distinct spatial and developmental patterns of occurrence of these epitopes in *Zea mays*. The hsGAX epitope is located mostly in unlignified cell walls and the unbranched xylan epitope in lignified cell walls (Suzuki *et al.*, 2000).

3.3.3 Proteoglycans and glycoproteins

As discussed above, some epitopes such as those bound by CCRC-M7 and LM6 occur both in RG-I polymers and complex glycan structures of AGP proteoglycans.

It is known that the core $(1\rightarrow3,1\rightarrow6)$ galactan structure of AGP glycans can also occur in pectic polysaccharides. Several panels of monoclonal antibodies to the glycan components of AGPs have been derived subsequent to immunization with cell extracts or isolated AGPs (see Table 3.1). In most cases the precise epitope structure is not known. Attempts at retrospective characterization of epitope structures have been made but these are often limited due to the lack of appropriate oligosaccharide samples (Yates *et al.*, 1996). Dedicated efforts towards the synthesis of oligosaccharide components of arabinogalactan components of AGPs and pectic polymers are now under way (Valdor and Mackie, 1997; Csávás *et al.*, 2001; Gu *et al.*, 2001; Hada *et al.*, 2001). These synthetic oligosaccharides will be useful to characterize existing antibodies and also to prepare immunogens to generate defined anti-AGP glycan antibodies.

Although the structural features bound by these anti-AGP antibodies are not known, they have revealed remarkable developmental regulation of AGPs during early development and are undoubtedly useful probes. In addition, these antibodies are a useful resource for the analysis of mutants with altered cell layers, such as the *scarecrow* mutant of *Arabidopsis* (Laurenzio *et al.*, 1996), and as molecular markers in conjunction with the biochemical characterization of AGPs (Gaspar *et al.*, 2001). Because of their heterogeneity, a key point in the study of AGP function is to focus on specific AGPs in specific systems. The core proteins of AGPs are attractive targets for antibody production and, once an AGP has been characterized at the gene/protein level, the generation of an antibody probe to the protein core should be straightforward and provide a highly specific and useful probe for that AGP, thereby overcoming many of the disadvantages of the anti-glycan probes. This has been achieved for LeAGP-1 from tomato (Gao and Showalter, 2000). Arising from similar approaches that led to the generation of panels of AGP probes, several panels of monoclonal antibodies have been generated to HRGPs (Table 3.1). These probes also indicate extensive patterns of developmental regulation of the HRGP epitopes, that in some cases are similar to the patterns of AGP epitopes (Knox, 1997).

Many of the anti-AGP and anti-HRGP antibodies were isolated from screens for antibodies binding to developmentally restricted antigens. Although the antigen class can generally be identified, a more precise characterization of epitope structure is much more difficult, as discussed above. A recent search for novel antigens related to vascular cell development has used an elegant approach. A *Zinnia* cell culture system, in which specific stages of cell development can be defined, was used to prepare cell wall material from an early stage of development that was used as an immunogen. Subsequent development of a phage display antibody library and selection of cell-specific antibodies using a subtractive approach led to a series of differentiation-specific phage antibodies (Shinohara *et al.*, 2000). For example, the CN8 antibody bound specifically to the tip of immature tracheary elements, indicating a cell wall component with distinct spatial localization and suggesting cell polarity. Analysis indicated that the antigen occurs in the hemicellulosic fractions of the *Zinnia* cell walls (Shinohara *et al.*, 2000). Further characterization of this epitope will be of considerable interest.

3.3.4 Phenolics and lignin

Phenolic components found in plant cell walls also present a considerable challenge to antibody generation. Some phenolic compounds, such as ferulic acid, appear to be the basis of important links within the polysaccharide networks of cell walls. Ferulic acid, attached to arabinose side chains, can dimerize in various ways and such links are likely to be key contributors to cell wall integrity. In the Chenopodiaceae, ferulic acid is associated with pectic polymers and in some species, such as those that belong to the Gramineae, it is attached to arabinoxylans. Probes for ferulic acid and various dimers would be useful to understand the contribution of these links to cell wall architecture. Steps towards this have been achieved with the development and use of an antiserum to 5-O-t-feruloyl-α-L-arabinofuranose using a protein conjugate as immunogen (Migné et al., 1998). The same study also generated an antiserum directed towards p-coumaric acid that is thought to be esterified to lignins in graminaceous cell walls (Migné et al., 1998).

The complex and variable lignin polymers result from the dehydrogenative polymerization of three monolignol precursors resulting in lignins containing varying amounts of p-hydroxyphenyl (H), guaiacyl (G) and syringyl (S) subunits (see Chapter 9). Different lignins vary in their subunit composition-making antibody probes important tools for analysis of lignins in situ. An interesting approach to probe development has used synthetic dehydrogenative products of H, G and mixed GS lignins directly as immunogens (without conjugation to protein) (Ruel et al., 1994). These antisera have been characterized with a novel technique in which antigens are embedded in resin and this has indicated that the antisera can have specificity to condensed or non-condensed interunit linkages of lignin (Joseleau and Ruel, 1997). These antisera have been used as immunogold probes to discriminate between lignins in cell walls in electron microscopy and have demonstrated the microheterogeneity of lignin deposition within a single cell wall (Ruel et al., 1999). They have also been used in studies on the structure and distribution of lignin in cell walls of maize coleoptiles (Müsel et al., 1997) and on the endodermal and hypodermal cell walls in maize roots (Zeier et al., 1999). Moreover, these antisera are key tools to uncover altered patterns of lignin deposition at cellular and subcellular levels in transgenic plants (Chabannes et al., 2001).

3.4 Extending antibody technologies: the way ahead

3.4.1 High throughput antibody characterization: microarrays

Generating antibodies is one challenge, detailed characterization is another. The generation of phage display antibodies involves the affinity selection of phage libraries against antigens and this bio-panning procedure rapidly (typically within two weeks) generates large numbers of putative antibodies with desired specificities. In our experience, the time limiting step in phage antibody production is the detailed analysis of each monoclonal phage population. Conventional assays, such as ELISAs and immunodot-assays, have the disadvantages that only a relatively small number of samples can be tested simultaneously, and large amounts of antibody are required for each assay. In order to alleviate this bottleneck with respect to the production of anti-glycan antibodies, we have recently used a novel microarray slide surface that has capacity to immobilize structurally and chemically diverse glycans without any derivatization of the slide surface, or the need to create reactive groups on the immobilized glycans (Willats *et al.*, 2002b). The slides are made of polystyrene and have the MaxiSorp™ surface that has been widely used in a microtitre plate format for ELISAs. We have generated microarrays of a range of cell wall components including pectic polysaccharides, arabinogalactan-proteins and cell wall extracts. Immobilization of the samples was assessed by probing the microarrays with monoclonal antibodies with known specificities as shown in Figure 3.2. Antibody binding indicated that highly reproducible arrays can be created using very low levels of sample antigens. In our analyses the detection limit of the arrays was as low as 1.6 µg/ml (using 50 pL) spots although of course this depends on the antigen/antibody pair. The use of microarrays for antibody analysis has the advantages that up to 10,000 samples can be rapidly assayed simultaneously using small volumes of antibody solution (~100 µL), and it is likely that such microarrays will be valuable tools for the high throughput generation of monoclonal antibodies against cell wall components.

3.4.2 Antibody engineering

The isolation of recombinant antibodies allows the manipulation of antibody specificity in a variety of ways. The anti-HG phage antibody PAM1 is based on the M13 phage and, to date, we have used the intact phage particle (known as PAM1$_{phage}$) as the antibody probe and anti-M13 secondary antibodies (with specificity for the M13 p8 major coat protein) in immunolabelling experiments (Willats *et al.*, 1999a, 2001b). Each M13 phage particle is about 800 nm long and is covered in several thousand copies of the p8 protein (Figure 3.1). The use of anti-p8 secondary antibodies therefore amplifies scFv binding with the result that the detection limit of PAM1$_{phage}$ is high (~ 10 ng on immuno-dot assays). However, the large size of PAM1$_{phage}$ is a disadvantage for immuno-localization studies because of the diffuse signal resulting for the large size of phage particles. In order to overcome this we have produced a soluble, non-phage bound version of PAM1 by cloning the PAM1 coding sequence into a bacterial expression vector and at the same time adding an N-terminal polyhistidine (HIS) tag to the protein (Figure 3.3). The soluble form of PAM1 (known as PAM1$_{scFv}$) can be readily isolated to high purity using a nickel column. PAM1$_{phage}$ and PAM1$_{scFv}$ have identical specificities although the detection limit of PAM1$_{scFv}$ is less than that of PAM1$_{phage}$ because the HIS tag presents only one binding site for each secondary antibody. However, the small size of PAM1$_{scFv}$ results in greatly improved resolution during immunolabelling (Figure 3.3).

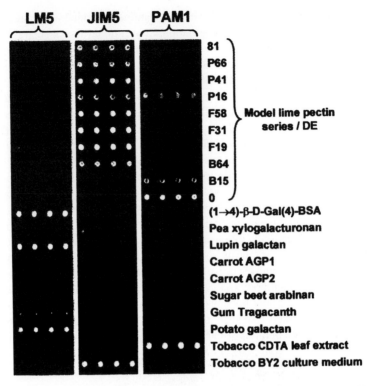

Figure 3.2 The integration of the characterization of anti-glycan probe specificities with micro-array systems. A series of identical carbohydrate arrays were created by immobilizing a series of polysaccharides, proteoglycans, neoglycoproteins and plant cell extracts. The arrays included a series of polysaccharides differing only in their extent/pattern of methyl-esterification. Five replicates are shown. Binding of monoclonal antibodies LM5, JIM5 and PAM1 was detected by probing with Cy3-conjugated secondary antibodies.

The generation of antibodies by phage display technology has several other advantages in addition to those already outlined. Since each phage particle carries the sequence encoding the displayed antibody fragment, the sequence is readily determined. The possibility then exists for manipulation of binding properties – either in a rational site-directed manner – or by the random alteration of sequence using error-prone PCR or mutator strains of host bacteria (Winter, 1998). Furthermore, antibody encoding sequences may be combined in order to generate bi-specific and bivalent synthetic antibodies (Conrath *et al.*, 2001). Recent work on scFv/green fluorescent protein (GFP) fusions has indicated that these fusions are often successful in the sense that both scFv and GFP functionalities are retained (Hink *et al.*, 2000). This brings into sight some exciting prospects for the generation of tailor-made

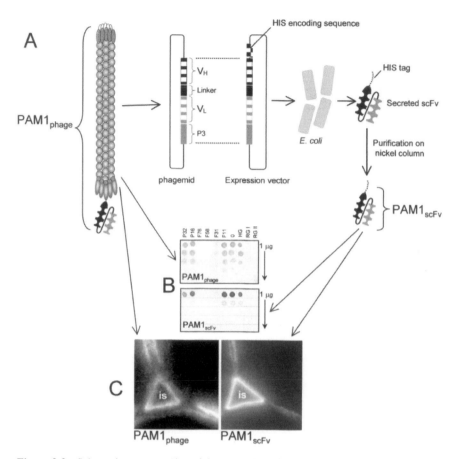

Figure 3.3 Schematic representation of the generation of soluble PAM1 scFv (A) and the use of this and the phage-bound form in immunodot assays against a range of pectic antigens (B) and for immunofluorescent labelling of the PAM1 HG epitope at an intercellular space (is) in TS of tobacco stem (C). PAM1$_{phage}$ has high sensitivity, for example in immunodot-assays, but low resolution in immunocytochemistry. Because each phagemid carries the scFv encoding sequence it is straightforward to generate soluble scFvs by cloning antibody sequences into bacterial expression vectors. PAM1$_{scFv}$ has lower sensitivity in immunodot-assays but due to its small size it is a vastly superior probe when used for immunocytochemistry. The immunodot-assay shows the binding of PAM1$_{phage}$ and PAM1$_{scFv}$ to examples of the series of model lime pectins with a range of extent/patterns of methyl-esterification (see text). Binding to HG, RG-I and RG-II is also shown. Samples were applied at 1 µg/dot to nitrocellulose and at 5-fold dilutions downwards.

inherently fluorescent probes. The expression of antibodies in plants has considerable potential for the disruption *in vivo* of the target epitopes. The development of this immunomodulation approach has been hampered in the past by the fact that many full-sized immunoglobulins are not efficiently assembled *in planta*. In contrast, scFvs with their undemanding folding requirements are relatively straightforward to express in plants in a way that retains functionality (Owen *et al.*, 1992). An intriguing future prospect will be the expression of scFv/GFP fusions with the ability to immunomodulate target antigens whilst providing direct positional information *in vivo*. If this approach were developed then the possibility also exists for the use of scFv/GFP fusions in fluorescence resonance energy transfer (FRET) studies to investigate intra- and intermolecular interactions (Arai *et al.*, 2000) within cell walls. The revolution in recombinant technology as applied to probes has extended beyond synthetic antibodies. For example, microbial non-catalytic carbohydrate binding modules (CBMs) have potential as a source of molecular probes for plant cell walls. Many microbial enzymes that catalyse the degradation of cell wall glycans consist of separate catalytic and binding domains or modules. The specificities of many CBMs for their glycan ligands have been extensively characterized and these are now recognized to include members that bind to celluloses, xylans and mannans (Freelove *et al.*, 2001). Recombinant CBMs with known specificities can readily be produced in large amounts using standard bacterial expression procedures. The addition of peptide tags (e.g. HIS) allows CBMs to have the potential to be used for the localization of cell wall glycans in just the same way as synthetic antibody fragments.

References

An, J., O'Neill, M.A., Albersheim, P. and Darvill, A.G. (1994) Isolation and structural characterization of β-D-glucosyluronic acid and 4-O-methyl β-D-glucosyluronic acid-containing oligosaccharides from the cell wall pectic polysaccharide, rhamnogalacturonan I. *Carbohydr. Res.*, 252, 235–243.

Andème-Onzighi, C., Lhuissier, F., Vicré, M., Yamada, H. and Driouich, A. (2000) A (1 →3,6)-β-D-galactosyl epitope containing uronic acids associated with bioactive pectins occurs in discrete cell wall domains in hypocotyl and root tissues of flax seedlings. *Histochem. Cell Biol.*, 113, 61–70.

Anderson, M.A., Sandrin, M.S. and Clarke, A.E. (1984) A high proportion of hybridomas raised to a plant extract secrete antibody to arabinose or galactose. *Plant Physiol.*, 75, 1013–1016.

Arai, R., Ueda, H., Tsumoto, K., Mahoney, W.C., Kumagai, I. and Nagamune, T. (2000) Fluorolabelling of antibody variable domains with green fluorescent protein variants: application to an energy transfer-based homogeneous immunoassay. *Protein Engng*, 13, 369–376.

Basile, D.V., Basile, M.R., Salama, N., Peart, J. and Roberts, K. (1999) Monoclonal antibodies to arabinogalactan-proteins (AGPs) released by *Gymnocolea inflata* when leaf and branch development is desuppressed. *Bryologist*, 102, 304–308.

Bush, M.S. and McCann, M.C. (1999) Pectic epitopes are differentially distributed in the cell walls of potato (*Solanum tuberosum*) tubers. *Physiologia Plantarum*, 107, 201–213.

Bush, M.S., Marry, M., Huxham, I.M., Jarvis, M.C. and McCann, M.C. (2001) Developmental regulation of pectic epitopes during potato tuberisation. *Planta*, 213, 869–880.

Catoire, L., Pierron, M., Morvan, C., Hervé du Penhoat, C. and Goldberg, R. (1998) Investigation of the action patterns of pectinmethylesterase isoforms through kinetic analyses and NMR spectroscopy. Implications in cell wall expansion. *J. Biol. Chem.*, 273, 33150–33156.

Chabannes, M., Ruel, K., Yoshinaga, A. *et al.* (2001) *In situ* analysis of lignins in transgenic tobacco reveals a differential impact of individual transformations on the spatial patterns of lignin deposition at the cellular and subcellular levels. *Plant J.*, 28, 271–282.

Conrath, K.E., Lauwereys, M., Wyns, L. and Muyldermans, S. (2001) Camel single-domain antibodies as modular building units in bispecific and bivalent antibody constructs. *J. Biol. Chem.*, 276, 7346–7350.

Csávás, M., Borbás, A., Jánossy, L., Batta, G. and Lipták, A. (2001) Synthesis of the α-L-Ara*f*-(1→2)-β-D-Gal*p*-(1→6)- β-D-Gal*p*-(→6)-[α-L-Ara*f*-(1→2)]- β-D-Gal*p*-(1→6)-D-Gal hexasaccharide as a possible repeating unit of the cell culture exudates of *Echinacea purpurea* arabinogalactan. *Carbohydr. Res.*, 336, 107–115.

Ermel, F.F., Follet-Gueye, M.L., Cibert, C. *et al.* (2000) Differential localization of arabinan and galactan side chains of rhamnogalacturonan 1 in cambial derivatives. *Planta*, 210, 732–740.

Freelove, A.C.J., Bolam, D.N., White, P., Hazlewood, G.P. and Gilbert, H.J. (2001) A novel carbohydrate binding protein is a component of the plant cell wall degrading complex of *Piromyces equi. J. Biol. Chem.*, 276, 43010–43017.

Freshour, G., Clay, R.P., Fuller, M.S., Albersheim, P., Darvill, A.G. and Hahn, M.G. (1996) Developmental and tissue-specific structural alterations of the cell wall polysaccharides of *Arabidopsis thaliana* roots. *Plant Physiol.*, 110, 1413–1429.

Fritz, S.E., Hood, K.R. and Hood, E.E. (1991) Localization of soluble and insoluble fractions of hydroxyproline rich glycoproteins during maize kernel development. *J. Cell Sci.*, 98, 545–550.

Gao, M. and Showalter, A.M. (2000) Immunolocalization of LeAGP-1, a modular arabinogalactan-protein, reveals its developmentally regulated expression in tomato. *Planta*, 210, 865–874.

Gasper, Y., Johnson, K.L., McKenna, J.A., Bacic, A. and Schultz, C.J. (2001) The complex structures of arabinogalactan-proteins and the journey towards understanding function. *Plant Mol. Biol.*, 47, 161–176.

Gu, G., Yang, F., Du, Y. and Kong, F. (2001) Synthesis of a hexasaccharide that relates to the arabinogalactan epitope. *Carbohydr. Res.*, 336, 99–106.

Hada, N., Ogino, T., Yamada, H. and Takeda, T. (2001) Syntheses of model compounds related to an antigenic epitope in pectic polysaccharides from *Bupleurum falcatum* L. *Carbohydr. Res.*, 334, 7–17.

Hink, M.A., Griep, R.A., Borst, J.W. *et al.* (2000) Structural dynamics of green fluorescent protein alone and fused with a single chain Fv protein. *J. Biol. Chem.*, 275, 17556–17560.

Huisman, M.M.H., Brüll, L.P., Thomas-Oates, J.E., Haverkamp, J., Schols, H.A. and Voragen, A.G.J. (2001) The occurrence of internal (1→5)-linked arabinofuranose and arabinopyranose residues in arabinogalactan side chains from soybean pectic substances. *Carbohydr. Res.*, 330, 103–114.

Jones, L., Seymour, G.B. and Knox, J.P. (1997) Localization of pectic galactan in tomato cell walls using a monoclonal antibody specific to (1→4)-β-D-galactan. *Plant Physiol.*, 113, 1405–1412.

Joseleau, J.P. and Ruel, K. (1997) Study of lignification by noninvasive techniques in growing maize internodes. An investigation by Fourier transform infrared cross-polarization-magic angle spinning ^{13}C-nuclear magnetic resonance spectroscopy and immunocytochemical transmission electron microscopy. *Plant Physiol.*, 114, 1123–1133.

Kikuchi, S., Ohinata, A., Tsumuraya, Y., Hashimoto, Y., Kaneko, Y. and Matsushima, H. (1993) Production and characterization of antibodies to the β-(1→6)-galactotetraosyl group and their interaction with arabinogalactan-proteins. *Planta*, 190, 525–535.

Knox, J.P. (1997) The use of antibodies to study the architecture and developmental regulation of plant cell walls. *Int. Rev. Cytol.*, 171, 79–120.

Knox, J.P., Day, S. and Roberts, K. (1989) A set of cell surface glycoproteins forms a marker of cell position, but not cell type, in the root apical meristem of *Daucus carota* L. *Development*, 106, 47–56.

Knox, J.P., Linstead, P.J., King, J., Cooper, C. and Roberts, K. (1990) Pectin esterification is spatially regulated both within cell walls and between developing tissues of root apices. *Planta*, 181, 512–521.

Knox, J.P., Linstead, P.J., Peart, J., Cooper, C. and Roberts, K. (1991) Developmentally-regulated epitopes of cell surface arabinogalactan-proteins and their relation to root tissue pattern formation. *Plant J.*, 1, 317–326.

Knox, J.P., Peart, J. and Neill, S.J. (1995) Identification of novel cell surface epitopes using a leaf epidermal strip assay system. *Planta*, 196, 266–270.

Kreuger, M. and van Holst, G.J. (1995) Arabinogalactan-protein epitopes in somatic embryogenesis of *Daucus carota* L. *Planta*, 197, 135–141.

Laurenzio, L.D., Wysocka-Diller, J., Malamy, J.E. *et al.* (1996) The *SCARECROW* gene regulates an asymmetric cell division that is essential for generating the radial organization of the Arabidopsis root. *Cell*, 86, 423–433.

Limberg, G., Körner, R., Bucholt, H.C., Christensen, T.M.I.E., Roepstorff, P. and Mikkelsen, J.D. (2000) Analysis of different de-esterification mechanisms for pectin by enzymatic fingerprinting using endopectin lyase and endopolygalacturonase II from *A. niger. Carbohydr. Res.*, 327, 293–307.

Liners, F., Letesson, J.-J., Didembourg, C. and van Cutsem, P. (1989) Monoclonal antibodies against pectin. Recognition of a conformation induced by calcium. *Plant Physiol.*, 91, 1419–1424.

Liners, F., Thibault, J.-F. and van Cutsem, P. (1992) Influence of the degree of polymerization of oligogalacturonates and of esterification pattern of pectin on their recognition by monoclonal antibodies. *Plant Physiol.*, 99, 1099–1104.

Matoh, T., Takasaki, M., Takabe, K. and Kobayashi, M. (1998) Immunocytochemistry of rhamnogalacturonan II in cell walls of higher plants. *Plant Cell Physiol.*, 39, 483–491.

McCartney, L., Ormerod, A.P., Gidley, M.J. and Knox, J.P. (2000) Temporal and spatial regulation of pectic (1→4)-β-D-galactan in cell walls of developing pea cotyledons: implications for mechanical properties. *Plant J.*, 22, 105–113.

Meikle, P.J., Bonig, I., Hoogenraad, N.J., Clarke, A.E. and Stone, B.A. (1991) The location of (1→3)-β-glucans in the walls of pollen tubes of *Nicotiana alata* using a (1→3)-β-glucan-specific monoclonal antibody. *Planta*, 185, 1–8.

Meikle, P.J., Hoogenraad, N.J., Bonig, I., Clarke, A.E. and Stone, B.A. (1994) A (1→3,1→4)-β-glucan-specific monoclonal antibody and its use in the quantiation and immunocytochemical location of (1→3,1→4)-β-glucans. *Plant J.*, 5, 1–9.

Migné, C., Prensier, G., Utille, J.-P., Angibeaud, P., Cornu, A. and Grenet, E. (1998) Immunocytochemical localisation of *para*-coumaric acid and feruloyl-arabinose in the cell walls of maize stem. *J. Sci. Food Agric.*, 78, 373–381.

Meyer, D.J., Afonso, C.L. and Galbraith, D.W. (1988) Isolation and characterization of monoclonal antibodies directed against plant plasma membrane and cell wall epitopes: identification of a monoclonal antibody that recognizes extensin and analysis of the process of epitope biosynthesis in plant tissues. *J. Cell Biol.*, 107, 163–144.

Müsel, G., Schindler, T., Bergfeld, R. *et al.* (1997) Structure and distribution of lignin in primary and secondary cell walls of maize coleoptiles analyzed by chemical and immunological probes. *Planta*, 201, 146–159.

Norman, P.M., Wingate, V.P.M., Fitter, M.S. and Lamb, C.J. (1986) Monoclonal antibodies to plant plasma membrane antigens. *Planta*, 167, 452–459.

O'Neill, M.A.., Eberhard, S., Albersheim, P. and Darvill, A.G. (2001) Requirement of borate cross-linking of cell wall rhamnogalacturonan II for *Arabidopsis* growth. *Sci.*, 294, 846–849.

Orfila, C. and Knox, J.P. (2000) Spatial regulation of pectic polysaccharides in relation to pit fields in cell walls of tomato fruit pericarp. *Plant Physiol.*, 122, 775–781.

Orfila, C., Seymour, G.B., Willats, W.G.T. *et al.* (2001) Altered middle lamella homogalacturonan and disrupted deposition of (1→5)-α-L-arabinan in the pericarp of *Cnr*, a ripening mutant of tomato. *Plant Physiol.*, 126, 210–221.

Owen, M., Gandecha, A., Cockburn, B. and Whitelam, G. (1999) Synthesis of a functional anti-phytochrome single-chain Fv protein in transgenic tobacco. *Bio/Technol.*, 10, 790–794.

Pennell, R.I., Knox, J.P., Scofield, G.N., Selvendran, R.R. and Roberts, K. (1989) A family of abundant plasma membrane-associated glycoproteins related to the arabinogalactan proteins is unique to flowering plants. *J.Cell Biol.*, 108, 1967–1977.

Pennell, R.I., Janniche, L., Kjellbom, P., Scofield, G.N., Peart, J.M. and Roberts, K. (1991) Developmental regulation of a plasma membrane arabinogalactan-protein in oilseed rape flowers. *Plant Cell*, 3, 1317–1380.

Pennell, R.I., Janniche, L., Scofield, G.N., Booji, H., de Vries, S.C. and Roberts, K. (1992) Identification of a transitional cell state in the developmental pathway to carrot somatic embryogenesis. *J. Cell Biol.*, 119, 1371–1380.

Pettolino, F.A., Hoogenraad, N.J., Ferguson, C., Bacic, A., Johnson, E. and Stone, B.A. (2001) A (1→4)-β-mannan-specific monoclonal antibody and its use in the immunocytochemical location of galactomannans. *Planta*, 214, 235–242.

Puhlmann, J., Bucheli, E., Swain, M.J. *et al.* (1994) Generation of monoclonal antibodies against plant cell wall polysaccharides. I Characterization of a monoclonal antibody to a terminal α-(1→2)-linked fucosyl-containing epitope. *Plant Physiol.*, 104, 699–710.

Ridley, B.L., O'Neill, M.A. and Mohnen, D. (2001) Pectins: structure, biosynthesis, and oligo-galacturonide-related signaling. *Phytochem.*, 57, 929–967.

Ruel, K., Faix, O. and Joseleau, J.-P. (1994) New immunogold probes for studying the distribution of the different lignin types during plant cell wall biogenesis. *J. Trace Microprobe Tech.* 12, 267–276.

Ruel, K., Burlat, V. and Joseleau, J-.P. (1999) Relationship between ultrastructural topochemistry of lignin and wood properties. *IAWA J.*, 20, 203–211.

Sakurai, M.H., Kiyohara, H., Matsumoto, T., Tsumuraya, Y., Hashimoto, Y. and Yamada, H. (1998) Characterization of antigenic epitopes in anti-ulcer pectic polysaccharides from *Bupleurum falcatum* L. using several carbohydrases. *Carbohydr. Res.*, 311, 219–229.

Serpe, M.D., Muir, A.J. and Keidel, A.M. (2001) Localization of cell wall polysaccharides in nonarticulated laticifers of *Asclepias speciosa* Torr. *Protoplasma*, 216, 215–226.

Shinohara, N., Demura, T. and Fukuda, H. (2000) Isolation of a vascular cell wall-specific monoclonal antibody recognizing a cell polarity by using a phage display subtraction method. *Proc. Natl. Acad. Sci. USA*, 97, 2585–2590.

Smallwood, M., Beven, A., Donovan, N. *et al.* (1994) Localization of cell wall proteins in relation to the developmental anatomy of the carrot root apex. *Plant J.*, 5, 237–246.

Smallwood, M., Martin, H. and Knox, J.P. (1995) An epitope of rice threonine- and hydroxyproline-rich glycoprotein is common to cell wall and hydrophobic plasma membrane glycoproteins. *Planta* 196, 510–522.

Smallwood, M., Yates, E.A., Willats, W.G.T., Martin, H. and Knox, J.P. (1996) Immunochemical comparison of membrane-associated and secreted arabinogalactan-proteins in rice and carrot. *Planta*, 198, 452–459.

Steffan, W., Kovàc, P., Albersheim, P., Darvill, A.G. and Hahn, M.G. (1995) Characterization of a monoclonal antibody that recognizes an arabinosylated (1→6)-β-D-galactan epitope in plant complex carbohydrates. *Carbohydr. Res.*, 275, 295–307.

Suzuki, K., Baba, K., Itoh, T. and Sone, L. (1998) Localization of the xyloglucan in cell walls in a suspension culture of tobacco by rapid-freezing and deep-etching techniques coupled with immunogold labelling. *Plant Cell Physiol.*, 39, 1003–1009.

Suzuki, K., Kitamura, S., Kato, Y. and Itoh, T. (2000) Highly substituted glucuronoarabinoxylans (hsGAXs) and low-branched xylans show a distinct localization pattern in the tissues of *Zea mays* L. *Plant Cell Physiol.*, 41, 948–959.

Valdor, J.F. and Mackie, W. (1997) Synthesis of a trisaccharide repeating unit related to arabinogalactan-protein (AGP) polysaccharides. *J. Carbohydr. Chem.*, 16, 429–440.

VandenBosch, K.A., Bradley, D.J., Knox, J.P., Perotto, S., Butcher, G.W. and Brewin, N.J. (1989) Common components of the infection thread matrix and the intercellular space identified by immunocytochemical analysis of pea nodules and uninfected roots. *EMBO J.*, 8, 335–342.

Vicré, M., Jauneau, A., Knox, J.P. and Driouich, A. (1998) Immunolocalization of β(1→4)- and β(1→6)-D-galactan epitopes in the cell wall and Golgi stacks of developing flax root tissues. *Protoplasma*, 203, 26–34.

Willats, W.G.T., Marcus, S.E. and Knox, J.P. (1998) Generation of a monoclonal antibody specific to (1→5)-α-L-arabinan. *Carbohydr. Res.*, 308, 149–152.

Willats, W.G.T., Gilmartin P.M., Mikkelsen, J.D. and Knox, J.P. (1999a) Cell wall antibodies without immunization: generation and use of de-esterified homogalacturonan block-specific antibodies from a naive phage display library. *Plant J.*, 18, 57–65.

Willats, W.G.T., Steele-King, C.G., Marcus, S.E. and Knox, J.P. (1999b) Side chains of pectic polysaccharides are regulated in relation to cell proliferation and cell differentiation. *Plant J.*, 20, 619–628.

Willats, W.G.T., Limberg, G., Bucholt, H.C. *et al.* (2000a) Analysis of pectic epitopes recognised by conventional and phage display monoclonal antibodies using defined oligosaccharides, polysaccharides and enzymatic degradation. *Carbohydr. Res.*, 327, 309–320.

Willats, W.G.T., Steele-King, C.G., McCartney, L., Orfila, C., Marcus, S.E. and Knox, J.P. (2000b) Making and using antibody probes to study plant cell walls. *Plant Physiol. Biochem.*, 38, 27–36.

Willats, W.G.T., McCartney, L.,Mackie, W. and Knox, J.P. (2001a) Pectin: cell biology and prospects for functional analysis. *Plant Mol. Biol.*, 47, 9–27.

Willats, W.G.T., Orfila, C., Limberg, G. *et al.* (2001b) Modulation of the degree and pattern of methyl-esterification of pectic homogalacturonan in plant cell walls: implications for pectin methyl esterase action, matrix properties and cell adhesion. *J. Biol. Chem.*, 276, 19404–19413.

Willats, W.G.T., Steele-King, C.G., Marcus, S.E. and Knox, J.P. (2002a) Antibody techniques. In *Molecular Plant Biology, Vol. 2: A Practical Approach*, Oxford University Press, Oxford, pp. 119–129.

Willats, W.G.T., Rasmussen, S.E., Kristensen, T., Mikkelsen, J.D. and Knox, J.P. (2002b) Sugar-coated microarrays: a novel slide surface for the high throughput analysis of glycans. *Proteomics* 2, 1666–1671.

Williams, M.N.V., Freshour, G., Darvill, A.G., Albersheim, P. and Hahn, M.G. (1996) An antibody Fab selected from a recombinant phage display library detects deesterified pectic polysaccharide rhamnogalacturonan II in plant cells. *Plant Cell*, 8, 673–685.

Winter, G. (1998) Synthetic human antibodies and a strategy for protein engineering. *FEBS Lett.*, 430, 92–94.

Yates, E.A., Valdor, J.F., Haslam, S.M. *et al.* (1996) Characterization of carbohydrate structural features recognized by anti-arabinogalactan-protein monoclonal antibodies. *Glycobiol.*, 6, 131–139.

Zeier, J., Ruel, K., Ryser, U. and Schreiber, L. (1999) Chemical analysis and immunolocalisation of lignin and suberin in endodermal and hypodermal/rhizodermal cell walls of developing maize (*Zea mays* L.) primary roots. *Planta*, 209, 1–12.

Zou, W., Mackenzie, R., Therien, L. *et al.* (1999) Conformational epitope of the type III group B *Streptococcus* capsular polysaccharide. *J. Immunol.*, 163, 820–825.

4 Non-enzymic cell wall (glyco)proteins

Kim L. Johnson, Brian J. Jones, Carolyn J. Schultz and
Antony Bacic

4.1 Introduction

Each of the 40 or so different types of cells in a plant is enclosed by a cell wall with distinct physico-chemical and functional properties. Even within the same cell type the wall can show subtle yet significant variation. Walls are modified during development and in response to biotic and abiotic stresses. The dynamic nature of the wall and its responsiveness is reflected in the complexity of its constituents. Wall proteins, which generally comprise less than 10% of the dry weight of the primary wall, are now recognized as critical components in maintaining both the physical and biological functions of the plant extracellular matrix (ECM). They are not restricted solely to the wall itself but also form structural and functional elements of the plasma membrane–cell wall continuum. This continuum is vital for the perception of signals from the external environment.

Historically, the wall has been viewed as an inert structure and the study of its components was largely influenced by this assumption. Thus, proteins extracted from the wall were initially seen as structural elements. What has now become clear is that many of these proteins fulfil a variety of roles in addition to that of structural components. Chemical analysis showed that the protein component is rich in the amino acids hydroxyproline (Hyp)/proline (Pro), serine/threonine (Ser/Thr) and glycine (Gly) (Table 4.1; Lamport, 1965, 2001). These are now known to be components of the abundant, hydroxyproline-rich glycoproteins (HRGPs) and the glycine-rich proteins (GRPs) (Table 4.2; Showalter, 1993; Cassab, 1998). The HRGPs

Table 4.1 Major amino acids (mol%) in walls of various plant species.

	tobacco[1]	tomato[1]	sycamore[1]	potato[1]	French bean[1]	Arabidopsis[2]
Hyp	21	16.2	12.4	8.8	6.3	0.6
Ser	10.7	10.6	11.8	8.8	9.8	5.8
Pro	7.1	6.9	5	6.2	8.7	4.8
Ala	4.6	4.5	5.4	6.9	4.9	9.9
Thr	6.3	5.8	4.1	8.5	4.1	5.4
Gly	5.6	9.7	6.1	9.1	7.2	9.5

[1]Walls were isolated from suspension cultured cells (adapted from Lamport, 1965).
[2]Walls were isolated from leaves (Zablackis *et al.*, 1995).

Table 4.2 Structural proteins of the plant cell wall and their characteristic motifs.

Protein class	Abundant aa[1]	aa motifs	% Sugar	Review
Extensins	O, S, K, Y, V, H	SOOOOSOSOOOOYYYK SOOOK SOOOOTOVYK	~50	Showalter, 1993; Cassab, 1998
PRPs	O, P, V, Y, K	PPVYK and PPVEK	0–20	Showalter, 1993; Cassab, 1998
AGPs	O, S, A, T, V	AO, SO, TP, TO	90–98	Fincher et al., 1983; Nothnagel, 1997
GRPs	G	GGGX, GGXXXGG and GXGX[2]	~0	Sachetto-Martins et al., 2000
TLRPs[3]	T, L, C	$C_{(2-3)}CC(X)_6C(X)_{2-3}CC$	~0	Domingo et al., 1999

[1]aa = amino acids.
[2]X = any amino acid.
[3]TLRPs = Tyr- and Lys-rich proteins.
Adapted from Showalter, 1993.

include the extensins, arabinogalactan-proteins (AGPs) and proline-rich proteins (PRPs). HRGP backbones can be highly glycosylated, with arabinofuranosyl (Araf) and galactopyranosyl (Galp) residues, being the most abundant sugars (Lamport, 1969; Clarke et al., 1979; Fincher et al., 1983). Throughout this review the term 'wall protein' will include the moderately glycosylated 'glycoproteins' and the extensively glycosylated 'proteoglycans'.

The term 'proteoglycan' is now defined in the mammalian literature and by IUPAC (International Union of Pure and Applied Chemistry) as a protein that contains a specific type of glycan chain (glycosaminoglycan), which is absent in plants. We use the term as originally conceived to infer a molecule that is primarily composed of carbohydrate with some protein (Gottschalk, 1972; Fincher et al., 1983).

The specific interactions of HRGPs and GRPs with themselves and with other wall components are still largely unknown. The proteins are proposed to interact with the major carbohydrate components of the wall to form a complex network (Bacic et al., 1988; Carpita and Gibeaut, 1993). Changes during development and exposure to abiotic and biotic stresses influence the nature of these interactions as well as altering the wall's composition and structure (Cassab and Varner, 1988). The wall also contains other families of proteins that are vital for wall assembly and remodelling during growth, development and stress responses. These include enzymes (hydrolases/proteases/glycosidases/peroxidases/esterases/etc.), expansins, and wall-associated kinases (WAKs), which are discussed in the other chapters of this book.

Modern analytical methods have greatly advanced our understanding of the structure and organization of wall proteins. The use of antibodies has provided a visual reference of the distribution of proteins in the wall and has highlighted the complex and dynamic nature of the extracellular matrix (Knox, 1997; Willats et al.,

2000; see Chapter 3). Cloning the genes that encode the protein backbones of wall proteins has been valuable and revealed full-length polypeptide sequences, expression patterns, and the potential sites of amino acid glycosylation. However, to date only a few aspects of the arrangement and sequence of the attached glycan chains have been completed.

This chapter will provide an overview of the HRGP and GRP wall protein families and discuss recent developments in the field. In particular, our review will focus on proteins from *Arabidopsis thaliana*, as the completion of the genome sequence allows us to look at the diversity within each family in a single species. The predicted proteins are discussed in terms of what their primary sequence tells us about location, hydroxylation and subsequent glycosylation of the protein. Amino acid modules that are common to the different classes will be discussed with regard to how they may influence function through interactions with other molecules and/or contribute to their insolubilization in the wall by cross-linking. Chimeric proteins with HRGP repeat motifs and additional domains will be presented as they often have specific expression patterns that may indicate distinct developmental roles. Finally, the putative functions of these wall proteins, whilst still mostly speculative, are discussed. For recent comprehensive reviews, readers should also consult Keller (1993); Showalter (1993); Knox (1995); Nothnagel (1997); Cassab (1998); Sommer-Knudsen *et al.* (1998); Josè-Estanyol and Puigdomènech (2000) and Wu *et al.* (2001).

4.2 Hydroxyproline-rich glycoproteins (HRGPs)

The nomenclature of the HRGPs is constantly evolving. This is not surprising since the molecular data derived from gene sequencing are outstripping the pace of biochemical studies. For consistency, we have used the existing biochemical/molecular data to delineate the major classical classes of HRGPs. The classification of new genes in each of these classes is not always straightforward. When gene sequences code for proteins that contain several distinct regions, we have used the following definitions: 'chimeric' is used if one region encodes a known classical HRGP motif (e.g. extensin) and the other regions contain unrelated motifs (e.g. Leu-rich repeats (LRRs)); and 'hybrid' is used when there are two different but known classical HRGP motifs (e.g. extensin motif and AGP motif). Using these definitions we have renamed the 'hybrid' PRPs (Josè-Estanyol *et al.*, 1992; Josè-Estanyol and Puigdomènech, 1998) as chimeric PRPs. Ultimately, the definition must rely on the structure of the mature protein, which in some cases (e.g. non-classical AGPs from tobacco (Mau *et al.*, 1995)) can change dramatically following post-translational proteolytic processing.

The HRGPs consist of the extensins (Showalter, 1993; Kieliszewski and Lamport, 1994; Cassab, 1998), AGPs (Fincher *et al.*, 1983; Nothnagel, 1997; Gaspar *et al.*, 2001; Showalter, 2001) and PRPs (Chen and Varner, 1985; Tierney *et al.*, 1988; Fowler *et al.*, 1999). Hybrid proteins that share repeat modules common to different classes of HRGPs suggest that there is a continuum from the minimally glycosylated to extensively glycosylated HRGPs (Sommer-Knudsen *et al.*, 1998). Despite this continuum, there are a large number of molecules that clearly reflect the original classification of extensins, AGPs and PRPs. The extent of post-translational processing is largely determined by amino acid repeat sequences in the protein backbone. Hydroxylation and glycosylation of HRGPs are important as they define the interactive surface of the molecule and hence should determine HRGP function, as is the case for extensively glycosylated proteoglycans in animals (Seppo and Tiemeyer, 2000).

4.2.1 Post-translational modification of HRGPs

An N-terminal signal sequence predicted by the gene sequence of all wall proteins targets them to the endoplasmic reticulum (ER)/Golgi apparatus for subsequent post-translational modification and secretion. The secretion signal is proteolytically removed in the ER. Many AGPs also possess a C-terminal hydrophobic signal sequence that directs addition of a glycosylphosphatidylinositol (GPI) membrane anchor (see section 4.2.3.1). Further modifications of the HRGPs include the hydroxylation of Pro residues and subsequent glycosylation in the ER/Golgi apparatus of both Hyp and Ser residues (Lamport *et al.*, 1973; Lamport, 1977; Wojtaszek *et al.*, 1999). In some cases, disulphide bond formation occurs in the ER, catalysed by protein disulphide isomerases (Frand *et al.*, 2000).

The amino acid motifs that direct the glycosylation of specific residues in extensins and AGPs have been studied. The characteristic repeat motif of extensins is contiguous Pro residues following a Ser residue (e.g. Ser-Pro$_4$ (SP$_4$)). The AGPs have XP repeats where X is commonly Ser, Thr or Ala, and predominantly non-contiguous Pro residues. The nature of the features that determine whether hydroxylation and glycosylation of these residues occurs has been elegantly addressed experimentally, using a variety of techniques, by Kieliszewski and colleagues (Kieliszewski and Lamport, 1994; Kieliszewski *et al.*, 1995; Shpak *et al.*, 1999; Goodrum *et al.*, 2000; Kieliszewski, 2001; Kieliszewski and Shpak, 2001; Shpak *et al.*, 2001; Zhou *et al.*, 2002) and is summarized below. Additional knowledge of the hydroxylation of AGPs comes from sequencing of AGP protein backbones (Chen *et al.*, 1994; Du *et al.*, 1994; Mau *et al.*, 1995; Gao *et al.*, 1999; Loopstra *et al.*, 2000; Schultz *et al.*, 2000).

4.2.1.1 Hydroxylation of proline

Initially it was believed that the conformation of the Pro-rich backbone controlled hydroxylation (Tanaka *et al.*, 1980). However, it now appears that peptide sequence rather than conformation determines hydroxylation of Pro residues and subsequent glycosylation (Kieliszewski and Lamport, 1994; Shpak *et al.*, 2001). Hydroxylation of Pro residues may involve promiscuous, but still sequence-specific, prolyl hydroxylases that modify both extensins and AGPs, or multiple prolyl hydroxylases that recognize a given 'window' of residues (Kieliszewski and Lamport, 1994). Different enzyme specificity for different plant species is also possible (Shpak *et al.*, 2001).

Despite the limited number of glycoproteins studied, some consistent patterns for hydroxylation have emerged. Prolyl hydroxylase does not hydroxylate all HRGP Pro residues; for example, Lys-Pro is never hydroxylated, wheras Pro-Val is always hydroxylated (Kieliszewski and Lamport, 1994) and Thr-Pro is hydroxylated only some of the time (Schultz *et al.*, 2000). Pro residues in extensins and AGPs, such as those occurring in SP_4, AP and SP repeats, are almost always hydroxylated (Shpak *et al.*, 2001). In many PRPs, the occurrence of Lys-Pro in the repeat motifs, and fewer contiguous Pro residues results in reduced hydroxylation (Kieliszewski *et al.*, 1995).

After hydroxylation of some, but not all, Pro residues by prolyl hydroxylases in the ER (Wojtaszek *et al.*, 1999), glycosylation of some, but not all, Hyp residues then occurs in the ER/Golgi by glycosyl transferases (Kieliszewski, 2001). The type of glycosylation is dependent on the contiguous or non-contiguous nature of the Hyp residues.

4.2.1.2 *Glycosylation of hydroxyproline*
Our knowledge of the glycosylation of HRGPs is well established for the extensins (Lamport, 1967; Kieliszewski and Lamport, 1994; Kieliszewski *et al.*, 1995) but remains incomplete for the AGPs (Fincher *et al.*, 1983; Serpe and Nothnagel, 1999; Schultz *et al.*, 2000). Sequencing the glycan chains and defining the site-specific glycosylation of these proteins is a major technical challenge; however, the use of a recombinant fusion-protein strategy (Shpak *et al.*, 1999; Kieliszewski and Shpak, 2001; Shpak *et al.*, 2001; Zhou *et al.*, 2002) has made the analyses of this important post-translational modification easier.

The Hyp-contiguity hypothesis suggests that blocks of contiguous Hyp residues, such as those that occur in extensins, are arabinosylated (Kieliszewski and Lamport, 1994). A revised version of the model predicts that Hyp (arabino)galactosylation occurs on the clustered non-contiguous Hyp residues that are commonly found in AGPs (Figure 4.1; Goodrum *et al.*, 2000). Evidence for this hypothesis was obtained by arabinosylation site mapping of a Pro- and Hyp-rich glycoprotein (PHRGP) isolated from Douglas fir (*Pseudotsuga menziesii*). This was done by partial alkaline hydrolysis of the glycoprotein to yield unique glycopeptides, which could be sequenced (Kieliszewski *et al.*, 1995). Modules containing the sequence Lys-Pro-Hyp-Val-Hyp were only arabinosylated 20% of the time (with a single Ara*f* residue) on the Hyp-5 residue, whereas in the sequence Lys-Pro-Hyp-Hyp-Val a triarabinoside was predominantly added to the Hyp-3 (70% of the time).

Biochemical techniques are difficult to apply to the study of many HRGPs as they can be highly glycosylated and difficult to purify to homogeneity. An alternative strategy using recombinant fusion proteins of simple HRGP repeats, based on the gum arabic glycoprotein (GAGP) with green fluorescent protein (GFP) has been devised (Shpak *et al.*, 1999). Recombinant heterologous proteins that were expressed and targeted for secretion in suspension cultured tobacco cells acted as substrates for endogenous prolyl hydroxylases and glycosyl transferases (GTs). Recombinant proteins with only non-contiguous Hyp residues contained exclusively arabinogalactan

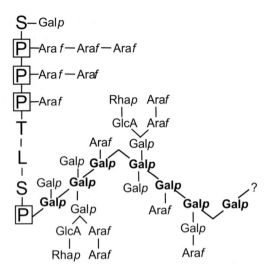

Figure 4.1 Schematic representation of the carbohydrate modifications of gum arabic glycoprotein (GAGP) which contains 'glycomodules' of both extensins and AGPs (Goodrum *et al.*, 2000).
Amino acid residues (single-letter code) of the protein backbone are shown with Hyp denoted by
a boxed P. The terminal and branching sugars in this model have been positioned arbitrarily. The
proposed pattern of glycosylation depends on whether Pro/Hyp residues are contiguous or non-
contiguous (Goodrum *et al.*, 2000). Contiguous Hyp residues have short Araf chains attached and
non-contiguous Hyp residues have larger AG chains (containing ca. 23 residues). In AGPs, the
AG chain may be ten or more repeats of the 7 β(1→3)-linked Galp backbone shown (bold), with
a periodate-sensitive sugar linkage (indicated by?) between them (Bacic *et al.*, 1987) and a large
degree of variation in the sugar side chains. Araf, arabinofuranose; Galp, galactopyranose; GlcA
([4-*O*-Me]±GlcpA), glucuronopyranose that may (+) or may not (-) be methylated at C(*O*)4; Rhap,
rhamnopyranose; *f*, furanose; *p*, pyranose. Modified from Goodrum *et al.* (2000).

(AG) polysaccharide chains, whereas recombinant proteins containing both non-
contiguous and contiguous Hyp residues showed both polysaccharide and arabino-
oligosaccharide addition, consistent with predictions (Shpak *et al.*, 1999). Further
constructs encoding repetitive blocks of Ser-Pro$_x$ (SP$_2$, SP$_3$ and SP$_4$) resulted in
contiguous Hyp-arabinosides, with the exception of the SP$_3$ glycoprotein that also
contained a Hyp-AG polysaccharide chain (Shpak *et al.*, 2001). The presence of AG
chains in the SP$_3$ motif was due to incomplete hydroxylation of the Pro residues.
This was probably a result of a low affinity of the tobacco prolyl hydroxylase for
this heterologous substrate, or alternatively that the plant cell suspension culture
had insufficient amounts of one of several prolyl hydroxylase co-factors, such as
oxygen, ascorbic acid, iron or α-ketoglutarate (Vuorela *et al.*, 1997). The value of
the Hyp-contiguity hypothesis to predict glycosylation patterns must now be tested
on an array of recombinant fusion proteins in different cell types since, for example,
Hyp-arabinosides are not common on AGPs (Bacic *et al.*, 1987, Du *et al.*, 1996a;
Shpak *et al.*, 1999).

Our current understanding of plant glycosyltransferases (GTs) at either the biochemical or molecular level is poor (Serpe and Nothnagel, 1999; Majewski-Sawka and Nothnagel, 2000; Perrin *et al.*, 2001). In animals, *O*-linked glycosylation requires a particular GT to add the first sugar to an amino acid in the core polypeptide (Ten Hagen *et al.*, 2001). Further glycosylation then occurs through the specific and hierarchical action of multiple GT members. There are likely to be many GTs involved in the assembly of the polysaccharide chains and arabino-oligosaccharides. These will include (1→3)β- and (1→6)β-galactosyltransferases (GalTs), (1→2)β-arabinosyltransferases (AraTs), as well as GTs for the terminal sugars, such as AraTs, FucTs, and RhaTs, etc. (Breton *et al.*, 1998).

The size, number and sequence of the glycosyl chains probably contribute to both physiochemical properties and function of glycoproteins, as extensins are generally moderately glycosylated with smaller arabinosides and are insoluble, whereas AGPs are heavily glycosylated with large polysaccharides and are soluble. Elucidation of Hyp glycosylation profiles will allow a better understanding of the functional and structural modelling of the wall (Kieliszewski, 2001).

4.2.2 Extensins

4.2.2.1 Extensin structure

Extensins are abundant wall glycoproteins and are the best characterized structural proteins. For comprehensive reviews readers should consult Showalter (1993); Cassab (1998); Lamport (2001) and references therein. In general, extensin genes encode proteins with a signal peptide followed by a repetitive region rich in Pro/Hyp, with the main repeat motif being SP_4. They are rich in Hyp and Ser, and moderately abundant in Val, Tyr, Lys and His. Extensins are basic proteins with isoelectric points of ~10 due to the Lys content.

Carbohydrate comprises approximately 50% of the mass of extensins and consists of *O*-linked side chains that include single galactopyranosyl (Gal*p*) residues attached to Ser, as well as mono-, di-, tri-(1,2)-β-linked arabinosides or a tetra-arabinoside (αAraf1–3βAraf1–2βAraf1–2βAraf1-) attached to Hyp (Figure 4.1). The addition of arabinosides to blocks of Hyp induces the extended polyproline-II conformation, a rod-like helical structure stabilized by glycan side chains, that characterizes extensins (van Holst and Varner, 1984).

There is confusion in the literature in defining extensins, particularly from the monocotyledons and the algae. Some proteins are classified as extensins based on SP_{2-5} motifs, yet they do not contain Val, Tyr, Lys and His and have Ser-Pro repeats more like the AGPs or Lys-Pro repeats like the PRPs. For example, the Thr and Hyp-rich glycoproteins (THRGPs) from maize (Kieliszewski *et al.*, 1990; Stiefel *et al.*, 1990) and rice (Caelles *et al.*, 1992), and a sugar beet extensin (Li *et al.*, 1990), have characteristics more appropriately classified as PRPs. This is also reflected in a different 3D structure of these proteins. Rather than a polyproline-II secondary conformation, they exist in a random coil that increases molecular flexibility, and decreases glycosylation (Kieliszewski *et al.*, 1990; Kieliszewski and Lamport, 1994).

HRGPs from *Chlamydomonas* (Woessner *et al.*, 1994; Ferris *et al.*, 2001) are often referred to as extensins; however, they also do not have Tyr, Val, Lys or His residues separating the SP_3 or SP_4 motifs and they also have AP, SP motifs common to AGPs. We refer to these algal wall proteins as hybrid HRGPs (see section 4.2.5).

In the dicotyledonous plant *Arabidopsis thaliana*, there is a large family of extensin genes (Merkouropoulos *et al.*, 1999; Yoshiba *et al.*, 2001). Using a computer program designed to search for proteins with biased amino acid compositions in the *Arabidopsis* annotated protein database, it was possible to identify twenty two putative extensins (Table 4.3; Schultz *et al.*, 2002). As a number of these match the same AGI (*Arabidopsis* genome initiative) number (see below), there may only be nineteen unique extensins. The program detected proteins with greater than 50% composition of Pro, Ala, Ser and Thr, which were then analysed for the presence of an N-terminal secretion signal sequence, and repeat motifs commonly found in extensins. We have numbered all of the new putative extensins, *Ext6–Ext22*, as five of the *Arabidopsis* genes have previously been characterized (Merkouropoulos *et al.*, 1999; Yoshiba *et al.*, 2001).

Although some *Arabidopsis* extensin cDNAs have been cloned (Merkouropoulos *et al.*, 1999; Yoshiba *et al.*, 2001), the comparison of these clones to the genomic sequences is difficult. For example, several supposedly different extensin cDNA clones match a single *Arabidopsis* genomic clone (*AtExt1=AtExt4*, *AtExt2=AtExt14*) however, they are predicted to have different protein sequences, and both *AtExt3* and *AtExt5* have some regions of sequence that match At1g21310 with 100% identity. As *AtExt1* was identified from genomic sequence (Landsberg erecta ecotype) and *AtExt4* from a cDNA (unknown ecotype), it is possible they are encoded by the same gene and that an artefact was introduced during the synthesis of the cDNA at the reverse transcription step and/or the genomic sequence is incorrectly annotated. The cDNA encoding *AtExt3* is almost certainly the result of a cloning artefact because its 5′ sequence matches a gene on chromosome 1 and 3′ sequence a gene on chromosome 2. As extensin genes are highly repetitive and GC-rich, errors can easily occur during sequencing and/or cloning. To determine if the *Arabidopsis* genes are correctly annotated, multiple independent cDNA clones for each gene should be sequenced.

The *Arabidopsis* extensins can be divided into different classes depending on their repeat motifs and using features of extensins characterised from other dicot species (Table 4.3). All 22 of the *Arabidopsis* extensins have repeats of SP_4 and/or SP_3, and these repeats are separated by short spacers (2–5 amino acid residues) rich in two or more of the amino acid residues Tyr, Val, Lys or His.

Interestingly, searches of the rice genome database (http://www.tigr.org/tigr-scripts/e2k1/irgsp.spl) have failed to find 'true' (i.e. non-chimeric) extensins, i.e. proteins containing both SP_{3-4} and some combination of the amino acids Val, Tyr, Lys and His, which are features of the *Arabidopsis* extensins. Much work will be required to determine how the different repeat motifs affect the overall structure and function of these putative extensins.

4.2.2.2 Chimeric extensins

A number of HRGPs have also been identified with characteristics common to the extensins as well as some unique regions. These genes often have more restricted expression patterns and may have specific development roles.

A chimeric extensin protein with a Leu-rich repeat (LRR) from *Arabidopsis* (LRX1) is specifically found in the walls of root hairs, based on immunolocalization studies of a tagged LRX protein (Baumberger *et al.*, 2001). The absence of LRX1 in transposon-tagged *lrx1* mutants causes defects in root hair development, such as abortion, swelling, and branching. The function of the extensin domain of LRX1 is possibly to direct the protein to a particular region of the wall and/or to insolubilize it (Baumberger *et al.*, 2001). LRX1 is suggested to contribute to establishing or stabilizing root hair polarization and tip growth by physically connecting the wall and the plasma membrane (Fowler and Quatrano, 1997). Alternatively, LRX1 could function in the regulation and organization of wall expansion at the root hair tip.

Another group of chimeric extensins containing LRRs associated with tip growth are the pollen extensin-like (Pex) proteins, that have an N-terminal globular domain and a C-terminal extensin-like domain. Pex proteins are encoded by two closely related genes in maize, *mPex1* and *mPex2*, while a single *tPex* gene has been identified in tomato. A number of putative *Pex* gene sequences have also been identified in *Arabidopsis*, sorghum and potato (Stratford *et al.*, 2001). An interesting feature of the Pex proteins is the absence of Tyr residues in the extensin-like domain; however, two Tyr residues are found in a conserved C-terminal region. The Pex proteins from maize exist in both soluble and insoluble forms, suggesting either two Tyr residues are sufficient to cross-link the proteins into the wall, or that a different form of cross-linking occurs, for example through Lys residues (see section 4.2.2.3).

The maize and tomato *Pex* genes are specifically expressed in pollen and the maize Pex proteins have been shown to be restricted to the intine of mature pollen and to the callosic sheath of the pollen tube wall (Rubinstein *et al.*, 1995; Stratford *et al.*, 2001). Possible functions of Pex proteins may be to provide structural support for the pollen tube necessary for its rapid growth and/or to mediate pollen-pistil interactions during pollination (Stratford *et al.*, 2001). All of the Pex proteins contain a conserved LRR, which is thought to be involved in specific protein-ligand interactions during pollen tube growth (Kobe and Deisenhofer, 1995).

The Cys-rich extensin-like proteins (CELPs), another class of chimeric extensins from *Nicotiana tabacum*, are encoded by a family of more than 10 related genes that are predominantly expressed in flowers and may be important for cell-cell recognition (Wu *et al.*, 1993, 2001). The predicted structure of these proteins has an N-terminal extensin-like region followed by a Cys-rich (8 residues) domain and a short, highly charged C-terminal region. The Cys-rich domain in CELPs is different from the Cys-rich domain in the hybrid HRGPs known as PELPs (see section 4.2.5; Bosch *et al.*, 2001).

Table 4.3 Characteristics of the *Arabidopsis thaliana* extensins.

Class	Extensin[1]	AGI locus[2]	aa length	aa composition (%)[3]						Repeat motifs	Related clones[4]
				P	Y	S	V	K	H		
A	AtExt1/	At1g76930	284	42	15	12	13	7	9	S(P)4VKH/YYS(P)3VYHS(P)4VHYS(P)3VVYH	S12022 extensin-rape[6] JT0754 extensin-like protein-potato[7]
	AtExt4		246	42	14	12	12	9	8		
B	AtExt3/	At1g21310	325	39	14	11	10	14	10	S(P)4VKHYS(P)3VYHS(P)4KKHYVYK	S12022 extensin-rape[6] 445130 HRGP-sunflower[8]
	AtExt5		203	38	16	11	9	12	11		
C	AtExt6	At2g24980	559	38	23	19	8	8	0	S(P)4YVYSS(P)4YYSPSPKVD/YYK	AF155232 extensin-pea[9]
	AtExt7	At4g08400	513	38	20	19	7	8	1		
	AtExt8	At4g08410	707	38	20	19	7	8	1		
	AtExt9	At5g06630	433	38	23	19	8	8	1		
	AtExt10	At5g06640	689	38	20	19	8	8	1		
	AtExt11	At5g49080	609	38	20	18	7	8	1		
	AtExt12	At4g13390	429	38	18	16	5	7	0		
	AtExt13	At5g35790	328	35	18	17	6	8	1		
D	AtExt2/	At3g54590	300	37	19	17	8	8	1	S(P)4YVYSS(P)4V/YYSPSPKV SPPHPVCVCPPPPPCY[5]	AF155232 extensin-pea[9] AJ271872 extensin-wood tobacco[10] AF271872, X70343 extensin-wood tobacco[11]
	AtExt14		743	39	20	19	8	8	1		
	AtExt15	At1g23720	895	41	19	19	6	8	0		
	AtExt16	At3g28550	1018	39	19	19	9	8	1		
	AtExt17	At3g54580	951	39	19	19	8	8	1		
E	AtExt18	At1g26250	443	47	22	14	10	4	0	S(P)3YVY	D86854 extensin-Madagascar periwinkle[12] X86030 extensin-like protein-cowpea[13] L22031 HRGP-soybean[14]
	AtExt19	At1g26240	478	48	20	16	9	5	0		
	AtExt20	At4g08370	350	35	15	15	7	2	1		

F	AtExt21	At2g43150	212	48	18	14	7	7	4	$S(P)_4VKS(P)_4YYYH$	227927 extensin-rape[15] AF239615 CRANTZ HRGP-cassava[16] S55036, L36982 Tyr-rich HRGP-parsley[17] M76670, X55684, X55685 extensin-tomato[18]
G	AtExt22	At4g08380	437	39	24	20	8	3	0	$SS(P)_3YAYS(P)_3SPYVYKSPPYVY$	

[1] Common name given to the extensins. For those that match the same AGI locus the GenBank accession numbers are AtExt1:U43627, AtExt4:AB031820, AtExt3:AB031819, AtExt5:AB031821, AtExt2:AB022782, AtExt14:AL138656.

[2] Locus identity number given by the Arabidopsis Genome Initiative.

[3] Amino acid (aa) composition calculated for the mature protein sequence.

[4] Related clones identified by standard protein-protein BLAST searches at NCBI; http://www.ncbi.nlm.nih.gov/BLAST/using repeat motifs shown: GenBank accession number, gene name and plant are provided.

[5] A conserved motif containing 3 Cys residues occurs once in the C-terminal region of the predicted protein backbone.

[6] Evans et al., 1990.
[7] Bown et al., 1993.
[8] Adams et al., 1992.
[9] Savenstrand et al., unpublished.
[10] Guzzardi and Jamet, unpublished.
[11] Parmentier et al., 1995.
[12] Ito et al., 1998.
[13] Arsenijevic-Maksimovic et al., 1997.
[14] Hong et al., 1994.
[15] Gatehouse et al., 1990.
[16] Han et al., unpublished.
[17] Kawalleck et al., 1995.
[18] Showalter et al., 1991.

4.2.2.3 Cross-linking of extensins into the wall

Two mechanisms of cross-linking extensins have been proposed: covalent and non-covalent. Ionic interactions with acidic pectin, for example, may be important for the precise positioning of extensins in the wall (Kieliszewski et al., 1992; Kieliszewski et al., 1994) and are known to occur because monomeric extensins can be eluted with salt. Covalent cross-linking probably then cements extensins in muro and the extent of cross-linking is controlled during development and in response to wounding and pathogen attack (Brisson et al., 1994; Otte and Barz, 1996).

An ionic interaction with chelator-extracted pectin from unripe tomato pericarp was investigated using synthetic peptides, based on sequences found in extensins, or a wound-induced native carrot extensin (MacDougall et al., 2001). A peptide containing the Tyr-Lys-Tyr-Lys motif was able to interact with rhamnogalacturonan-I (RG-I). Increasing amounts of carrot extensin interfered with calcium cross-linking of the pectin chains, forming weaker gels, suggesting a role for extensins in modulating the pectin-gel networks of walls. The Tyr-Lys-Tyr-Lys motif is also thought to be involved in covalent cross-linking and may reflect an alternative function for this sequence (MacDougall et al., 2001).

Most of the extensins are insolubilized in the wall by covalent links with themselves and/or other cell wall components (Table 4.4; Qi et al., 1995; Schnabelrauch et al., 1996; Wojtaszek et al., 1997; Otte and Barz, 2000). The cross-link site must contain an amino acid susceptible to peroxidatic oxidation, which suggests Tyr or Lys (Schnabelrauch et al., 1996), both of which are directly involved in oxidative and peroxidative intermolecular cross-links in animals (Waite, 1990).

The chemical nature of extensin cross-linking is unclear and was initially thought to occur by an oxidatively coupled dimer of Tyr, isodityrosine (Idt), that can be generated quickly and leads to both inter- and intra-polypeptide cross-links (Figure 4.2; Fry, 1986). However, intermolecular cross-linked extensin has not been detected and

Table 4.4 Proposed cross-links for wall proteins.

aa involved in cross-link	Proposed cross-link	Wall molecules	Reference
Covalent			
Tyr	Idt, Di-Idt	Extensin-extensin	(Brady et al., 1996)
		Extensin-PRP	(Brady and Fry, 1997)
		Extensin-pectin	(Qi et al., 1995)
		PRPs	(Otte and Barz, 2000)
		GRPs	(Ringli et al., 2001)
Val-Tyr-Lys	Intermolecular	Extensin-extensin	(Kieliszewski and Lamport, 1994)
Cys	Disulfide bond	TLRP	(Domingo et al., 1999)
		PRPs	(Fowler et al., 1999)
		GRPs-WAKs	(Park et al., 2001)
Non-covalent	Ionic	Extensin-pectin	(MacDougall et al., 2001)
		AGP-pectin	(Baldwin et al., 1993)

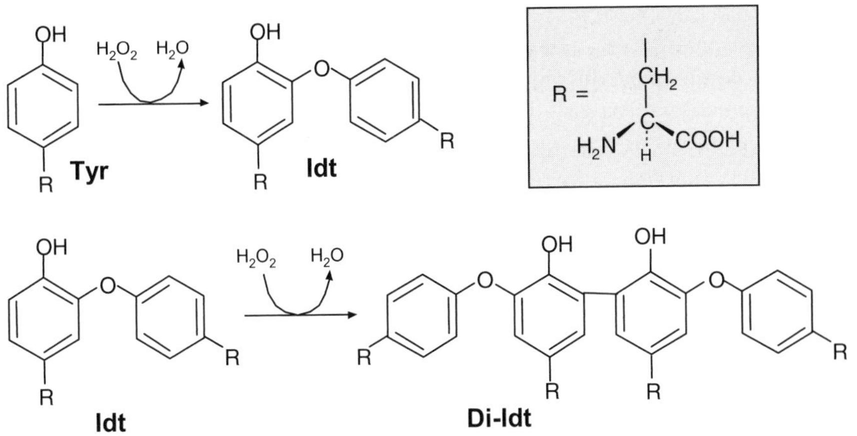

Figure 4.2 Chemical structures and possible *in vivo* reaction for the formation of Idt (isodityrosine) and Di-Idt (di-isodityrosine) from Tyr (tyrosine). Modified from Brady and Fry (1997).

only a few examples of intramolecular cross-linked Idt have been shown (Epstein and Lamport, 1984; Fong *et al.*, 1992; Zhou *et al.*, 1992). A tetrameric derivative of Tyr, Di-Idt formed by the oxidation of Idt, could possibly form intermolecular cross-links between extensins (Figure 4.2; Brady *et al.*, 1996; Brady and Fry, 1997).

Direct evidence that covalent cross-linking of extensins occurs through Tyr or Lys residues has not yet been obtained experimentally; however, the extractability of extensins decreases in plant cells challenged with H_2O_2 and elicitors and extensin monomers from suspension-cultured cells can be cross-linked *in vitro* by the action of peroxidases and H_2O_2 (Brisson *et al.*, 1994; Schnabelrauch *et al.*, 1996; Dey *et al.*, 1997; Wojtaszek *et al.*, 1997; Otte and Barz, 2000; Jackson *et al.*, 2001). A selective extensin peroxidase, pI 4.6, from tomato cell suspension cultures was shown to only cross-link extensins with the motif Val-Tyr-Lys (Schnabelrauch *et al.*, 1996). Interestingly, cross-linking of extensins with this specific peroxidase only occurs with native glycosylated extensins (Schnabelrauch *et al.*, 1996).

Extensins are also proposed to form covalent links with pectins, possibly via a Type II pectic arabino-3,6-linked galactan (Keegstra *et al.*, 1973) or a phenolic cross-link from a feruloylated sugar in the pectin to an amino acid in the extensin. This latter cross-linking would be restricted to the Caryophyllales (e.g. sugar beet and spinach), the only dicotyledon group to contain feruloylated pectins (Bacic *et al.*, 1988; Brownleader and Dey, 1993; Harris, 2000). Extensin fragments from walls of cotton suspension cultures co-purified with RG-I following trypsin digestion and the extensin peptides ran independently on an SDS-PAGE gel only after hydrogen fluoride deglycosylation, suggesting a covalent association between the extensins and RG-I (Qi *et al.*, 1995).

4.2.2.4   Extensin function

The expression of several extensin genes, for example the tomato extensins P1, P2 and P3, is developmentally regulated (Showalter, 1993; Ito *et al.*, 1998). Developmental expression of extensin-like genes has also been found during formation of tracheary elements in the *Zinnia* mesophyll cell system (Milioni *et al.*, 2001) and loblolly pine (Bao *et al.*, 1992). Extensins are induced by stresses such as wounding (Showalter *et al.*, 1991; Parmentier *et al.*, 1995; Wycoff *et al.*, 1995; Hirsinger *et al.*, 1997; Merkouropoulos *et al.*, 1999), water stress (Yoshiba *et al.*, 2001), elicitors and pathogen attack (Mazau and Esquerré-Tugayé, 1986; Corbin *et al.*, 1987; Kawalleck *et al.*, 1995; Garcia-Muniz *et al.*, 1998). Three putative extensin genes are on the *Arabidopsis* functional genomics consortium (AFGC) microarray (Wisman and Ohlrogge, 2000) and some microarray analysis results are shown in Table 4.5. The higher levels of expression in roots observed for Ext5 is supported by a large number of ESTs (29) for this gene from a root-specific cDNA library. It is not known why these extensins are more highly expressed in normal atmospheric levels of CO_2 (360 ppm) compared to elevated CO_2 (1000 ppm).

Table 4.5   Expression patterns of extensin ESTs on the *Arabidopsis* microarray.

Microarray experiment[1]	Expt ID	genes on the AFGC array[2]		
		AtExt5 143G3T7	AtExt9 142J1T7	AtExt12 4B2T7P
Whole plant to root	7203	**0.373**	0.582	0.666
Root to whole plant	7205	**3.22**	1.881	**2.213**
360ppm to 1000ppm CO_2	10847	**3.753**	**4.09**	**2.231**
1000ppm to 360ppm CO_2	10848	0.973	0.657	**0.419**
Mutant (*cch*) to wild-type (chlorophyll starvation)	11604	0.668	0.53	0.707
Wild-type to mutant (*cch*)	11605	**3.157**	**2.864**	NA[4]
TMV systemic leaves to mock inoculation[3]	7342	**0.494**	1.204	NA[4]
Mock inoculation to TMV systemic leaves	7343	**2.811**	1.03	NA[4]
Transcription inhibitor 120 to 0 min	11333	**3.028**	**2.028**	1.261
Transcription inhibitor 0 to 120 min	11375	**0.48**	NA[4]	0.559

[1]Microarray websites: http://afgc.stanford.edu/afgc_html/site2.htm. See Schultz *et al.* (2002) for an explanation of how this data was generated.
[2]The gene name and the expressed sequence tag (EST) number are provided. With the AFGC array data, experiments are considered 'significant' (shown in bold) if BOTH the test and the reciprocal experiments are significant (i.e. one is greater than 2 and the other is less than 0.5 or vice versa).
[3]TMV = tobacco mosaic virus.
[4]NA = not available.

Structural roles

Extensin cross-linking is thought to provide additional rigidity to the wall during development. Strengthening the wall in response to wounding and infection of tissue also creates a physical barrier against pathogens. Certain Tyr-rich extensins are activated in response to stress (Corbin *et al.*, 1987; Kawalleck *et al.*, 1995) and appear to be substrates for cross-linking and insolubilization into the wall.

The *HRGP4.1* gene from bean (*Phaseolus vulgaris* L.) can be activated as part of a defence response, both by wounding and by elicitation. Using *HRGP4.1* promoter-::GUS fusion studies it was shown that wounding causes local activation in the phloem, suggesting a supporting role for vascular tissues (Wycoff *et al.*, 1995). Enhanced wound-induced expression was observed in addition to the normal tissue-specific developmental expression in stem nodes and root tips, suggesting that HRGP4.1 has specific structural roles in development as well as protective functions in defence (Wycoff *et al.*, 1995).

Extensins are cross-linked in response to aluminium (Al)-stress and this may aid in the development of a rigid insoluble matrix. The wall proteins (KCl- and SDS-extractable) studied under Al-stress conditions for two wheat lines, an Al-sensitive and an Al-tolerant line, differed mostly in their extensins (Kenzhebaeva *et al.*, 2001). The untreated plants of both lines were low in covalently bound extensins. When the seedlings were treated with Al the extensin content increased in both wheat lines, especially in the Al-sensitive line and was correlated with inhibition of root elongation. Extensins from pea root tips are also proposed to bind Al both *in vivo* and *in vitro*. This may result in structural changes to the extensins that is related to the increased wall rigidity under Al stress (Kenjebaeva *et al.*, 2001).

Developmental roles

Cross-linking of extensins by an extensin peroxidase may be an important determinant that restricts plant growth (Brownleader *et al.*, 2000). This is suggested by the action of an inhibitor of extensin peroxidase, derived from suspension cultured tomato cells, that is capable of completely inhibiting peroxidase-mediated extensin cross-linking *in vitro*. Brownleader *et al.* (2000) showed that in the presence of the inhibitor seedling growth is increased by up to 15%. This suggests that inhibition of tomato hypocotyl growth is mediated, at least partially, by cross-linking of wall extensin. Little peroxidase-mediated cross-linking of extensin in the wall is thought to occur in etiolated seedlings, thereby facilitating active cell expansion (Jackson *et al.*, 1999). The role of cross-linking peroxidases in restricting plant growth has also been demonstrated by under- or over-expressing peroxidase resulting in taller and shorter transformed plants, respectively (Lagrimini *et al.*, 1997). Several processes are thought to contribute to wall rigidification, with extensin cross-linking being only one of these. The role of expansins, XTHs and other wall proteins in cell expansion/rigidification and their relative contributions must also be considered (see Chapter 8 this book).

In a recent publication, one of the *Arabidopsis* extensin genes, *At1g21310*, is apparently involved in embryo development, based on the root-shoot-hypocotyl-defective (*RSH*) phenotype of a knockout mutant (Hall and Cannon, 2002). It is unclear from their work whether the tagged gene is most similar to *Ext3*, *Ext5* or *Ext1* and this almost certainly arises from the cloning/sequencing/ecotype differences with *Ext3*, *Ext5* and *Ext1* (see section 2.2.1). Expression of RSH is critical for the correct positioning of the cell plate during cytokinesis in embryo cells; however, the role of this extensin is not yet known (Hall and Cannon, 2002).

4.2.3 Arabinogalactan-proteins (AGPs)

4.2.3.1 Structure

AGPs are highly glycosylated proteoglycans of the wall and plasma membrane and have been extensively reviewed in recent times (Fincher *et al.*, 1983; Knox, 1995; Bacic *et al.*, 1996; Nothnagel, 1997; Serpe and Nothnagel, 1999; Bacic *et al.*, 2000; Nothnagel *et al.*, 2000; Gaspar *et al.*, 2001; Showalter, 2001). As a consequence, only a brief overview is provided here. AGPs are generally defined by their ability to react with a synthetic chemical dye, the β-glycosyl Yariv reagent (Yariv *et al.*, 1967). Yariv requires both the protein and carbohydrate component of the molecule and is variable in the strength of binding (Gleeson and Clarke, 1980; Bacic *et al.*, 2000). The secondary structure of AGPs is not well understood. Two models are proposed, the 'wattle blossom' and the 'twisted hairy rope', as both spheroidal and rod-like AGPs have been imaged microscopically (Figure 4.3; Fincher *et al.*, 1983; Qi *et al.*, 1991; Baldwin *et al.*, 1993).

Protein backbones of classical AGPs are rich in the amino acids Ser, Pro, Thr and Ala. The main repeat motifs of AGPs are SP, AP and TP, with the majority of Pro (> 85%) residues predicted to be modified to Hyp. Genes encoding AGPs have been identified in a variety of plant species including pine (Loopstra and Sederoff, 1995), pear (Chen *et al.*, 1994; Mau *et al.*, 1995), tobacco (Du *et al.*, 1994; Mau *et al.*, 1995; Du *et al.*, 1996b; Gilson *et al.*, 2001), tomato (Pogson and Davies, 1995; Li and Showalter, 1996), rice (http://www.tigr.org/tdb/e2k1/osa1/) and *Arabidopsis* (Schultz *et al.*, 2000; Gaspar *et al.*, 2001; Schultz *et al.*, 2002) and are probably found throughout the plant kingdom (reviewed in Nothnagel, 1997). In *Arabidopsis*, the GPI-anchored 'classical' AGPs can be divided into three subclasses: the classical AGPs, those with Lys-rich domains, and AG-peptides with short protein backbones (Table 4.6; Figure 4.4A–C; Schultz *et al.*, 2002).

Proteins encoded by classical AGP genes have a hydrophobic C-terminal domain which directs the addition of a glycosylphosphatidylinositol (GPI) membrane anchor (Youl *et al.*, 1998; Oxley and Bacic, 1999; Sherrier *et al.*, 1999; Svetek *et al.*, 1999). The C-terminal signal for addition of a GPI-anchor contains small aliphatic amino acids, followed by a short, possibly charged, region and terminates with a hydrophobic domain (Udenfriend and Kodukula, 1995; Schultz *et al.*, 1998). It is likely that processing of the GPI-anchored AGPs occurs in the ER as it does for mammalian, yeast and protozoan GPI-anchored proteins (Udenfriend and Kodukula, 1995;

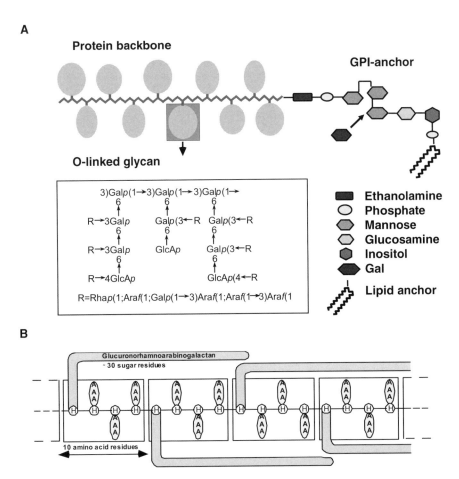

Figure 4.3   Two models of AGP structure. A. The 'wattle blossom' model of the structure of AGPs (Fincher *et al.*, 1983) with a glycosylphosphatidylinositol-(GPI-) membrane anchor attached (Oxley and Bacic, 1999). In this model, there may be 25 Hyp residues in an AGP, each of which may bear an arabinogalactan (AG) chain. Each AG chain may contain 15 or more repeats of a β(1→3)–linked Gal oligosaccharide. The molecule as a whole is spheroidal. The structure of the GPI-anchor shows an ethanolamine (EtNH₂)-phosphate (P) between the anchor and the C-terminus of the protein backbone, which is common to all GPI-anchors. The core oligosaccharide of the GPI shown is based on PcAGP1 (Oxley and Bacic, 1999) which comprises 2- and 6-linked Man*p* residues, a 4-linked glucosamine residue, and a mono-substituted inositol. An additional Gal*p* residue linked to C(*O*)4 of the 6-linked Man*p* residue occurs in some GPI-anchors. The lipid moiety is a ceramide composed primarily of a phytosphingosine base and tetracosanoic acid. Glc*p*A, glucuronopyranose; Rha*p*, rhamnopyranose; Ara*f*, arabinofuranose. B. The 'twisted-hairy rope' model of the structure of the gum arabic glycoprotein (Qi *et al.*, 1991). In this model, a hypothetical block size of 7 kDa contains 10 amino acid residues (1 kDa), 30 sugar residues (4.4 kDa), and 3 Hyp-triarabinosides (1.32 kDa). The glucuronorhamnoarabinogalactan has a galactan backbone with Glc*p*A, Rha*p* and Ara*f* side chains. Modified from Fincher *et al.* (1983); Qi *et al.* (1991) and Oxley and Bacic (1999).

Table 4.6 Genes encoding AGP protein backbones in *Arabidopsis*.

Class	AGP	AGI Locus[1]	Total ESTs[2]	Characteristics	Related clones
A	AtAGP1	At5g64310	30	Pro-rich backbone	PcAGP 1 – pear[3]
	AtAGP2	At2g22470	14	85–151 aa in length	NaAGP1 – tobacco[4]
	AtAGP3	AL161596	11	GPI-anchored	BnSta39-3,4 – rape[5]
	AtAGP4	At5g10430	35		PtX3H6 – pine[6]
	AtAGP5	At1g35230	5		
	AtAGP6	At5g14380	6		
	AtAGP7	At5g65390	3		
	AtAGP9	At2g14890	48		
	AtAGP10	At4g09030	6		
	AtAGP11	At3g01700	4		
	AtAGP25	At5g18690	1		
	AtAGP26	At2g47930	5		
	AtAGP27	At3g06360	0		
B	AtAGP17	At2g23130	2	Pro-rich backbone with	LeAGP-1 – tomato[7]
	AtAGP18	At4g37450	4	Lys-rich domain (15–20 aa)	NaAGP4 – tobacco[8]
	AtAGP19	At1g68725	0	GPI-anchored	
C	AtAGP12	At3g13520	14	AG-peptide;	AG-peptide – wheat[9]
	AtAGP13	At4g26320	2	Pro-rich backbone	
	AtAGP14	At5g56540	8	10–13 aa in length	
	AtAGP15	At5g11740	20	Most GPI-anchored	
	AtAGP16	At2g46330	21		
	AtAGP20	At3g61640	3		
	AtAGP21	At1g55330	3		
	AtAGP22	At5g53250	5		
	AtAGP23	At3g57690	1		
	AtAGP24	At5g40730	7		
D	AtFLA1	At5g55730	15	Fascilin-like AGPs; One or two	PtX14A9 – pine[6]
	AtFLA2	At4g12730	44	Pro-rich domains and one or two fasciclin-like domains Most GPI-anchored	

AtFLA3	At2g24450	0		
AtFLA4	At3g46550	0		
AtFLA5	At4g31370	0		
AtFLA6	At2g20520	5		
AtFLA7	At2g04780	27		
AtFLA8	At2g45470	44		
AtFLA9	At1g03870	29		
AtFLA10	At3g60900	3		
AtFLA11	At5g03170	10		
AtFLA12	At5g60490	5		
AtFLA13	At5g44130	4		
AtFLA14	At3g12660	0		
AtFLA15	At3g52370	9		
AtFLA16	At2g35860	8		
AtFLA17	At5g06390	0		
AtFLA18	At3g11700	4		
AtFLA19	At1g15190	0		
AtFLA20	At5g40940	0		
AtFLA21	At5g06920	0		
E	*AtAGP28*	At1g03820	2	Chimeric/non-classical; Asn-rich C-terminal domain No GPI-anchor	NaAGP2 – tobacco[10] PcAGP2 – pear[10]

[1]Locus identity number given by the Arabidopsis Genome Initiative.
[2]Expression of total ESTs from all libraries were obtained from the *A. thaliana* gene index at http://www.tigr.org/tdb/agi/
[3]Chen *et al.*, 1994.
[4]Du *et al.*, 1994.
[5]Gerster *et al.*, 1996.
[6]Loopstra and Sederoff, 1995.
[7]Pogson and Davies, 1995; Gao *et al.*, 1999.
[8]Gilson *et al.*, 2001.
[9]This putative AG-peptide was identified by its small size and carbohydrate composition in a preparation of purified AGP from wheat (Fincher *et al.*, 1974). No cDNA clone or protein has been sequenced.
[10]Mau *et al.*, 1995.

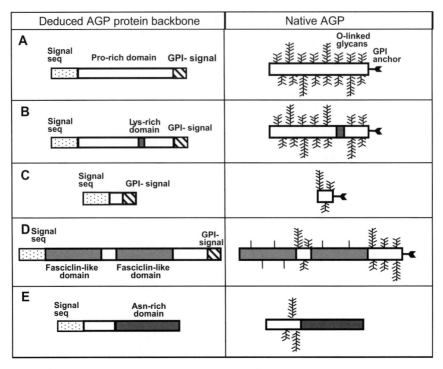

Figure 4.4   Schematic representation of the different classes of AGPs deduced from DNA sequence (left-hand panels; see Table 4.6 for accession numbers), and the predicted structures of the native AGPs after processing and post-translational modification (right-hand panels). Not drawn to scale. Processing involves removal of the predicted N-terminal signal sequence, removal of the predicted C-terminal GPI-signal sequence (if present), followed by the attachment of the GPI-anchor to the C-terminus. Pro residues are hydroxylated to Hyp and O-linked glycans (indicated by feathers) are added to most of the Hyp residues. A-C. Classical AGPs contain Pro/Hyp residues throughout the mature protein backbone and can be divided into three subclasses, with classical AGPs, those with Lys-rich domains, and AG-peptides with short protein backbones. D. There are four subclasses of fasciclin-like AGPs (FLAs) with only one represented. The protein backbone of FLAs can contain one or two fasciclin-like domains and one or two AGP domains (Gaspar *et al.*, 2001). Fasciclin domains contain motifs for the addition of N-glycans (indicated by vertical lines, up to four sites per domain). Not all FLAs are predicted to have a GPI-anchor (Schultz *et al.*, 2002). E. A putative chimeric (non-classical) AGP containing an Asn-rich domain at the C-terminus has been identified in *Arabidopsis*. Modified from Gaspar *et al.* (2001).

Thompson and Okuyama, 2000). Putative *Arabidopsis* orthologs of GPI-anchor synthesis and processing genes have been found (Gaspar *et al.*, 2001).

  The structure of the lipid moiety of the GPI-anchor of two AGPs, from pear (Oxley and Bacic, 1999) and rose (Svetek *et al.*, 1999) suspension cultured cells, has been fully characterized, with both having a phosphoceramide lipid anchor (Figure 4.3). The glycan moiety of the pear GPI anchor has also been fully sequenced and contains the conserved minimal structure found in all eukaryotes (α-D-Man(1$\rightarrow$

2) α-D-Man(1→6) α-D-Man(1→4) α-D-GlcNH$_2$-inositol) with a plant-specific substitution of a β-D-Galp (1→4) residue on the third Man residue (Figure 4.3; Oxley and Bacic, 1999). GPI-anchored AGPs can be released into the ECM from the membrane and this may occur through either phospholipase C or phospholipase D action (Svetek *et al.*, 1999; Takos *et al.*, 2000). A study of radioactively labelled ethanolamine incorporation into the GPI-anchor of AGPs in an *Arabidopsis* cell culture system revealed that 85% of AGPs were in the culture medium while 15% were recoverable from the cells (Darjania *et al.*, 2000). Despite synthesis of different sized GPI-anchored AGPs at one time, large AGPs were the earliest to be released into the culture medium with AGP species of decreasing size detected as time progressed (Darjania *et al.*, 2002).

Large type II AG chains are *O*-glycosidically linked to the Hyp residues in the protein backbone, resulting in the total mass of the molecule consisting of 90–99% carbohydrate (reviewed in Nothnagel, 1997; Serpe and Nothnagel, 1999; Bacic *et al.*, 2000). These type II AGs have (1→3) β-D-linked Galp residues that form a backbone substituted at C(O)6 by side-chains of (1→6) β-D-linked Galp (Figure 4.3). The side chains often terminate in α-L-Ara*f*, and other sugars, such as Fucp, Rhap and GlcpA (with or without 4-O-methyl ether), are common in some AGPs (reviewed in Nothnagel, 1997; Bacic *et al.*, 2000). Some AG chains may consist of as many as 120 sugar residues (Gane *et al.*, 1995). The (1→3)-β-galactan backbone contains repeat blocks of about 7 Galp residues interrupted by a periodate-sensitive linkage that is postulated to be either (1→5) α-L-Ara*f* or (1→6) β-D-Galp (Figure 4.1; Churms *et al.*, 1981; Bacic *et al.*, 1987). Expression of a major tomato AGP, LeAGP-1, as a fusion glycoprotein with GFP in tobacco (Zhou *et al.*, 2002), enabled its purification and carbohydrate analysis. Hyp-glycoside profiles showed that 54% of the total Hyp had polysaccharide substituents with a median size of 20 residues; however, some had as many as 52 residues. The rest of the Hyp was either non-glycosylated or had Hyp-arabinosides.

4.2.3.2 Chimeric AGPs

A distinct subclass of AGPs are the fasciclin-like AGPs (FLAs) that, in addition to AGP motifs, have fasciclin-like domains (Table 4.6; Figure 4.4D; Gaspar *et al.*, 2001). Fasciclins, first described in *Drosophila*, are cell adhesion molecules that have a role in axon guidance (Zinn *et al.*, 1988; Elkins *et al.*, 1990). A *Volvox* glycoprotein with fasciclin-like domains, Algal-CAM, is required to obtain proper contact between neighbouring cells during the formation of daughter colonies (Huber and Sumper, 1994). As FLAs contain domains with the potential for both protein-protein and/or protein-carbohydrate interactions, they are likely to have important roles in development. Twenty-one genes encoding FLAs have been identified in *Arabidopsis* (Table 4.6; Schultz *et al.*, 2002) that contain one or two AGP domains and one or two fasciclin-like domains (Gaspar *et al.*, 2001).

Genes encoding chimeric 'non-classical' AGPs have also been characterized in several species including tobacco, cotton and pear (Sheng *et al.*, 1991; John and Crow, 1992; Chen *et al.*, 1993; Cheung *et al.*, 1993; Mau *et al.*, 1995; Du *et al.*, 1996b;

Gaspar *et al.*, 2001). The gene structure predicts a protein backbone with an N-terminal secretion signal, a Pro-rich region, and a C-terminal hydrophilic region that is highly variable (Figure 4.4E). There is experimental evidence that at least some of these chimeric AGPs (e.g. NaAGP2 and PcAGP2 (Mau *et al.*, 1995)) are proteolytically processed to produce a mature AGP with a 'classical' protein backbone.

Chimeric AGPs have been identified in *Arabidopsis* (Table 4.6). AtAGP28 has an Asn-rich C-terminal domain similar to chimeric AGPs from tobacco and pear (Table 4.6; Mau *et al.*, 1995; Gaspar *et al.*, 2001) and AtAGP30 (van Hengel and Roberts, 2001) is a putative homolog of the Gal-rich stylar glycoprotein (GaRSGP) from *Nicotiana alata* (Sommer-Knudsen *et al.*, 1996). GaRSGP has both *O*- and *N*-linked glycans and reacts very weakly with β-Glc Yariv reagent (Sommer-Knudsen *et al.*, 1996). The deduced primary structure of GaRSGP has a Pro-rich N-terminal domain reminiscent of both PRPs and AGPs and a Cys-rich C-terminal region. Therefore, in our suggested terminology AtAGP30 and GaRSGP are hybrid HRGPs, and accordingly they are discussed further in section 4.2.5.

Another chimeric AGP with a specific role is protodermal factor 1 (PDF1), that is expressed exclusively in the L1 layer of shoot apices and protoderm of organ primordia (Abe *et al.*, 1999, 2001). The AGP-like region of PDF1 has repeats of PSHTPTP which is distinctly different from other *Arabidopsis* AGPs (Table 4.6). The predicted C-terminal region of PDF1, despite being Pro-poor, is not Cys-rich and suggests a more specialized role for this chimeric AGP.

4.2.3.3 AGP function

It is difficult to assign specific roles to the AGPs as the diversity of backbones suggests many functions. AGPs that are not plasma membrane bound via a GPI-anchor are readily soluble and only a small percentage are wall bound (Svetek *et al.*, 1999; Darjania *et al.*, 2000). There is some evidence to suggest that secreted AGPs may bind to pectic fractions of the wall (Baldwin *et al.*, 1993; Carpita and Gibeaut, 1993). AGPs could have a 'structural' role by providing 'bulk' around which cross-linking occurs and in this way regulate pore size. Alternatively, they may act as a buffer between the rigid wall and the plasma membrane.

It has also been proposed that AGPs have a signalling or communication role. GPI-anchors on the classical AGPs (Youl *et al.*, 1998; Svetek *et al.*, 1999) provide a plausible mechanism for this to occur, as animal GPI-anchored glycoproteins are involved in signal transduction (Peles *et al.*, 1997; Selleck, 2000). The GPI-anchor may confer properties such as:

1.   increased lateral mobility in the lipid bilayer;
2.   regulated release from the cell surface;
3.   polarized targeting to different cell surfaces; and
4.   inclusion in lipid rafts

(Hooper, 1997; Muñiz and Riezman, 2000; Chatterjee *et al.*, 2001).

AGPs have been implicated in such diverse roles as cell-cell recognition, cell fate, embryogenesis and xylem development. There are several reports of vascular-specific or preferential localization of AGPs (Loopstra and Sederoff, 1995; Stacey *et al.*, 1995; Casero *et al.*, 1998; Loopstra *et al.*, 2000) and these AGPs are proposed to function in secondary cell wall thickening and programmed cell death, although a discrete functional relationship is yet to be shown. An AGP-like molecule associated with secondary development has been identified from the transdifferentiation of isolated mesophyll cells of *Zinnia elegans* L. into tracheary elements (Fukuda, 1997). Local intercellular communication involved in tracheary element differentiation is thought to be mediated by a 'xylogen' (Motose *et al.*, 2001a). 'Xylogen' preparations have the ability to bind Yariv reagent, suggesting the presence of AGPs. However, unlike most classical AGPs, the 'xylogen' AGPs are also susceptible to protease cleavage, indicating the presence of additional domains (Motose *et al.*, 2001b). Interestingly, the chimeric AGPs (FLAs and 'non-classical' AGPs) are also susceptible to proteolysis.

A large-scale sequencing project of cDNAs expressed in the process of tracheary element formation identified a transcript, *TED3* (Milioni *et al.*, 2001), that encodes a protein whose sequence shows similarity to two chimeric AGPs, *PcAGP2* and *NaAGP2* from pear and tobacco, respectively (Mau *et al.*, 1995). *TED3* transcripts accumulate 12 to 24 hours before the beginning of secondary wall thickening in the *in vitro Zinnia* system (Milioni *et al.*, 2001).

Two AGPs preferentially expressed in differentiating xylem of loblolly pine (PtX3H6 and PtX14A9; Table 4.6) are hormonally regulated (No and Loopstra, 2000). They are also differentially regulated during seedling development, which may be mediated by hormonal signalling. Hormonal and developmentally regulated AGPs are also found on the AFGC microarray (Schultz *et al.*, 2002).

Embryogenic cell cultures provide a model system to identify signals that affect embryogenesis and molecules with AGP epitopes have been implicated in somatic embryogenesis (Kreuger and van Holst, 1995; McCabe *et al.*, 1997; Toonen *et al.*, 1997). An AGP epitope recognized by the monoclonal antibody JIM8 exhibits polarity in terms of its spatial localization in the plasma membrane of cells destined to become embryogenic (McCabe *et al.*, 1997). Certain AGPs involved in embryogenesis contain N-acetylglucosamine (GlcNAc) and cleavage of this residue by chitinase is proposed to produce a cellular signal (van Hengel *et al.*, 2001). Classical AGPs do not appear to contain GlcNAc, but FLAs, chimeric AGPs and hybrid HRGPs (see section 4.2.5) with N-linked glycans are potential candidates.

A reverse genetics approach to determine the function of individual AGPs is not as direct an approach as first envisioned, possibly due to the redundancy that often occurs in large gene families (Pickett and Meeks-Wagner, 1995). The generation of double and triple mutants and observing growth and development of mutants under stress conditions may overcome this problem (Krysan *et al.*, 1999; Meissner *et al.*, 1999). It is possible that individual family members have both overlapping and unique roles and that the unique roles are only observable under specific conditions (Krysan *et al.*, 1999). This seems to be the case for the only *Arabidopsis*

AGP mutants associated with a specific phenotype; agp17 and fla4. Neither of these mutants has a phenotype when grown under standard conditions; however, agp17 is resistant to *Agrobacterium tumefaciens* mediated transformation (*rat1*) (Nam *et al.*, 1999; Gaspar *et al.*, 2001), and fla4 is salt overly sensitive (*sos5*) (Shi *et al.*, 2003). When *sos5* (*fla4*) plants are grown on high salt media, they have a reduced growth compared to wild-type and they exhibit a root-swelling phenotype.

4.2.4 Proline-rich proteins (PRPs)

4.2.4.1 Structure of PRPs

PRPs and 'hybrid' proline-rich proteins (HyPRPs) represent another family of HRGPs that accumulate in the wall (reviewed in Carpita and Gibeaut, 1993; Showalter, 1993; Cassab, 1998). The term PRP has been given to proteins with an N-terminal secretion signal, followed by a coding sequence that is substantially enriched in Pro residues, often in blocks of KPPVY(K) that can be repeated over forty times in one polypeptide. Because PRPs generally have a K residue preceding PP, PRPs are thought to be either minimally glycosylated with arabino-oligosaccharides, or may not be glycosylated at all (see section 4.2.1.2; Kieliszewski *et al.*, 1995).

Based on their primary structure, genes encoding PRPs have been placed into several sub-classes (Figure 4.5; Showalter, 1993; Fowler *et al.*, 1999). Members of the first class (A), including PRPs from carrot (Chen and Varner, 1985) and soybean (Hong *et al.*, 1990), have Pro-rich repeats for the entire length of the mature protein (Figure 4.5A) and do not appear to be highly glycosylated. A second class (B) comprises proteins with a predicted Pro-rich N-terminal domain and a C-terminal domain that lacks Pro-rich or repeat sequences, and are generally hydrophilic and Cys-rich (Figure 4.5B). These have previously been termed HyPRPs (Josè-Estanyol *et al.*, 1992; Josè-Estanyol and Puigdomènech, 1998); however, for consistency with the definitions used in this review they will be referred to as chimeric PRPs. A third class (C) has been identified (Fowler *et al.*, 1999) that includes proteins that are also chimeric, with a unique N-terminal domain that is non-repetitive, and a PRP-like C-terminal region (Figure 4.5C).

In *Arabidopsis*, four genes encoding chimeric PRPs (*AtPRP1, AtPRP2, AtPRP3, and AtPRP4*) have been characterized (Fowler *et al.*, 1999). A computer program was used to identify other putative PRPs in the *Arabidopsis* annotated protein database (Schultz *et al.*, 2002). Two searches were performed, one for proteins with greater than 49% Pro, Lys, Val and Tyr and another for those with greater than 47% Pro, Lys, Val and Leu. Table 4.7 shows the large number of putative PRPs that were found but includes only those that are predicted to have an N-terminal secretion signal and PRP-like repeats. Some of these *Arabidopsis* putative PRPs contain TP, AP and SP motifs, in addition to KP, and may contain AGP-like polysaccharides based on the Hyp-contiguity hypothesis (see section 4.2.1.2; Kieliszewski and Lamport, 1994). This suggests that they could also be hybrid HRGPs (see section 4.2.5) but their accurate classification will require analysis of the isolated mature (glyco)proteins.

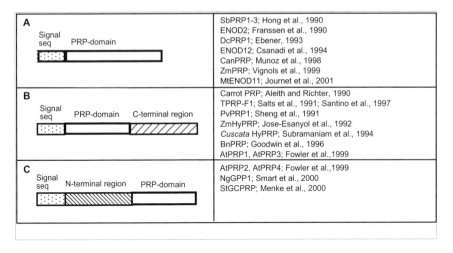

A		
Signal seq	PRP-domain	

SbPRP1-3; Hong et al., 1990
ENOD2; Franssen et al., 1990
DcPRP1; Ebener, 1993
ENOD12; Csanadi et al., 1994
CanPRP; Munoz et al., 1998
ZmPRP; Vignols et al., 1999
MtENOD11; Journet et al., 2001

B		
Signal seq	PRP-domain	C-terminal region

Carrot PRP; Aleith and Richter, 1990
TPRP-F1; Salts et al., 1991; Santino et al., 1997
PvPRP1; Sheng et al., 1991
ZmHyPRP; Jose-Esanyol et al., 1992
Cuscata HyPRP; Subramaniam et al., 1994
BnPRP; Goodwin et al., 1996
AtPRP1, AtPRP3; Fowler et al.,1999

C		
Signal seq	N-terminal region	PRP-domain

AtPRP2, AtPRP4; Fowler et al.,1999
NgGPP1; Smart et al., 2000
StGCPRP; Menke et al., 2000

Figure 4.5 Schematic representation of the different classes of PRPs deduced from DNA sequences. Not drawn to scale. Processing of PRPs involves removal of the predicted N-terminal signal sequence. A. PRPs which predict a Pro-rich backbone, often with the repeat sequence KPPVY(K), for the entire length of the mature protein backbone. These PRPs are present in both monocots and dicots. B. 'Chimeric' PRPs previously called hybrid PRPs (HyPRPs), predict a protein backbone with Pro-rich repeats at the N-terminal of the mature protein and a C-terminal that is often Pro-poor, hydrophobic and contains several Cys residues. C. Another class of chimeric PRPs has been identified which contains a unique, non-repetitive domain at the N-terminal and a basic domain containing Pro-rich repeats at the C-terminal.

4.2.4.2   PRP function

PRPs from class A were first identified in carrot roots as proteins accumulating in the wall in response to wounding (Chen and Varner, 1985). PRPs are also up-regulated during seedling, leaf, stem, root hair, and seed coat development, and during the early stages of pea fruit development (Rodríguez-Concepción *et al.*, 2001). PRPs are also responsive to phytohormones including gibberelic acid (Rodríguez-Concepción *et al.*, 2001), abscisic acid (Josè-Estanyol and Puigdomènech, 1998), ethylene (Bernhardt and Tierney, 2000), auxin, and cytokinins (Ebener *et al.*, 1993) and environmental cues such as heat (Györgyey *et al.*, 1997) and cold (Goodwin *et al.*, 1996). As with the extensins and AGPs, cDNAs encoding PRPs are likely to be on the AFGC microarray and this resource should be analysed for expression patterns during development and in response to different stimuli.

Several chimeric PRPs also have specific expression patterns, for example, in early fruit growth in tomato (Salts *et al.*, 1991; Santino *et al.*, 1997), embryogenesis in maize (Josè-Estanyol *et al.*, 1992) and somatic embryogenesis of carrot (Aleith and Richter, 1990). Two chimeric PRPs from class C, *Solanum tuberosum* guard cell PRP (StGCPRP; Menke *et al.*, 2000) and *Nicotiana glauca* guard cell PRP

Table 4.7 Characteristics of putative PRPs from *Arabidopsis*.

Class	AGI Locus[1]	aa length	aa composition (%)[2]		Repeat motifs	Related clones/reference[3]
			Pro	Other major aas		
A	At5g09520	99	29	E(16), K(14), L(12)	PKPE(M/L)PK(L/V)PE	T07173 hypothetical protein senescence down-regulated-tomato[4]
	At5g09530	336	31	E,K(17), L(11)		
	At5g15780	375	26	L(12), S(10)		
	At5g59170	263	42	K,Y(14)	KYPPP EQYPPPIKKYPPP	M76546 HRGP-sunflower[5]
	At2g27380	761	45	T,K(10), Y,V(7), I(6)	SPPIKPPPVHKPPTPTY	
B	At1g54970 (AtPRP1)	313	23	T(11), K,Y(10), Y(9)	KPTLSPPVYT	
	At3g62680 (AtPRP3)	313	22	T(12), Y(11), K,V(9)		
	At3g22120	310	38	T(14), K(11), V(8)	PTPPVVTPPT	X94976 cell wall-plasma membrane linker protein-rape[6]
	At2g45180	109	14	L(16), K(11), T(10), V(9)		
	At3g50570	167	14	S(11), N(9), K,I(8)	PNI	AU091308 unknown protein-rice[7]
	At2g33790	213	16	K(14), V(10), L(9)	PPAKAPIKLP	S31096 PRP-tobacco[8]
	At1g28290	335	27	K(15), V(12), T(7)	KPPVKPPVYPP	AAB28459 ENOD2 nodule-specific PRP-broadbean[9]
C	At4g15160	403	25	K(9), V(7)	KPPPP	S06733 HRGP-tobacco[10]
	At2g10940	264	27	V(15), L(14), K(9), T(7)	PKLPVPPVTV	
	At5g14920	254	32	T(16), K(10), V(9)	VQPPTYKPP	T05717 HRGP-barley
	At2g21140 (AtPRP2)	320	20	K(13), V(9), I(8)	PPPKIEHPPPVPVYK	
	At4g38770 (AtPRP4)	419	33	K(14), V(10)		

[1]Locus identity number given by the Arabidopsis Genome Initiative, common name is also provided for known PRPs (Fowler *et al.*, 1999).

[2]Amino acid (aa) composition calculated from predicted mature protein sequence.

[3]Related clones identified by standard protein-protein BLAST searches at NCBI; http://www.ncbi.nlm.nih.gov/BLAST/using repeat motifs shown: GenBank accession number, gene name and plant are provided.

[4]John *et al.*, 1997.

[5]Adams *et al.*, 1992.

[6]Goodwin *et al.*, 1996.

[7]Buell *et al.*, unpublished.

[8]Chen *et al.*, 1993.

[9]Protein identification number is provided (Perlick and Puhler, 1993).

[10]Keller and Lamb, 1989.

(NgGPP1; Smart *et al.*, 2000), are guard cell-specific proteins that are down- and up-regulated, respectively, by drought stress.

PRPs are rapidly insolubilized by oxidative cross-linking into the hypocotyl wall upon wounding or elicitor treatment (Bradley *et al.*, 1992; Showalter, 1993; Brisson *et al.*, 1994). As proposed for extensins, covalent linkages through Tyr and/or Lys residues are thought to be the mechanism that cross-links PRPs into the wall (Table 4.4; see section 4.2.2.3). Accordingly, PRPs have many Tyr residues as this amino acid is found in the repeat motif.

AtPRP1 and AtPRP3 have a C-terminal domain enriched in Tyr and have been localized to the seedling wall using specific antibodies. AtPRP3 is concentrated in regions of active wall synthesis and is subsequently cross-linked within the mature wall of root hairs (Hu and Tierney, 2001). T-DNA knockout lines for AtPRP3 show a defective root hair branching phenotype and AtPRP3 is proposed to play a role in tailoring the structure of the newly formed root hair wall (Bernhardt and Tierney, 2000).

In contrast, there are PRPs such as MtENOD11 and MtENOD12 from *Medicago truncatula* that have a low overall abundance of Tyr residues and a virtual absence of Tyr in the PPXXX Pro-rich repeat sequences (Figure 4.5A; Journet *et al.*, 2001). The early nodulins (ENODs) include many different types of proteins that are expressed by the plant during symbiotic nodule formation (Scheres *et al.*, 1990; Pichon *et al.*, 1992; Albrecht *et al.*, 1999; Journet *et al.*, 2001), some of which are chimeric AGPs (e.g. ENOD5 from broad bean (Frühling *et al.*, 2000)). It has been suggested that the lack of Tyr residues may result in lower levels of cross-linking and greater porosity of the wall, thereby enabling penetration by symbiotic bacteria (Journet *et al.*, 2001).

Carpita and Gibeaut (1993) suggest that PRPs are cross-linked to extensins, forming a heteropeptide framework that locks the cellulose microfibrils within the 3D network of the wall. Evidence to support the theory that PRPs are involved in wall stiffening and strengthening has come from a number of species such as chick pea (*Cicer arietinum*) (Muñoz *et al.*, 1998), soybean (Averyhart-Fullard *et al.*, 1988; Kleis-San Francisco and Tierney, 1990), alfalfa (Wilson and Cooper, 1994), and carrot (Brisson *et al.*, 1994), where the proteins can become insolubilized into the wall. Interestingly, net defect, a net-like pattern of seed cracking, and other soybean seed mutations that lead to seed coat cracking, is associated with reduced levels of soluble PRPs (Nicholas *et al.*, 1993; Percy *et al.*, 1999). The lack of soluble PRPs may be the result of more rapid insolubilization into the wall, resulting in inflexibility of the wall at an earlier stage of seed development (Percy *et al.*, 1999).

4.2.5 Hybrid HRGPs

A number of proteins have been identified which share modules of two different HRGP families. HRGPs from volvocine walls such as GP1, ZSP1, VSP1 and a2 contain both SP_{3-4} 'extensin-like' repeats and AGP motifs (Woessner *et al.*, 1994; Ferris *et al.*, 2001). These hybrid HRGPs lack Tyr and other amino acids that occur between the SP_x repeat motifs of 'true' extensins (see section 4.2.2.1). This is particularly relevant given that the majority of the *Chlamydomonas* wall polymers are postulated to be bound into the wall by non-covalent linkages (Woessner *et al.*, 1994). Interestingly, GP1 from the green alga *Chlamydomonas reinhardtii* has novel repeat motifs, yet adopts a polyproline-II helical conformation (Ferris *et al.*, 2001). A PPSPX repeat domain of GP1 forms the rod-like structure and this is interrupted by a flexible kink due to a Pro-rich sequence (PPPPPRPPFPANTPM) that is not hydroxylated. It is likely that the kink exposes amino acids capable of binding negatively charged partner molecules or enables interchain, non-covalent interactions (Ferris *et al.*, 2001).

Another abundant hybrid HRGP is the 120-kDa glycoprotein from *Nicotiana alata* that is found in the stylar transmitting tissue wall of the flower (Lind *et al.*, 1994; Schultz *et al.*, 1997). The predicted protein backbone includes the SP_{3-4} motifs common to extensins and AP, SP motifs common to AGPs (Schultz *et al.*, 1997). Purification of the native 120-kDa glycoprotein showed that it contains both AG chains and short arabinosides (Lind *et al.*, 1994). Several functions, such as defence, cell-cell communication, cell growth and development, maintaining structural integrity of the wall, and providing nutrient resources for pollen tubes have all been suggested as possible roles for this 120-kDa glycoprotein (Wu *et al.*, 2001).

Interestingly, the C-terminal Cys-containing domain of the 120-kDa glycoprotein has similarity to several different classes of hybrid HRGPs. This domain has 76% identity to the C-terminal domain of the pistil-specific extensin-like proteins (PELPIII) from *Nicotiana tabacum*, that despite their name also contain AGP-like motifs (Bosch *et al.*, 2001). It also has approximately 55% amino acid identity to the C-terminal domain of GaRSGP (Sommer-Knudsen *et al.*, 1996) and the transmitting tissue-specific (TTS) proteins from *N.tabacum* (Cheung *et al.*, 1995; Wu *et al.*, 2000). These proteins are abundant in sexual tissues and react, to varying degrees, with β-Glc Yariv reagent. These glycoproteins have been proposed to contribute chemical and/or physical factors from the female sporophytic tissue to pollen tube growth *in vivo* (Bosch *et al.*, 2001). This C-terminal domain is less well conserved in non-solanaceous plants where it is 40% identical to the C-terminal domain of DcAGP1 (Baldwin *et al.*, 2001) from carrot and AtAGP30 from *Arabidopsis* (van Hengel and Roberts, 2001).

AtEPR1 from *Arabidopsis* is specifically expressed in the endosperm with expression controlled by gibberelic acid. The repeat unit is $YSPPX_a(Y/K)PPPX_bX_cX_dPPTPT$, where X_a can be any amino acid, X_b can be Ile or Val; X_c Gln, His or Lys and X_d Lys, Met, Val or Pro (Dubreucq *et al.*, 2000). AtEPR1 is more closely related to several dicot and monocot proteins, including the THRGPs from maize and rice (see section 4.2.2.1), than to the *Arabidopsis* extensins (Dubreucq *et al.*, 2000). The predicted protein backbone of AtEPR is Thr-rich and has characteristics of both extensins and PRPs. The spatial and temporal regulation of *AtEPR1* gene expression suggests a specific role for the protein in modifying the wall structure during seed germination, thus facilitating radicle protrusion (Dubreucq *et al.*, 2000).

A putative hybrid HRGP, Hvex1, has been isolated from developing barley grains and is differentially expressed in the coenocytic endosperm and the surrounding sporophytic tissues (Sturaro *et al.*, 1998). Hvex1 has a similar overall amino acid composition to extensins, with a Pro and Lys content of 42% and 13%, respectively. It is distinguished by four distinct domains and a lower Thr content (9%) compared to the THRGP monocotyledon 'extensins' (see section 4.2.2.1). Features such as an N-terminal signal peptide, motifs found in the maize THRGP, as well as some PRPs (Chen and Varner, 1985), and a single SPPPP site located near the C-terminus, suggest that it is a HRGP. A number of AGP-like motifs are also found in the predicted protein backbone, further complicating the classification of this hybrid HRGP. The presence of many different HRGP motifs in one protein may reflect distinct functional sites for interactions *in muro* between Hvex1 and other wall components (Sturaro *et al.*, 1998).

4.3   Glycine-rich proteins (GRPs)

4.3.1   GRP structure

Unlike most other plant structural proteins, GRPs generally do not contain Pro-rich sequences and are not known to be glycosylated (reviewed in Keller, 1993; Showalter, 1993; Sachetto-Martins *et al.*, 2000). The Gly-rich domains of plant GRPs consist of sequence repeats that can be summarized by the formula $(Gly)_n X$, where X can be any amino acid, and n is generally 1 to 5. A number of different types of Gly-rich repeats have been identified, including GGGX, GGXXGG and GXGX (reviewed in Sachetto-Martins *et al.*, 2000).

A large number of Gly-rich proteins are present in the *Arabidopsis* genome. Using a biased amino acid searching criterion (Schultz *et al.*, 2002), hundreds of putative GRPs were identified with >37% Gly, Ala, Leu and Phe. Increasing the stringency to >50% Gly, Ala, Leu and Phe identified 41 putative GRPs. These were then analysed for the presence of a signal peptide and Gly repeat motifs (e.g. rather than runs of continuous Gly residues). Proteins that met these requirements and previously identified wall GRPs (de Oliveira *et al.*, 1990) are shown in Table 4.8. The most frequently observed motifs in *Arabidopsis* and other dicotyledons and monocotyledons (Sachetto-Martins *et al.*, 2000) are repeats of GGGX and $G_{2-5}X$. These motifs are usually present in GRPs with a signal peptide and Gly content of 40–75%. A novel repeat GGXXGG is observed in a subclass (C) of *Arabidopsis* GRPs with a Gly content of 28–38% (Table 4.8).

The repeat GGXXGG is present in GRPs from several species and some of these share similarity to the soybean nodulin 24 (Pawlowski *et al.*, 1997; Sachetto-Martins *et al.*, 2000). In *Arabidopsis* only AtGRP-3 belongs to this subclass (D). Other Gly-rich proteins with unusual repeats and non-Gly domains (i.e. chimeric GRPs) are present in *Arabidopsis*. Characterization of these proteins was considered beyond the scope of this review.

Table 4.8 Putative and known GRPs from *Arabidopsis*.

Class	AGI Locus[1]	aa length	Gly	Other major aa's	Repeat motif	Related clones[3]
A	At3g17050 (AtGRP1)	349	74	A(10), F,L(4)	GGGX	AJ293726 putative GRP-rape[4]
	At3g23450	452	68	I(10), K(7), F(4)		AP003047 putative GRP-rice[5]
	At2g36120	255	60	A(10), S(6), Y(5)		P09789 GRP-petunia[6]
	At5g46730	268	57	A(11), S(9), Y(7)		
	At4g29030	115	40	L(8), S(7), V(6)		
B	At2g32690	201	63	L(10), F,K(5), H(4)	$G_{2-5}X$	T09608 stress-induced protein-alfalfa[7]
	At2g05580	302	60	Q(12), K(8), E(4)		AB007818 GRP-citrus unshiu[8]
	At3g20470	174	56	A(8), F(6), L(5)		A42844 ABA and stress-induced protein-alfalfa[7]
	At2g05440	127	53	H(13), Y(6), E(5)		Q07202 cold and drought-regulated protein-alfalfa[9]
	At1g11850	108	53	L(17), A(6)		
	At2g05510	127	50	H(15), Y(6)		
	At1g04800	200	49	K(9), F(8), D(6)		
	At1g04660	212	47	V(10), L(7), A,I(6)		
	At1g62240 (AtGRP5)	227	46	A(8), F(6), L(5)		
C	At2g05380 (AtGRP3S)	116	38	Q,Y(8), H,C(6.5)	GGXXGG	AJ27650934 GRP-bread wheat[10]
	At5g07540 (AtGRP6)	244	31	A(15), P(10), S(9), K(7)		P09789 GRP-petunia[6]
	At5g07510 (AtGRP4)	185	28	A(18), L(11), P(10), T(8)		
D	At2g05520 (AtGRP3)	145	37	Y(11), Q(9), R,N(7)	GGGGXXXGGGG	S14980, S14977, Q01157 GRP-tomato[11]

[1] Locus identity number given by the Arabidopsis Genome Initiative, common name is also provided for known GRPs.
[2] Amino acid (aa) composition calculated from predicted mature protein sequence.
[3] Related clones identified by standard protein-protein BLAST searches at NCBI; http://www.ncbi.nlm.nih.gov/BLAST/using Gly repeat motifs: GenBank accession number, gene name and plant are provided.
[4] Bowers and Trick, unpublished.
[5] Sasaki et al., unpublished.
[6] Condit and Meagher, 1986.
[7] Luo et al., 1992.
[8] Hisada et al., 1999.
[9] Laberge et al., 1993.
[10] Ringli et al., 2001.
[11] Showalter et al., 1991.

The presence of an N-terminal signal peptide in many GRPs suggests a wall localization. This has been confirmed by immunolocalization studies and there is some evidence of GRPs at the membrane/wall interface (Condit, 1993). Two models for the secondary structure of Gly-rich segments have been proposed based on studies of GRPs in mammals and both may exist for different GRPs (Sachetto-Martins et al., 2000). Gly loops (Steinert et al., 1991) are proposed for GRPs such as the Arabidopsis anther-specific GRP (de Oliveira et al., 1993) and epidermis-specific GRP (Sachetto-Martins et al., 1995), allowing non-Gly residues or domains to interact with each other. The second 'β-sheets' model, proposed for PtGRP-1 (Condit and Meagher, 1986), suggests the bulky side chains of non-Gly amino acids project on the same side of the protein to generate hydrophobic regions. However, this model should be considered with caution as computer modelling algorithms do not support this prediction (Sachetto-Martins et al., 2000). Future studies with purified proteins are necessary to determine the correct conformation of Gly-rich segments in plant GRPs.

4.3.2 GRP function

Sachetto-Martins et al. (2000) provide an extensive review of the expression patterns of GRPs, showing that they are developmentally regulated and are induced by physical, chemical and biological factors. Up-regulation of GRP genes in response to water stress, wounding and pathogen attack suggest a protective role.

There is evidence that a GRP, GRP1.8 of French bean (Phaseolus vulgaris), is partially soluble in early stages of protoxylem development, but insolublized at later stages (Keller et al., 1989). To analyse the interaction of GRP1.8 with other wall components, Ringli et al. (2001) used a reporter-protein system whereby GRP1.8 domains were added to a soluble chitinase. This enabled hydrophobic interactions and insolubilization of the fusion protein in the wall to be studied (Ringli et al., 2001). Tyr-residues were proposed to enable the GRP fusion protein to form high molecular mass complexes in the presence of H_2O_2 and peroxidase. This was shown using a second construct with a wheat GRP (wGRP1) that has a similar overall amino acid sequence to GRP1.8, except that it lacks Tyr and contains Phe (Ringli et al., 2001). This protein remained soluble.

GRP1.8 may also play a role in cellular repair as it is synthesized by the xylem parenchyma and transported into the modified wall of protoxylem elements (Ryser et al., 1997). The deposition of the hydrophobic GRP1.8 during protoxylem development suggests a role in preventing water loss and in wall strengthening (Ringli et al., 2001).

An Arabidopsis GRP, AtGRP-3 is predicted to be a wall protein that interacts with wall associated kinases (WAKs, see Chapter 5), specifically binding to WAK1 both in vitro and in vivo (Park et al., 2001). Both AtGRP-3 and WAKs have conserved Cys residues, and it is possible these are responsible for the interaction between AtGRP-3, WAK1 and other proteins (Table 4.4; Park et al., 2001). WAKs are bound to the wall in part by a covalent association with pectin (Anderson et al., 2001). The ability of WAKs to bind both GRPs and pectins may be related to their role in regulating cell expansion (see Chapters 7 and 8 of this book).

4.4   Other wall proteins

A novel wall protein from tobacco, the Tyr- and Lys-rich protein (TLRP), is another structural protein that is not Hyp-rich. TLRP is related to a TLRP from tomato and both have a Cys-rich domain (CD) at the C-terminus (Domingo *et al.*, 1999). TLRP from tobacco is likely to be cross-linked into the wall, as TLRP antibody binding is restricted to the walls of lignified cells. A fusion protein of the CD with a highly soluble pathogenesis-related protein, PR1, expressed in transgenic tobacco plants was used to investigate cross-linking of the protein *in vitro*. Domingo *et al.* (1999) show that the presence of the CD is sufficient to insolubilize the PR1 fusion protein into the wall. TLRPs are proposed to interact with lignin and function in the differentiation of xylem vessels (Domingo *et al.*, 1999). The synthesis and deposition of lignin must be tightly coordinated in order to reinforce secondary walls so that they can withstand the negative pressures generated in the xylem during water transport (McCann, 1997). Since this chapter was intended to be an overview rather than a comprehensive review, it is possible that there are other wall proteins that have been characterized, but have been overlooked.

4.5   Conclusion

Bioinformatic analysis of genomic sequences has revealed that many of the wall proteins belong to multigene families. This in itself implies that wall proteins must play critical roles in development and in response to biotic and abiotic stresses. Despite a large number of 'gain-of-function' and 'loss-of-function' genomics techniques, defining the function of individual genes from large multigene families remains a significant technical challenge. It is possible that many of these wall proteins have a structural role, with only a small subset having developmental roles. It is also worth recalling that there will be glycoproteins that are phylogenetically restricted. For example, no *Arabidopsis* homolog has been found for the 120-kDa glycoprotein backbone from *Nicotiana alata* (Schultz *et al.*, 1997). Furthermore, cereal genomics may not truly reflect the entire monocotyledon group as monocotyledons can be divided into two subgroups based on their wall compositions (Bacic *et al.*, 1988).

   Post-translational processing of secreted proteins (e.g. hydroxylation, glycosylation, proteolysis, GPI-anchor addition, etc.) provides the potential for a multitude of products from a single gene. A particular gene product may be differently processed in different cell types and/or different stages of development. It has been estimated that there could be as many as 200 times more proteins than there are genes, giving rise to a new paradigm: 'the static genome and dynamic proteome' (Missler and Sudhof, 1998). In order to gain a better understanding of function, it will be necessary to isolate these glycoproteins for chemical and physical characterization from specific wall types, at specific stages of development and following different environmental stimuli. Unravelling the glycosylation machinery that gives rise to the enormous heterogeneity, and the unlimited scope for biological specificity, will also be critical. We are at the very early stages of this new and exciting journey of functional genomics of 'cell wall (glyco)proteins'.

Acknowledgements

We are grateful for funding from the Australian Research Council Large Grant (A10020017) to support this work. K.J. is a recipient of a University of Melbourne Research Scholarship. We thank Yolanda Gaspar and Edward Newbigin for critical comments on this manuscript.

References

Abe, M., Takahashi, T. and Komeda, Y. (1999) Cloning and characterization of an L1 layer-specific gene in *Arabidopsis thaliana*. *Plant Cell Physiol.*, 40, 571–580.

Abe, M., Takahashi, T. and Komeda, Y. (2001) Identification of a *cis*-regulatory element for L1 layer-specific homeodomain protein. *Plant J.*, 26, 487–494.

Adams, C.A., Nelson, W.S., Nunberg, A.N. and Thomas, T.L. (1992) A wound-inducible member of the hydroxyproline-rich glycoprotein gene family in sunflower. *Plant Physiol.*, 99, 775–776.

Albrecht, C., Geurts, R. and Bisseling, T. (1999) Legume nodulation and mycorrhizae formation: two extremes in host specificity meet. *EMBO J.*, 18, 281–288.

Aleith, F. and Richter, G. (1990) Gene expression during induction of somatic embryogenesis in carrot cell suspensions. *Planta*, 183, 17–24.

Anderson, C.M., Wagner, T.A., Perret, M., He, Z.-H., He, D. and Kohorn, B.D. (2001) WAKs: cell wall-associated kinases linking the cytoplasm to the extracellular matrix. *Plant Mol. Biol.*, 47, 197–206.

Arsenijevic-Maksimovic, V., Broughton, W.J. and Krause, A. (1997) Rhizobia modulate root-hair-specific expression of extensin genes. *Mol. Plant–Microbe Interact.*, 10, 95–101.

Averyhart-Fullard, V., Datta, K. and Marcus, A. (1988) A hydroxyproline-rich protein in the soybean cell wall. *Proc. Natl Acad. Sci. USA*, 85, 1082–1085.

Bacic, A., Churms, S.C., Stephen, A.M., Cohen, P.B. and Fincher, G.B. (1987) Fine structure of the arabinogalactan-protein from *Lolium multiflorum*. *Carbohydr. Res.*, 162, 85–93.

Bacic, A., Harris, P.J. and Stone, B.A. (1988) Structure and function of plant cell walls. In *The Biochemistry of Plants*, Vol. 14 (ed. J. Priess), Academic Press, New York, pp. 297–371.

Bacic, A., Du, H., Stone, B.A. and Clarke, A.E. (1996) Arabinogalactan-proteins: a family of cell surface/extracellular matrix plant proteoglycans. In *Essays in Biochemistry*, Vol. 31 (ed. D.K. Apps), Portland Press, London, pp. 91–101.

Bacic, A., Currie, G., Gilson, P. *et al.* (2000) Structural classes of arabinogalactan-proteins. In *Cell and developmental biology of arabinogalactan-proteins* (eds E.A. Nothnagel, A. Bacic and A.E. Clarke), Kluwer Academic/Plenum Publishers, New York, pp. 11–23.

Baldwin, T.C., McCann, M.C. and Roberts, K. (1993) A novel hydroxyproline-deficient arabinogalactan protein secreted by suspension-cultured cells of *Daucus carota*. Purification and partial characterization. *Plant Physiol.*, 103, 115–123.

Baldwin, T.C., Domingo, C., Schindler, T., Seetharaman, G., Stacey, N. and Roberts, K. (2001) DcAGP1, a secreted arabinogalactan protein, is related to a family of basic proline-rich proteins. *Plant Mol. Biol.*, 45, 421–435.

Bao, W., O'Malley, D.M. and Sederoff, R.R. (1992) Wood contains a cell-wall structural protein. *Proc. Natl. Acad. Sci. USA*, 89, 6604–6608.

Baumberger, N., Ringli, C. and Keller, B. (2001) The chimeric leucine-rich repeat/extensin cell wall protein LRX1 is required for root hair morphogenesis in *Arabidopsis thaliana*. *Genes Devel.*, 15, 1128–1139.

Bernhardt, C. and Tierney, M.L. (2000) Expression of AtPRP3, a proline-rich structural cell wall protein from *Arabidopsis*, is regulated by cell-type-specific developmental pathways involved in root hair formation. *Plant Physiol.*, 122, 705–714.

Bosch, M., Sommer-Knudsen, J., Derksen, J. and Mariani, C. (2001) Class III pistil-specific extensin-like proteins from tobacco have characteristics of arabinogalactan proteins. *Plant Physiol.*, 125, 2180–2188.

Bowers, N.L. and Trick, M. (unpublished) Microsynteny at the FCA region between *Arabidopsis thaliana* and *Brassica napus*.

Bown, D.P., Bolwell, G.P. and Gatehouse, J.A. (1993) Characterisation of potato (*Solanum tuberosum* L.) extensins: a novel extensin-like cDNA from dormant tubers. *Gene*, 134, 229–233.

Bradley, D.J., Kjellbom, P. and Lamb, C.J. (1992) Elicitor- and wound-induced oxidative cross-linking of a proline-rich plant cell wall protein: a novel, rapid defence response. *Cell*, 70, 21–30.

Brady, J.D. and Fry, S.C. (1997) Formation of di-isodityrosine and loss of isodityrosine in the cell walls of tomato cell-suspension cultures treated with fungal elicitors or H_2O_2. *Plant Physiol.*, 115, 87–92.

Brady, J.D., Sadler, I.H. and Fry, S.C. (1996) Di-isodityrosine, a novel tetrameric derivative of tyrosine in plant cell wall proteins: a new potential cross-link. *J. Biochem.*, 315, 323–327.

Breton, C., Bettler, E., Joziasse, D.H., Geremia, R.A. and Imberty, A. (1998) Sequence-function relationships of prokaryotic and eukaryotic galactosyltransferases. *J. Biol. Chem.*, 123, 1000–1009.

Brisson, L.F., Tenhaken, R. and Lamb, C. (1994) Function of oxidative cross-linking of cell wall structural proteins in plant disease resistance. *Plant Cell*, 6, 1703–1712.

Brownleader, M.D. and Dey, P.M. (1993) Purification of extensin from cell walls of tomato (hybrid of *Lycopersicon esculentum* and *L. peruvianum*) cells in suspension culture. *Planta*, 191, 457–469.

Brownleader, M.D., Hopkins, J., Mobasheri, A., Dey, P.M., Jackson, P. and Trevan, M. (2000) Role of extensin peroxidase in tomato (*Lycopersicon esculentum* Mill.) seedling growth. *Planta*, 210, 668–676.

Buell, C.R., Yuan, Q., Moffat, K.S. *et al.* (unpublished) *Oryza sativa* chromosome 3 BAC OSJNBa0091J19 genomic sequence.

Caelles, C., Delsney, M. and Puigdomènech, P. (1992) The hydroxyproline-rich glycoprotein gene from *Oryza sativa*. *Plant Mol. Biol.*, 18, 617–619.

Carpita, N.C. and Gibeaut, D.M. (1993) Structural models of primary cell walls in flowering plants: consistency of molecular structure with the physical properties of the walls during growth. *Plant J.*, 3, 1–30.

Casero, P.J., Casimiro, I. and Knox, J.P. (1998) Occurrence of cell surface arabinogalactan-protein and extensin epitopes in relation to pericycle and vascular tissue development in the root apex of four species. *Planta*, 204, 252–259.

Cassab, G.I. (1998) Plant cell wall proteins. *Annu. Rev. Plant Physiol. Plant Mol. Biol.*, 49, 281–309.

Cassab, G.I. and Varner, J.E. (1988) Cell wall proteins. *Annu. Rev. Plant Physiol. Plant Mol. Biol.*, 39, 321–353.

Chatterjee, S., Smith, E.R., Hanada, K., Stevens, V.L. and Mayor, S. (2001) GPI anchoring leads to sphingolipid-dependent retention of endocytosed proteins in the recycling endosomal compartment. *EMBO J.*, 20, 1583–1592.

Chen, C.-G., Mau, S.-L. and Clarke, A.E. (1993) Nucleotide sequence and style-specific expression of a novel proline-rich protein gene from *Nicotiana alata. Plant Mol. Biol.*, 21, 391–395.

Chen, C.-G., Pu, Z.-Y., Moritz, R.L., *et al.* (1994) Molecular cloning of a gene encoding an arabinogalactan-protein from pear (*Pyrus communis*) cell suspension culture. *Proc. Natl Acad. Sci. USA*, 91, 10305–10309.

Chen, J. and Varner, J.E. (1985) Isolation and characterization of cDNA clones for carrot extensin and a proline-rich 33-kDa protein. *Proc. Natl Acad. Sci. USA*, 82, 4399–4403.

Cheung, A.Y., May, B., Kawata, E.E., Gu, Q. and Wu, H.-W. (1993) Characterization of cDNAs for stylar transmitting tissue-specific proline-rich proteins in tobacco. *Plant J.*, 3, 151–160.

Cheung, A.Y., Wang, H. and Wu, H.-M. (1995) A floral transmitting tissue-specific glycoprotein attracts pollen tubes and stimulates their growth. *Cell*, 82, 383–393.

Churms, S.C., Stephen, A.M. and Siddiqui, I.R. (1981) Evidence for repeating sub-units in the molecular structure of the acidic arabinogalactan from rapeseed (*Brassica campestris*). *Carbohydr. Res.*, 94, 119–122.

Clarke, A.E., Anderson, R.L. and Stone, B.A. (1979) Form and function of arabinogalactans and arabinogalactan-proteins. *Phytochem.*, 18, 521–540.

Condit, C.M. (1993) Developmental expression and localization of petunia glycine-rich protein 1. *Plant Cell*, 5, 277–288.

Condit, C.M. and Meagher, R.B. (1986) A gene encoding a novel glycine-rich structural protein of petunia. *Nature*, 323, 178–181.

Corbin, D.R., Sauer, N. and Lamb, C.J. (1987) Differential regulation of a hydroxyproline-rich glycoprotein gene family in wounded and infected plants. *Mol. Cell. Biol.*, 7, 4337–4344.

Csanadi, G., Szecsi, J., Kalo, P., Kiss, P., Endre, G., Kondorosi, A., Kondorosi, E. and Kiss, G.B. (1994) *ENOD12*, an early nodulin gene, is not required for nodule formation and efficient nitrogen fixation in alfalfa. *Plant Cell*, 6, 201–213.

Darjania, L., Ichise, N., Ichikawa, S., Okamoto, T., Okuyama, H. and Thompson Jr, G.A. (2000) Metabolism of glycosylphosphatidylinositol-anchored proteins in *Arabidopsis. Biochem. Soc. Trans.*, 28, 725–727.

Darjania, L., Ichise, N., Ichikawa, S., Okamoto, T., Okuyama, H. and Thompson Jr, G.A. (2002) Dynamic turnover of arabinogalactan proteins in cultured *Arabidopsis* cells. *Plant Physiol. Biochem.*, 40, 69–79.

de Oliveira, D.E., Franco, L.O., Simoens, C. *et al.* (1993) Inflorescence-specific genes from *Arabidopsis thaliana* encoding glycine-rich proteins. *Plant J.*, 3, 495–507.

de Oliveira, D.E., Seurinck, J., Inzé, D., Van Montagu, M. and Botterman, J. (1990) Differential expression of five *Arabidopsis* genes encoding glycine-rich proteins. *Plant Cell*, 2, 427–436.

Dey, P.M., Brownleader, M.D., Pantelides, A.T., Trevan, M., Smith, J.J. and Saddler, G. (1997) Extensin from suspension-cultured potato cells: a hydroxyproline-rich glycoprotein, devoid of agglutinin activity. *Planta*, 202, 179–187.

Domingo, C., Sauri, A., Mansilla, E., Conejero, V. and Vera, P. (1999) Identification of a novel peptide motif that mediates cross-linking of proteins to cell walls. *Plant J.*, 20, 563–570.

Du, H., Simpson, R.J., Moritz, R.L., Clarke, A.E. and Bacic, A. (1994) Isolation of the protein backbone of an arabinogalactan-protein from the styles of *Nicotiana alata* and characterization of a corresponding cDNA. *Plant Cell*, 6, 1643–1653.

Du, H., Clarke, A.E. and Bacic, A. (1996a) Arabinogalactan-proteins: a class of extracellular matrix proteoglycans involved in plant growth and development. *Trends Cell Biol.*, 6, 411–414.

Du, H., Simpson, R.J., Clarke, A.E. and Bacic, A. (1996b) Molecular characterization of a stigma-specific gene encoding an arabinogalactan-protein (AGP) from *Nicotiana alata. Plant J.*, 9, 313–323.

Dubreucq, B., Berger, N., Vincent, E., *et al.* (2000) The *Arabidopsis AtEPR1* extensin-like gene is specifically expressed in endosperm during seed germination. *Plant J.,* 23, 643–652.

Ebener, W., Fowler, T.J., Suzuki, H., Shaver, J. and Tierney, M.L. (1993) Expression of DcPRP1 is linked to carrot storage root formation and is induced by wounding and auxin treatment. *Plant Physiol.,* 101, 259–265.

Elkins, T., Hortsch, M., Bieber, A.J., Snow, P.M. and Goodman, C.S. (1990) *Drosophila* fasciclin I is a novel homophilic adhesion molecule that along with fasciclin III can mediate cell sorting. *J. Cell Biol.,* 110, 1825–1832.

Epstein, L. and Lamport, D.T.A. (1984) An intramolecular linkage involving isodityrosine in extensin. *Phytochem.,* 23, 1241–1246.

Evans, I.M., Gatehouse, L.N., Gatehouse, J.A., Yarwood, J.N., Boulter, D. and Croy, R.R.D. (1990) The extensin gene family in oilseed rape (*Brassica napus* L.): characterisation of sequences of representative members of the family. *Mol. Genl Genet.,* 223, 273–287.

Ferris, P.J., Woessner, J.P., Waffenschmidt, S., Kilz, S., Drees, J. and Goodenough, U.W. (2001) Glycosylated polyproline II rods with kinks as a structural motif in plant hydroxyproline-rich glycoproteins. *Biochem.,* 40, 2978–2987.

Fincher, G.B., Sawyer, W.H. and Stone, B.A. (1974) Chemical and physical properties of an arabinogalactan-peptide from wheat endosperm. *Biochem. J.,* 139, 535–545.

Fincher, G.B., Stone, B.A. and Clarke, A.E. (1983) Arabinogalactan-proteins: structure, biosynthesis and function. *Annu. Rev. Plant Physiol.,* 34, 47–70.

Fong, C., Kieliszewski, M.J., de Zacks, R., Leykam, J.F. and Lamport, D.T.A. (1992) A gymnosperm extensin contains the serine-tetrahydroxyproline motif. *Plant Physiol.,* 99, 548–552.

Fowler, J.E. and Quatrano, R.S. (1997) Plant cell morphogenesis: plasma membrane interactions with the cytoskeleton and cell wall. *Annu. Rev. Cell. Devel. Biol.,* 13, 697–743.

Fowler, T.J., Bernhardt, C. and Tierney, M.L. (1999) Characterization and expression of four proline-rich cell wall protein genes in *Arabidopsis* encoding two distinct subsets of multiple domain proteins. *Plant Physiol.,* 121, 1081–1091.

Frand, A.R., Cuozzo, J.W. and Kaiser, C.A. (2000) Pathways for protein disulphide bond formation. *Trends Cell Biol.,* 10, 203–210.

Franssen, H.J., Thompson, D.V., Idler, K., Kormelink, R., Kammen, A.V. and Bisseling, T. (1990) Nucleotide sequence of two soybean ENOD2 early nodulin genes encoding Ngm-75. *Plant Mol. Biol.* 14, 103–106.

Frühling, M., Hohnjec, N., Schröder, G., Küster, H., Pühler, A. and Perlick, A.M. (2000) Genomic organization and expression properties of the *VfENOD5* gene from broad bean (*Vicia faba* L.). *Plant Sci.,* 155, 169–178.

Fry, S.C. (1986) Cross-linking of matrix polymers in the growing cell walls of angiosperms. *Annu. Rev. Plant Physiol.,* 37, 165–186.

Fukuda, H. (1997) Tracheary element differentiation. *Plant Cell,* 9, 1147–1156.

Gane, A.M., Craik, D., Munro, S.L.A., Howlett, G.J., Clarke, A.E. and Bacic, A. (1995) Structural analysis of the carbohydrate moiety of arabinogalactan-proteins from stigmas and styles of *Nicotiana alata. Carbohydr. Res.,* 277, 67–85.

Gao, M., Kieliszewski, M.J., Lamport, D.T.A. and Showalter, A.M. (1999) Isolation, characterization and immunolocalization of a novel, modular tomato arabinogalactan-protein corresponding to the LeAGP-1 gene. *Plant J.,* 18, 43–55.

Garcia-Muniz, N., Martinez-Izquierdo, J.A. and Puigdomènech, P. (1998) Induction of mRNA accumulation corresponding to a gene encoding a cell wall hydroxyproline-rich glycoprotein by fungal elicitors. *Plant Mol. Biol.,* 38, 623–632.

Gaspar, Y.M., Johnson, K.L., McKenna, J.A., Bacic, A. and Schultz, C.J. (2001) The complex structures of arabinogalactan-proteins and the journey towards a function. *Plant Mol. Biol.,* 47, 161–176.

Gatehouse, L.N., Evans, I.M., Gatehouse, J.A. and Croy, R.R.D. (1990) Characterisation of a rape (*Brassica napus* L.) extensin gene encoding a polypeptide relatively rich in tyrosine. *Plant Sci.*, 71, 223–231.

Gerster, J., Allard, S. and Robert, L.S. (1996) Molecular characterization of two *Brassica napus* pollen-expressed genes encoding putative arabinogalactan proteins. *Plant Physiol.*, 110, 1231–1237.

Gilson, P., Gaspar, Y., Oxley, D., Youl, J.J. and Bacic, A. (2001) NaAGP4 is an arabinogalactan-protein whose expression is suppressed by wounding and fungal infection in *Nicotiana alata*. *Protoplasma*, 215, 128–139.

Gleeson, P.A. and Clarke, A.E. (1980) Antigenic determinants of a plant proteoglycan, the *Gladiolus* style arabinogalactan-protein. *Biochem. J.*, 191, 437–447.

Goodrum, L.J., Patel, A., Leykam, J.F. and Kieliszewski, M.J. (2000) Gum arabic glycoprotein contains glycomodules of both extensin and arabinogalactan-glycoproteins. *Phytochem.*, 54, 99–106.

Goodwin, W., Pallas, J.A. and Jenkins, G.I. (1996) Transcripts of a gene encoding a putative cell wall-plasma membrane linker protein are specifically cold-induced in *Brassica napus*. *Plant Mol. Biol.*, 31, 771–781.

Gottschalk, A. (1972) Definition of glycoproteins and their delineation from other carbohydrate-protein complexes. In *Glycoproteins. Their Composition, Structure and Function*, Vol. 5 (ed. A. Gottschalk), Elsevier, Amsterdam, pp. 24–30.

Guzzardi, P. and Jamet, E. (unpublished) The *Nicotiana sylvestris Ext1.2A* gene is expressed in the root transition zone, upon wounding of stems or ribs and in proliferating cells.

Györgyey, J., Németh, K., Magyar, Z. *et al.* (1997) Expression of a novel-type small proline-rich protein gene of alfalfa is induced by 2,4-dichlorophenoxiacetic acid in dedifferentiated callus cells. *Plant Mol. Biol.*, 34, 593–602.

Hall, Q. and Cannon, M.C. (2002) The cell wall hydroxyproline-rich glycoprotein RSH is essential for normal embryo development in Arabidopsis. *Plant Cell*, 14, 1161–1172.

Han, Y., Gomez-Vasquez, R., Reilly, K., Tohme, J., Cooper, R.M. and Beeching, J.R. (unpublished) Hydroxyproline-rich glycoproteins expressed during stress responses in cassava.

Harris, P.J. (2000) Compositions of monocotyledon cell walls: implications for biosystematics, in *Monocots: Systematics and Evolution* (eds K.L. Wilson and D.A. Morrison), CSIRO, Melbourne, pp. 114–126.

van Hengel, A.J. and Roberts, K. (2001) AtAGP30, an arabinogalactan-protein from the cell wall of the primary root modulates seed germination. In *9th International Cell Wall Meeting*, Toulouse, France, p. 223.

van Hengel, A.J., Tadesse, Z., Immerzeel, P., Schols, H., van Kammen, A. and de Vries, S.C. (2001) N-acetylglucosamine and glucosamine-containing arabinogalactan proteins control somatic embryogenesis. *Plant Physiol.*, 125, 1880–1890.

Hirsinger, C., Parmentier, Y., Durr, A., Fleck, J. and Jamet, E. (1997) Characterization of a tobacco extensin gene and regulation of its gene family in healthy plants under various stress conditions. *Plant Mol. Biol.*, 33, 279–289.

Hisada, S., Kita, M., Endo-Inagaki, T., Omura, M. and Moriguchi, T. (1999) Refinement of cDNA clone expression analysis in random sequencing from the rapid cell development phase of citrus fruit. *Plant Physiol.*, 155, 699–705.

van Holst, G.-J. and Varner, J.E. (1984) Reinforced polyproline II conformation in a hydroxyproline-rich cell wall glycoprotein from carrot root. *Plant Physiol.*, 74, 247–251.

Hong, J.C., Nagao, R.T. and Key, J.L. (1990) Characterization of a proline-rich cell wall protein gene family of soybean. *J. Biol. Chem.*, 265, 2470–2475.

Hong, J.C., Cheong, Y.H., Nagau, R.T., Bahk, J.D., Cho, M.J. and Key, J.L. (1994) Isolation and characterization of three soybean extensin cDNAs. *Plant Physiol.*, 104, 793–796.

Hooper, N.M. (1997) Glycosyl-phosphatidylinositol anchored membrane enzymes. *Clinica Chimica Acta*, 266, 3–12.

Hu, J. and Tierney, M.L. (2001) *Arabidopsis* lines lacking expression of a root hair-specific proline-rich cell wall protein (AtPRP3) are altered in root hair shape. In *9th International Cell Wall Meeting*, Toulouse, France, p. 65.

Huber, O. and Sumper, M. (1994) Algal-CAMs: isoforms of a cell adhesion molecule in embryos of the alga *Volvox* with homology to *Drosophila* fasciclin I. *EMBO J.*, 13, 4212–4222.

Ito, M., Kodama, H., Komamine, A. and Watanabe, A. (1998) Expression of extensin genes is dependent on the stage of the cell cycle and cell proliferation in suspension-cultured *Catharanthus roseus* cells. *Plant Mol. Biol.*, 36, 343–351.

Jackson, P., Paulo, S., Brownleader, M., Freire, P.O. and Ricardo, C.P.P. (1999) An extensin peroxidase is associated with white-light inhibition of lupin (*Lupinus albus*) hypocotyl growth. *Austral. J. Plant Physiol.*, 26, 29–36.

Jackson, P.A.P., Galinha, C.I.R., Pereira, C.S. *et al.* (2001) Rapid deposition of extensin during the elicitation of grapevine callus cultures is specifically catalyzed by a 40-kilodalton peroxidase. *Plant Physiol.*, 127, 1065–1076.

John, I., Hackett, R., Cooper, W., Drake, R., Farrell, A. and Grierson, D. (1997) Cloning and characterization of tomato leaf senescence-related cDNAs. *Plant Mol. Biol.*, 33, 641–651.

John, M.E. and Crow, L.J. (1992) Gene expression in cotton (*Gossypium hirsutum* L.) fiber: cloning of the mRNAs. *Proc. Natl Acad. Sci. USA*, 89, 5769–5773.

Josè-Estanyol, M. and Puigdomènech, P. (1998) Developmental and hormonal regulation of genes coding for proline-rich proteins in female inflorescences and kernels of maize. *Plant Physiol.*, 116, 485–494.

Josè-Estanyol, M. and Puigdomènech, P. (2000) Plant cell wall glycoproteins and their genes. *Plant Physiol. Biochem. (Paris)*, 38, 97–108.

Josè-Estanyol, M., Ruiz-Avila, L. and Puigdomènech, P. (1992) A maize embryo-specific gene encodes a proline-rich and hydrophobic protein. *Plant Cell*, 4, 413–423.

Journet, E.-P., El-Gachtouli, N., Vernoud, V. *et al.* (2001) *Medicago truncatula* ENOD11: a novel RPRP-encoding early nodulin gene expressed during mycorrhization in arbuscule-containing cells. *Mol. Plant–Microbe Interact.*, 14, 737–748.

Kawalleck, P., Schmelzer, E., Hahlbrock, K. and Somssich, I.E. (1995) Two pathogen-responsive genes in parsley encode a tyrosine-rich hydroxyproline-rich glycoprotein (HRGP) and an anionic peroxidase. *Mol. General Genet.*, 247, 444–452.

Keegstra, K., Talmadge, K., Bauer, W.D. and Albersheim, P. (1973) The structure of plant cell walls. III. A model of the walls of suspension-cultured sycamore cells based on the interconnections of the macromolecular components. *Plant Physiol.*, 51, 188–196.

Keller, B. (1993) Structural cell wall proteins. *Plant Physiol.*, 101, 1127–1130.

Keller, B. and Lamb, C.J. (1989) Specific expression of a novel cell wall hydroxyproline-rich glycoprotein gene in lateral root initiation. *Genes Devel.*, 3, 1639–1646.

Keller, B., Templeton, M.D. and Lamb, C.J. (1989) Specific localization of a plant cell wall glycine-rich protein in protoxylem cells of the vascular system. *Proc. Natl Acad. Sci. USA*, 86, 1529–1533.

Kenjebaeva, S., Yamamoto, Y. and Matsumoto, H. (2001) The impact of aluminium on the distribution of cell wall glycoproteins of pea root tip and their Al-binding capacity. *Soil Sci. Plant Nutr.*, 47, 629–636.

Kenzhebaeva, S.S., Yamamoto, Y. and Matsumoto, H. (2001) Aluminium-induced changes in cell-wall glycoproteins in the root tips of Al-tolerant and Al-sensitive wheat lines. *Russ. J. Plant Physiol.*, 48, 441–447.

Kieliszewski, M.J. (2001) The latest hype on Hyp-*O*-glycosylation codes. *Phytochem.*, 57, 319–323.

Kieliszewski, M.J. and Lamport, D.T.A. (1994) Extensin: repetitive motifs, functional sites, posttranslational codes, and phylogeny. *Plant J.*, 5, 157–172.

Kieliszewski, M.J. and Shpak, E. (2001) Synthetic genes for the elucidation of glycosylation codes for arabinogalactan-proteins and other hydroxyproline-rich glycoproteins. *Cell. Mol. Life Sci.,* 58, 1386–1398.

Kieliszewski, M., Leykam, J.F. and Lamport, D.T.A. (1990) Structure of the threonine-rich extensin from *Zea mays. Plant Physiol.,* 85, 823–827.

Kieliszewski, M.J., Kamyab, A., Leykam, J.F. and Lamport, D.T.A. (1992) A histidine-rich extensin from *Zea mays* is an arabinogalactan protein. *Plant Physiol.,* 99, 538–547.

Kieliszewski, M.J., Showalter, A.M. and Leykam, J.F. (1994) Potato lectin: a modular protein sharing sequence similarities with the extensin family, the havein lectin family, and snake venom disintegrins (platelet aggregation inhibitors). *Plant J.,* 5, 849–861.

Kieliszewski, M.J., O'Neill, M., Leykam, J. and Orlando, R. (1995) Tandem mass spectrometry and structural elucidation of glycopeptides from a hydroxyproline-rich plant cell wall glycoprotein indicate that contiguous hydroxyproline residues are the major sites of hydroxyproline *O*-arabinosylation. *J. Biol. Chem.,* 279, 2541–2549.

Kleis-San Francisco, S.M. and Tierney, M.L. (1990) Isolation and characterization of a proline-rich cell wall protein from soybean seedlings. *Plant Physiol.,* 94, 1897–1902.

Knox, J.P. (1995) The extracellular matrix in higher plants: developmentally regulated proteoglycans and glycoproteins of the plant cell surface. *FASEB J.,* 9, 1004–1012.

Knox, J.P. (1997) The use of antibodies to study the architecture and developmental regulation of plant cell walls. *Int. J. Cytol.,* 171, 79–120.

Kobe, B. and Deisenhofer, J. (1995) A structural basis of the interactions between leucine-rich repeats and protein ligands. *Nature,* 374, 183–186.

Kreuger, M. and van Holst, G.-J. (1995) Arabinogalactan-protein epitopes in somatic embryogenesis of *Daucus carota* L. *Planta,* 197, 135–141.

Krysan, P.J., Young, J.C. and Sussman, M.R. (1999) T-DNA as an insertional mutagen in *Arabidopsis. Plant Cell,* 11, 2283–2290.

Laberge, S., Castonguay, Y. and Vezina, L.P. (1993) New cold- and drought-regulated gene from *Medicago sativa. Plant Physiol.,* 101, 1411–1412.

Lagrimini, L.M., Gingas, V., Finger, F., Rothstein, S. and Liu, T.-T.Y. (1997) Characterization of antisense transformed plants deficient in the tobacco anionic peroxidase. *Plant Physiol.,* 114, 1187–1196.

Lamport, D.T.A. (1965) The protein component of primary cell walls. *Adv. Botan. Res.,* 2, 151–218.

Lamport, D.T.A. (1967) Hydroxyproline-*O*-glycosidic linkage of the plant cell wall glycoprotein. *Nature,* 216, 1322–1324.

Lamport, D.T.A. (1969) The isolation and partial characterization of hydroyproline-rich glycopeptides obtained by enzymic degradation of primary cell walls. *Biochem.,* 8, 1155–1163.

Lamport, D.T.A. (1977) Structure, biosynthesis and significance of cell wall glycoproteins. *Rec. Adv. Phytochem.,* 11, 79–115.

Lamport, D.T.A. (2001) Life behind cell walls: paradigm lost, paradigm regained. *Cell. Mol. Life Sci.,* 58, 1363–1385.

Lamport, D.T.A., Katona, L. and Roerig, S. (1973) Galactosylserine in extensin. *J. Biochem.,* 133, 125–131.

Li, S.-X. and Showalter, A.M. (1996) Cloning and developmental/stress-regulated expression of a gene encoding a tomato arabinogalactan protein. *Plant Mol. Biol.,* 32, 641–652.

Li, X.-B., Kieliszewski, M. and Lamport, D.T.A. (1990) A chenopod extensin lacks repetitive tetrahydroxyproline blocks. *Plant Physiol.,* 92, 327–333.

Lind, J.L., Bacic, A., Clarke, A.E. and Anderson, M.A. (1994) A style-specific hydroxyproline-rich glycoprotein with properties of both extensins and arabinogalactan proteins. *Plant J.,* 6, 491–502.

Loopstra, C.A. and Sederoff, R.R. (1995) Xylem-specific gene expression in loblolly pine. *Plant Mol. Biol.,* 27, 277–291.

Loopstra, C.A., Puryear, J.D. and No, E.-G. (2000) Purification and cloning of an arabinogalactan-protein from xylem of loblolly pine. *Planta,* 210, 686–689.

Luo, M., Liu, J.H., Mohapatra, S., Hill, R.D. and Mohapatra, S.S. (1992) Characterization of a gene family encoding abscisic acid- and environmental stress-inducible proteins of alfalfa. *J. Biol. Chem.,* 267, 15367–15374.

MacDougall, A.J., Brett, G.M., Morris, V.J., Rigby, N.M., Ridout, M.J. and Ring, S.G. (2001) The effect of peptide-pectin interactions on the gelation behaviour of a plant cell wall pectin. *Carbohydr. Res.,* 335, 115–126.

Majewski-Sawka, A. and Nothnagel, E.A. (2000) The multiple roles of arabinogalactan proteins in plant development. *Plant Physiol.,* 122, 3–9.

Mau, S.-L., Chen, C.-G., Pu, Z.-Y. *et al.* (1995) Molecular cloning of cDNAs encoding the protein backbones of arabinogalactan-proteins from the filtrate of suspension-cultured cells of *Pyrus communis* and *Nicotiana alata. Plant J.,* 8, 269–281.

Mazau, D. and Esquerré-Tugayé, M.-T. (1986) Hydroxyproline-rich glycoprotein accumulation in the cell walls of plants infected by various pathogens. *Physiol. Mol. Plant Pathol.,* 29, 147–157.

McCabe, P.F., Valentine, T.A., Forsberg, L.S. and Pennell, R.I. (1997) Soluble signals from cells identified at the cell wall establish a developmental pathway in carrot. *Plant Cell,* 9, 2225–2241.

McCann, M. (1997) Tracheary element formation: building up to a dead end. *Trends Plant Sci.,* 2, 333–338.

Meissner, R.C., Jin, H., Cominelli, E. *et al.* (1999) Function search in a large transcription factor gene family in *Arabidopsis*: assessing the potential of reverse genetics to identify insertional mutations in R2R3 *MYB* genes. *Plant Cell,* 11, 1827–1840.

Menke, U., Renault, N. and Mueller-Roeber, B. (2000) StGCPRP, a potato gene strongly expressed in stomatal guard cells, defines a novel type of repetitive proline-rich proteins. *Plant Physiol.,* 122, 677–686.

Merkouropoulos, G., Barnett, D.C. and Shirsat, A.H. (1999) The *Arabidopsis* extensin gene is developmentally regulated, is induced by wounding, methyl jasmonate, abscisic and salicylic acid, and codes for a protein with unusual motifs. *Planta,* 208, 212–219.

Milioni, D., Sado, P.-E., Stacey, N.J., Domingo, C., Roberts, K. and McCann, M.C. (2001) Differential expression of cell-wall-related genes during the formation of tracheary elements in the *Zinnia* mesophyll system. *Plant Mol. Biol.,* 47, 221–238.

Missler, M. and Sudhof, T.C. (1998) Three genes and 1001 products. *Trends Genet.,* 14, 20–26.

Motose, H., Fukuda, H. and Sugiyama, M. (2001a) Involvement of local intercellular communication in the differentiation of *Zinnia* mesophyll cells into tracheary elements. *Planta,* 213, 121–131.

Motose, H., Sugiyama, M. and Fukuda, H. (2001b) An arabinogalactan protein(s) is a key component of a fraction that mediates local intercellular communication involved in tracheary element differentiation of *Zinnia* mesophyll cells. *Plant Cell Physiol.,* 42, 129–137.

Muñiz, M. and Riezman, H. (2000) Intracellular transport of GPI-anchored proteins. *EMBO J.,* 19, 10–15.

Muñoz, F.J., Dopico, B. and Labrador, E. (1998) A cDNA encoding a proline-rich protein from *Cicer arietinum.* Changes in expression during development and abiotic stresses. *Physiol. Plant.,* 102, 582–590.

Nam, J., Mysore, K.S., Zheng, C., Knue, M.K., Matthysse, A.G. and Gelvin, S.B. (1999) Identification of T-DNA tagged *Arabidopsis* mutants that are resistant to transformation by *Agrobacterium. Mol. General Genet.,* 261, 429–438.

Nicholas, C.D., Lindstrom, J.T. and Vodkin, L.O. (1993) Variation of proline rich cell wall proteins in soybean lines with anthocyanin mutations. *Plant Mol. Biol.,* 21, 145–156.

No, E.-G. and Loopstra, C.A. (2000) Hormonal and developmental regulation of two arabinogalactan-proteins in xylem of loblolly pine (*Pinus taeda*). *Physiol. Plant.,* 110, 524–529.

Nothnagel, E.A. (1997) Proteoglycans and related components in plant cells. *Int. Rev. Cytol.*, 174, 195–291.

Nothnagel, E.A., Bacic, A. and Clarke, A.E. (2000) *Cell and Developmental Biology of Arabinogalactan-Proteins*, Kluwer Academic/Plenum Publishing Corp, New York.

Otte, O. and Barz, W. (1996) The elicitor-induced oxidative burst in cultured chickpea cells drives the rapid insolubilization of two cell wall structural proteins. *Planta*, 200, 238–246.

Otte, O. and Barz, W. (2000) Characterization and oxidative in vitro cross-linking of an extensin-like protein and a proline-rich protein purified from chickpea cell walls. *Phytochem.*, 53, 1–5.

Oxley, D. and Bacic, A. (1999) Structure of the glycosyl-phosphatidylinositol membrane anchor of an arabinogalactan-protein from *Pyrus communis* suspension-cultured cells. *Proc. Natl Acad. Sci. USA*, 6, 14246–14251.

Park, A.R., Cho, S.K., Yun, U.J. *et al.* (2001) Interaction of the *Arabidopsis* Receptor Protein Kinase Wak1 with a glycine-rich protein, AtGRP-3. *J. Biol. Chem.*, 276, 26688–26693.

Parmentier, Y., Durr, A., Marbach, J. *et al.* (1995) A novel wound-inducible extensin gene is expressed early in newly isolated protoplasts of *Nicotiana sylvestris. Plant Mol. Biol.*, 29, 279–292.

Pawlowski, K., Twigg, P., Dobritsa, S., Guan, C.H. and Mullin, B.C. (1997) A nodule-specific gene family from *Alnus glutinosa* encodes glycine- and histidine-rich proteins expressed in the early stages of actinorhizal nodule development. *Mol. Plant–Microbe Interact.*, 10, 656–664.

Peles, E., Nativ, M., Lustig, M. *et al.* (1997) Identification of a novel contactin-associated transmembrane receptor with multiple domains implicated in protein-protein interactions. *EMBO J.*, 16, 978–988.

Percy, J.D., Philip, R. and Vodkin, L.O. (1999) A defective seed coat pattern (Net) is correlated with the post-transcriptional abundance of soluble proline-rich cell wall proteins. *Plant Mol. Biol.*, 40, 603–613.

Perlick, A.M. and Puhler, A. (1993) A survey of transcripts expressed specifically in root nodules of broadbean (*Vicia faba* L.). *Plant Mol. Biol.*, 22, 957–970.

Perrin, R., Wilkerson, C. and Keegstra, K. (2001) Golgi enzymes that synthesize plant cell wall polysaccharides: finding and evaluating candidates in the genomic era. *Plant Mol. Biol.*, 47, 219–224.

Pichon, M., Journet, E.-P., De Billy, F. *et al.* (1992) *Rhizobium meliloti* elicits transient expression of the early nodulin gene *ENOD*12 in the differentiating root epidermis of transgenic alfalfa. *Plant Cell*, 4, 1199–1211.

Pickett, F.B. and Meeks-Wagner, D.R. (1995) Seeing double: appreciating genetic redundancy. *Plant Cell*, 7, 1347–1356.

Pogson, B.J. and Davies, C. (1995) Characterization of a cDNA encoding the protein moiety of a putative arabinogalactan protein from *Lycopersicon esculentum. Plant Mol. Biol.*, 28, 347–352.

Qi, W., Fong, C. and Lamport, D.T.A. (1991) Gum arabic glycoprotein is a twisted hairy rope: a new model based on *O*-galactosylhydroxyproline as the polysaccharide attachment site. *Plant Physiol.*, 96, 848–855.

Qi, X., Behrens, B.X., West, P.R. and Mort, A.J. (1995) Solubilization and partial characterization of extensin fragments from cell walls of cotton suspension cultures. Evidence for a covalent cross-link between extensin and pectin. *Plant Physiol.*, 108, 1691–1701.

Ringli, C., Hauf, G. and Keller, B. (2001) Hydrophobic interactions of the structural protein GRP1.8 in the cell wall of protoxylem elements. *Plant Physiol.*, 125, 673–682.

Rodríguez-Concepción, M., Pérez-García, A. and Beltrán, J.P. (2001) Up-regulation of genes encoding novel extracellular proteins during fruit set in pea. *Plant Mol. Biol.*, 46, 373–382.

Rubinstein, A.L., Broadwater, A.H., Lowrey, K.B. and Bedinger, P.A. (1995) *Pex1*, a pollen-specific gene with an extensin-like domain. *Proc. Natl Acad. Sci. USA*, 92, 3086–3090.

Ryser, U., Schorderet, M., Zhao, G.-F. *et al.* (1997) Structural cell-wall proteins in protoxylem development: evidence for a repair process mediated by a glycine-rich protein. *Plant J.,* 12, 97–111.

Sachetto-Martins, G., Fernandes, L.D., Felix, D. and de Oliveira, D.E. (1995) Preferential transcriptional activity of a glycine-rich protein gene from *Arabidopsis thaliana* in protoderm-derived cells. *Int. J. Plant Sci.,* 156, 460–470.

Sachetto-Martins, G., Franco, L.O. and de Oliveira, D.E. (2000) Plant glycine-rich proteins: a family or just proteins with a common motif? *Biochim. Biophys. Acta,* 1492, 1–14.

Salts, V., Wachs, R., Gruissem, W. and Barg, R. (1991) Sequence coding for a novel proline-rich protein preferentially expressed in young tomato fruit. *Plant Mol. Biol.,* 17, 149–150.

Santino, C.G., Stanford, G.L. and Connor, T.W. (1997) Developmental and transgenic analysis of two tomato fruit enhanced genes. *Plant Mol. Biol.,* 33, 405–416.

Sasaki, T., Matsumoto, T. and Yamamoto, K. (unpublished) *Oryza sativa* nipponbare (GA3) genomic DNA, chromosome 1, PAC clone: P0666G04.

Savenstrand, H., Brosche, M. and Strid, A. (unpublished) Stress-induced extensin cDNA from *Pisum sativum.*

Scheres, B., van de Wiel, C., Zalensky, A. *et al.* (1990) The ENOD12 gene product is involved in the infection process during the pea-rhizobium interaction. *Cell,* 60, 281–294.

Schnabelrauch, L.S., Kieliszewski, M.J., Upham, B.L., Alizedeh, H. and Lamport, D.T.A. (1996) Isolation of pI 4.6 extensin peroxidase from tomato cell suspension cultures and identification of Val-Tyr-Lys as putative intermolecular cross-link site. *Plant J.,* 9, 477–489.

Schultz, C.J., Hauser, K., Lind, J.L. *et al.* (1997) Molecular characterisation of a cDNA sequence encoding the backbone of a style-specific 120kDa glycoprotein which has features of both extensins and arabinogalactan-proteins. *Plant Mol. Biol.,* 35, 833–845.

Schultz, C.J., Gilson, P., Oxley, D., Youl, J.J. and Bacic, A. (1998) GPI-anchors on arabinogalactan-proteins: implications for signalling in plants. *Trends Plant Sci.,* 3, 426–431.

Schultz, C.J., Johnson, K., Currie, G. and Bacic, A. (2000) The classical arabinogalactan protein gene family of *Arabidopsis. Plant Cell,* 12, 1–18.

Schultz, C.J., Rumsewicz, M.P., Johnson, K.L., Jones, B.J., Gaspar, Y.M. and Bacic, A. (2002) Using genomics resources to guide research directions: the arabinogalactan-protein gene family as a test case. *Plant Physiol.,* 129, 1448–1463.

Selleck, S.B. (2000) Proteoglycans and pattern formation: sugar biochemistry meets developmental genetics. *Trends Genet.,* 16, 206–212.

Seppo, A. and Tiemeyer, M. (2000) Function and structure of *Drosophila* glycans. *Glycobiol.,* 10, 751–760.

Serpe, M.D. and Nothnagel, E.A. (1999) Arabinogalactan-proteins in the multiple domains of the plant cell surface. *Adv. Botan. Res.,* 30, 207–289.

Sheng, J., D'Ovidio, R. and Mehdy, M.C. (1991) Negative and positive regulation of a novel proline-rich protein mRNA by fungal elicitor and wounding. *Plant J.,* 1, 345–354.

Sherrier, D.J., Prime, T.A. and Dupree, P. (1999) Glycosylphosphatidylinositol-anchored cell-surface proteins from *Arabidopsis. Electrophoresis,* 20, 2027–2035.

Shi, H., Kim, Y., Guo, Y., Stevenson, B. and Zhu, J.-K. (2003) The Arabidopsis SOS5 locus encodes a putative cell surface adhesion protein and is required for normal cell expansion. *Plant Cell,* 15, 19–32.

Showalter, A.M. (1993) Structure and function of plant cell wall proteins. *Plant Cell,* 5, 9–23.

Showalter, A.M. (2001) Arabinogalactan-proteins: structure, expression and function. *Cell. Mol. Life Sci.,* 58, 1399–1417.

Showalter, A.M., Zhou, J., Rumeau, D., Worst, S.G. and Varner, J.E. (1991) Tomato extensin and extensin-like cDNAs: structure and expression in response to wounding. *Plant Mol. Biol.,* 16, 547–565.

Shpak, E., Leykam, J.F. and Kieliszewski, M.J. (1999) Synthetic genes for glycoprotein design and the elucidation of hydroxyproline-O-glycosylation codes. Proc. Natl Acad. Sci. USA, 96, 14736–14741.

Shpak, E., Barbar, E., Leykam, J.F. and Kieliszewski, M.J. (2001) Contiguous hydroxyproline residues direct hydroxyproline arabinosylation in Nicotiana tabacum. J. Biol. Chem., 276, 11272–11278.

Smart, L.B., Cameron, K.D. and Bennett, A.B. (2000) Isolation of genes predominantly expressed in guard cells and epidermal cells of Nicotiana glauca. Plant Mol. Biol., 42, 857–869.

Sommer-Knudsen, J., Clarke, A.E. and Bacic, A. (1996) A galactose-rich, cell-wall glycoprotein from styles of Nicotiana alata. Plant J., 9, 71–83.

Sommer-Knudsen, J., Bacic, A. and Clarke, A.E. (1998) Hydroxyproline-rich plant glycoproteins. Phytochem., 47, 483–497.

Stacey, N.J., Roberts, K., Carpita, N.C., Wells, B. and McCann, M.C. (1995) Dynamic changes in cell surface molecules are very early events in the differentiation of mesophyll cells from Zinnia elegans into tracheary elements. Plant J., 8, 891–906.

Steinert, P.M., Mack, J.W., Korge, B.P., Gan, S.Q., Haynes, S.R. and Steven, A.C. (1991) Glycine loops in proteins, their occurrence in certain intermediate filament chains loricrins and single-stranded RNA binding proteins. International J. Biol. Macromolecules, 13, 130–139.

Stiefel, V., Ruiz-Avila, L., Raz, R. et al. (1990) Expression of a maize cell wall hydroxyproline-rich glycoprotein gene in early leaf and root vascular differentiation. Plant Cell, 2, 785–793.

Stratford, S., Barnes, W., Hohorst, D.L. et al. (2001) A leucine-rich repeat region is conserved in pollen extensin-like (Pex) proteins in monocots and dicots. Plant Mol. Biol., 46, 43–56.

Sturaro, M., Linnestad, C., Kleinhofs, A., Olsen, O.-A. and Doan, D.N.P. (1998) Characterization of a cDNA encoding a putative extensin from developing barley grains (Hordeum vulgare L.). J. Exper. Bot., 49, 1935–1944.

Subramaniam, K., Ranie, J., Srinivasa, B.R., Sinha, A.M. and Mahadevan, S. (1994) Cloning and sequence of a cDNA encoding a novel hybrid proline-rich protein associated with cytokinin-induced haustoria formation in Cuscuta reflexa. Gene, 141, 207–210.

Svetek, J., Yadav, M.P. and Nothnagel, E.A. (1999) Presence of a glycosylphosphatidylinositol lipid anchor on rose arabinogalactan proteins. J. Biol. Chem., 274, 14724–14733.

Takos, A.M., Dry, I.B. and Soole, K.L. (2000) Glycosyl-phosphatidylinositol-anchor addition signals are processed in Nicotiana tabacum. Plant J., 21, 43–52.

Tanaka, M., Shibata, H. and Uchida, T. (1980) A new prolyl hydroxylase acting on poly-L-proline from suspension cultured cells of Vinca rosea. Biochim. Biophys. Acta, 616, 188–198.

Ten Hagen, K.G., Bedi, G.S., Tetaert, D. et al. (2001) Cloning and characterization of a ninth member of the UDP-GalNAc:polypeptide N-Acetylgalactosaminyltransferase family, pp-GaNTase-T9. J. Biol. Chem., 276, 17395–17404.

Thompson, G.A. and Okuyama, H. (2000) Lipid-linked proteins of plants. Progr. Lipid Res., 39, 19–39.

Tierney, M.L., Wiechert, J. and Pluymers, D. (1988) Analysis of the expression of extensin and p33-related cell wall proteins in carrot and soybean. Mol. General Genet., 211, 393–399.

Toonen, M.A.J., Schmidt, E.D.L., Vankammen, A. and Devries, S.C. (1997) Promotive and inhibitory effects of diverse arabinogalactan proteins on Daucus carota L somatic embryogenesis. Planta, 203, 188–195.

Udenfriend, S. and Kodukula, K. (1995) How glycosyl-phosphatidylinositol-anchored membrane proteins are made. Annu. Rev. Biochem., 64, 563–591.

Vignols, F., José-Estanyol, M., Caparrós-Ruiz, D., Rigau, J. and Puigdomènech, P. (1999) Involvement of maize proline-rich protein in secondary cell wall formation as deduced from its specific mRNA localization. Plant Mol. Biol., 39, 945–952.

Vuorela, A., Myllyharju, J., Nissi, R., Pihlajaniemi, T. and Kivirikko, K.I. (1997) Assembly of human prolyl 4-hydroxylase and type III collagen in the yeast Pichia pastoris: formation of a stable enzyme tetramer requires coexpression with collagen and assembly of a stable collagen requires coexpression with prolyl 4-hydroxylase. *EMBO J.,* 16, 6702–6712.

Waite, J.H. (1990) The phylogeny and chemical diversity of quinone-tanned glues and varnishes. *Compar. Biochem. Physiol.,* 97B, 19–29.

Willats, W.G.T., Steele-King, C.G., McCartney, L., Orfila, C., Marcus, S.E. and Knox, J.P. (2000) Making and using antibody probes to study plant cell walls. *Plant Physiol. Biochem.,* 38, 27–36.

Wilson, R.C. and Cooper, J.B. (1994) Characterization of PRP1 and PRP2 from *Medicago truca-tula. Plant Physiol.,* 105, 445–446.

Wisman, E. and Ohlrogge, J. (2000) *Arabidopsis* microarray service facilities. *Plant Physiol.,* 124, 1468–1471.

Woessner, J.P., Molendijk, A.J., Egmond, P.V., Klis, F.M., Goodenough, U.W. and Haring, M.A. (1994) Domain conservation in several volvocalean cell wall proteins. *Plant Mol. Biol.,* 26, 947–960.

Wojtaszek, P., Trethowan, J. and Bolwell, G.P. (1997) Reconstitution in vitro of the components and conditions required for the oxidative cross-linking of extracellular proteins in French bean (*Phaseolus vulgaris* L.). *FEBS Lett.,* 405, 95–98.

Wojtaszek, P., Smith, C.G. and Bolwell, G.P. (1999) Ultrastructural localisation and further bio-chemical characterisation of prolyl 4-hydroxylase from *Phaseolus vulgaris*: comparative analysis. *Int. J. Biochem. Cell Biol.,* 31, 463–477.

Wu, H.-M., Zou, J., May, B., Gu, Q. and Cheung, A.Y. (1993) A tobacco gene family for flower cell wall proteins with a proline-rich domain and a cysteine-rich domain. *Proc. Natl. Acad. Sci. USA,* 90, 6829–6833.

Wu, H.-M., Wong, E., Ogdahl, J. and Cheung, A.Y. (2000) A pollen tube growth-promoting ara-binogalactan protein from *Nicotiana alata* is similar to the tobacco TTS protein. *Plant J.,* 22, 165–176.

Wu, H., de Graaf, B., Mariani, C. and Cheung, A.Y. (2001) Hydroxyproline-rich glycoproteins in plant reproductive tissues: structure, functions and regulation. *Cell.Mol. Life Sci.,* 58, 1418–1429.

Wycoff, K.L., Powell, P.A., Gonzales, R.A., Corbin, D.R., Lamb, C. and Dixon, R.A. (1995) Stress activation of a bean hydroxyproline-rich glycoprotein promoter is superimposed on a pattern of tissue-specific developmental expression. *Plant Physiol.,* 109, 41–52.

Yariv, J., Lis, H. and Katchalski, E. (1967) Precipitation of arabic acid and some seed polysac-charides by glycosylphenylazo dyes. *Biochem. J.,* 105, 1c-2c.

Yoshiba, Y., Aoki, C., Iuchi, S. *et al.* (2001) Characterization of four extensin genes in *Arabidopsis thaliana* by differential gene expression under stress and non-stress conditions. *DNA Res.,* 8, 115–122.

Youl, J.J., Bacic, A. and Oxley, D. (1998) Arabinogalactan-proteins from *Nicotiana alata* and *Pyrus communis* contain glycosylphosphatidylinositol membrane anchors. *Proc. Natl. Acad. Sci. USA,* 95, 7921–7926.

Zablackis, E., Huang, J., Muller, B., Darvill, A.G. and Albersheim, P. (1995) Characteriza-tion of the cell-wall polysaccharides of *Arabidopsis thaliana* leaves. *Plant Physiol.,* 107, 1129–1138.

Zhou, J., Rumeau, D. and Showalter, A.M. (1992) Isolation and characterization of two wound-regulated tomato extensin genes. *Plant Mol. Biol.,* 20, 5–17.

Zhou, Z.D., Tan, L., Showalter, A.M., Lamport, D.T.A. and Kieliszewski, M.J. (2002) Tomato LeAGP-1 arabinogalactan-protein purified from transgenic tobacco corroborates the Hyp contiguity hypothesis. *Plant J.,* 31, 431–444.

Zinn, K., McAllister, L. and Goodman, C.S. (1988) Sequence analysis and neuronal expression of fasciclin I in grasshopper and *Drosophila. Cell,* 53, 577–587.

5 Towards an understanding of the supramolecular organization of the lignified wall

Alain-M. Boudet

5.1 Introduction

Primary plant cell walls are complex materials essentially made of polysaccharides. They can be described for example by the type I model of Carpita and Gibeaut (1993) as being composed of three networks: a cellulose/xyloglucan network (>50% dry weight) embedded in a pectin matrix (25–40% dry weight) locked into shape by cross-linked glycoproteins (extensin) (1–10% dry weight). The supramolecular arrangement of polymers in the cell walls implies steric constraints so that favoured conformations may be different from those in solution. This primary cell wall is a porous structure with hydrophilic/hydrophobic domains. Weak bonds, H bonds and hydrophobic interactions seem to be driving forces for some of the associations of wall constituents. Covalent linkages are also involved for stronger associations of wall components.

After cell growth has ceased, hydroxycinnamic acid-mediated cross-linking may occur between cell wall constituents and, after a secondary wall has been formed, the walls are often reinforced by the deposition of other polymers such as lignins and suberin. These changes lead to maximum wall strength and rigidity for the plant body and, in addition, the hydrophobicity of lignin waterproofs the conducting cells of the xylem. Further 'decoration' of the wall may involve the deposition of lower molecular weight components, such as phenolic acids, flavonoids, tannins, stilbenes and lignans, which are important during the last stages of wall differentiation, for example, in the heartwood of many tree species (Beritognolo *et al.*, 2002). All these processes which increase the mechanical resistance of the walls reduce the susceptibility of the plant to abiotic and biotic stress factors (Nicholson and Hammerschmidt, 1992). These complex forms of differentiated cell walls and particularly the lignified walls represent the major proportion of the plant biomass and an immense reservoir of carbon in the form of lignocelluloses. Various data are available to demonstrate that the chemical composition of lignocellulosics is variable depending on the species, the developmental stage and the environmental conditions. This gross composition has a dramatic impact on the technological value of raw materials, including wood fibres and forage material, but also fruits and vegetables, and numerous strategies have been and are being developed to optimize the composition of plant cell walls for different agro-industrial purposes.

These specific structural and chemical wall patterns result from the coordinated expression of numerous genes involved in the biosynthesis, assembly and deposition of the different cell wall constituents (Plomion *et al.*, 2001) and, at the moment, large-scale projects involving cDNA sequencing of woody species are being made to identify genes that are related to the secondary wall formation and wood biosynthesis (Allona *et al.*, 1998; Sterky *et al.*, 1998). In addition, model systems such as *Arabidopsis* or *Zinnia* are exploited for similar purposes. Finally transcript profiling using cDNA microarrays (Hertzberg *et al.*, 2001) and proteomics are being increasingly applied to track the genes/proteins of interest in the differentiation of cell walls and the advances made will provide new targets for future interventions (see Chapter 10).

However, cell walls are composite materials resulting from the assembly of different polymers. In addition to the basic chemical composition, the nature of the interactions between wall constituents is undoubtedly a crucial parameter for determining the wall properties '*in planta*' and the technological characteristics of plant products. This aspect is becoming particularly important since we are entering an era where plants will be frequently considered as sources of renewable carbon in the context of sustainable development. Consequently, wall polymers are likely to be increasingly used starting from conventional resources or from redesigned plant materials with optimized cell wall composition. In this way, a better knowledge of the supramolecular organization of the plant cell wall, and particularly of the lignified plant wall – the major carbon sink in the biosphere, will become necessary for optimized breeding, genetic manipulation and processing of plants and plant products.

Such knowledge is still in its infancy and various reviews (Iiyama *et al.*, 1994; Burlat *et al.*, 2001) have emphasized that our understanding of how cell wall macromolecular assembly is controlled is very limited. Such complex problems require interdisciplinary approaches and integrated processing of results from different sources. In this review, we attempt to provide the most recent information available.

5.2   The dynamics of lignification: chemical and ultrastructural aspects

Lignins result from the polymerization of phenoxy radicals that are essentially derived from three hydroxycinnamyl alcohols, termed the monolignols. Coupling these radicals with monomeric or oligomeric molecules builds up the lignin polymer by a non-enzymatic process involving at least eleven kinds of intermonomeric linkages. Lignin polymers, which are deposited within the carbohydrate matrix of the cell wall, represent about a third of the terrestrial biomass. Typically the lignified cell wall consists of a thin primary layer, a thicker multilamellar secondary layer and sometimes a tertiary layer. The secondary wall layer is rich in cellulose and the non-cellulose polysaccharides are qualitatively different from those of the primary wall.

Although the major structural features of lignin are reasonably well understood, its chemical composition, unlike that of other cell wall polymers, still cannot be precisely defined. One major problem is the lack of a selective procedure for the quantitative isolation of lignin in a pure and unaltered form. Destructive analyses based on strong chemical treatments of the firmly wall-bound lignin polymers do not quantitatively convert native lignins into monomeric or oligomeric fragments, nor do they provide information on the three-dimensional structure of lignin. Moreover, *in vitro* formation of dehydrogenative polymers (DHPs) from monolignols, which do provide useful lignin structural models, does not reproduce the actual polymerization conditions in the cell wall, including the effect of preformed polysaccharides on lignification. Various non-destructive methods (reviewed in Terashima *et al.*, 1998) may give a more general view of native lignins *in situ*. We will report here selected features which seem important within the scope of this review, bearing in mind that the various assumptions mentioned essentially constitute, for the time being, working hypotheses which may well be modified in the future. Non-destructive methods have revealed that the structure of lignin is heterogeneous at cell and subcellular levels with respect to its monomer composition, inter-unit linkages and also, as discussed later, with its association with polysaccharides. In addition, lignin deposition does not occur randomly in the cell wall but chronologically from the external part of the wall to the inside of the cell. These aspects have been described in detail by Terashima *et al.* (1998) and Donaldson (2001) in two excellent reviews.

The first stage of lignification usually occurs at the cell corners and cell-corner/compound middle lamella regions in a pectic-substance-rich environment. In layers S_2 and S_3 the main lignification process develops slowly in a xylan or mannan matrix embedding cellulose microfibrils. The lignin deposited at the cell corners and in the compound middle lamella during the early stages is enriched in *p*-hydroxyphenyl (H) and guaiacyl (G) units and may be of a highly condensed type. There is a less lignified secondary wall containing guaiacyl units in gymnosperms or a mixture of guaiacyl and syringyl units in angiosperms. In angiosperm lignins, depending on the cellular function of the lignified elements, the lignin patterns may differ. Vessels involved in water conduction exhibit a highly lignified secondary wall containing guaiacyl units (Fergus and Goring, 1970) and fibres with structural functions are characterized by a less highly lignified secondary wall containing syringyl units.

Phenolic hydroxyl groups represent among the most reactive sites of lignins (Lai *et al.*, 1999). In softwoods, the lignin in the secondary wall generally contains higher levels of phenolic hydroxyl groups and fewer condensed units. It is also clear that very few syringyl units in the hardwood lignin are present as reactive free phenolic structures.

The content and composition of lignins in wood are also influenced by its age and growth conditions as reflected by significant differences between juvenile wood and mature wood, earlywood and latewood, and normal wood and reaction wood (Lai *et al.*, 1999).

The distribution of various inter-unit linkages during the building of lignin macromolecules has been reported by Terashima *et al.* (1998). The first products of the coupling of phenoxy radicals are β-O-4′, β-5′ and β-β′ dimers; and in the next stage of polymerization bulk-type oligomers are formed. As the size of the polymerization

product increases, endwise polymerization becomes prevalent, due to steric parameters and monolignol availability. As a result, the globular lignin macromolecule is
composed of a bulk polymer inside and an endwise polymer in the outer part. In the
last stages of lignin assembly, new linkages (particularly 5–5′ between β-O-4′ type
endwise chains) may be formed between globular polylignols. Beyond the discovery
of new original structures in lignins such as dibenzodioxocines, Brunow's group has
suggested a role for oxidation potentials and local concentrations of monolignols in
the polymerization process. They have shown that cross-coupling between lignin
precursors and the lignin polymer occurs only within a restricted range of oxidation
potentials of the phenols and should not be regarded as a random process (Syrjänen
and Brunow, 1998). The same group (Brunow *et al.*, 1998) has also emphasized the
role of local concentrations of coniferyl alcohol at the sites of lignification in the
control of lignin structure.

These studies point out the importance of radical cross-coupling conditions in
the lignification process. At this stage, it is still difficult to conclude if the process is
mostly under chemical control (oxidation potentials, local concentrations of monolignols, monolignol availability, etc.) or significantly regulated by template effects
or protein-dependent mechanisms. Future studies will determine if the polymer
represents a product of combinatorial-type chemistry with no regularly repeating
macro structures or if there are 'limits to ramdomness' in the organization of cell
wall lignin.

5.3   Interactions and cross-linking between non-lignin components of the cell wall

Lignins are deposited within a pre-formed polysaccharide network where interactions and chemical linkages are established between wall constituents. This section
provides an overview of the relevant knowledge in this area.

There is good evidence that the cellulose microfibril surfaces are coated with non-
cellulosic polysaccharides hydrogen bonded to cellulose which, possibly depending
on their length, may act as an adhesive between microfibrils (see Chapter 1). The
strength of heterogeneous fibres such as wood fibres is indeed largely assumed to
be determined by the strength of association between hemicelluloses and cellulose.
Hemicelluloses such as xylan, glucomannan and xyloglucan bind tightly to cellulose in specific and non-specific patterns, the driving force behind the interactions
being hydrogen bonds. Some of these interactions are highly specific, setting high
demands on structural complementarity, but it seems that the binding is mainly
non-specific. Cello-, manno- and xylo-oligosaccharides all interact differently with
the cellulose surface and tetrasaccharides were found to be the shortest cellulose-
binding oligosaccharides. During pulping, the structures of xylan and mannan are
modified and thus interactions between hemicelluloses and cellulose are different
from those in native wood. Preliminary results suggest that adsorbed xylan may
make cellulose more flexible, which may further increase the strength of the paper.

Other potential linkages occurring between polysaccharides have been reported (Iiyama *et al.*, 1994): glycosidic linkages of polysaccharide chains, ester linkages between carboxyl groups of uronic acid residues and hydroxyls on neighbouring polysaccharides (see Chapter 1). As an example, Thompson and Fry (2000) have recently reported covalent linkage between xyloglucan and acidic pectins in suspension-cultured rose cells.

Cell wall structural proteins can associate with themselves and with other cell wall constituents, as reported in more detail in section 5.5 and in Chapter 4. As an example, Saulnier *et al.* (1995) have demonstrated that protein-polysaccharide linkages might be the main cause of insolubility of maize bran heteroxylans. In this case, the actual nature of the protein-polysaccharide linkages is not known.

Cross-linking through hydroxycinnamic acids is an important mechanism which may already occur at the primary wall stage in the commelinoid orders of monocotyledons and secondarily in Chenopodiaceae. Ferulic acid plays a major role in these cross-linking phenomena even though coumaric and sinapic acids have also been found to be involved. Ferulic acid is found esterified to various wall constituents and the formation of dimers (diferulic bridges) through oxidative reactions enables covalent inter-molecular cross-linking between hemicelluloses (Saulnier *et al.*, 1999), hemicelluloses and lignins (Jacquet *et al.*, 1995) and potentially between proteins and lignins (Fry, 1986). This cross-linking decreases cell wall extensibility (Wakabayashi *et al.*, 1997) and contributes to its mesh-like network (Iiyama *et al.*, 1994), potentially affecting the digestibility and mechanical properties of plant tissues (Kamisaka *et al.*, 1990). Ralph *et al.* (2000) have characterized up to 9 different ferulate dimers which are now routinely being found in a variety of samples and they also suggested the occurrence of disinapate bridges in wild rice.

Dehydrodiferulates have been identified and quantified in various plant tissues but the isolation of dehydrodiferulates linked to neutral sugars has only been reported in a limited number of cases: bamboo shoots (Ishii, 1991) and maize-bran (Saulnier *et al.*, 1999). In this last material for example, dehydrodiferuloylated oligosaccharides were isolated and linkages between 5–5′ diferulate and arabinofuranoside moieties were clearly demonstrated. The same group (Saulnier *et al.*, 1999) has calculated that, on average, each heteroxylan molecule is cross-linked through about 15 dehydrodiferulate bridges.

Arabinoxylan has been suggested to be feruloylated intra-protoplasmically (Myton and Fry, 1994) or *in muro* (Yamamoto and Towers, 1985). Obel *et al.* (personal communication) have proposed that feruloylCoA, the most likely substrate for feruloylation, is a precursor of a feruloylated protein which acts as an intermediate in feruloylation of arabinoxylan in wheat. Many proteins are known to contain arabinose, some of which are cell wall proteins and could be candidates for carrying the ferulic acid residue intra-or extracellularly. In this context, a further requirement would be the presence of trans-glucosidases transferring ferulic acid from proteins to arabinoxylan. However, there is also evidence for the involvement of feruloylglucose in the feruloylation of arabinoxylan. The intra-protoplasmic ferulic acid dimer formation observed by Fry *et al.* (2000) seems to be confined to specific dimers.

According to Obel *et al.* (personal communication), on the basis of kinetic experiments, 8–5´-diferulic acid bridges are formed intracellularly, in contrast to the 5–5´ dimer and 8-O-4´ dimer, which are formed extracellularly.

Beyond the classical occurrence of hydroxycinnamate dimers, Fry (2000) has recently demonstrated in the primary wall of cultured maize and spinach cells that trimers and larger products may make the largest contribution to ferulate coupling and therefore possibly to wall tightening in cultured cells. The discovery that trimers and larger oligomers are the predominant coupling products of feruloyl-arabinoxylans argues in favour of inter-polysaccharide cross-links rather than intra-polysaccharide loops.

Cross-coupling of hydroxycinnamates is assumed to affect cell-cell adhesion since dehydrodiferulates are often concentrated in cell-cell junctions. These covalent bridges may thus influence the commercial and agronomic value of plant products including, for example, wheat grains processing, forage digestibility and doughing in bread making. In addition, diferulates have recently been the subject of an increasing interest as potential protective nutraceutical antioxidants.

One potential means to probe the putative roles of dehydrodiferulates, and their industrial applications, is the characterization of feruloyl-esterases which may hydrolyse phenolic cross-links. Such enzymes have been described in different microorganisms (Garcia-Conesa *et al.*, 1999) and recently in plants (Sancho *et al.*, 1999). Such esterases will be useful tools in plant cell wall research and could provide solutions to existing problems and new opportunities in the agri-food industry (Kroon, 2000). The use of specific enzymes for modelling cell wall composition and structure, either *in planta* or during commercial processes, is indeed a promising area. As an example, Vincken *et al.* (personal communication) have significantly modified the sugar composition of potato walls following the wall-targeted ectopic expression of a rhamnogalacturonan lyase from *Aspergillus aculeatus*.

5.4   Integration of lignins in the extracellular matrix

5.4.1 Ultrastructural aspects

Fujino and Itoh (1998) have studied the lignification process in *Eucalyptus tereticornis Sm.* using rapid-freeze deep-etching electron microscopy to provide a three-dimensional perspective of lignin deposition. This study demonstrates the loss of wall porosity as lignin deposition proceeds, illustrating the likely reasons for differences in lignin concentration among wall regions in mature wood. The highly porous carbohydrate matrix in the middle lamella and primary wall are thought to allow greater lignin deposition by space filling than the more densely packed secondary cell wall.

More recent additional experiments (Itoh, 2002) on lignifying and lignified cell walls of *Pinus thunbergii* wood fibres have confirmed the role of lignins in reducing the porosity of the cell wall. In unlignified S_2 walls, the cellulose microfibrils are unidirectional and many pores and cross-linking are visible (Figure 5.1a). When fully lignified, the S$_2$ wall is highly compacted and the encrustation of lignin seems to seal the 'pore system' in the cell structure, which is difficult to distinguish in the xylem (Figure 5.1b). The microfibril diameter in the lignified wall is higher than in unlignified walls. These interpretations are confirmed by experiments involving extensive delignification of the walls by sodium chlorite. The general appearance of these treated cell walls is similar to that of unlignified walls with slit-like pores, cross-links between microfibrils and reduced diameters of microfibrils (~7.7 nm) (Figure 5.1c).

It is concluded that the unlignified secondary wall is a highly porous structure, the pores being involved in the transport of water and solutes in the cell wall. This porosity pattern is dramatically altered in lignified walls in agreement with the water-proofing and compaction-associated properties of lignins.

5.4.2 Interactions and potential linkages with polysaccharides

Lignins are deposited in a preformed network of polysaccharides which may play an important role as template for the formation of the lignin macromolecule, since the polysaccharides and lignins are thought to be bound by both covalent and non-covalent interactions to form a lignin-polysaccharide complex (Freudenberg, 1968; Sarkanen, 1998). As underlined by Terashima et al. (1998), lignin deposition has not been observed in any location without prior deposition of polysaccharides and the structure of carbohydrate-free lignin released from suspension-cultured cells is different from that of typical lignins.

Lignification is influenced by the carbohydrate matrix in which it occurs (Donaldson, 1994; Salmen and Olsson, 1998) and the carbohydrate components of the cell wall exert a mechanical influence on the expansion of lignin, causing the formation of either spherical (middle lamella and primary wall) or elongated (secondary wall) structures (Donaldson, 1994). Using a Raman microprobe, Atalla and Agarwal (1985) and Houtman and Atalla (1995) have shown that the aromatic rings of lignin are often oriented within the plane of the cell as a result of the mechanical or chemical influence of the carbohydrate wall components.

Different types of cross-linking and interactions have been proposed between lignin and polysaccharides and have been discussed in detail by Iiyama et al. (1994).

Briefly, they include:

1. a link between uronic acids and hydroxyl groups on lignin surfaces to give α or γ esters on monolignol side chains;
2. a direct ether linkage between polysaccharides and lignin; and
3. hydroxycinnamic acid mediated linkages between hydroxycinnamoyl residues on polysaccharides and hydroxycinnamic acids directly esterified or etherified to lignin surfaces.

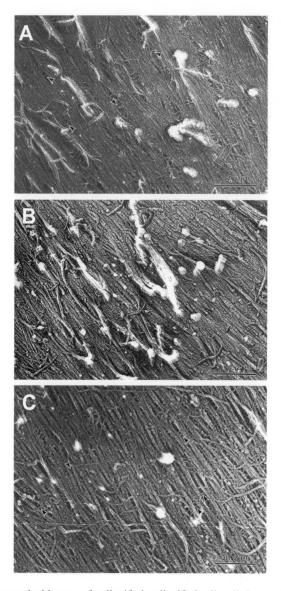

Figure 5.1 Deep-etched images of unlignified or lignified cell walls in wood fibres of *Pinus thunbergii*. (a) Unlignified S$_2$ wall in the differentiating wood fibres. A dense structure with scattered oval pores and cross-links is shown (cross-links are indicated by arrowheads). (b) Fully lignified S$_2$ wall in the mature wood fibre. Neither pores nor cross-links are visible. The deep-etched image shows a radial view of the wood sample. (c) Delignified wall (corresponding to the material shown in (b). Cross-links are visible between the microfibrils (indicated by arrowheads) and pores appear as gaps. In all three pictures a scale bar is shown representing 0.2 μm. Reproduced with permission from Dr Itoh, from 'Deep-etching electron microscopy and 3-dimensional cell wall architecture' in *Wood Formation in Trees*, published by Taylor and Francis, 2002.

Furthermore, hydrogen bonds may also be involved and, beyond chemical linkages, physical cross-links (entanglement) could also potentially associate lignins and other cell wall components.

The occurrence of some of these linkages has been confirmed through chemical analyses of wall components (Scalbert *et al.*, 1985) or indirectly through the incorporation of feruloylarabinose ester into a synthetic lignin (DHP) (Ralph *et al.*, 1992).

Phenolic acids (coumaric and ferulic acids) esterified with carbohydrates might act as lignin anchors by their participation in polymerization of the monolignols (Jacquet *et al.*, 1995; Ralph *et al.*, 1995). These last authors have demonstrated that ferulates in the wall seem to only be coupling with lignin monomers and not pre-formed lignin oligomers, and therefore would appear to be sites where lignification is initiated. In maize bran, typical lignin structures were found to be tightly associated to the alkali-extracted heteroxylans (Lapierre *et al.*, 2001). In a critical analysis of their results, the authors point out that, at this stage, it is not known whether these linkages exist *ab initio* and/or are formed during the alkaline extraction of heteroxylan. However, their results argue for the occurrence of covalent linkages between heteroxylan chains and lignin and that lignin acts in the organization of polysaccharide networks.

Analysis of residual lignins in the pulps resulting from chemical pulp manufacturing is a means to identify the chemical linkages between lignins and polysaccharides which resist the dissolution procedure (Tamminen and Hortling, 1999). The most stable bonds are considered to be the α-ether bonds (Gierer and Wännstrom, 1986). Resistant bonds between lignin and galactose originating from pectic substances (Minor, 1991), and also between lignins and glucose through both glycosidic and ether-type bonds (suggesting linkages with cellulose) (Kosikova and Ebringerova, 1994), have been also characterized in pulp. However, one problem with the characterization of lignin-carbohydrate linkages in residual lignins isolated from pulp is that these bonds may be formed during the pulping processes and may not occur in native lignins.

It is very likely that the supramolecular organization of the lignified cell wall differs with the morphological regions within the cell wall, the type of cell and the plant species. It is well known that different kinds of polysaccharides are deposited in each cell wall layer being associated with the cellulose microfibrils, which are orientated in different ways. For example, using a specific polyclonal antibody against glucomannans, the most abundant hemicelluloses in softwoods, Maeda *et al.* (2000) have shown that labelling was restricted to the secondary wall of differentiating and differentiated *Chamaecyparis obtusa* tracheids, and did not detect glucomannans in their primary cell wall or compound middle lamella. As the labelling decreases in the outer and middle layers of the secondary wall during cell wall formation, the authors suggest that the epitopes of glucomannans could be masked by the deposition of lignins as the cell wall differentiate.

Wall polysaccharides are proposed to act as a templates for macromolecular lignin assembly and hence affect lignin molecular weight and higher order structure. It was proposed by Terashima *et al.* (1998) that the compound middle lamella lignin is a thick mass with a high molecular weight, while the secondary wall lignin, associated intimately with hemicelluloses, is a thin film which surrounds the cellulose microfibrils like a twisted honeycomb. The thin film may contain 4–7 layers of monolignol units associated with hemicelluloses through covalent and non-covalent interactions.

Additional hypotheses result from the synthesis of artificial lignins (dehydrogenation polymers; DHP) which lead to polymerization products of lower masses than those of the natural extracted lignins. These results suggest a potential role for other cell wall components in controlling the lignin polymer molecular weight. Higuchi *et al.* (1971) reported that the preparation of coniferyl alcohol DHP in the presence of hemicelluloses and pectin increased their molecular weight. Furthermore, Grabber *et al.* (1996) have performed polymerization of DHPs into an actual cell wall matrix. When this is done, the resulting lignins are very similar to those in the native plant material as determined mainly by thioacidolysis analysis. More recently, Cathala and Monties (2001) and Cathala *et al.* (2001) polymerized coniferyl alcohol in the presence of pectin. A pectin-DHP complex was isolated in which DHP was covalently linked by ester bonds to the pectin. This lignin carbohydrate complex (LCC) was only formed following the slow addition of coniferyl alcohol to the pectin solution and DHPs' solubility increases with the pectin concentration, since the LCC acts as a surfactant molecule to keep the unbound DHPs in solution by the formation of aggregate or micelle-like structures. These data illustrate the potential the microenvironment has to affect lignin synthesis.

In addition, it was demonstrated (Cathala *et al.*, 2001) that the complex exhibited a lower hydrophilicity than the individual components, providing a possible mechanism for the exclusion of water associated with lignification in the outer part of the cell wall (Terashima *et al.*, 1993). Indeed, different observations support the view that lignification proceeds from the outer to the inner parts of differentiating cell walls (Terashima *et al.*, 1993), with a corresponding displacement of water from the hydrophobic lignified areas to the still hydrophilic water-swollen polysaccharide gel in the unlignified inner part of the wall. This removal of water results in the shrinkage of the cell wall and may have a significant effect on the mechanical and permeability properties of the cell wall.

5.5 New insights gained from analysis of transgenic plants and cell wall mutants

Tobacco and poplar have been used as model systems for the antisense and sense suppression of a considerable number of genes that regulate individual steps in phenylpropanoid and monolignol biosynthesis, with varying consequences for lignin composition and abundance. These approaches have essentially induced three main effects:

- quantitative reductions in the levels of lignins;
- qualitative effects on the monomers incorporated without an overall reduction in total phenolic content; and
- cumulative changes both in content and composition of lignins.

In addition to their interest in biotechnological applications and in contributing to a reappraisal of the fundamental understanding of the lignification pathway, these studies have shown both that lignification is a very flexible process in term of chemical composition and also that lignin deposition within the wall is a highly organized process. Particularly, they have confirmed the differential spatial deposition of different types of lignins within different cell types and between the different layers of the wall.

5.5.1   Tobacco lines down-regulated for enzymes of monolignol synthesis

In tobacco stems, the down-regulation of cinnamoyl CoA reductase (CCRH line) which decreases the lignin content up to 50%, induced an extensive disorganization of sub-layers S_2 (in fibres) and 2 (in vessels) when compared to wild-type, and the S_3, and 3 sub-layers were hardly visible. This general loosening of the secondary wall was due to disorganization of the cellulose framework, where cellulose microfibrils appeared individualized (Figure 5.2). Immunolabelling experiments with the CCRH line were also carried out, using transmission electron microscopy with antibodies directed against condensed or non-condensed (GS) lignin sub-units (Chabannes *et al.*, 2001). The results obtained specifically for fibres are discussed here. In the wild-type (WT), non-condensed GS lignin sub-units were homogeneously distributed in the different sub-layers of the fibres, but were absent from the middle lamellae and cell corners, showing that these anatomical zones did not harbour the same type of lignin as the secondary walls of fibres (Figure 5.3a). In the case of the CCRH down-regulated line, the non-condensed GS epitopes had become concentrated in the S_1 layer and the outer part of S_2 (Figure 5.3b). In contrast, there was little difference in the patterning of condensed units between WT and CCRH. In both cases, these condensed motifs were principally distributed in the S_1 sub-layer of the secondary wall. These first observations suggested that non-condensed GS lignin sub-units could play a role in the cohesion and assembly of lignified secondary walls.

The crucial role of this particular type of lignin was indirectly confirmed by a careful examination of another transgenic tobacco line down-regulated for both cinnamoyl CoA reductase (CCR) and cinnamyl alcohol dehydrogenase (CAD) (Chabannes *et al.*, 2001). This double transformant (DT) exhibits a normal vascular system despite a dramatic reduction of gross lignin content in the stems (up to 50%). The deposition of non-condensed lignins in fibres from the DT line was not affected (Figure 5.3c), whereas condensed GS lignin sub-units are practically absent (Figure 5.3f) in comparison to the control (Figure 5.3d). In contrast, the deposition of condensed lignin sub-units is similar in the CCRH line and in the control (Figure 5.3e). Consequently, this last type of lignin motif does not seem to exert a major role

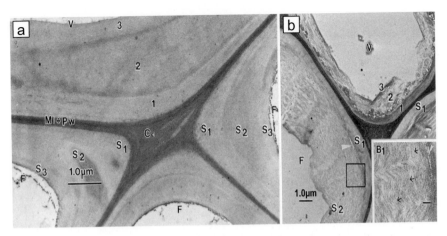

Figure 5.2   Ultrastructural organization of tobacco xylem cell walls in (a) the wild-type and (b) the CCRH line (down-regulated for cinnamoyl COA reductase). (a) Wild-type. Fibre and vessel walls exhibit a general staining covering the different layers and sublayers. In fibre walls, S_1 (outer layer), S_2 (middle layer) and S_3 (innermost layer), are identifiable. In vessel wall, three concentric sublayers are visible and are noted 1, 2, 3 from outer layer to inner layer. Cell corner and middle lamellae are strongly reactive to the PATAg staining. (b and inset b_1) CCRH depressed line. A dramatic loosening in the cell wall architecture of S_2 can be seen both in fibres and vessels ± in fibres, with a concentric sublayering appearing, ending at a clear separation of cellulose microfibrils in S_2. Arrowheads indicate weak points between S_1 and S_2. The inset (b_1) shows an enlarged view of S_2, underlying the unmasking of cellulose microfibrils which appear individualized (small arrows). Reproduced with permission from Dr Chabannes 'In situ analysis of lignins in transgenic tobacco reveals a differential impact of individual transformations on the spatial patterns of lignin deposition at the cellular and subcellular levels'. *Plant J.*, 28(3), 271–282, published by Blackwell, 2001.

in maintaining the structural and functional integrity of the conducting elements (Chabannes *et al.*, 2001). Similar patterns were observed for vessels.

The depletion of lignification in S_2 and 2 sub-layers from CCRH xylem walls, which is associated with an alteration of the ultrastructure, constitutes a strong indication that lignin plays an active role in secondary wall assembly. More specifically, these results show that cohesion between cellulose-hemicellulose elements of the secondary wall involves a specific type of lignin: GS non-condensed unit enriched lignins.

With the same reduction in total lignin content, the double transformant (DT) resulting from the cross between individual homozygous CCR and CAD down-regulated lines did not show loosening of its xylem wall. This constitutes clear evidence that, in addition to the global amount of lignin, the type of lignins may be important in maintaining the cohesion of secondary wall. The basis for the potential adhesive role of these particular lignins is not currently clear, but the fact that it is the non-condensed form, rather than the condensed form of lignin, that plays a significant role in secondary wall assembly agrees with the view that the restricted space between the

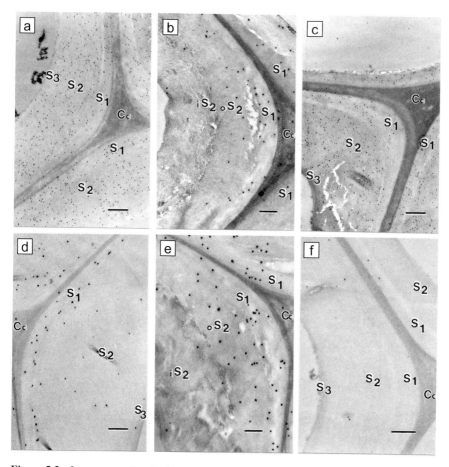

Figure 5.3 Immunocytochemical localization of non-condensed and condensed mixed guaiacyl-syringyl lignin subunits in fibres from wild-type (WT), CCRH depressed and DT (CAD.H 3 × CCRH) lines. (a–c) Labelling for non-condensed GS lignin subunits. (a) In WT, gold particles cover the three sublayers of the secondary wall of fibres but are absent from the middle lamella and cell corner. (b) In the CCRH depressed line, gold particles have become mostly localized in S_1 and partly in outer part of S_2 ($_oS_2$). Non-condensed epitopes are absent in the loosened inner part of S_2 ($_iS_2$). (c) In the DT, an abundant and homogeneous deposit of gold particles is seen on the entire S_2. S_1 is less labelled, particularly in the area adjacent to the cell corner. As in WT, gold particles are absent from the middle lamella and cell corner. (d–f) Labelling for condensed GS lignin subunits. (d) In WT, gold deposits are restricted to the S_1 sublayer. (e) In the CCRH depressed line, gold labelling has become concentrated in S_1 and in the outer part of S_2 ($_oS_2$). A few gold particles are deposited in the cell corner. (f) In the DT, no gold deposits can be detected, either in S_1 or in S_2, or in the middle lamella or cell corner. Cc, cell corner; F, fibre; Ml, middle lamella; S_1, S_2 and S_3, outer-, middle and innermost layers of the secondary wall; $_iS_2$, $_oS_2$, inner and outer parts of S_2 sublayer; V, vessel; 1, 2, 3, outer, middle and innermost layers of the secondary wall of vessels. Bars in a–f = 0.5 μm. Reproduced with permission from Dr Chabannes 'In situ analysis of lignins in transgenic tobacco reveals a differential impact of individual transformations on the spatial patterns of lignin deposition at the cellular and subcellular levels'. *Plant J.*, 28(3), 271–282, published by Blackwell, 2001.

cellulose-hemicellulose in the template favours the extended macromolecular conformation of the non-condensed lignin molecule over the bulky form of condensed
lignin (Joseleau and Ruel, 1997; Ruel *et al.*, 2001).

Variations in the distribution of lignin substructures within the different domains
of the cell wall were also observed in the WT plant. This could be related to the
influence of the polysaccharide matrix and to the fact that the various sugars of the
secondary wall interact differently with the various lignin precursors (Houtman and
Atalla, 1995).

The physical-chemical factors such as the matrix effect due to the polysaccharide
environment (Siegel, 1957) and its polyelectrolyte structure (Houtman and Atalla,
1995) are likely important during lignin polymerization. It is already known that
the type of lignins synthesized in cell corners and secondary walls is strongly influenced by the geometry of the randomly arranged carbohydrate polymers present in
the former and the orientated arrangement of microfibrils in the latter. One direct
consequence of the random environment is to give rise to bulk polymerization of
lignin monomers, between which strong covalent linkages are favoured. On the
other hand, the narrow microfibril secondary wall environment induces an endwise
type of polymerization that leads to the predominance of non-condensed linkages
(Roussel and Lim, 1995).

Overall, these results further confirm that the spatio-temporal deposition of specific lignins is likely dependent on the chemical microenvironment. These results
also confirm that, when the availability of monolignols is reduced, the deposition of
lignins preferentially occurs in the external domain of the wall, where putative initiation sites of lignification have been localized (see section 5.5.2 for more details).
It is interesting to note the parallel between the specific occurrence of lignins in the
S_1 sub-layer of the secondary wall of CCRH tobacco lines (Chabannes *et al.*, 2001)
and the specific occurrence in the same S_1 sub-layer of dirigent (monomer binding)
sites (Burlat *et al.*, 2001).

5.5.2   Cell wall mutants

In addition to transgenic lines, the recent isolation of different cell wall mutants,
particularly in *Arabidopsis thaliana*, has provided new arguments for the individual
roles of the different cell wall components in the extracellular matrix and for tightly
coupled interactions between them.

As an example of a primary cell wall mutant, *Korrigan 'kor'* has defective expression of a member of the endo-1,4-β-D-glucanase family. In addition to reduced
growth, it exhibits pronounced changes in the cellulose-xyloglucan network within
the primary wall and an alteration of the physicochemical properties of the walls
(Nicol *et al.*, 1998). The authors suggest that this phenotype could be due to the
absence of correct cleavage of xyloglucans necessary to permit the incorporation of
new microfibrils into the existing cellulose-xyloglucan network. These data confirm
the high dynamic state of the elongating wall.

A series of *Arabidopsis* mutants deficient in secondary cell wall cellulose deposition has been described by Turner *et al.* (2001). These mutations, termed *irregular xylem (irx1, 2* and *3)* caused the collapse of mature xylem cells in the inflorescence stems. There were no apparent differences in either the deposition of lignin or in the composition of the non-cellulosic carbohydrate fraction of the walls in these mutants. Consequently the observed phenotype is due to the decrease in cellulose content. This polymer was not replaced by other polysaccharides and is apparently crucial for the integrity of xylem cells.

Other *Arabidopsis* mutants have been characterized by an abnormal lignification. In stems of the *elp 1* mutant, lignin was ectopically deposited in the walls of pith parenchyma and this deposition of lignin appears to be independent of secondary wall thickening, as it is the case in certain responses to wounding or infection (Zhong *et al.*, 2000). In contrast, in the *eli 1* mutant, the ectopic lignification of cells throughout the plant that never normally lignify seems to be related with an inappropriate initiation of secondary wall formation and a cell elongation defect (Caño-Delgado *et al.*, 2000).

Among the mutants with a decreased lignification, a severe lignin mutant, *irx 4*, has been identified as a result of its collapsed xylem phenotype (Jones *et al.*, 2001). Interesting observations have been reported on the effect of the *irx4* mutation on the ultrastructure of the cell wall. The decreased lignin levels resulted in hugely expanded cell walls, which often occupy a large proportion of the cell interior, but also in a dramatic alteration of the mechanical properties, including a decrease of stem stiffness and stem strength. These characteristics are in part similar to those observed for the CCR down-regulated tobacco lines (see section 5.5.1). Jones *et al.* (2001) concluded that lignin is a crucial component of the wall and is needed to maintain structural integrity and anchor the cellulose and hemicelluloses together. They refer to cross-linking studies carried out using the cellulose synthesizing bacterium *Acetobacter xylinum* which indicate that, whilst xyloglucans (primary cell wall hemicelluloses) exhibit extensive binding and cross-linking to cellulose, xylans (the major cross-linking glucan in lignified secondary cell wall) do not appear to bind strongly to cellulose (Whitney *et al.*, 1995). It is thus suggested that lignin acts by directly cross-linking xylans and cellulose or by altering the physical environment of the wall (removal of water) and consequently influencing the ability of cellulose to cross-link with the xylan (Jones *et al.*, 2001).

The future exploitation of different transgenic lines and cell wall mutants in combination with ultrastructural and mechanical studies should reveal in more detail the role of lignins and of specific lignin components in the cohesion and supramolecular organization of the secondary wall. As an example, *Arabidopsis* mutants and transgenic lines have been obtained with completely different lignin composition (S/G ratios) (Marita *et al.*, 1999). A further exploitation of these modified lines should highlight the role of lignin composition in the structural and physical properties of the wall.

5.6 Cell wall proteins: their structural roles and potential involvement in the initiation of lignification and wall assembly

In addition to the major polysaccharidic and phenolic polymers, which have been subjected to experimental analysis for some time, the so-called structural protein moiety of the cell wall has only relatively recently been studied in any detail (see Chapter 4), and the results reveal an unexpected complexity.

This protein complement not only seems to play a crucial role in the supramolecular organization of the cell wall but is likely involved in the different functional aspects of the wall (Showalter, 1993). Major classes of wall glycoproteins have been characterized for a long time, including hydroxyproline-rich proteins (HRGP), arabinogalactan proteins (AGP), glycine-rich proteins (GRP), cysteine-rich proteins and proline-rich proteins (PRP). Our understanding of these proteins has benefited from genomic analyses, which have revealed the occurrence of different multigene families even though the functions of the corresponding proteins are often only partially known at best (Jose-Estanyol and Puigdomenech, 2000). The fact that some of the corresponding genes are stringently regulated during development suggests that these proteins play a role in determining the extracellular matrix structure for specific cell types and secondarily cell differentiation and morphogenesis (Cordewener et al., 1991; Bernhardt and Tierney, 2000; Garcia Gomez et al., 2000; Baumberger et al., 2001). Several examples are given in Chapter 4.

In addition, biochemistry and enzymology studies have shown that a wide range of enzymes, including oxidases, hydrolases, transglycosylases and non-enzymatic proteins such as expansin (McQueen-Mason et al., 1992) or dirigent proteins (Davin and Lewis, 2000) are associated with the cell wall with varying affinities (see also Chapters 8 and 9). The nature of these interactions is only partly known, but a cysteine-rich peptide motif has been identified that mediates the insolubilization of proteins in cell walls (Domingo et al., 1999). For the lignified secondary wall at least, it has been suggested that peroxidases become covalently bound to lignin, whilst catalysing its polymerization, through linkages between tyrosine residues and the growing lignin polymer (Evans and Himmelsbach, 1991). More recent data confirm this insolubilization of peroxidase and also laccase during the in vitro polymerization of monolignols (McDougall, 2001).

The present knowledge of cell wall proteins is likely far from being complete due to the strong covalent binding of some molecular species to polysaccharidic or phenolic components, or to the low concentration of minor species. Proteomics represents a unique opportunity to characterize in more detail the protein complement of the cell wall but until now only a limited amount of data has been obtained through this powerful approach (see Chapter 10). Robertson et al. (1997) have characterized, after sequential extractions and separation on SDS-PAGE, the N-terminal amino-acid sequence of a set of readily solubilized primary wall proteins from cell suspension cultures of different plant species. The protein populations differed according to species and sequence data revealed a huge diversity of structures and potential enzymatic functions. In addition, a significant proportion of these proteins have not been identified by database search.

More recently, the same group (Blee *et al.*, 2001) was able to identify (again on the basis of N-terminal sequences) dramatic differences in the subset of wall extractable proteins between primary and secondary walls of tobacco cell suspension cultures. In the same way, in our hands, a proteomic analysis, through a direct proteolysis of isolated *Arabidopsis* walls, revealed a range of unknown firmly bound proteins (Rossignol *et al.*, unpublished).

Most of the structural proteins are very difficult to extract from the cell wall since they become insoluble after their secretion, possibly as a result of cross-linking (Fry, 1986). For example, the isodityrosine intramolecular cross-link between HRGP molecules is well established (Fry, 1986), but its putative intermolecular counterpart remains elusive, even though the discovery of a trimer and tetramer of tyrosine makes the hypothesis much more convincing (Brady *et al.*, 1997). Other wall proteins which contain Tyr-rich repeated sequences that could be involved in isodityrosine cross-links may have structural functions (PRPs, GRPs). Speculative proposals for covalent cross-links between hydroxycinnamic acids esterified to wall polysaccharides, and tyrosine or cysteine residues on wall proteins through dehydrogenative polymerization, have been reported, although this has not been confirmed (Bacic *et al.*, 1988).

There is also evidence that both HRGPs and GRPs are associated with lignin and possibly act as *foci* for lignin polymerization (see Iiyama *et al.*, 1993), but the types of linkages have not been identified. Since the predominant localization of GRPs in vascular tissue suggests that they play a role in the reinforcement of cell walls and/or increase wall hydrophobicity to reduce friction for transported fluid, the recent data obtained for these proteins will be reported here in more detail.

Structural cell wall glycine-rich proteins contain up to 70% glycine residues, a characteristic shared by other proteins of animal origin where they are considered to strengthen biological structures (Ringli *et al.*, 2001b). In plants, these proteins are particularly abundant in cells of xylem tissues, as demonstrated by the analysis of GUS expression driven by the grp10 promoter in transgenic tobacco, or by *in situ* hybridization of the *grp1.8* gene in different species (Ye *et al.*, 1991). In addition, immunolocalization experiments found the GRPs in xylem elements (Keller *et al.*, 1988; Ryser and Keller, 1992). In the immunocytochemistry studies, GRPs were consistently found to be associated with protoxylem cells and particularly with the elongating protoxylem primary walls (Figure 5.4a). A range of observations suggests that at least $GRP_{1.8}$ might be part of a repair mechanism, reinforcing the hydrolysed (modified) primary walls of the protoxylem. This would mechanically stabilize the water conducting elements and prevent water flow between adjacent protoxylem vessels in rapidly growing organs (Ryser *et al.*, 1997; Ringli *et al.*, 2001b). Additional studies on *in vitro* cross-linking of two cell wall GRPs, bean $GRP_{1.8}$ (containing tyrosine) and wheat $GRP_{1.0}$ (containing phenylalanine but not tyrosine) have shown dramatic differences between the two GRPs. Incubation of $GRP_{1.8}$ with a peroxidase converted it to high-molecular-weight aggregates, whereas the phenylalanine-

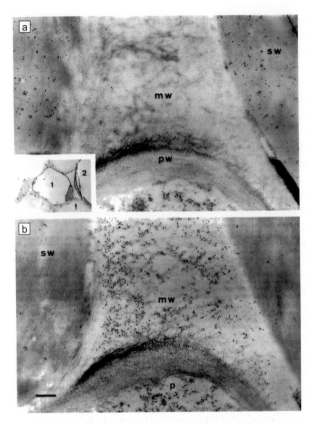

Figure 5.4 Differential localization of the PRP$_2$ and GRP$_{1.8}$ structural cell wall proteins in protoxylem cells of *Phaseolus vulgaris*. (a) PRP$_2$ antibodies label the lignified secondary walls (sw)
of two adjacent protoxylem elements; pw = primary wall of xylem parenchyma cell. Inset: the
enlarged region is indicated with an arrow and the two protoxylem cells are numbered. (b) GRP$_{1.8}$
antibodies strongly label the modified primary walls (mw) of the protoxylem elements; p = xylem
parenchyma cell. Bar = 250 nm. Reproduced with permission from Dr Ryser 'Structural cell-wall
proteins in protoxylem development: evidence for a repair process mediated by a glycine-rich
protein'. *Plant J.*, 12, 97–111.

containing GRP was not affected by peroxidase incubation (Ringli *et al.*, 2001a).
This strongly implies the involvement of tyrosine residues in peroxidase-mediated
inter-molecular cross-linking.

The interaction of GRP$_{1.8}$ with the extracellular matrix was also studied by
protein extraction experiments of different fusion proteins obtained from different
domains of GRP$_{1.8}$ and expressed in vascular tissue of tobacco (Ringli *et al.*, 2001a).
The analyses demonstrated that GRP$_{1.8}$ undergoes hydrophobic interactions in the
cell wall. These interesting data reveal that GRP$_{1.8}$ might be capable of establishing
different types of linkages and interactions with different components of the extracellular matrix.

Another class of cell wall proteins, proline-rich proteins (PRPs), seems to be involved in a different way in protoxylem differentiation. Antibodies raised against a 33-kDa rich protein of soybean (PRP_2) labelled epitopes in the lignified secondary wall thickenings of protoxylem, but did not react with modified primary walls labelled with $GRP_{1.8}$ antibodies (Figure 5.4b). The secretion of PRPs correlated with the lignification of the secondary walls and preceded the secretion of $GRP_{1.8}$ in protoxylem development (Ryser et al., 1997; Ringli et al., 2001b). These interesting studies pointed out that in protoxylem cells two structural cell wall proteins GRPs and PRPs are deposited in distinct cell wall domains and are secreted at different points in protoxylem development.

Because lignification starts far from the protoplast at the cell corners in the middle lamella (Donaldson, 2001), it has been suggested that there are initiation sites at specific regions of the cell wall which begin the polymerization process although their exact nature is not known. Extensin-like proteins (Bao et al., 1992), proline-rich protein (Müsel et al., 1997) and dirigent (non-enzymatic) proteins (Davin and Lewis, 2000) have been either spatially or temporally correlated with the onset of lignification.

The dirigent proteins, which are ubiquitous throughout the plant kingdom, display no catalytic activity but are capable of binding monolignols in a specific manner. In the presence of oxidases they engender stereospecific coupling which oxidases alone cannot catalyse in vitro. Recent data (Burlat et al., 2001) have shown, through immunolabelling techniques, that within lignified secondary xylem, dirigent proteins were primarily localized in the S_1 sub-layer and compound middle lamella. This localization is coincident with the site for initiation of lignin biosynthesis. According to these authors, putative arrays of dirigent (coniferyl-alcohol-derived radical binding) sites could locally initiate the lignification process and the expansion of lignification would occur through an iterative process of lignin assembly in a template-guided manner (Sarkanen, 1998).

While nucleation or directing sites are likely involved in the lignification process, the role of proteins in controlling the coupling of monomers is still a matter of debate. The metabolic plasticity of lignins and the substitution of monomers in different transgenic lines is not in good agreement with a tight coupling specificity. For example, COMT down-regulated plants lacking the ability to perform a final methylation step to make sinapyl alcohol, used 5-hydroxyconiferyl alcohol instead in striking quantities, giving rise to novel benzodioxane structures as major components of the lignin (Ralph et al., 2001).

In addition to the potential involvement of dirigent proteins, peroxidases and laccases are both candidates for initiation and progression of the polymerization. Laccases constitute, with ascorbate oxidases and ceruloplasmins, a large and diverse family of copper-containing metalloproteins called 'blue copper oxidases' that are able to catalyse the 4-electron reduction of oxygen to water. Plant laccases are cell-wall-associated enzymes that are relatively difficult to purify. Due to their capacity to oxidize phenolic compounds, laccases were postulated to be involved in the final step of lignification, but at the present time, an unambiguous demonstration of this potential role is lacking (see Ranocha *et al.*, 2000). Using both the sequence data obtained from purified poplar laccase, and heterologous laccase cDNA as probes, we have identified five different laccase genes in poplar, which exhibit a relatively low degree of sequence homology amongst themselves. We have transformed poplar with four different laccase genes (*lac1, lac3, lac90* and *lac110*) in the antisense orientation under the control of a strong constitutive promoter, 35S CAMV. However, no significant differences in growth or development, nor in lignin content and monomeric composition, were observed between antisense laccase and control poplars (Ranocha *et al.*, 2002). One potential difficulty in determining the physiological roles of individual members of multigene families using this approach is the potential functional compensation by another member of the gene family that is insufficiently similar in terms of sequence homology to be affected by the transgene. This may be especially problematic in the case of poplar laccases, owing to their relatively low homology. Thus, it is difficult for the moment to draw any clear-cut conclusions as to the role of laccases in lignification.

Nevertheless, in one line of antisense laccase transformants (poplar underexpressing the *lac3* gene) a two-to-three-fold increase in soluble phenolics was observed. Moreover, examination of the xylem by fluorescence microscopy revealed an irregular contour of the cell, an apparently looser frame around the cells and a loss of cohesion between cells in the antisense line (Figure 5.5b) when compared to the control (Figure 5.5a). Moreover, in the antisense plants, the fluorescence emission was not homogeneous throughout the entire width of the wall, but was negligible in the middle lamella/primary wall region between adjacent fibres. As a consequence, the cells appeared to be detached from one another. The intensity of phloroglucinol staining was not altered in *lac3* antisense (AS) plants and this is in agreement with the fact that no quantitative or qualitative differences in lignin were detected in these plants.

At the electron microscope level, defects in cell wall structure of xylem fibres were clearly visible. In controls (Figure 5.5c), as expected, the secondary walls were firmly laid down on the primary walls of xylem fibre cells. In contrast, in *lac3* AS plants adhesion defects occured within the secondary wall of a given cell (Figure 5.5d).

An increase in cell wall fragility in *lac3* down-regulated plants suggested that mechanical properties of stems had also been altered. To this end, microtome-cutting tests were performed and demonstrated that the modification of *lac3* gene expression leads to significant alterations of the mechanical properties of wood (Ranocha *et al.*, 2002).

Laccases have been proposed to be involved in cross-linking phenomena (e.g. formation of diferulic bridges) and a decrease in the activity of a specific laccase could reduce the mobilization of simple phenolics for such linkages. This would induce this loose organization of the wall and indirectly an accumulation of soluble phenolics.

It is clear that the characterization of the type of reaction catalysed by *lac3* and the corresponding substrates through *in vitro* and *in vivo* approaches should highlight

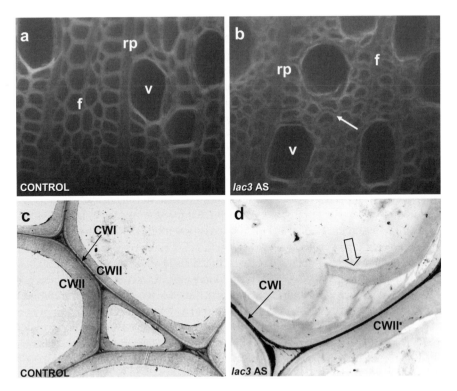

Figure 5.5 UV fluorescence and electron microscopy analysis of wild-type and *lac3* antisense poplar xylem tissue. (a) Transverse stem sections of wild-type poplar xylem visualized by UV fluorescence microscopy. (b) Similar analysis of a *lac3* antisense poplar line. v: vessels, rp: ray parenchyma, f: fibres. Note the deformed cell contour of xylem fibres and an absence of fluorescence emission at the middle lamella/primary wall region in the transgenic plant. (c) Electron micrographs of xylem fibres in a wild-type poplar plant and (d) *lac3* antisense plant. Note detachment of secondary wall strips (open arrow). CWI: primary wall, CWII: secondary wall.

our knowledge of cell wall assembly and of the supramolecular organization of the secondary cell wall.

Future experiments based on molecular genetics, proteomics, transgenesis and immunocytochemistry should help to identify the proteins involved in the supramolecular organization of the lignified wall and how they function.

5.7   Conclusions

Xylogenesis and lignin deposition in the cell wall represent an example of cell differentiation in an exceptionally complex form driven by the coordinated expression of hundreds of genes.

Lignification of pre-existing polysaccharide matrices adds a degree of chemical complexity and results in new composite materials. Previous studies have demonstrated the high compositional heterogeneity and the large chemical flexibility of the lignin moiety. Through molecular biology and transgenesis, lignin composition and content can be manipulated with correlative impacts on the structure and properties of the wall even though clear evidence is scarce to demonstrate the preferential role of specific lignin, structures or compositions, on the mechanical and physical characteristics of the wall.

As with other composite materials, the association between the different macromolecular components of the lignified wall is likely of the utmost importance in determining the physical and mechanical properties of the wall and, secondly, the technological properties of plant products and plant-derived materials. This field has to date been poorly explored but it is known that the heterogeneity of the lignified cell wall relies on the occurrence of sub-domains with specific chemical composition and potentially specific patterns of interactions between wall constituents. Up to now, and for obvious technical reasons, the focus has been on the chemical characteristics of the cell wall, but molecular biology transgenesis and mutant plants can be of great help to elucidate the interactions between wall components. In addition to previous successful programs aiming to manipulate lignin content and composition through lignin genetic engineering (see Boudet and Grima-Pettenati, 1996), new programs are being initiated to manipulate matrix polysaccharides. There is already some evidence that the synthesis and deposition of lignin and hemicelluloses during secondary wall formation are linked and co-regulated (Taylor and Haigler, 1993). The expected results may have important general implications in understanding the co-regulation of the synthesis of polymeric cell wall components and in stimulating coupled approaches aimed at manipulating both hemicelluloses and lignins. Furthermore, the introduction into transgenic plants of microbial genes encoding polysaccharides modifying or degrading enzymes targeted to the cell wall can help to understand the role of individual components in the cohesion and stability of the final lignified network. However, it is clear that, in this complex area, combinatorial approaches will be increasingly necessary. The resulting products of transgenesis experiments or other specific plant materials resulting from mutations should be investigated by a range of complementary techniques.

This has been partly illustrated for immunocytochemistry in this review which reveals the complexity, the dynamics and the domain specificity of the cell wall. In a more general way, the advantages of this technique for expanding the knowledge of the cell wall has been highlighted at the last International Cell Wall Meeting (Toulouse, September 2001).

Other sophisticated microscopy techniques such as FTIR-Raman microscopy or atomic force microscopy, possibly coupled with specific enzymatic pretreatment of the walls (Kroon, 2000), will likely provide new and important insights. These *in situ* studies also need to be coupled to recent developments in instrumental chemical analysis (different forms of NMR, MS, etc.) in order to correlate structures with chemical changes as discussed in depth by Morris *et al.* in Chapter 2 of this book.

The coupling and combination of these different techniques will provide new opportunities to understand the supramolecular organization of the cell wall. Beyond this set of techniques, that address the composition and interactions between cell wall constituents, methodologies are also necessary to probe mechanical properties of the composite (e.g. resistance, elasticity, rigidity) in order to associate changes in linkages, bonds, cohesion between wall constituents to bulk mechanical properties (Ha *et al.*, 1997). Finally modelling studies will be also complementary and necessary for a full understanding of these interactions (Cathala *et al.*, 2000).

The development of linkages between lignins and other wall polymers is a final step in the process of wall strengthening in the context of the adaptive significance of lignification in plants. This level of organization will likely be an important area for future investigations aiming to control and improve plant material for practical applications. For example, it is particularly difficult to correlate the digestibility of forage crops with a particular chemical composition. We can speculate that the key factors are most likely related to the nature of the interactions between wall components than to the constituents themselves. For example, in the 'rubbery wood' disease which induces a decrease of the rigidity and mechanical resistance of woody stems, it has been demonstrated that, in addition to a lowered lignin content and altered lignin monomer composition, there are also changes in the association of lignins with wall polysaccharides (see Raven, 1977).

Lignin also has negative effect in pulp industry where it has to be removed from carbohydrate constituents. These carbohydrates would also be valuable as sources of sugar in fermentation of 'green chemicals'. However, the need to separate the lignin in order to achieve effective hydrolysis (saccharification) is again costly. In this context, research programs have been developed aiming, through genetic engineering, to design plants with a modified lignin content or with lignin of different compositions, more suited to specific agricultural and industrial uses. Genetically modified plants, with, for example, more easily extractable lignins have been obtained which could be of immediate utility (Grima-Pettenati and Goffner, 1999).

As the age of oil changes to the age of biomass, a better understanding of the polymers involved in the cell wall, the major storage compartment of renewable carbon in the biosphere, will become increasingly important. In addition, knowledge of the assembly mechanisms of the different molecules involved in this composite material would undoubtedly facilitate industrial processes aiming to produce, from individual polymers, chemicals and materials that society needs in the framework of sustainable development.

Acknowledgements

I wish to thank the members of my research group 'Expression and modulation of lignification' for their contribution to some of the results presented in this chapter.

Thanks are due to Dr Ruel and Dr Joseleau for their efficient involvement in the microscopic analysis of transgenic tobaccos. I am grateful to Dr C. Lapierre, Dr J.

Grima-Pettenati, Dr S. Fry and Dr J. Ralph for critically reading the manuscript and for very helpful suggestions. The support of the European Community (TIMBER project –Fair – CT 95–0424 and COPOL project – QL SK – CT- 2000–01493), of the CNRS and of the University Paul Sabatier is also gratefully acknowledged.

References

Allona, I., Quinn, M., Shoop, E. *et al.* (1998) Analysis of xylem formation in pine by cDNA sequencing. *Proc. Natl. Acad. Sci. USA*, 95, 9693–9698.

Atalla, R.H. and Agarwal, U.P. (1985) Raman microprobe evidence for lignin orientation in cell walls of native woody tissue. *Sci.*, 227, 636–638.

Bacic, A., Harris, P.J. and Stone, B.A. (1988) Structure and function of plant cell walls. In *The Biochemistry of Plants: a comprehensive treatise. Carbohydrates* (eds P.K. Stumpf and E.E. Conn), Academic Press, New York, pp. 297–371.

Bao, W., O'Malley, D.M. and Sederoff, R.R. (1992) Wood contains a cell-wall structural protein. *Proc. Natl. Acad. Sci. USA*, 89, 6604–6608.

Baumberger, N., Ringli, C. and Keller, B. (2001) The chimeric leucine-rich repeat/extension cell wall protein I. is required for root hair morphogenesis in *Arabidopsis thaliana. Genes Devel.*, 15, 1128–1139.

Beritognolo, I., Magel, E., Abdel-Latif, E., Charpentier, J.P., Jay-Allemand, C. and Breton, C. (2002) Expression of genes encoding chalcone synthase, flavonone 3-hydroxylase and dihydroflavonol 4-reductase correlates with flavonol accumulation during heartwood formation in *Juglans nigra L. Tree Physiol.*, 22, 291–300.

Bernhardt, C. and Tierney, M.L. (2000) Expression of At PRP3, a proline rich structural cell wall protein from arabidopsis is regulated by cell type-specific developmental path in root hair formation. *Plant Physiol.*, 122, 705–714.

Blee, K.A., Wheatley, E.R., Bonham, V.A. *et al.* (2001) Proteomic analysis reveals a novel set of cell wall proteins in a transformed tobacco cell culture that synthesizes secondary walls as determined by biochemical and morphological parameters. *Planta*, 212, 404–415.

Boudet, A.M. and Grima-Pettenati, J. (1996) Lignin genetic engineering. *Mol. Breeding*, 2, 25–39.

Brady, J.D., Sadler, I.H. and Fry, S.C. (1997) Pulcherosine, an oxidative coupled trimer of tyrosine in plant cell walls: its role in cross-link formation. *Phytochem.*, 47, 349–353.

Brunow, G., Kilpeläinen, I., Sipilä, J. *et al.* (1998) Oxidative coupling of phenols and the biosynthesis of lignin. In *Lignin and Lignan Biosynthesis* (eds N.G. Lewis and S. Sarkanen), American Chemical Society, New Orleans, LA, pp. 131–147.

Burlat, V., Kwon, M., Davin, L.B. and Lewis, N.G. (2001) Dirigent proteins and dirigent sites in lignifying tissues. *Phytochem.*, 57, 883–897.

Caño-Delgado, A.I., Metzlaff, K. and Bevan, M.W. (2000) The *eli1* mutation reveals a link between cell expansion and secondary cell wall formation in *Arabidopsis thaliana. Devel.*, 127, 3395–3405.

Carpita, N.C. and Gibeaut, D.N. (1993) Structural models of primary cell walls in flowering plants: consistency of molecular structure with the physical properties of the walls during growth. *Plant J.*, 3, 1–30.

Cathala, B. and Monties, B. (2001) Influence of pectins on the solubility and the molar mass distribution of dehydrogenative polymers (DHPs, lignin model compounds). *Int. J. Biol. Macromolecules*, 29, 45–51.

Cathala, B., Lee, L.T., Aguie-Beghin, V., Douillard, R. and Monties, B. (2000) Organization behaviour of guaiacyl and guaiacyl/syringyl dehydrogenation polymers (lignin model compounds) at the air water interface. *Langmuir*, 16, 10444–10448.

Cathala, B., Chabbert, B., Joly, C., Dole, P. and Monties, B. (2001) Synthesis, characterization and water sorption properties of pectin-dehydrogenation polymer (lignin model compound) complex. *Phytochem.*, 56, 195–202.

Chabannes, M., Ruel, K., Yoshinaga, A. *et al.* (2001) *In situ* analysis of lignins from tobacco transgenic lines down-regulated for several enzymes of monolignol synthesis reveals a differential impact of transformation on the spatial patterns of lignin deposition at the cellular and subcellular levels. *Plant J.*, 28, 271–282.

Cordewener, J., Booit, H., Van der Zandt, H., Van Engelen, F., Van Kammen, A. and de Vries, S. (1991) Tunicamycin-inhibited carrot somatic embryogenesis can be restored by secreted cationic peroxidase isoenzymes. *Planta*, 184, 478–486.

Davin, L.B. and Lewis, N.G. (2000) Dirigent proteins and dirigent sites explain the mystery of specificity of radical precursor coupling in lignan and lignin biosynthesis. *Plant Physiol.*, 123, 453–461.

Domingo, C., Sauri, A., Mansilla, E., Conejero, V. and Vera, P. (1999) Identification of a novel peptide motif that mediates cross-linking of proteins to cell wall. *Plant J.*, 20, 563–570.

Donaldson, L.A. (1994) Mechanical constraints on lignin deposition during lignification. *Wood Sci. Technol.*, 28, 111–118.

Donaldson, L.A. (2001) Lignification and lignin topochemistry – an ultrastructural view. *Phytochem.*, 57, 859–873.

Evans, J.J. and Himmelsbach, D.S. (1991) Incorporation of peroxidase into synthetic lignin. *J. Agric. Food Chem.*, 39, 830–832.

Fergus, B.J. and Goring, D.A.I. (1970) The location of guaiacyl and syringyl lignins in birch xylem tissue. *Holzforsch.*, 24, 118–124.

Freudenberg, K. (1968) The constitution and biosynthesis of lignin. In *Constitution and Biosynthesis of Lignin* (eds K. Freudenberg and A.C. Neish), Springer-Verlag, New York, pp. 47–122.

Fry, S.C. (1986) Cross-linking of matrix polymers in the growing cell walls of angiosperms. *Annu. Rev. Plant. Physiol.*, 37, 165–186.

Fry, S.C. (2000) Ménage à trois: oxidative coupling of feruloyl polysaccharides *in vivo* proceeds beyond dimers. *Polyphenols Actualités*, 19, 8–12.

Fry, S.C., Willis, S.C. and Paterson, A.J. (2000) Intraprotoplasmic and wall-localised formation of arabinoxylan-bound diferulates and larger ferulate coupling-products in maize cell-suspension cultures. *Planta*, 211, 679–692.

Fujino, T. and Itoh, T. (1998) Changes in the three-dimensional architecture of the cell wall during lignification of xylem cells in *Eucalyptus tereticornis. Holzforsch.*, 52, 111–116.

Garcia Gomez, B.I., Campos, F., Hernandez, M. and Covarrubias, A.A. (2000) Two bean cell wall proteins more abundant during water deficit are high in proline and interact with a plasma membrane protein. *Plant J.*, 22, 277–288.

Garcia-Conesa, M.T., Kroon, P.A., Ralph, J. *et al.* (1999) A cinnamoyl esterase from *Aspergillus niger* can break plant cell wall cross-links without release of free diferulic acids. *Eur. J. Biochem.*, 266, 644–652.

Gierer, J. and Wannström, S. (1986) Formation of ether bonds between lignin and carbohydrates during alkaline pulping processes. *Holzforsch.*, 40, 347.

Grabber, J.H., Ralph, J., Hatfield, R.D., Quideau, S., Kuster, T. and Pell, A.N. (1996) Dehydrogenation polymer-cell wall complexes as a model for lignified grass walls. *J. Agric. Food Chem.*, 44, 1453–1459.

Grima-Pettenati, J. and Goffner, D. (1999) Lignin genetic engineering revisited. *Plant Sci.*, 145, 51–65.

Ha, M.A., Apperley, D.C. and Jarvis, M.C. (1997) Molecular rigidity in dry and hydrated onion cell walls. *Plant Physiol.*, 115, 593–598.

Hertzberg, M., Aspebord, H., Schrader, J. *et al.* (2001) A transcriptional roadmap to wood formation. *Proc. Natl. Acad. Sci. USA*, 98, 14732–14737.

Higuchi, T., Ogino, K. and Tanahashi, M. (1971) Effect of polysaccharides on dehydropolymeriza-
    tion of coniferyl alcohol. *Wood Res.*, 51, 1–11.
Houtman, C.J. and Atalla, R.H. (1995) Cellulose lignin interactions. A computational study. *Plant
    Physiol.*, 107, 977–984.
Iiyama, K., Lam, T.B.T., Meikle, P.J., Ng, K., Rhodes, D. and Stone, B.A. (1993) Cell wall biosyn-
    thesis and its regulation. In *Forage cell wall structure and digestibility* (eds H. Jung, D. Bux-
    ton, R. Hatfield and J. Ralph), American Society of Agronomy, Madison, WI, pp. 621–683.
Iiyama, K., Lam, T.B.T. and Stone, B.A. (1994) Covalent cross-links in the cell wall. *Plant Physiol.*,
    104, 315–320.
Ishii, T. (1991) Isolation and characterization of a diferuloyl arabinoxylan hexasaccharide from
    bamboo shoot cell walls. *Carbohydr. Res.*, 219, 15–22.
Itoh, T. (2002) Deep-etching electron microscopy and 3-dimensional cell wall architecture. In
    Wood formation in trees (ed. N. Chaffey), Taylor and Francis, London and New York, pp.
    83–98.
Jacquet, G., Pollet, B. and Lapierre, C. (1995) New ether-linked ferulic acid coniferyl alcohol dim-
    ers identified in grass straws. *J. Agr. Food Chem.*, 43, 2746–2751.
Jones, L., Ennos, A.R. and Turner, S.R. (2001) Cloning and characterization of irregular xylem
    4(irx4): a severely lignin-deficient mutant of Arabidopsis. *Plant J.*, 26, 205–216.
Jose-Estanyol, M. and Puigdomenech, P. (2000) Plant cell wall glycoproteins and their genes.
    Plant Physiol. Biochem., 38, 97–108.
Joseleau, J.P. and Ruel, K. (1997) Study of lignification by nominative techniques in growing maize
    internodes – an investigation by Fourier transform infrared, cross-polarisation – magic angle
    spinning ^{13}C nuclear magnetic resonance spectroscopy and immunocytochemical transmis-
    sion electron microscopy. *Plant Physiol.*, 114, 1123–1133.
Kamisaka, S., Takeda, S., Takahashi, K. and Shibata, K. (1990) Diferulic acid and ferulic acid
    in the cell wall of Avena coleoptiles: their relationship to mechanical properties of the cell
    wall. *Physiol. Plant,* 78, 1–7.
Keller, B., Sauer, N. and Lamb, C.J. (1988) Glycine-rich cell wall proteins in bean: gene structure
    and association of the protein with the vascular system. *EMBO J.*, 12, 3625–3633.
Kosikova, B. and Ebringerova, A. (1994) Lignin carbohydrate bonds in a residual soda spruce pulp
    lignin. *Wood Sci. Technol.*, 28, 291.
Kroon, P.A. (2000) What role for feruloyl esterases today? *Polyphenols Actualités*, 19, 4–5.
Lai, Y.Z., Funaoka, M. and Chen, H.T. (1999) Chemical heterogeneity in woody lignins. In *Ad-
    vances in Lignocellulosics Characterization* (ed. D.S. Argyropoulos), TAPPI Press, Atlanta,
    pp. 43–53.
Lapierre, C., Pollet, B., Ralet, M.C. and Saulnier, L. (2001) The phenolic fraction of maize bran:
    evidence for lignin-heteroxylan association. *Phytochem.*, 57, 765–772.
McDougall, G.J. (2001) Cell wall proteins from sitka spruce xylem are selectively insolubilised
    during formation of dehydrogenation polymers of coniferyl alcohol. *Phytochem.*, 57,
    157–163.
McQueen-Mason, S.J., Durachko, D.M. and Cosgrove, D.J. (1992) Two endogenous proteins that
    induce cell wall extension in plants. *Plant Cell*, 4, 1425–1433.
Maeda, Y., Awano, T., Takabe, K. and Fujita, M. (2000) Immunolocalization of glucomannans
    in the cell wall of differentiating tracheids in *Chamaecyparis obtusa. Protoplasma,* 213,
    148–156.
Marita, J.M., Ralph, J., Hatfield, R.D., and Chapple, C. (1999) NMR characterization of lignins
    in *Arabidopsis* altered in the activity of ferulate 5-hydroxylase. *Proc. Natl. Acad. Sci. USA*,
    96, 12328–12332.
Minor, J.L. (1991) Location of lignin-bonded pectic polysaccharides. *J. Wood. Chem. Technol.*,
    11, 159.

Müsel, G., Schindler, T., Bergfeld, R. *et al.* (1997) Structure and distribution of lignin in primary and secondary cell walls of maize coleoptiles analyzed by chemical and immunological probes. *Planta*, 201, 146–159.

Myton, K.E. and Fry, S.C. (1994) Intraprotoplasmic feruloylation of arabinoxylans in *Festuca arundinacea* cell cultures. *Planta*, 193, 326–330.

Nicholson, R.L. and Hammerschmidt, R. (1992) Phenolic compounds and their role in disease resistance. *Annu. Rev. Phytopath.*, 30, 369–389.

Nicol, F., His, I., Jauneau, A., Vernhettes, S., Canut, H. and Hofte, H. (1998) A plasma membrane-bound putative endo-1,4-bêta-D-glucanase is required for normal wall assembly and cell elongation in *Arabidopsis*. *EMBO J.*, 17, 5563–5576.

Plomion, C., Leprovost, G. and Stokes, A. (2001) Wood formation in trees. *Plant Physiol.*, 127, 1513–1523.

Ralph, J., Helm, R.F., Quideau, S. and Hatfield, R.D. (1992) Lignin-feruloyl ester cross-links in grasses. Part I. Incorporation of feruloyl esters into coniferyl alcohol dehydrogenation polymers. *J. Chem. Soc. Perkin Trans.*, 2961–2968.

Ralph, J., Grabber, J.H. and Hatfield, R.D. (1995) Lignin ferulate crosslinks in grasses: active incorporation of ferulate polysaccharide esters into ryegrass lignins. *Carbohydr. Res.*, 275, 167–178.

Ralph, J., Bunzel, M., Marita, J.M. *et al.* (2000) Diferulates analysis: new diferulates and disinapates in insoluble cereal fiber. *Polyphenols Actualités*, 19, 13–17.

Ralph, J., Lapierre, C., Marita, J.M. *et al.* (2001) Elucidation of new structures in lignins of CAD- and COMT- deficient plant by NMR. *Phytochem.*, 57, 993–1003.

Ranocha, P., Goffner, D. and Boudet, A.M. (2000) Plant laccases: are they involved in lignification? In *Cell and Molecular Biology of Wood Formation* (eds R. Savidge, J. Barnett and R. Napier), Bios Scientific Publishers Ltd, Oxford, pp. 397–410.

Ranocha, P., Chabannes, M., Danoun, S., Jauneau, A., Boudet, A.M. and Goffner, D. (2002) Laccase down-regulation causes alterations in phenolic metabolism and cell wall structure in poplar. *Plant Physiol.*, 129, 145–155.

Raven, J.A. (1977) The evolution of vascular land plants in relation to supracellular transport processes. *Adv. Bot. Res.*, 5, 153–219.

Ringli, C., Hauf, G. and Keller, B. (2001a) Hydrophobic interactions of the structural protein GRP1.8 in the cell wall of protoxylem elements. *Plant Physiol.*, 125, 673–682.

Ringli, C., Keller, B. and Ryser, U. (2001b) Glycine-rich proteins as structural components of plant cell walls. *Cell. Mol. Life Sci.*, 58, 1430–1441.

Robertson, D., Mitchell, G.P., Gilroy, J.S., Gerrish, C., Bolwell, G.P. and Slabas, A.R. (1997) Differential extraction and protein sequencing reveals major differences in patterns of primary cell wall proteins from plants. *J. Biol. Chem.*, 272, 15841–15848.

Roussel, M.R. and Lim, C. (1995) Dynamic model of lignin growing in restricted spaces. *Macromolecules*, 28, 370–376.

Ruel, K., Chabannes, M., Boudet, A.M., Legrand, M. and Joseleau, J.P. (2001) Reassessment of qualitative changes in lignification of transgenic tobacco plants and their impact on cell wall assembly. *Phytochem.*, 57, 875–882.

Ryser, U. and Keller, B. (1992) Ultrastructural localization of a bean glycine-rich protein in unlignified primary walls of protoxylem cells. *Plant Cell*, 4, 773–783.

Ryser, U., Shorderet, M., Zhao, G.F. *et al.* (1997) Structural cell-wall proteins in protoxylem development: evidence for a repair process mediated by a glycine-rich protein. *Plant J.*, 12, 97–111.

Salmen, L. and Olsson, A.M. (1998) Interactions between hemicelluloses, lignin and cellulose: structure property relationships. *J. Pulp Paper Sci.*, 24, 99–103.

Sancho, A.I., Faulds, C.B., Bartolome, B. and Williamson, G. (1999) Characterization of feruloyl esterase activity in barley. *J. Sci. Food Agric.*, 79, 447–449.

Sarkanen, S. (1998) Template polymerization in lignin biosynthesis. In *Lignin and Lignan Biosynthesis* (eds N.G. Lewis and S. Sarkanen), American Chemical Society Symposium, Washington, DC, pp. 194–208.

Saulnier, L., Marot, C., Chanliaud, E. and Thibault, J.F. (1995) Cell wall polysaccharide interactions in maize bran. *Carbohydr. Polymers*, 26, 279–287.

Saulnier, L., Crepeau, M.J., Lahaye, M. *et al.* (1999) Isolation and structural determination of two 5,5′-diferuloyl oligosaccharides indicate maize heteroxylans are covalently cross-linked by oxidatively coupled ferulates. *Carbohydr. Res.*, 320, 82–92.

Scalbert, A., Monties, B., Lallemand, J.Y., Guittet, E. and Rolando, C. (1985) Ether linkage between phenolic acids and lignin fractions from wheat straw. *Phytochem.*, 24, 1359–1362.

Showalter, A.M. (1993) Structure and function of plant cell wall proteins. *Plant Cell*, 5, 9–23.

Siegel, S. (1957) Non-enzymatic molecules as matrices in biological synthesis: the role of polysaccharides in peroxidase-catalysed lignin polymer from eugenol. *J. Am. Chem. Soc.* 79, 1628–1632.

Sterky, F., Regan, S., Karlsson, J. *et al.* (1998) Gene discovery in the wood-forming tissues of poplar: analysis of 5,692 expressed sequence tags. *Proc. Natl. Acad. Sci. USA*, 95, 13330–13335.

Syrjänen, K. and Brunow, G. (1998) Oxidative cross coupling of *p*-hydroxycinnamic alcohols with dimeric arylglycerol β-aryl ether lignin model compounds. The effect of oxidation potentials. *J. Chem. Soc., Perkin Trans.* 1, 3425–3429.

Tamminen, T.L. and Hortling, B.R. (1999) Isolation and characterization of residual lignin. In *Advances in Lignocellulosics Characterization* (ed. D.S. Argyropoulos), TAPPI Press, Atlanta, pp. 1–42.

Taylor, J.G. and Haigler, C.H. (1993) Patterned secondary cell-wall assembly in tracheary elements occurs in a self perpetuating cascade. *Acta Bot. Neerland.*, 42, 153–163.

Terashima, N., Fukushima, K., He, L.F. and Takabe, K. (1993). In *Forage Cell Wall Structure and Digestibility* (eds H.G. Jung, D.R. Buxton, R.D. Hatfield and J. Ralph), ASA-CSSA-SSSA, Madison, WI, pp. 247–270.

Terashima, N., Nakashima, J. and Takabe, K. (1998) Proposed structure for protolignin in plant cell wall. In *Lignin and Lignan Biosynthesis* (eds N.G. Lewis and S. Sarkanen), American Chemical Society, New Orleans, LA, pp. 180–193.

Thompson, J.E. and Fry, S.C. (2000) Evidence for covalent linkage between xyloglucan and acidic pectins in suspension-cultured rose cells. *Planta*, 211, 275–286.

Turner, S.R., Taylor, N. and Jones, L. (2001) Mutations of the secondary cell wall. *Plant Mol. Biol.*, 47, 209–219.

Wakabayashi, K., Hoson, T. and Kamisaka, S. (1997) Osmotic stress suppresses cell wall stiffening and the increase in cell wall bound ferulic and diferulic acids in wheat coleoptiles. *Plant Physiol.*, 113, 967–973.

Whitney, S.E.C., Brigham, J.E., Darke, A.H., Reid, J.S.G., and Gidley, M.J. (1995) In vitro assembly of cellulose/xyloglucan networks: ultrastructural and molecular aspects. *Plant J.*, 8, 491–504.

Yamamoto, E. and Towers, G.H.N. (1985) Cell wall bound ferulic acid in barley seedlings during development and its photoisomerization. *J. Plant Physiol.*, 117, 441–449.

Ye, Z.H., Song, Y.R., Marcus, A. and Varner, J.E. (1991) Comparative localization of three classes of cell wall proteins. *Plant J.*, 1, 175–183.

Zhong, R., Ripperger, A. and Ye, Z.H. (2000) Ectopic deposition of lignin in the pith of stems of two arabidopsis mutants. *Plant Physiol.*, 123, 59–69.

6 Plant cell wall biosynthesis: making the bricks

Monika S. Doblin, Claudia E. Vergara, Steve Read,
Ed Newbigin and Antony Bacic

6.1 Introduction

6.1.1 Importance of polysaccharide synthesis

Each cell in a plant is surrounded by an extracellular matrix called the cell wall that is responsible for determining cell size and shape. Primary walls surround growing cells, while thickened secondary walls confer mechanical strength after cell growth has ceased. Walls are also involved in cell adhesion, cell-cell communication and defence responses (Bacic *et al.,* 1988; Carpita and Gibeaut, 1993; Gibeaut and Carpita, 1994a). Polysaccharides comprise the bulk of wall structural components, being ~90% by dry weight of primary walls and ~60% by dry weight of lignified secondary walls (Bacic *et al.,* 1988). In most cell types, the wall consists of three structurally independent but interwoven polymer networks – cellulose microfibrils coated with branched non-cellulosic polysaccharides ('hemicelluloses' such as xyloglucans, glucuronoarabinoxylans or glucomannans), a gel-like matrix of pectin, and cross-linked structural proteins. Other chapters in this volume provide detailed descriptions of the polysaccharides and structural proteins that make up the wall (Chapters 1 and 4), as well as the proteins and enzymes that modify various wall components after their deposition in the wall (Chapters 8 and 9).

The focus of this chapter is plant cell wall biosynthesis, and in particular the synthesis of wall polysaccharides. Polysaccharide synthesis and wall assembly occurs through a complex and intricate series of steps that often begins in an intracellular compartment and ends in the wall itself, after the polysaccharides have been deposited. The great structural diversity of wall polysaccharides is due to the large number of possible constituent sugars and non-sugar components, the variety of ways these can be linked together, and the many ways in which linear polymers can be branched or modified. Regulation of these steps is central to cell development, because the polysaccharide composition of the wall changes during cell division, elongation and differentiation. At a larger scale, for the plant to assemble its mature organs and tissues in an orderly manner, the stages of wall synthesis in the range of different cell types must be integrated through time and space and must also respond to internal and external developmental cues; how this coordinate regulation of wall synthesis is achieved is unknown and remains a major future research question in plant biology.

6.1.2 General features of plant cell wall biosynthesis

The wall is initially laid down at the membranous cell plate (phragmoplast) that forms after nuclear division. The cell plate fuses with the plasma membrane of the parental cell and divides it into two new cells. The wall material first added to the cell plate persists as the thin pectin-rich 'middle lamella' between adjacent cells. Meristematic cells are initially small and undifferentiated, and the different cell types of mature tissues are produced through subsequent cell expansion and differentiation. Expanding cells are surrounded by a thin (100 nm or less) and highly hydrated (~60% of wet weight) primary wall able to yield to the hydrostatic forces exerted by the protoplast that drive growth. Once the cell has reached its mature size, the primary wall is modified so that it is no longer extensible, and deposition of new wall material produces a rigid secondary wall up to several micrometres thick. The secondary wall, which may also contain lignin, may completely surround a cell or may be formed only in localized regions producing, for example, spirally thickened tracheary elements.

Figure 6.1 shows where different polysaccharides are made in an idealized plant cell. Regardless of where it occurs, polysaccharide synthesis can be broken down into four distinct stages: the production of activated nucleotide-sugar donors, the initiation of polymer synthesis, polymer elongation, and the termination of synthesis (Delmer and Stone, 1988). The key enzymes in wall biogenesis are the polysaccharide (glycan) synthases and glycosyl transferases that catalyse formation of the bonds between adjacent monosaccharides from activated nucleotide-sugar donors. The specificity of these enzymes determines the sequence of sugar residues within a polysaccharide, and their branching pattern.

The simple (unbranched) polysaccharides cellulose and callose are synthesized by processive glycan synthases directly at the plasma membrane (Kudlicka and Brown, 1997; Delmer, 1999). These enzymes are defined as processive because they repeatedly transfer the same type of sugar residue to a growing glycan polymer: for example, a cellulose synthase transfers a Glc residue in β-linkage to the C(O)4 position on the Glc residue at the non-reducing end of the growing cellulose polymer, forming a $(1,4)$-β-glucosidic bond. In contrast, complex (branched) non-cellulosic and pectic polysaccharides are synthesized within the endoplasmic reticulum and Golgi apparatus (Fincher and Stone, 1981; Gibeaut and Carpita, 1994b) using both glycan synthases and glycosyl transferases: first, processive glycan synthases repeatedly transfer single sugar residues to form the growing polysaccharide backbone, then a number of non-processive glycosyl transferases decorate this backbone by each transferring a single sugar residue onto the growing branch. The completed complex polysaccharides are then transported to the cell surface via secretory vesicles and incorporated into the wall. A few unbranched polysaccharides are also synthesized in the Golgi apparatus, such as a mixed-linkage $(1-3),(1-4)$-β-glucan found in grasses and some closely related families, and a glucomannan found in primary and secondary walls. It has been estimated that well over 2,000 different gene products are involved in making and maintaining the cell wall (Carpita et al., 2001).

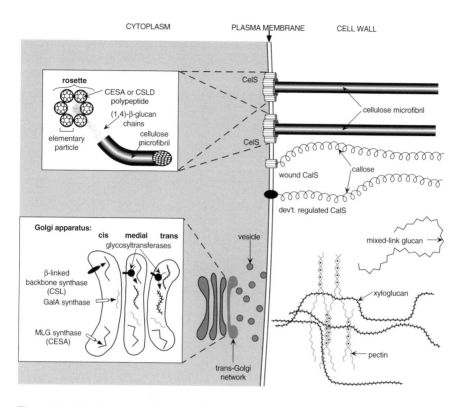

Figure 6.1   Gene families involved in wall polysaccharide synthesis in plants. This figure shows an idealized primary wall and the proposed arrangement and location of some of the major enzymes involved in polysaccharide biosynthesis. The wall shown is idealized because it contains polysaccharides from the type I (xyloglucan) and type II (mixed-linkage glucan, MLG) walls that are not normally found together. Also, in a real wall, these polysaccharides would be interwoven with the cellulose microfibrils and not independent as they are shown here. Chains of (1,4)-β-glucan (cellulose) are produced by cellulose-synthesizing (CelS) complexes, or rosettes, at the plasma membrane that associate with each other to form microfibrils (see Figure 6.2). Members of the CESA family and possibly of the CSLD family are implicated in cellulose synthesis and form part of the CelS complex. Callose, a (1,3)-β-glucan, is also synthesized at the plasma membrane by a callose-synthesizing (CalS) complex. The callose made by wounded cells (wound callose) may be synthesized by a deregulated CelS complex whereas CalS in cell types such as pollen tubes that normally make callose (developmentally regulated callose) may contain members of the GSL family. All other polysaccharides (pectins, xyloglucans and the mixed-linkage glucan are illustrated) are synthesized within the Golgi. CESAs, CSLs and glycosyltransferases (GTs) have all been suggested as being backbone synthases for these polysaccharides. The sugar side chains found on many of these polymers are most likely added by Golgi-localized GTs. The completed polysaccharides are transported to the plasma membrane in vesicles for secretion into the wall. Within the wall, further modification of the polysaccharides can occur either during or after deposition.

This review will cover current work that is elucidating the genes and gene families required for synthesis of individual wall polysaccharides, and will focus on the glycan synthases and glycosyl transferases and enzymes producing their nucleotide sugar substrates. The processes for the modification of polysaccharides by non-glycosyl constituents (e.g. methylation, acetylation and the addition of phenolic groups), although of immense importance in determining the physicochemical properties of the polysaccharides, are outside of the scope of this review.

6.2   Synthesis at the plasma membrane

Only two wall polysaccharides are made at the plant plasma membrane: the β-glucans cellulose and callose. Cellulose is the main fibrillar component of plant walls and is made by all cells. Callose, in contrast, is synthesized transiently at a variety of sites during development, such as cell plates, and as a response to wounding, but is only a permanent component of the wall of one cell type, the pollen tube (Stone and Clarke, 1992).

6.2.1   Use of cytoplasmic UDP-glucose in glucan synthesis at the plasma membrane

Both cellulose and callose are synthesized from the cytoplasmic substrate UDP-Glc which is present in plant cells at millimolar levels (Carpita and Delmer, 1981; Wagner and Backer, 1992; Schlüpmann et al., 1994). The labelling kinetics and rate of turnover of UDP-Glc in excised cotton fibres incubated in medium containing [^{14}C]Glc were consistent with UDP-Glc being the precursor for cellulose synthesis in these cells (Carpita and Delmer, 1981). A similar result was obtained for callose synthesis in cultured pollen tubes incubated in medium containing [^{14}C]sucrose (Schlüpmann et al., 1994). These are the two cell types exhibiting the highest rates of glucan synthesis, and the results have been taken to be generally true for all plant cells. However, interactions of cellulose and callose synthases with membrane-bound forms of sucrose synthase (SuSy; Amor et al., 1995) or other UDP-Glc transferases (Hong et al., 2001b) suggest that there may in addition be specific mechanisms for channeling sugar residues into the active sites of these glucan synthases.

6.2.2   General features of cellulose biosynthesis

Cellulose is a (1,4)-β-glucan, a polymer of (1,4)-β-linked Glc residues, and is the most ubiquitous and abundant plant wall polysaccharide. In general, primary walls contain 10–40% cellulose, whereas secondary walls contain 40–60% cellulose (Bacic et al., 1988). Cellulose is synthesized by terminal complexes that are embedded in the plasma membrane; in higher plants, these take the form of hexagonal 'rosettes' roughly 25 nm across (Brown 1996; Kimura et al., 1999a).

A major question in cellulose synthesis concerns the nature and organization of the proteins that comprise the rosette, the functional cellulose synthase complex which we will refer to as CelS. This structure can presumably be divided into catalytic subunits responsible for $(1,4)$-β-glucan polymer synthesis and regulatory components. The active site that binds substrate, UDP-Glc, is probably located on the cytoplasmic face of the plasma membrane, and conventionally the growing cellulose polymer has been visualized as extruding into the wall through a pore structure made of components of the rosette (Delmer, 1999; Brown and Saxena, 2000). Recently, however, it has been suggested that lipid-linked Glc or oligoglucan precursors move across the plasma membrane, at least during the process of initiation of a cellulose chain, with polymer formation subsequently occurring on the outer face of the membrane (Peng *et al.*, 2002; Read and Bacic, 2002).

A primary wall cellulose microfibril contains about 36 individual $(1,4)$-β-glucan chains. Therefore, if a single hexagonal rosette made a microfibril, each of the 6 elementary particles comprising the rosette could contain 6 polypeptides each capable of polymerizing a single glucan chain (Brown and Saxena, 2000). This model is illustrated schematically in Figure 6.2. An alternative view is that wall cellulose microfibrils are assembled from smaller 'elementary fibrils' each composed of 10 to 15 glucan chains (Ha *et al.*, 1998), and if a rosette only produces an elementary fibril then each rosette subunit need only contain no more than 2–3 CelS polypeptides. A model with more than 36 CelS polypeptides in a rosette would be required if the rosette makes a microfibril of 36 $(1,4)$-β-glucan chains and more than one CelS polypeptide were required to make a single $(1,4)$-β-glucan chain, either for the polymer elongation process itself or if separate CelS polypeptides were required for polymer initiation and elongation (Read and Bacic, 2002). Whatever the case, all models point to a complex structure for the rosette.

6.2.3   *First identification of a cellulose synthase: the* CESA *genes*

The road towards identifying components of the plant CelS complex has been long and arduous. Conventional biochemical approaches to purification are impossible because the enzyme retains virtually no activity *in vitro* despite high levels of CelS activity *in vivo*. Instead, when incubated with the substrate UDP-Glc, isolated cell membranes from numerous plant species produce large amounts of the $(1,3)$-β-glucan, callose, as a wound response (Delmer, 1987; Read and Delmer, 1990; Stone and Clarke, 1992; Kudlicka *et al.*, 1995). Probing plant DNA libraries with the bacterial cellulose synthase catalytic subunit (*CESA*) genes (Lin and Brown, 1989; Lin *et al.*, 1990; Saxena *et al.*, 1990, 1994) was also unsuccessful, as the bacterial probes were not sufficiently similar to hybridize to homologous plant genes (Delmer, 1999). The plant *CESA* genes were finally identified by sequencing and inspection of random clones from a cotton-fibre cDNA library (Pear *et al.*, 1996). The two cotton *CESA* genes (*GhCESA1* and *GhCESA2*) first identified have many of the domains found in the bacterial *CESA* genes and also regions of high sequence similarity, although overall the plant and bacterial genes have less than 30% nucleotide identity. In addition, levels of *GhCESA1* and *GhCESA2* transcripts rise at the onset of second-ary wall synthesis when cotton fibres deposit massive amounts of cellulose, and the bacterially expressed central cytoplasmic region of *GhCESA1* binds UDP-Glc in a Mg^{2+}-dependent but not Ca^{2+}-dependent manner (that is, under conditions that favour synthesis of cellulose rather than callose). Together the data indicated that the cotton *CESA* genes were probably functional homologues of the bacterial *CESA* genes, and Pear *et al.* (1996) concluded that these genes encode the CelS catalytic subunit. This conclusion has subsequently been substantiated by genetic analysis of plants deficient in cellulose deposition (Table 6.1).

Prominent among the regions conserved between the plant and bacterial CESA proteins (that is, the proteins encoded by the *CESA* genes) is the D,D,D,QXXRW motif (where X is any amino acid). This motif was originally identified in sequence alignments by Saxena *et al.* (1995). Enzymes that catalyse transfer of a single sugar residue (non-processive glycosyl transferases: see section 6.3) were suggested to contain just the first two invariant D residues ('domain A'). Enzymes that catalyse multiple sugar additions (processive glycan synthases that make polysaccharide backbones) were suggested to contain the first two invariant D residues ('domain A') plus a third invariant D residue and the QXXRW signature ('domain B') and are classified as members of glycosyl transferase (GT) family 2 (Saxena *et al.*, 1995;

Table 6.1   *Arabidopsis CESA* genes, mutants and their expression patterns

Gene name[1]	Synonym[2]	Vegetative expression pattern[3]
AtCESA1	*RSW* (root swelling) *1*	Primary wall in cotyledons, leaves, hypocotyl, stem, root and vascular system throughout plant
AtCESA2		As for *AtCESA1*
AtCESA3	*IXR* (isoxaben-resistant) *1*	As for *AtCESA1*
AtCESA4	*IRX* (irregular xylem) *5*	Secondary wall in vascular system throughout plant
AtCESA5		Primary wall in cotyledons, leaves, hypocotyl, stem, root, vascular system throughout plant and vegetative meristem
AtCESA6	*PRC* (Procuste) *1 / IXR2*	As for *AtCESA5*
AtCESA7	*IRX3*	As for *AtCESA4*
AtCESA8	*IRX1*	As for *AtCESA4*
AtCESA9		Petiole of rosette leaves
AtCESA10		Petiole of rosette leaves

[1]http://cellwall.stanford.edu/

[2] Synonyms are shown for genes in which mutations have been characterized. The *rsw1* mutation was characterized by Arioli *et al.* (1998a); *ixr1* by Scheible *et al.* (2001); *irx5* by Taylor *et al.* (2003); *prc1* by Fagard *et al.* (2000) and *ixr2* by Desprez *et al.* (2002); *irx3* by Taylor *et al.* (1999); and *irx1* by Taylor *et al.* (2000).

[3]Expression data are from Doblin *et al.* (2003). Associations with primary or secondary wall synthesis are based on mutant analyses. Floral expression data not shown.

Coutinho and Henrissat, 1999). In plants, the GT 2 family comprises essentially the *CESA* genes and related cellulose-synthase-like (*CSL*) genes, but, in other organisms, the GT2 family includes chitin synthases and hyaluronan synthases. In this review we shall maintain this distinction between the two classes of enzyme (glycan synthases and glycosyl transferases) even though, as more examples are described, the simple allocation of these activities to different gene families may become less clear.

Site-directed mutagenesis studies, as well as resolution of the crystal structure of a member of the GT 2 family (SpsA from *Bacillus subtilis*), have shown the critical nature of the D,D,D,QXXRW motif for substrate binding and catalysis (Charnock and Davies, 1999; Charnock *et al.*, 2001; Saxena *et al.*, 2001), and the *GhCESA1* gene product cannot bind UDP-Glc when a region containing the first conserved D residue is deleted (Pear *et al.*, 1996). The results of Jing and DeAngelis (2000) and Yoshida *et al.* (2000) suggest that various hyaluronan synthases, which make a linear polysaccharide of the repeating disaccharide $(1–4)$-β-D-GlcA-$(1–3)$-β-D-GlcNAc, may have separate active sites for GlcA transfer and GlcNAc transfer, and it is possible that some of the *CESA* genes also have a second active site.

The *Arabidopsis* genome has 10 *CESA* genes numbered sequentially from *AtCESA1* to *AtCESA10* (Delmer, 1999; see http://cellwall.stanford.edu/). All the *CESA*-encoded proteins are integral membrane proteins (type III), with two transmembrane domains at the N-terminal end and 5 or 6 transmembrane domains at the C-terminal end (Figure 6.2). If this topology is correct, then the D,D,D,QXXRW motif is on the cytoplasmic side of the plasma membrane. Pear *et al.* (1996) also identified a number of other motifs in the plant CESA proteins. Figure 6.3 shows some of these including the N-terminal 'zinc-finger' domain, thought to be involved in interactions with other protein partners, the 'conserved region-plant' (CR-P) that contains sequences not found in other GT 2 family members, and the 'hypervariable' (HVR) region.

Direct experimental evidence that the *CESA* genes were involved in cellulose synthesis came with the characterization of *Arabidopsis* plants with mutations in these genes. Mutations in some *CESA* genes specifically affect primary wall synthesis. *RSW1/AtCESA1* codes for a 122-kDa polypeptide with 67% amino-acid identity to the GhCESA1 protein (Arioli *et al.*, 1998a). Plants with *rsw1* mutations have reduced levels of crystalline cellulose in their primary walls and increased radial expansion of all cell types, especially roots. Two observations have linked *AtCESA1/RSW1* to cellulose synthesis: the plasma membrane cellulose synthase rosettes disassemble when plants with temperature-sensitive alleles of *rsw1* are grown at non-permissive temperatures (Arioli *et al.*, 1998a), and rosette proteins can be labelled with antibodies to the *RSW1* gene product (Kimura *et al.*, 1999a). Treating plants with the cellulose synthesis inhibitor 2,6-dichlorobenzonitrile (DCB) also mimics the *rsw1* phenotype (Peng *et al.*, 2000; Williamson *et al.*, 2001a). There is thus direct evidence for the protein encoded by the *AtCESA1/RSW1* allele being both in the plasma membrane and part of the CelS complex.

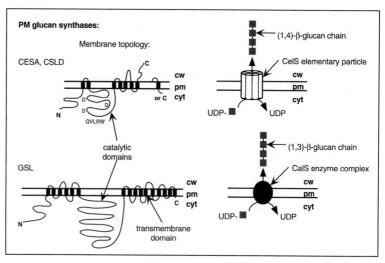

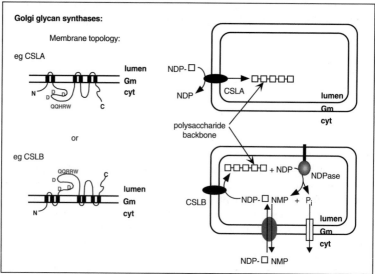

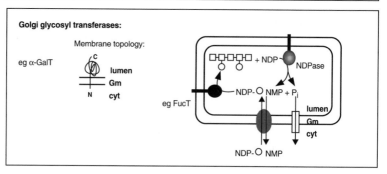

Mutations in another *Arabidopsis* gene, *AtCESA7/IRX3*, are also characterized by reduced cellulose levels, but in this case only in secondary walls. Plants with *irx3* mutations display a 'collapsed xylem' phenotype, and have reduced levels of secondary wall cellulose in xylem cells (Turner and Somerville, 1997; Taylor *et al.*, 1999). Thus it appears that the *CESA* genes are involved in synthesis of cellulose in both primary and secondary walls.

The discovery of the *CESA* genes led to identification of a relatively large number of related *Arabidopsis* and other plant sequences in public databases (Richmond and Somerville, 2000; Hazen *et al.*, 2002). A compilation of these synthase sequences (http://cellwall.stanford.edu/) is maintained by Todd Richmond and Chris Somerville, and gives excellent and up-to-date information on the number of genes in a particular organism and their expression level. In addition to the 'true' *CESA* genes, the *Arabidopsis* genome also contains six groups of cellulose synthase-like (*CSL*) genes, the *CSLA, B, C, D, E* and *G* families (Richmond and Somerville, 2000). There is little evidence for the functional role of the *CSL* genes, but all except the *CSLD* family may encode different polysaccharide synthases located in the endomembrane system (see section 6.3.3); the *CSLD* genes have been suggested to encode cellulose synthases similar to the *CESA* gene products (see section 6.2.6).

Figure 6.2   Topology of the plant polysaccharide synthases.
Upper panel: Glucan synthases found at the plasma membrane (PM). The CESAs and possibly the CSLDs are components of the CelS rosette that have their N-terminal end and large central region (containing the D,D,D,QVLRW motif) on the cytoplasmic (cyt) side of the membrane. Depending on the number of transmembrane helices present (indicated by black vertical bars), the C-terminal end is either cytoplasmic or on the cell wall (cw) side of the membrane. UDP-Glc in the cytoplasm is the activated Glc donor used for cellulose synthesis. GSLs are larger than CESAs and contain ~16 transmembrane helices. The N-terminal and large central domains are both on the cytoplasmic side of the membrane. The structure of the CalS complex is unknown but UDP-Glc is the donor sugar nucleotide.
Middle panel: Glycan synthases found in the Golgi membrane (GM). Some CSL family members have an even number of transmembrane helices at their N-terminal end and therefore have their central catalytic domains on the cytoplasmic side of the membrane. This domain is on the luminal side in CSLs with an odd number of N-terminal membrane helices. The former can utilize NDP-sugars made in the cytoplasm whereas the latter uses NDP-sugars in the Golgi lumen. NDP-sugars are produced in the cytoplasm by either the *de novo* or salvage pathways and then enter the Golgi lumen via specific sugar transporters.
Bottom panel: Glycosyl transferases found in the Golgi membrane. GTs are type II membrane proteins with a short N-terminal domain on the cytoplasmic side of the membrane and large catalytic C-terminal domain on the luminal side of the membrane. These enzymes utilize NDP-sugars that enter the Golgi via specific sugar transporters.

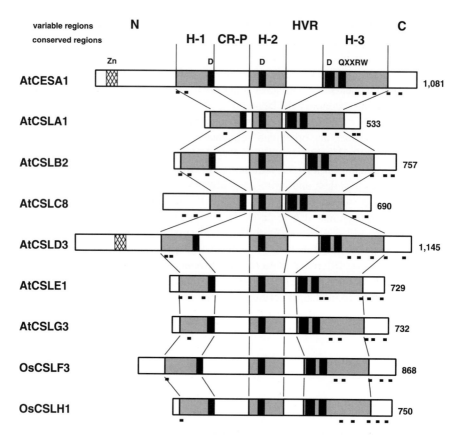

Figure 6.3 Organization of the CESA and CSL gene products. Representative gene products (drawn to scale) of the *CESA* superfamily of *Arabidopsis* (*AtCESA1 – AtCSLG3*) and rice (*OsCSLF3* and *OsCSLH1*) are shown. The boxed regions indicate the variable N-terminal, hypervariable (HVR) and C-terminal regions, the three conserved homology regions (H-1 to 3) that contain the D,D,D,QXXRW motif, and the conserved region – plant (CR-P). Vertical black boxes indicate the U1 to U4 regions that contain the D,D,D,QXXRW sequences, respectively. The zinc-finger domain of the CESAs and the CSLDs is also shown. Numbers to the right indicate the size of each polypeptide and the transmembrane helices are indicated by dashes.

6.2.4 Roles of different CESA family members

CESA expression studies have been undertaken in a number of species (see for instance Holland *et al.*, 2000; Dhugga 2001) in attempts to deduce the role of a particular *CESA* from information on where and when it is expressed. The *CESA* family of *Arabidopsis* has been the most thoroughly characterized (Table 6.1), and different members are expressed to varying levels and in different cell types. For example, *AtCESA1* is much more highly expressed than *AtCESA9*, not only because *AtCESA1* is expressed in many more cells and cell-types than is *AtCESA9*, but also because *AtCESA1* is more highly expressed in those cells (Doblin *et al.,* 2002). In contrast, the expression of *AtCESA7* is limited to the xylem but it is expressed there as highly as is *AtCESA1* in other cell types (Taylor *et al.,* 1999; Doblin *et al.,* 2002). Some *CESA* genes are expressed in several different tissues, but in a cell type that is common to these tissues (Table 6.1). Thus, epidermal cells of all plant organs are affected in the *AtCESA1* (*rsw1*) mutant (Arioli *et al.,* 1998a), and the *AtCESA7* (*irx3*) mutant shows a collapsed xylem phenotype in leaves, hypocotyls, stem and root (Taylor *et al.,* 1999).

Generally, most cell types express two or more *CESA* genes (Table 6.1). Thus, *At-CESA1, 3* and *6* are all expressed in root and hypocotyl cells undergoing expansion, whereas *AtCESA4, 7* and *8* are all expressed in xylem cells. This is consistent with *AtCESA1, 3* and *6* all being involved in synthesis of cellulose in primary walls, and *AtCESA4, 7* and *8* in synthesis of cellulose in secondary walls. Primary wall CESA isoforms are more similar to each other than they are to secondary wall CESA isoforms, and phylogenetic analysis shows that this grouping is repeated for a taxonomically broad range of plants (Holland *et al.,* 2000). This specialization of function may thus have an ancient origin that predates the origin of flowering plants.

Although the expression patterns of different *CESA* genes overlap, mutant analysis indicates that they do not have overlapping functions. Mutations in any of six *Arabidopsis CESA* genes have observable phenotypes that are related to impaired cellulose production (Table 6.1) and, even though at least two other *CESA* genes are expressed in affected cells, the non-mutated genes do not complement the mutant genes. One possibility is that different but functionally identical CESA isoforms are randomly assembled into rosettes, with single defective subunits affecting rosette stability or function and reducing cellulose production. However, this should give either dominant or semi-dominant mutations, while all known *CESA* mutations that lead to reduced cellulose accumulation are recessive, only having a phenotype when in the homozygous state (Arioli *et al.,* 1998a; Fagard *et al.,* 2000; Taylor *et al.,* 2000). Furthermore, if the CESA isoforms were interchangeable, then it should be possible to complement a mutation in one by over-expression of the wild-type gene for another. This, however, does not appear to be the case, as mutations in *AtCESA1/RSW1* cannot be complemented by a wild-type *AtCESA3* gene (Burn *et al.,* 2002), even though both genes are involved in synthesizing the primary wall (Table 6.1). The best explanation for these data is that incorporation of CESA proteins into rosettes is non-random, and that at least three primary wall and three secondary wall *CESA* genes are not functionally redundant (Fagard *et al.,* 2000; Perrin, 2001).

The major question thus becomes the different functions of these various CESA isoforms. However, it is not clear how many different CESA isoforms are required for cellulose production within a single cell. Promoter-GUS fusion experiments indicate that certain *Arabidopsis* cell-types express only two *CESA* genes: for example, only *AtCESA9* and *AtCESA10* are expressed in cells at the base of the rosette leaves where the inflorescence stem is attached (Doblin *et al.,* 2003). There are also cell-types in which many *CESA* genes are expressed: the primary wall *CESA* genes *AtCESA1, 3* and *6* are expressed in xylem in addition to *AtCESA4, 7* and *8*, and *AtCESA2* and

AtCESA5 are expressed in the vascular bundles in expanding tissues in addition to *AtCESA1, 3* and *6* (Table 6.1). It is also not clear how many different CESA proteins are needed to form a functional rosette. Taylor *et al.* (2000) showed that 'solubilized' *AtCESA7* and *AtCESA8* 'co-purify', which is consistent with them being part of the same rosette, and a variety of models of rosette structure and function propose more than one type of CESA protein in each rosette (Perrin, 2001; Scheible *et al.,* 2001; Read and Bacic, 2002).

One explanation for the number of CESA proteins required is that the processes of cellulose chain initiation and chain elongation may be different (Delmer, 1999; Brown and Saxena, 2000; Williamson *et al.,* 2001b). There is evidence for the existence of a specific, lipid-linked primer for chain initiation (Peng *et al.,* 2002). Sitosterol-β-glucoside (SG) is synthesized from sitosterol and UDP-Glc at the inner face of the plasma membrane by the enzyme UDP-Glc:sterol glucosyltransferase (Warnecke *et al.,* 1997; Cantatore *et al.,* 2000). When cotton-fibre membranes are incubated with SG and UDP-Glc, they synthesize short (1,4)-β-D-gluco-oligosaccharides linked to sitosterol, known as sitosterol cellodextrins (SCDs). This glycosyl transferase activity is due to a *CESA* gene product, because yeast membranes expressing non-mutated *GhCESA1* synthesize SCDs when incubated with the same substrates, whereas yeast membranes expressing a mutated form of GhCESA1 not containing the first conserved D residue do not carry out this reaction (Peng *et al.,* 2002). Two supportive pieces of evidence come from studies with herbicides that specifically inhibit cellulose synthesis. Levels of SG and SCD are reduced upon treatment of cultured cotton ovules with DCB, and the resultant cellulose-deficient phenotype is reversed by adding SG (Peng *et al.,* 2002); and cellulase digestion of the non-crystalline cellulose that accumulates after treatment of cells with the herbicide CGA 325'615 releases not only CESA proteins but also small amounts of sitosterol linked to Glc (Peng *et al.,* 2001, 2002).

Elongation of the initial SCDs to form a polymeric cellulose molecule may first require cleavage from the sitosterol primer. This would explain the involvement in cellulose synthesis of the membrane-bound (1,4)-β-glucanase *KORRIGAN (KOR)* (Peng *et al.,* 2001, 2002). Some CESA isoforms may thus be required for synthesis of SCDs during chain initiation, while others would be involved in elongating the glucan chains after cleavage from the lipid (Read and Bacic, 2002). All CESA isoforms would utilize UDP-Glc to form a (1,4)-β-glucosidic linkage, the only difference being in the acceptor molecule, and each cellulose synthase rosette may contain different CESA isoforms responsible for initiation and elongation (Read and Bacic, 2002).

In order to account for the ~180° rotation of adjacent Glc residues in (1,4)-β-glucan chains, some models of glucan polymerization have proposed that each CESA polypeptide has more than one substrate binding site (Albersheim *et al.,* 1997; Carpita and Vergara, 1998; Buckeridge *et al.,* 1999). This appears to be the case for a mouse hyaluronan synthase, with transfer of GlcNAc residues occurring at the site defined by the D,D,D,QXXRW motif, and transfer of GlcA residues occurring at a second glycosyl transfer site (Yoshida *et al.,* 2000). However, the crystal structure of the SpsA GT 2 Family synthase shows that this enzyme has only a single glycosyl transfer site (Charnock and Davies, 1999; Charnock *et al.*, 2001). It is possible that two adjacent CESA polypeptides, each with one catalytic site, could in combination add a cellobiose unit (that is, two adjacent (1,4)-β-linked Glc residues) to the growing chain (Vergara and Carpita, 2001). CESA dimerization may also favour the formation of a channel through which the growing glucan chain could be secreted (Vergara and Carpita, 2001). However, if a rosette makes the 36 glucan chains of a microfibril, then this model would imply that there are 72 CESA proteins in the cellulose synthase rosette, or even more if separate CESA polypeptides are needed for chain initiation! This number is reduced if some CESAs have two catalytic sites.

Associations between CESA polypeptides may be mediated by N-terminal sequences. As mentioned above, all plant *CESA* genes have an N-terminal zinc-finger domain similar to the RING finger motif known to mediate protein-protein interactions in eukaryotic systems (Figure 6.3). Yeast two-hybrid experiments have shown that the Zn-finger domain of GhCESA1 is able to interact with itself to form homodimers, and will also form heterodimers with the Zn-finger domain of GhCESA2 in a redox-dependent manner (Kurek *et al.*, 2002). CESA monomers, dimers and tetramers were also detected in membranes of yeast cells expressing the full-length *GhCESA1* cDNA, using an antibody to the GhCESA1 Zn-finger motif (Kurek *et al.*, 2002). Further characterization of rosette components and their interactions, as well as data on which CESA proteins are expressed in particular cell-types, is required before the process of selection and assembly of CESA proteins into a rosette is understood.

6.2.5   *Other components of the cellulose synthase machinery*

The KOR endo-(1,4)-β-D-glucanase is present in all plant organs, with highest levels in rapidly expanding tissues like dark-grown hypocotyls (Brummell *et al.*, 1997; Nicol *et al.*, 1998; Zuo *et al.*, 2000). The phenotypes seen in *kor* mutants, such as a short hypocotyl with increased radial expansion, wavy cell plates and incomplete cell walls, are reminiscent of the *rsw1* phenotype and of wild-type plants grown in the presence of herbicides that inhibit cellulose deposition (Nicol *et al.*, 1998; Zuo *et al.*, 2000; Lane *et al.*, 2001; Sato *et al.*, 2001). This suggests a specific role for KOR in cellulose biosynthesis, and Peng *et al.* (2001, 2002) have proposed that KOR is responsible for cleavage of lipid-linked (1,4)-β-glucan chains during initiation of polymerization.

Immunolocalization studies show that sucrose synthase (SuSy), which catalyses the formation of UDP-Glc from sucrose, associates with sites of cellulose synthesis. SuSy is located between the cortical microtubules and wall in regions of cotton fibres depositing cellulose (Amor *et al.*, 1995; Haigler *et al.*, 2001), and is enriched near the plasma membrane underlying *Zinnia* tracheary element thickenings (Amor *et al.*, 1995; Salnikov *et al.*, 2001). SuSy is a peripheral membrane protein that has been proposed to provide UDP-Glc directly to the CESA proteins, thus channeling the Glc moiety of sucrose into cellulose (Amor *et al.*, 1995). A 93-kDa polypeptide, similar in size to the membrane-bound form of SuSy, has been identified in cotton fibre membrane extracts purified by a CelS 'product entrapment' procedure (Li *et al.*, 1993), and antisense downregulation of SuSy in potato tubers leads to a reduction in cellulose levels (Haigler *et al.*, 2001). SuSy is thus associated with cellulose synthesis, but may or may not be an integral component of the CelS rosette, and protein-protein interaction screens using plant-specific regions of GhCESA1 have not provided evidence for a direct interaction between CESA polypeptides and SuSy (D. Delmer, personal communication).

The cytoskeleton has long been viewed as playing a role in cellulose biosynthesis, with the cortical microtubule network being proposed to align the cellulose microfibrils as these are deposited into the wall (reviewed in Baskin, 2001). However, in both cotton fibres and *Zinnia* tracheary elements, a disorganized pattern of cellulose deposition results from treatment with compounds that disrupt either microtubules or actin microfilaments, indicating that actin is also involved (Kobayashi *et al.*, 1988; Seagull 1990). Indeed, actin may play a role in setting the pattern of cortical microtubules, which in turn directs the pattern of cellulose microfibril deposition (Delmer and Amor, 1995). The most recent alignment model, 'templated incorporation', proposes that a scaffold of wall polysaccharides, and plasma membrane proteins in contact with microtubules, directs the orientation of the cellulose microfibril as it is deposited into the wall (Baskin, 2001). A novel CelS purification method suggests, however, that the CelS is able to interact with tubulin directly: eluate from an anti-tubulin column loaded with 'solubilized' plasma membranes from azuki beans contained 10-nm granules that produced $(1,4)$-β-glucan in the presence of UDP-Glc and Mg^{2+}, with synthesis being inhibited by DCB (Mizuno, 1998). Dissociation of these granular structures released eight polypeptides, including α- and β-tubulin (50 and 51 kDa), polypeptides at 120 kDa and 93 kDa which may represent CESA and SuSy subunits respectively (Mizuno, 1998), and an 18-kDa polypeptide that may be the same as the uncharacterized 18-kDa protein in cotton fibre extracts that was specifically labelled with a photoreactive analogue of DCB (Delmer *et al.*, 1987). A possible interaction of SuSy with actin has also been proposed (Winter *et al.*, 1998; Haigler *et al.*, 2001).

A small GTPase, Rac13, has been implicated in regulation of cellulose synthesis. Expression of the *Rac13* gene increases in the transition phase from primary to secondary wall synthesis in cotton fibres (Delmer *et al.*, 1995), coinciding with the production of hydrogen peroxide (H_2O_2) which has been shown to stimulate cellulose synthesis during fibre differentiation (Potikha *et al.*, 1999). A dominant-active form of Rac13 constitutively activates H_2O_2 production in soybean and *Arabidopsis* cell cultures (Potikha *et al.*, 1999). Kurek *et al.* (2002) have shown that the formation of dimers and higher-order aggregates of CESA proteins (and thus possibly the formation of functioning rosettes) is favoured under oxidative conditions. The cellulose synthesis inhibitor CGA 325'615 causes rosette disassembly and accumulation of non-crystalline glucan in cotton ovules and *Arabidopsis* seedlings, but these effects are completely reversed by H_2O_2, indicating that the herbicide could act by reducing the ability of CESA proteins to be oxidized and incorporated into active rosettes (Peng *et al.*, 2001; Kurek *et al.*, 2002).

A few mutations that affect genes involved in synthesizing or modifying N-gly-cans also lead to reduced levels of cellulose accumulation. *Arabidopsis* embryos with the lethal *cyt-1* mutation have strongly reduced levels of cellulose, and are characterized by radial swelling and abnormal wall structures (Nickle and Meinke, 1998; Lukowitz *et al.*, 2001). The *CYT1* gene encodes a Man-1-P guanylyltransferase, an enzyme required for the production of GDP-Man, GDP-Fuc, ascorbic acid, glycosyl-phosphatidylinositol membrane anchors, and the core N-glycan of glycoproteins. A second embryo-lethal mutation, *knf*, also produces embryos that are radially swollen and have a strongly reduced cellulose content (Mayer *et al.*, 1991; Boisson *et al.*, 2001; Gillmor *et al.*, 2002). KNF encodes α-glucosidase I, an enzyme that removes terminal Glc from the N-linked glycan precursor $Glc_3Man_9GlcNAc_2$. Together these mutant phenotypes indicate a role for N-glycosylation at some stage in the process of cellulose synthesis, although there is no evidence to indicate that the CESA proteins themselves are glycoproteins even though they contain the consensus sequence for N-glycosylation (Lukowitz *et al.*, 2001).

The genetic lesions in cellulose-deficient mutants such as *tbr* (trichome birefrin-gence, impaired in its ability to synthesize secondary wall cellulose in trichomes and other cell types; Potikha and Delmer 1995) and the *brittle culm* lines of barley (which have reduced ability to produce secondary wall cellulose; Kokubo *et al.*, 1991; Kimura *et al.*, 1999b) still remain to be characterized.

6.2.6   *Involvement of* CSLD *genes in cellulose biosynthesis*

As stated earlier, the *Arabidopsis* genome contains six families of *CSL* genes: *CSLA*, *B*, *C*, *D*, *E* and *G* (Richmond and Somerville, 2000, 2001). As Figure 6.3 shows, each *CSL* family is characterized by specific sequence features, but all have the D,D,D,QXXRW motif and thus are members of the GT-2 family. We will discuss these genes further in section 6.3.3 and here restrict ourselves to discussing just one of these families, the *CSLD* genes, which have been suggested to encode cellulose synthases.

Of all the *CSL* gene families, the *CSLD* genes are the most closely related in sequence to the *CESA* genes (Richmond and Somerville, 2000). All CSLD proteins have the homology (H) and plant-specific regions (CR-P and HVR) found in the CESA proteins (Figure 6.3), probably also the same plasma membrane topology (Doblin *et al.*, 2001; Figure 6.2), and a modified Zn-finger motif at their N-terminal end. The *CSLD* genes appear to be expressed by tip-growing cells such as pollen tubes and root hairs, and may encode proteins involved in cellulose synthesis in these highly specialized cells. Mutations in *AtCSLD3/KOJAK* affect the ability of *Arabi-dopsis* root hairs to elongate, with a phenotype that indicates *AtCSLD3* is required for synthesis of the primary wall at the root hair tip, and the reduced wall strength in mutant root hairs suggesting the affected polysaccharide is cellulose (Favery *et al.*, 2001; Wang *et al.*, 2001). Similarly, *NaCSLD1* appears responsible for the synthesis of cellulose in *Nicotiana alata* pollen tubes, cells that do not appear to express a 'true' *CESA* gene yet make cellulose (Doblin *et al.*, 2001). The observation that *AtCSLD3*, when transiently expressed as a fusion protein with green fluorescent protein (GFP), localizes to the endoplasmic reticulum argues against the involvement of CSLD proteins in cellulose synthesis (Favery *et al.*, 2001). Further experiments that determine where the CSLD proteins reside in the cell will help resolve this issue.

AtCSLD3 is expressed not only in roots but also in other tissues such as leaves, stems, flowers and siliques (Favery *et al.*, 2001; Wang *et al.*, 2001; Doblin *et al.*, 2003), but the *KOJAK* mutant only has a phenotype in root hairs, suggesting that there is functional redundancy in other tissues (Wang *et al.*, 2001). The recessive nature of the *csld3-1* mutation supports this hypothesis (Wang *et al.*, 2001). At least two other *CSLD* genes (*AtCSLD2* and *AtCSLD5*) and possibly *AtCESA1, 3* and *6*, are also expressed in root hairs (Doblin *et al.*, 2003). If all three CSLD proteins are involved in cellulose synthesis, there may be as many as six genes encoding CelS catalytic subunits of non-identical function within this cell-type. Since a root-hair phenotype was not observed in the *rsw1, ixr1* or *procuste* mutants, the CSLD proteins could be redundant to the CESA isoforms, but alternatively the CESA isoforms in root hairs may not be incorporated into rosettes or be involved in cellulose synthesis in this cell-type. Detailed wall analyses of the four *AtCSLD3* mutant alleles should be able to determine their exact wall deficiency, aiding in the task of assigning function to the *CSLD* genes.

6.2.7   *Callose, callose synthases, and the relationship between callose deposition and cellulose deposition*

Given that six of the ten *Arabidopsis CESA* genes have now been specifically linked with defects in cellulose synthesis, it would seem likely that all CESA proteins are involved in synthesis of cellulose, rather than another polysaccharide. However, there are suggestions that a *CESA* gene may be responsible for synthesis of a mixed-linkage (1,3),(1,4)-β-glucan in the Golgi apparatus (see section 6.3.3), and there are observations that may best be explained if some of these *CESA* genes are also able to synthesize the (1,3)-β-glucan callose at the plasma membrane.

Callose is widely distributed within plant tissues, but in much lower quantities than cellulose. Callose is composed of a triple helix of a linear homopolymer of (1,3)-β-linked D-Glc residues with occasional (1,6)-β-linked branches (Stone and Clarke, 1992), and the different linkage between the Glc residues gives callose quite different physico-chemical properties to cellulose. Callose can form either fibrils or a more space-filling, gel-like structure, and has structural as well as protective and possibly other functions (Stone and Clarke, 1992). Callose is deposited transiently in various specialized walls or wall-associated structures at particular stages of cell growth and differentiation (Fincher and Stone, 1981; Stone and Clarke, 1992), for example at the cell plate in dividing cells (Samuels *et al.*, 1995; Scherp *et al.*, 2001), at plasmodesmata and sieve plates, in cotton fibres (Meinert and Delmer, 1977) and various other hairs, and in development of pollen and eggs, that is, microsporogenesis and megasporogenesis (Rodkiewicz, 1970). In higher plants, callose deposition is also induced under certain physiological conditions in which the plasma membrane is perturbed. These conditions include mechanical wounding, chemical treatment, physiological stress or pathogen attack (Fincher and Stone, 1981; Read and Delmer, 1990; Stone and Clarke, 1992). The capacity for deposition of 'wound callose' is widespread within plant tissues and is thought to be a protection mechanism, creating a physical barrier at the plasma membrane.

CelS and callose synthase (CalS) enzymes both reside in the plasma membrane, utilize UDP-Glc as substrate, and have similar maximum activities (Read and Delmer, 1990). However, these enzymes have opposing modes of regulation, with progressive cell disruption causing an increasing rate of callose synthesis in proportion to the decreasing rate of cellulose synthesis and disappearance of plasma membrane rosettes (Read and Delmer, 1990). Isolated membranes from a range of plant tissues have a high level of CalS activity that uses UDP-Glc as substrate to make long, insoluble chains of (1,3)-β-glucan. This CalS activity requires micromolar levels of Ca^{2+} (Hayashi *et al.*, 1987), and release of intracellular Ca^{2+} on cell disruption is one factor that can cause its rapid activation (Kohle *et al.*, 1985); other factors include release of vacuolar furfuryl-β-glucoside (Ohana *et al.*, 1993) and changes in the lipid environment (Kauss and Jeblick, 1986).

The opposing regulation of CelS and CalS has led to the suggestion that the CalS enzyme activated post-translationally upon membrane disruption is a deregulated form of CelS (Figure 6.1; Delmer, 1977; Jacob and Northcote, 1985; Delmer, 1987; Girard and Maclachlan, 1987; Read and Delmer, 1990). There is no direct evidence for this conversion in higher plants, but corroborative evidence comes from other organisms. First, the *CRDS* gene of *Agrobacterium* is a bacterial member of the GT Family 2 and related to the *CESA* genes, but synthesizes (1,3)-β-glucan (Stasinopoulos *et al.*, 1999). Second, the slime-mold *Dictyostelium discoideum* contains a single *CESA* gene and synthesizes only cellulose in a developmentally regulated manner *in vivo*; however, isolated membranes from *D. discoideum* make (1,3)-β-glucan when incubated with UDP-Glc, and only do so if the single *CESA* gene is not disrupted (Blanton and Northcote, 1990; Blanton *et al.*, 2000). Third, grasses and related species make a mixed-linkage (1–3),(1–4)-β-glucan *in vivo* but isolated Golgi membranes synthesize (1,3)-β-glucan, consistent with the suggestion that the mixed-linkage glucan synthase is encoded by a *CESA* gene (Dhugga, 2001; Vergara and Carpita, 2001).

The pollen tube is the only cell type in which callose is the main structural polysaccharide throughout growth and development, rather than transiently, and it synthesizes callose and cellulose simultaneously (Stone and Clarke, 1992; Ferguson *et al.*, 1998; Li *et al.*, 1999). Pollen tubes are also distinct in that their extracts contain a Ca^{2+}-independent CalS activity, which is consistent with callose being deposited behind the growing tube tip where Ca^{2+} concentrations are not elevated (Schlüpmann *et al.*, 1993). In addition, pollen tubes do not appear to express any *CESA* genes, which may explain their lack of a Ca^{2+}-dependent wound-activated CalS if this is indeed a deregulated *CESA* gene product; instead a *CSLD* gene product has been linked to cellulose deposition in pollen tubes (Doblin *et al.*, 2001; see section 6.2.6).

6.2.8   *Identification of callose synthases: the* GSL *genes*

While CalS is active after cell breakage, purification of this large, integral membrane enzyme to homogeneity has not been achieved, although there are very many reports of its partial purification and characterization of associated polypeptides that may or may not represent the catalytic subunit (examples include Frost *et al.*, 1990; Delmer *et al.*, 1991; Dhugga and Ray, 1994; Kudlicka and Brown, 1997; McCormack *et al.*, 1997; Turner *et al.*, 1998). However, in the last two years, a family of glucan synthase genes have been described from plants that are candidates for encoding at least some CalS enzymes. These genes have been named glucan-synthase-like, or *GSL*, genes, because of their similarity to the *FKS* family of fungal genes that encode (1,3)-β-glucan synthases (Douglas *et al.*, 1994), but they have no sequence similarity with the *CES/CSL* superfamily. *GSL* genes have been linked to deposition of the plant (1,3)-β-glucan callose in pollen tubes (*NaGSL1*: Doblin *et al.*, 2001) and cotton fibres (*CFL1*: Cui *et al.*, 2001), at cell plates (*CALS1*: Hong *et al.*, 2001a), and generally during systemic-acquired resistance (Østergaard *et al.*, 2002); other reports relate to suspension-cultured *Lolium* endosperm cells (Wardak *et al.*, 1999), *Hieracium* ovules (Paech *et al.*, 1999) and barley (Li *et al.*, 2001). A compilation of the *GSL* genes is maintained at http://cellwall.stanford.edu/gsl/index.shtml.

The *GSL* genes all encode large, integral membrane polypeptides: for example, *NaGSL1* is predicted to encode a polypeptide of 1931 amino acids with a molecular mass of 221 kDa (Doblin *et al.*, 2001), close to the size of the *FKS*-encoded protein in yeast at 215 kDa. The plant sequences also resemble the yeast sequences in having two groups of transmembrane helices, one toward the N-terminus and one at the C-terminus, and two major cytoplasmic regions, an N-terminal region of approximately 500 amino acids and a central region of approximately 700 amino acids (Figure 6.2). Overall, the deduced GSL and FKS proteins are about 20% identical (40% similar) at the amino-acid sequence level, which is close to the degree of similarity found between plant and bacterial CESA proteins; as a group the plant *GSL* genes are more closely related to each other than to any of the fungal *FKS* sequences (Doblin *et al.*, 2001).

The highest percentage identity between the deduced GSL and FKS proteins lies in their central cytoplasmic region which presumably contains the active site, although the consensus amino-acid motif D,D,D,QXXRW for binding UDP-Glc in the GT 2 family (Campbell *et al.*, 1997; Coutinho and Henrissat, 1999; Henrissat *et al.*, 2001) is not present. Østergaard *et al.* (2002) report that the putative UDP-Glc binding sequence RXTG is present in *AtGSL5*, and Hong *et al.* (2001a) report the similar sequence KSGG in *CALS1* (now *AtGSL6*) although they did not find sufficient evidence to support this representing the UDP-Glc binding site of the *GSL* genes. A large number of other possible functional domains have been deduced to lie in the GSL sequences (Cui *et al.*, 2001; Doblin *et al.*, 2001; Hong *et al.*, 2001a;

Østergaard *et al.*, 2002). *Arabidopsis* contains 12 identified *GSL* genes (http: //cellwall.stanford.edu/gsl/index.shtml), presumably related to deposition of callose in a number of specialized cell types and structures. Two of the genes, *AtGSL1* and *AtGSL5,* are distinct from the remainder of the family, being 120–150 amino acids shorter and with a different gene organization, containing only two and three introns respectively compared to the approximately 40 introns of the other 10 family members.

Doblin *et al.* (2001) used RT-PCR with pollen-tube RNA and primers designed to regions conserved between *FKS1*, *FKS2* and homologous plant ESTs to detect, then clone, transcripts of *NaGSL1* in *Nicotiana alata*. *NaGSL1* is abundantly expressed in late-stage anthers, pollen grains and developing pollen tubes, but expression was not detected in any other tissue. A second gene, *NaGSL2*, is expressed at low levels in immature floral organs. The *NaGSL1* gene was predicted to encode the catalytic subunit of the pollen-tube CalS enzyme. *NaGSL1* is the orthologue of the *Arabidopsis* gene *AtGSL2* (Doblin *et al.*, 2001).

CalS activity appears and callose deposition commences within 1 hour of hydration of *N. alata* pollen, just as the grains germinate, and increases during tube growth (Li *et al.*, 1999). Mature pollen grains contain large amounts of the *NaGSL1* transcript (Doblin *et al.*, 2001), and translation of this mRNA as well as new transcription may contribute to the increase in CalS activity at pollen germination, then during tube growth. There are also substantial amounts of latent CalS activity in isolated, permeabilized pollen-tube membranes, with this latent CalS being activated by protease treatment or specific detergents that may alter lipid-protein interactions (Schlüpmann *et al.*, 1993; Li *et al.*, 1997). Localized activation of a zymogen form of CalS may be responsible for synthesis of thick callose plugs that 'wall off' older sections of the tube (Li *et al.*, 1999).

Cui *et al.* (2001) used 5′-RACE, again with primers designed to regions conserved between *FKS1* and homologous plant ESTs, to identify the *CFL1* gene expressed in cotton fibres. *CFL1* is expressed in young roots and seedlings as well as in cotton fibres during primary wall development. The N-terminal region of the expressed protein was able to bind calmodulin, but it was not clear whether this was related to any Ca^{2+}-dependence of the encoded CalS activity. Antibodies raised against *CFL1* sequences detected a polypeptide larger than 200 kDa in solubilized cotton fibre membranes, and antigenically cross-reactive material was also found in the callose produced during purification of CalS by product-entrapment (Cui *et al.*, 2001). *CFL1* is most similar to the *Arabidopsis* gene *AtGSL10* (http: //cellwall.stanford.edu/gsl/index.shtml).

Hong *et al.* (2001a) linked the activity of a gene they named *CALS1*, but now called *AtGSL6* (http://cellwall.stanford.edu/gsl/index.shtml), to deposition of callose at the cell plate between dividing cells. *CALS1* was discovered because it is adjacent in the *Arabidopsis* genome to *UGT1*, a cell-plate-specific UDP-Glc transferase (Hong *et al.*, 2001b). A GFP::CALS1 fusion protein expressed in transgenic tobacco BY-2 cells was located at growing cell plates during cell division, and the cells had more callose at their cell plates and a 40–60% increase in CalS activity levels (Hong *et al.,* 2001a); the GFP-tagged CALS1 protein also co-purified with endogenous CalS activity. When the experiment was repeated with a truncated CALS1 protein missing the C-terminal set of transmembrane helices, the resultant GFP fusion protein still located to cell plates but CalS activity was not elevated.

Some *GSL* genes may also respond to environmental conditions. *AtGSL5* is expressed in flowers (Østergaard *et al.,* 2002) and (from EST data) in roots (http://cellwall.stanford.edu/gsl/index.shtml). However, *AtGSL5* expression in rosette leaves was increased after treatment of plants with salicylic acid, which induces systemically acquired resistance to pathogens, and was also increased in the *Arabidopsis mpk4* mutant that displays constitutive systemically acquired resistance and elevated levels of callose and CalS activity (Østergaard *et al.,* 2002). Induction of *AtGSL5* and possibly other *GSL* genes is thus part of plant responses to pathogen attack.

Direct evidence that the *GSL* gene family encodes CalS enzymes is still being collected, to support the deduction made from their similarity to the yeast *FKS* genes. Supportive evidence is that Østergaard *et al.* (2002) reported partial complementation of a yeast *fks1* mutant by *AtGSL5*; Hong *et al.* (2001a) showed that transgenic expression of *CALS1 (AtGSL6)* elevated CalS activity; and Li *et al.* (2001) reported that antibodies raised against the expressed cytoplasmic domain of a barley *GSL* gene detected a 220-kDa polypeptide in fractions enriched in CalS activity, with mass-spectrometric analysis of tryptic peptides from these enriched fractions showing several matches with peptides deduced from the barley *GSL* and another *Arabidopsis GSL* gene. To date, however, no *Arabidopsis GSL* mutants have been identified or characterized.

As for the *CESA* genes, the distinct roles and expression patterns of most of the 12 *Arabidopsis GSL* genes remain unknown, as does the number of GSL proteins in a CalS complex or the possibility that various *GSL* genes have overlapping functions. One difficulty is that developmentally regulated CalS activity cannot often be reliably assayed as it is overwhelmed by the wound-activated CalS of isolated plasma membranes; this wound-activated activity may be due to one particular *GSL* gene product, may be a property of the sum of the *GSL* genes expressed in a tissue, or may not be due to a *GSL* gene product at all.

6.2.9   Other components of the callose synthase machinery

Many polypeptides (reviewed in Turner *et al.,* 1998) have been implicated as being components of the wound-activated CalS enzyme complex, by virtue of co-purification with enzyme activity or labelling with analogues of substrate UDP-Glc, but few of these have been identified or proven to have a functional role in synthesis of callose and there is no genetic evidence to support their role. There are occasional reports of labelling or co-purification with CalS of large polypeptides of the size of the *GSL* gene products (Delmer *et al.,* 1991; Turner *et al.,* 1998) but even there the possibility of contamination, for example with the 190-kDa polypeptide of clathrin (Turner *et al.,* 1998), cannot be ruled out. Components that have been linked to callose synthesis include the 65-kDa polypeptide of annexin which localizes with CalS at plasmodesmata and may control metal-ion activation of CalS (Andrawis *et al.,* 1993; Delmer *et al.,* 1993), and SuSy which also colocalizes with CalS at the cell plate and may channel a Glc residue from sucrose into a UDP-Glc molecule committed to transfer it directly to the elongating glucan chain (Amor *et al.,* 1995).

Hong *et al.* (2001a, b) showed a direct linkage between the cell-plate-specific *CALS1 (AtGSL6)* and two other polypeptides, namely the cell-plate structural protein phragmoplastin, and a novel 60-kDa UDP-Glc transferase named UGT1. This transferase may represent the 57-kDa polypeptide that co-purifies with the wound-activated CalS and that can be labelled with 5-azido-UDP-Glc (Frost *et al.,* 1990). Hong *et al.* (2001b) proposed that UGT1 channels Glc residues from SuSy to CalS at the growing cell plate.

UGT1 also bound GTP-Rop1, a Rho-like protein (Hong *et al.,* 2001b). Rho-like proteins have a number of roles but one of these, in yeast, is to regulate (1,3)-β-glucan synthase activity (Qadota *et al.,* 1996). The observation of Hong *et al.* (2001b) may indicate that the plant *GSL* genes have similar regulatory properties to the fungal *FKS* gene products, in addition to their sequence similarity. The yeast enzyme, unlike plant CalS, requires added GTP for activity, and this binds to a separate, soluble regulatory subunit now known to be a prenylated version of Rho. In addition, sequences similar to the yeast *GNS1* gene are also found within the *GSL* genes (Hong *et al.,* 2001a); yeast *GNS1* mutants confer resistance to anti-fungal echinocandins and have defects in (1–3)-β-glucan synthase activities (El-Sherbeini and Clemas, 1995). However, the pollen tube CalS is echinocandin insensitive.

6.3   Synthesis in the Golgi apparatus

6.3.1   General features of polysaccharide synthesis in the Golgi

The plant Golgi apparatus is responsible for synthesis of wall matrix polysaccharides that bind to and surround cellulose microfibrils. There are two main classes of matrix polysaccharides, non-cellulosic polysaccharide (cross-linking glycans also known as hemicelluloses) and pectins, and these polymers form two distinct domains within the cell wall matrix (Bacic *et al.,* 1988; Carpita and Gibeaut, 1993). The detailed structures of the non-cellulosic polysaccharides are covered elsewhere in this volume.

The non-cellulosic cross-linking glycans have (1–4)-β-linked backbones with a variable degree of substitution by side branches, and they form non-covalent interactions with cellulose. The major cross-linking glycans in the primary walls of flowering plants are xyloglucans and glucuronoarabinoxylans, with mannans and galacto(gluco)mannans being less abundant but widely distributed. The (1–3),(1–4)-β-glucan, also called a mixed-linkage glucan, found in grasses and related families, is an example of a cross-linking glycan with an unsubstituted backbone; it contains runs of (1–4)-β-D-linked glucosyl units connected by single (1–3)-β-D-linkages. The major cross-linking glycan in secondary walls of flowering plants are xylans (Bacic *et al.,* 1988; Carpita and Gibeaut, 1993).

Pectins are structurally complex polysaccharides rich in D-galacturonic acid (GalA) residues. The three types of pectic polysaccharides are homogalacturonan (HGA), rhamnogalacturonan I (RG-I), and rhamnogalacturonan II (RG-II). HGAs are homopolymers of $(1-4)$-α-D-GalA; in some plants, portions of HGA contain a branching β-D-Xyl linked to GalA residues forming a xylogalacturonan (Mohnen, 1999; Ridley *et al.,* 2001). The RG-I backbone contains repeats of the disaccharide $(1-2)$-α-L-Rha-$(1-4)$-α-D-GalA, with Rha residues further substituted with other linear and branched polysaccharides such as $(1-5)$-α-L-arabinans and $(1-4)$-β-D-(arabino)galactans. RG-II is a substituted galacturonan with a backbone of $(1-4)$-α-D-GalA residues substituted with four structurally different oligosaccharide side chains at different locations (Ridley *et al.,* 2001).

The plant Golgi apparatus has a dynamic organization that changes throughout development and during the cell cycle (Nebenführ and Staehelin, 2001). The functional unit of the Golgi is the individual dictyosome, its associated *trans*-Golgi network, and the Golgi matrix that surrounds both structures (Staehelin and Moore, 1995). Plant dictyosomes are formed by five to eight cisternae that display a distinct *cis*-to-*trans* polarity, with the *cis*-side receiving vesicles from the endoplasmic reticulum and the *trans*-side exporting the contents of these vesicles after processing and sorting.

The complexity of polysaccharide synthesis in the plant Golgi apparatus has led to the suggestion that enzyme activities are compartmentalized into distinct cisternae (Moore *et al.,* 1991). This is certainly true in mammalian cells, where different signals within Golgi-localized proteins determine either their Golgi retention or their targeting to specific portions of the Golgi apparatus (Grabenhorst and Conradt, 1999). Immunocytochemical and biochemical fractionation studies suggest specific targeting also occurs in plants. Antibodies to carbohydrate epitopes have been used to locate xyloglucans in the *trans*-Golgi cisternae and *trans*-Golgi network (Moore *et al.,* 1991; Zhang and Staehelin, 1992; Puhlmann *et al.,* 1994) and unesterified pectin in the *cis* and medial Golgi cisternae (Lynch and Staehelin, 1992). Methylesterified pectin was found in *medial* and *trans* cisternae, suggesting that backbone synthesis is initiated in the *cis*-Golgi cisternae and methyl esterification takes place in later cisternae (Sherrier and VandenBosch, 1994).

6.3.2 *Nucleotide sugar precursors for polysaccharide synthesis in the Golgi*

The sugar units used for synthesis of polysaccharides come from nucleotide sugars made in the cytoplasm (Feingold and Avigad, 1980; Coates *et al.,* 1980; Bonin *et al.,* 1997; see Figure 6.4). Initial reactions in the *de-novo* pathway of sugar nucleotide synthesis and interconversion produce UDP-Glc from UTP and Glc-1-P, and GDP-Man from GTP and Man-1-P. Further modifications of the Glc and Man sugar units then result in the formation of all the substrates needed by the various glycan

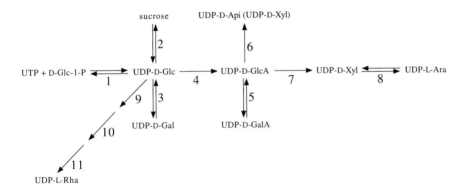

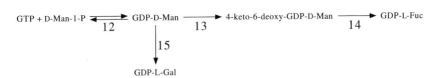

Figure 6.4 Enzymes and genes in the sugar nucleotide interconversion pathway of *Arabidopsis*. The pathway for *de novo* generation of nucleotide sugars is shown. The *myo*-inositol and salvage pathways are not shown. (1) UDP-Glc pyrophosphorylase; (2) Sucrose synthase (SuSy); (3) UDP-D-Glc 4-epimerase encoded by genes *UGE1–4* (Dörmann and Benning, 1996). Mutants in *UGE4* are *rhd1* (Schiefelbein and Somerville, 1990), *reb1* (Baskin *et al.*, 1992); (4) UDP-D-Glc dehydrogenase encoded by *UGD* (Seitz *et al.*, 2000); (5) UDP-D-GlcA 4-epimerase; (6) UDP-D-Api(UDP-D-Xyl) synthase; (7) UDP-D-GlcA decarboxylase; (8) UDP-D-Xyl 4-epimerase encoded by *MUR4* (Burget and Reiter, 1999); (9) 4,6-dehydratase; (10) 3,5-epimerase; (11) 4-reductase; (12) GDP-D-Man pyrophosphorylase encoded by *VTC1* (Conklin *et al.*, 1999); (13) GDP-D-Man 4,6-dehydratase encoded by *GMD1* and *GMD2/MUR1* (Bonin *et al.*, 1997); (14) 3,5-epimerase 4-reductase encoded by *GER1* and *GER2* (Bonin and Reiter, 2000); (15) GDP-D-Man 3,5-epimerase (Wolucka *et al.*, 2001).

synthases and glycosyl transferases (Feingold and Avigad, 1980; Reiter and Vanzin, 2001). Two alternative pathways for nucleotide sugar production can operate:

1. the salvage pathway which recycles sugars released from the wall during assembly and turnover by sequential action of monosaccharide kinases and nucleotide sugar pyrophosphorylases (Feingold and Avigad, 1980); and

2. the *myo*-inositol pathway, whereby *myo*-inositol is oxidized to GlcA then activated to UDP-GlcA (Feingold and Avigad, 1980).

Figure 6.4 identifies some of the cloned genes from the sugar interconversion pathway of *Arabidopsis*. In many cases, gene identification has relied on the discovery in the *Arabidopsis* genome of sequences related to bacterial, yeast or mammalian sugar interconversion genes. Evidence that the cloned *Arabidopsis* gene encodes a particular enzyme is often based on its ability to complement an appropriate yeast mutant, or on biochemical studies with the heterologously expressed protein.

UDP-Glc dehydrogenase (UGD) catalyses the first committed step in the conversion of UDP-Glc to UDP-GlcA (Figure 6.4). The epimerization of UDP-GlcA gives UDP-GalA, the activated precursor for incorporation of GalA into the backbone of pectic polysaccharides; UDP-GlcA is also the precursor for the synthesis of UDP-Xyl, UDP-Ara, and UDP-Api. Reporter gene studies with the *Arabidopsis UGD* promoter showed expression is correlated with growth, and is highest in the primary root and expanding leaves (Seitz *et al.*, 2000). Low levels of *UGD* expression in cotyledons and hypocotyls may be because these tissues use the *myo*-inositol pathway to produce UDP-GlcA.

UDP-Glc epimerase (UGE) catalyses the reversible epimerization of UDP-Glc to UDP-Gal, the activated precursor for incorporation of Gal into non-cellulosic polysaccharides. Dörmann and Benning (1996) isolated an *Arabidopsis* cDNA (*UGE1*) for a functional UDP-Glc epimerase. *UGE1* is one of four *UGE* genes in *Arabidopsis* and EST surveys suggest it is the most highly expressed; *UGE1* transcripts accumulate in all tissues, particularly stems (Dörmann and Benning, 1996; Reiter and Vanzin, 2001). All of the *Arabidopsis UGE* genes encode cytosolic enzymes (Dörmann and Benning, 1996; Reiter and Vanzin, 2001). Varying levels of *UGE1* expression by transgenic means has no visible effect on *Arabidopsis* plants grown in soil, and even plants with 10% of the wild-type level of UGE activity appear to grow normally (Dörmann and Benning, 1998). Plants with *UGE1* levels raised by transgenic expression were able to grow on agar containing 1% Gal, whereas untransformed plants and plants with lowered *UGE1* levels could not. However, *Arabidopsis* lines overexpressing *UGE1* and grown in the presence of Gal had less Gal in their walls than either untransformed plants or those with reduced UGE activity, suggesting that galactan biosynthesis is controlled by the level of expression of the relevant galactan synthase/galactosyl transferase or the flux through the Golgi; the exogenous Gal may raise the pool of UDP-Gal but the raised UGE activity converts this excess UDP-Gal into UDP-Glc and thus prevents more Gal from being incorporated into the walls (Dörman and Benning, 1998). Mutations in *UGE4* (*rhd1*, the root hair development locus 1, and *reb1*, the root epidermal bulging locus 1) are, however, associated with aberrant root-hair development in *Arabidopsis* (Schiefelbein and Somerville, 1990; Baskin *et al.*, 1992; Reiter and Vanzin, 2001), possibly through an effect on synthesis of galactosylated wall proteins controlling root epidermal cell expansion (Ding and Zhu, 1997).

The role of sugar interconversion in wall biosynthesis is also being addressed by the analysis of the '*mur*' collection of *Arabidopsis* mutant lines. The 11 *MUR* loci (*MUR1* to *MUR11*) were identified by screening 5200 chemically mutagenized *Arabidopsis* plants for changes in the monosaccharide composition of hydrolysed leaf wall polysaccharides (Reiter *et al.*, 1997). Three *mur* mutants have defects in the *de novo* synthesis of Fuc and their walls either contain less (*mur2* and *mur3*) or little or no Fuc (*mur1*), although only *mur1* plants are visibly affected, being slightly dwarfed with inflorescence meristems more brittle than those of wild-type plants (Reiter *et al.,* 1997).

The *MUR1 (AtGMD2)* gene encodes an isoform of GDP-D-Man-4,6-dehydratase, the enzyme that catalyses the first step in the interconversion of GDP-D-Man to GDP-L-Fuc (Figure 6.4; Bonin *et al.,* 1997). Mutations in this gene markedly reduce the amounts of Fuc present on pectin and fucosylated xyloglucan in shoot organs, but has less effect on the Fuc content of roots (Reiter *et al.,* 1997), presumably because a second GDP-D-Man-4,6-dehydratase isoform supplies these tissues with L-Fuc (Bonin *et al.,* 1997). The xyloglucan of *mur1* plants has no detectable structural or conformational changes, presumably since the missing L-Fuc is partly replaced by the closely related sugar, L-Gal (Zablackis *et al.,* 1996). L-Gal is a relatively rare sugar in plants and is not normally found in walls (Feingold and Avigad, 1980); the source of the L-Gal in *mur1* plants does not appear to be the 3,5-epimerase-4-reductase encoded by the *GER1* gene as on its own this will not use GDP-Man as substrate (Bonin and Reiter, 2000), but could be a GDP-Man 3,5-epimerase from the ascorbic acid pathway if this is expressed in the same tissues as *MUR1* (Wolucka *et al.,* 2001).

Nucleotide sugars are transported into the lumen of the Golgi by specific transporters (Gibeaut, 2000; Keegstra and Raikhel, 2001), where they are used for biosynthesis of not just wall polysaccharides but also glycolipids and glycoproteins. UDP-Glc is transported into the lumen by a UDP-Glc transporter in exchange for UMP, with the UMP being produced by a luminal UDPase that hydrolyses the UDP resulting from use of UDP-Glc in synthetic reactions (Muñoz *et al.,* 1996; Orellana *et al.,* 1997; Neckelmann and Orellana, 1998; see Figure 6.2). Norambuena *et al.* (2002) cloned an *Arabidopsis* gene, *AtUTR1*, that codes for a hydrophobic protein with high similarity to the human UDP-Gal transporter-related gene (*hUGTrel1*). Expression of *AtUTR1* in yeast mutant cells resistant to ricin because of a defect that blocks transport of UDP-Gal into the Golgi (ricin is a cytotoxic protein that binds to Gal residues) restores transport of UDP-Gal into the Golgi and the cells are consequently as sensitive to ricin as wild-type cells (Norambuena *et al.,* 2002). Interestingly, in addition to UDP-Gal, the AtUTR1 transporter also appears able to transport UDP-Glc (Norambuena *et al.,* 2002).

The GDP-Fuc needed for xyloglucan biosynthesis is transported into the Golgi by a distinct transporter that has a similar mechanism to the UDP-Glc transporter, and functions as a GMP/GDP-Fuc antiporter (Wulff *et al.,* 2000). A Golgi-localized GDP-Man transporter has also recently been cloned from *Arabidopsis* (Baldwin *et al.,* 2001); GDP-Man is used in the synthesis of (galacto)mannans of secondary cell walls.

6.3.3   Synthesis of non-cellulosic polysaccharide backbones: possible role of CSL and CESA genes

Biochemical studies have provided evidence that supports the Golgi lumen location of the reactions involved in synthesis of the backbone of non-cellulosic polysaccharides, but have not led to identification of the relevant gene families. Typically, enriched Golgi membrane preparations are incubated with radioactively labelled nucleotide sugars, and the labelled sugar residue is transferred to a polysaccharide product. Activities detected include the synthases for mixed-linkage (1,3),(1,4)-β-glucan (Henry and Stone, 1982; Gibeaut and Carpita, 1993; Becker *et al.*, 1995), xyloglucan (Gordon and Maclachlan, 1989), glucuronoarabinoxylans (Hobbs *et al.*, 1991), glu-comannan (Piro *et al.*, 1993), and galactomannan (Reid *et al.*, 1995). The synthesis of some of these polysaccharides requires sequential action of many enzymes; for example, synthesis of glucuronoarabinoxylan requires a xylan synthase to form the backbone, and an arabinosyl transferase and a glucuronosyl transferase to attach the branch units. Subcellular fractionation studies using pea epicotyls indicate that both the xylan synthase and the glucuronyl transferase activities are located in the Golgi membranes, with the glucuronosyl transferase activity concentrated in the *cis*-Golgi and stimulated by UDP-Xyl, suggesting that these enzymes are co-located (Waldron and Brett, 1987; Hobbs *et al.*, 1991).

Current thinking identifies the *CSL* genes as one set of candidates for encoding the enzymes responsible for processive synthesis of the β-linked backbones of these non-cellulosic polysaccharides. The *Arabidopsis* genome contains six families of *CSL* genes (*CSLA, B, C, D, E* and *G*) containing 9, 6, 5, 6, 1 and 3 members, re-spectively (Richmond and Somerville, 2000, 2001). *CESA* and *CSL* genes are also present in the incomplete rice (*Oryza sativa*) genome, but no rice *CSLB* or *CSLG* genes have been identified to date. Instead, rice appears to contain two other *CSL* families, *CSLF* and *CSLH* (Hazen *et al.*, 2002); the *CSLF* family is most closely related to the *CSLD* family, and the *CSLH* family is most closely related to the *CSLB* family. At present, rice has 10, 9, 4, 5, 7 and 2 members respectively of the *CSLA, C, D, E, F* and *H* families, although these numbers are likely to increase when the complete rice genome becomes available.

The *CESA/CSL* gene superfamily is much more highly diversified in plants than in other cellulose-synthesizing organisms such as cyanobacteria, bacteria and fungi (Saxena and Brown, 1995; Blanton *et al.*, 2000; Nobles *et al.*, 2001). Phylogenetic analyses using *CESA* and *CSL* sequences from various plant species have shown that orthologous genes (similar genes across different species) are usually more similar than paralogous genes (similar genes within a species), which indicates genes in the *CESA/CSL* superfamily diverged early in angiosperm evolution before the diver-gence of the monocot and dicot lineages (Fagard *et al.*, 2000; Holland *et al.*, 2000; Doblin *et al.*, 2001; Vergara and Carpita 2001).

While it is thought this divergence reflects functional specialization, no explicit activity has yet been unequivocally assigned to any *CSL* gene, nor have any of their products been localized. The current view is that it is likely that the *CSLD* genes are involved in cellulose biosynthesis at the plasma membrane (see section 6.2.6), while other *CSL* gene families are involved in synthesis of the backbone of vari-ous non-cellulosic polysaccharides in the Golgi apparatus (Cutler and Somerville, 1997; Arioli *et al.*, 1998b; Delmer, 1999; Richmond and Somerville, 2000, 2001; Vergara and Carpita, 2001). It is also possible that individual *CESA* genes may be involved in synthesis of the (1–3),(1–4)-β-glucan in the Golgi apparatus of grasses and related species. Since the *CSLF* and *CSLH* families are found in rice but not in *Arabidopsis*, these genes may be responsible for the synthesis of polysaccharides found predominantly or only within the walls of grasses (Richmond and Somerville, 2001; Hazen *et al.*, 2002). Conversely, the absence of *CSLB* or *CSLG* genes from the genomes of rice and maize implicates these genes in the synthesis of polysaccharides predominantly found in dicots (Richmond and Somerville, 2001). The *CSLA, C, D* and *E* families should thus be involved in the synthesis of polysaccharides common to both dicots and monocots.

Each *CSL* family is characterized by specific sequence features, but all have the D,D,D,QXXRW motif, are members of the GT 2 family, and are presumed to be involved in processive synthesis of β-linked polysaccharides (Campbell *et al.*, 1997; Coutinho and Henrissat, 1999; Henrissat *et al.*, 2001; see Table 6.2). A role in synthesis of α-linked polysaccharides, such as the pectic components HGA, RG-I and RG-II, is less likely, since all members of the GT 2 family with known products invert rather than retain the anomeric configuration of their α-linked nucleotide sugar substrate (Charnock *et al.*, 2001; Henrissat *et al.*, 2001). However, such a possibility cannot be discounted until the control of substrate and reaction specificity is better understood. Similarly, a role in catalysis of single sugar additions to polysaccharide side chains cannot be ruled out, as the factors controlling reaction processivity are also unknown. Discounting these possibilities for the moment, however, and depending on the plant species, Golgi-localized *CSL* and *CESA* genes may be involved in the synthesis of β-linked non-cellulosic polysaccharides (Cutler and Somerville, 1997; Delmer, 1999; Richmond and Somerville, 2000, 2001; Vergara and Carpita, 2001). Initial reports using antibody labelling or GFP-fusion experiments suggest that *Arabidopsis* CSLA, CSLB, CSLE and CSLG proteins are located in the Golgi (Richmond and Somerville, 2000, 2001), although these locations need to be confirmed.

All *CSL* genes are predicted to encode integral membrane proteins, with 0–3 transmembrane domains in their N-terminal region and 3–6 transmembrane domains in their C-terminal region (Richmond and Somerville, 2000; see Figure 6.2). The CSLA and CSLC proteins are the smallest, and have the lowest overall amino-acid identity to the CESA proteins (Figure 6.3). No signal peptides are apparent on any of the proteins but all N-termini are believed to be cytoplasmic. Although the membrane topologies are not experimentally confirmed, CSLs with an odd number (e.g. 1 or 3) N-terminal transmembrane domains are predicted to reside in the Golgi with the presumed catalytic region containing the D,D,D,QXXRW motif in the lumen (Figure 6.2). In contrast, CSLs with an even number (e.g. 2 or 4) N-terminal helices are predicted to reside in the Golgi (Figure 6.2).

Based on EST analyses and other work, the *CSLA* and *CSLC* genes are the most abundantly expressed of the CSL families, both in *Arabidopsis* and other plant species, although they are not as highly expressed as the *CESA* genes (Richmond and Somerville, 2001; http://cellwall.stanford.edu/cesa/index.shtml). The *CSLD* genes are also well represented in EST libraries, while the *CSLB* genes are poorly represented. The *CSLG* genes are expressed at low levels in *Arabidopsis*, but are more abundant in other dicots such as *Glycine max* (soybean) and *Medicago truncatula* (Richmond and Somerville, 2001; http://cellwall.stanford.edu/cesa/index.shtml). It is not clear whether these data reflect differences in polysaccharide abundance or some other factor such as enzyme specific activity or rate of protein turnover.

Only a handful of *CSL* mutations are currently known, and reports on many of these are still to be published. The best characterized to date are the *AtCSLD3/KOJAK* mutants discussed in section 6.2.6. The *Arabidopsis rat4* mutant has a T-DNA insertion in *AtCSLA9*, and is phenotypically normal except that plants are more resistant to *Agrobacterium tumefaciens* infection than are wild-type plants (Nam *et al.*, 1999; Richmond and Somerville, 2001). *Agrobacterium* attachment to roots is unaffected but the genetic lesion affects an early stage of infection (Nam *et al.*, 1999). If the AtCSLA9 gene product plays a role in the synthesis of a polysaccharide in the Golgi apparatus of root cells, then analysis of the wall composition of *rat4* roots should provide an insight into which wall component is affected. There are also reports of an insertional mutant in *AtCSLC4* that specifically affects xylan synthesis, indicating that this gene encodes a xylan synthase, which are supported by the observation that sense suppression of *AtCSLC4* in transgenic plants also reduces xylan levels (Somerville and Cutler, 1998). To date only brief reports of these experiments have been published and further investigations of the *CSL* gene families are required to determine what roles these genes play in wall synthesis.

The Golgi apparatus of grasses and related families synthesizes large amounts of the mixed-linkage (1,3),(1,4)-β-glucan during endosperm development as well for deposition into primary walls of elongating coleoptiles (Figure 6.1). The mixed-linkage glucan synthase uses UDP-Glc as substrate, requires Mg^{2+} or Mn^{2+} as co-factor, and is sensitive to disruption or permeabilization of the Golgi membranes (Henry and Stone, 1982; Gibeaut and Carpita, 1993; Becker *et al.*, 1995). In addition, unlike plants that do not make this polysaccharide, a callose synthase is associated with the Golgi apparatus of grasses and related families, and increases in Golgi callose synthase activity coincide with loss of mixed-linkage glucan synthase activity (Gibeaut and Carpita, 1993, 1994b). In this feature, the Golgi mixed-linkage glucan synthase resembles the plasma membrane cellulose synthase, and it has therefore been suggested that the Golgi mixed-linkage glucan synthase is encoded by a *CESA* gene. Comparative and phylogenetic sequence analysis and expression profiling have been used to search for candidate genes (Buckeridge *et al.*, 1999; Dhugga, 2001; Vergara and Carpita, 2001), and EST surveys of developing maize endosperms have identified *ZmCESA5* as a *CESA* gene that does not have an obvious orthologue in *Arabidopsis* and that therefore may encode the mixed-linkage glucan synthase (Dhugga, 2001). Alternatively, it cannot be entirely ruled out that some of these polysaccharides are synthesized by Golgi-resident type II membrane bound glycosyl transferases (as described below) since such enzymes are involved in the elaboration of polysaccharide chains on mammalian proteoglycans (Schwartz, 2000).

6.3.4   Synthesis of branches on non-cellulosic polysaccharides: role of glycosyl transferases

The high diversity and structural complexity of non-cellulosic polysaccharides and pectins is due in large part to the decoration of their backbones with short side chains of sugars by Golgi-resident glycosyl transferases (GTs). These catalyse the transfer of individual sugar residues from an activated nucleotide sugar substrate to an acceptor molecule (the growing polysaccharide) with the stereochemistry at the donor sugar C1 position being either retained or inverted. Although they have very different amino-acid sequences, all Golgi-localized GTs are type II membrane proteins and consist of a short N-terminal cytoplasmic domain, a single transmembrane segment, a stem region of variable length, and a large C-terminal globular catalytic domain located inside the Golgi lumen (Paulson and Colley, 1989; Figure 6.2). Very little is known about the molecular basis of reaction specificity (that is, donor selectivity and acceptor preference), but several GT families (both inverting and retaining) contain a DXD motif following a stretch of hydrophobic amino acids predicted to form a β-strand (Breton and Imberty, 1999). The motif occurs in GTs where catalysis requires either Mn^{2+} or Mg^{2+} and is involved in coordinating the metal ion during nucleotide-sugar binding (Charnock and Davies, 1999; Gastinel *et al.*, 1999; Ünligil *et al.*, 2000). Based on amino-acid similarities, in particular the presence or absence of specific motifs, as well as substrate/product stereochemistry (inverting or retaining), 60 different families of GTs have been described (Coutinho and Henrissat, 1999). GTs from different families generally show limited sequence identity, and moreover the identity among members of the same GT family is often restricted to a few short motifs (Breton and Imberty, 1999). However, using a bioinformatic approach, it is possible to divide the various GT sequences into subfamilies based on these conserved motifs: for example, the (1–3)-β-GalTs belong to Family 31, and the (1–4)-β-GalTs belong to Family 7.

The *Arabidopsis* genome contains 414 different GT genes representing 34 of the 60 GT families described above (Coutinho and Henrissat, 1999). These genes also code for enzymes involved in protein and lipid glycosylation and in metabolite catabolism, but the fact that as many as 21 GTs may be required simply to produce the linkages in the pectic polysaccharide RG-II (The Arabidopsis Genome Initiative, 2000) indicates that a very large number of GT genes will eventually be shown to be involved in polysaccharide biosynthesis. It is presumed that GTs from several different families will be involved in matrix polysaccharide biosynthesis, to add the diverse range of sugars to the various polymer backbones.

To date, only two plant genes have been reported that encode GTs involved in non-processive synthesis of polysaccharide side chains: a xyloglucan fucosyltransferase (XyGFucT) from *Arabidopsis* (Perrin *et al.*, 1999), and a galactomannan galactosyl transferase (GalT) from fenugreek (Edwards *et al.*, 1999). Each is a member of a multigene family (see below).

Xyloglucans consist of linear chains of (1–4)-β-Glc with Xyl units attached at the C(O)6 position of 3 out of every 4 Glc residues; some of the Xyl units are further substituted by Ara, Gal and Fuc (Bacic *et al.*, 1988). The XyFucT transfers single Fuc units from GDP-Fuc to terminal Gal residues in a (1–2)-α-linkage, and Perrin *et al.* (1999) purified adequate quantities of this enzyme from pea microsomal membranes to obtain partial amino-acid sequence. The relevant *Arabidopsis* gene (*AtFUT1*) was then cloned and characterized, and the function of its encoded protein was confirmed by heterologous expression. The *Arabidopsis* genome contains a total of nine sequences with homology to *AtFUT1*, but their function is currently unknown (Sarria *et al.*, 2001). Positional cloning of the *MUR2* gene indicated that it corresponded to this XyFucT, and *AtFUT1* successfully complemented a *mur2* mutation (Vanzin *et al.*, 2002).

Two enzyme activities are involved in the coordinate synthesis of galactomannan, a mannan synthase that processively transfers (1–4)-β-linked Man residues from GDP-Man to the growing mannan backbone, and a GalT that adds single (1–6)-α-linked Gal side-branches from UDP-Gal. The activities are linked in that, in membrane preparations, GalT activity requires a growing mannan backbone and only functions if the mannan synthase is simultaneously active (Edwards *et al.*, 1989, 1992). Membrane preparations from endosperm of developing seeds of *Trigonella foenum-graecum* (fenugreek), a tissue with high galactomannan synthesis activity, were used to identify a 51-kDa polypeptide that displayed high levels of galactomannan-dependent GalT activity, and amino-acid sequence information was used to clone the relevant gene (Edwards *et al.*, 1999). Extracts from *Pichia pastoris* yeast cells expressing the *GALT* gene were capable of adding Gal units onto a mannan acceptor molecule *in vitro*, producing authentic galactomannan as determined by enzymatic hydrolysis of the products (Edwards *et al.*, 1999). The *Arabidopsis* genome contains eight sequences with homology to the fenugreek *GALT* gene.

6.4   Future directions

The past 10 years has seen some major breakthroughs in our understanding of polysaccharide biosynthesis. At the molecular level, a number of multigene families have been identified, but the specific functions are still to be allotted to many of the individual genes in these families. Both gain-of-function and loss-of-function experiments, in association with biochemical and ultrastructural work, will be needed to allocate these functions, and to understand the degrees of specialization between isoforms catalysing the same reaction (e.g. the *CESA* gene family). It is also likely that some synthetic steps will require genes from families not currently implicated in polysaccharide biosynthesis, and that the different gene families will give insight into mechanisms of glycosidic bond formation and specificity of transfer. For example, the Family 7 GalTs identified in mammalian genomic databases are absent from the *Arabidopsis* genome in spite of the fact that plants are known to synthesize (1–4)-β galactans. The data will also allow development of an understanding of the molecular machinery of wall synthesis, ranging from rosette structure to the mechanisms of polysaccharide chain initiation (primers), elongation, and termination, and thus of the regulation of this process through cell development and differentiation. Another intriguing observation is the apparent coordinate regulation between wall components to ensure that the structural integrity of the wall is maintained. In a number of reports, it has been observed that perturbation of cellulose biosynthesis (either through gene knockout or enzyme inhibitors) leads to enhanced biosynthesis of pectins with a lower degree of methyl esterification in primary walls (Shedletzky *et al.*, 1992; His *et al.*, 2001). Alternatively, the down-regulation of lignin biosynthesis in secondary walls leads to enhanced levels of cellulose (Hu *et al.*, 1999). All of these factors will need to be understood if the proposed genetic approaches to rational design of plant cell walls for targeted agro-industrial applications can become a reality.

References

Albersheim, P., Darvill, A., Roberts, K., Staehelin, L.A. and Varner, J.E. (1997) Do the structures of cell wall polysaccharides define their mode of synthesis? *Plant Physiol.*, 113, 1–3.

Amor, Y., Haigler, C.H., Johnson, S., Wainscott, M. and Delmer, D.P. (1995) A membrane-associated form of sucrose synthase and its potential role in synthesis of cellulose and callose in plants. *Proc. Natl. Acad. Sci. USA*, 92, 9353–9357.

Andrawis, A., Solomon, M. and Delmer, D.P. (1993) Cotton fiber annexins: a potential role in the regulation of callose synthase. *Plant J.*, 3, 763–772.

Arioli, T., Peng, L., Betzner, A.S. *et al.* (1998a) Molecular analysis of cellulose biosynthesis in *Arabidopsis. Sci.*, 279, 717–720.

Arioli, T., Burn, J.E., Betzner, A.S. and Williamson, R.E. (1998b) ... response: how many cellulose synthase-like gene products actually make cellulose? *Trends Plant Sci.*, 3, 165–166.

Bacic, A., Harris, P.J. and Stone, B.A. (1988) Structure and function of plant cell walls. In *The Biochemistry of Plants* (ed. J. Priess), Academic Press, New York, pp. 297–371.

Baldwin, T.C., Handford, M.G., Yuseff, M.I., Orellana, A. and Dupree, P. (2001) Identification and characterization of *GONST1*, a Golgi-localized GDP-Mannose transporter in *Arabidopsis. Plant Cell*, 13, 2283–2295.

Baskin, T.I. (2001) On the alignment of cellulose microfibrils by cortical microtubules: a review and a model. *Protoplasma*, 215, 150–171.

Baskin, T.I., Betzner, A.S., Hoggart, R., Cork, A. and Williamson, R.E. (1992) Root morphology mutants in *Arabidopsis thaliana. Austral. J. Plant Physiol.*, 19, 427–437.

Becker, M., Vincent, C. and Reid, J.S.G. (1995) Biosynthesis of (1,3)(1,4)-beta-glucan and (1,3)-beta-glucan in barley (*Hordeum vulgare* L) – Properties of the membrane-bound glucan synthases. *Planta*, 195, 331–338.

Blanton, R.L. and Northcote, D.H. (1990) A 1,4-β-D-glucan-synthase system from *Dictyostelium discoideum. Planta*, 180, 324–332.

Blanton, R.L., Fuller, D., Iranfar, N., Grimson, M.J. and Loomis, W.F. (2000) The cellulose synthase gene of *Dictyostelium. Proc. Natl. Acad. Sci. USA*, 97, 2391–2396.

Boisson, M., Gomord, V., Audran, C. *et al.* (2001) *Arabidopsis* glucosidase I mutants reveal a critical role of N-glycan trimming in seed development. *EMBO J.*, 20, 1010–1019.

Bonin, C.P. and Reiter, W.-D. (2000) A bifunctional epimerase reductase acts downstream of the *MUR1* gene product and completes the *de novo* synthesis of GDP-L-fucose in *Arabidopsis. Plant J.*, 21, 445–454.

Bonin, C.P., Potter, I., Vanzin, G.F. and Reiter, W.-D. (1997). The *MUR1* gene of *Arabidopsis thaliana* encodes an isoform of GDP-D-mannose-4,6-dehydratase, catalyzing the first step in the *de novo* synthesis of GDP-L-fucose. *Proc. Natl. Acad. Sci. USA*, 94, 2085–2090.

Breton, C. and Imberty, A. (1999) Structure/function studies of glycosyltransferases. *Curr. Op. Struc. Biol.*, 9, 563–571.

Brown, R.M. Jr (1996) The biosynthesis of cellulose. *J. Macromol. Sci. Pure App. Chem.*, VA33, 1345–1373.

Brown, R.M. Jr and Saxena, I.M. (2000) Cellulose biosynthesis: a model for understanding the assembly of biopolymers. *Plant Physiol. Biochem.*, 38, 57–67.

Brummell, D.A., Catala, C., Lashbrook, C.C. and Bennett, A.B. (1997) A membrane-anchored E-type endo-1,4-β-glucanase is localized on Golgi and plasma membranes of higher plants. *Proc. Natl. Acad. Sci. USA*, 94, 4794–4799.

Buckeridge, M.S., Vergara, C.E. and Carpita, N.C. (1999) The mechanism of synthesis of a cereal mixed-linkage (1→3),(1→4)-β-D-glucan: evidence for multiple sites of glucosyl transfer in the synthase complex. *Plant Physiol.*, 120, 1105–1116.

Burget, E.G. and Reiter, W.-D. (1999) The *mur4* mutant of *Arabidopsis* is partially defective in the *de novo* synthesis of uridine diphospho L-arabinose. *Plant Physiol.*, 121, 383–389.

Burn, J.E., Hocart, C.H., Birch, R.J., Cork, A.C. and Williamson, R.E. (2002) Functional analysis of the cellulose synthase genes *CesA1*, *CesA2*, and *CesA3* in *Arabidopsis*. *Plant Physiol.*, 129, 797–807.

Campbell, J.A., Davies, G.J., Bulone, V. and Henrissat, B. (1997) A classification of nucleotide-diphospho-sugar glycosyltransferases based on amino acid sequence similarity. *Biochem. J.*, 326, 929–942.

Cantatore, J.L., Murphy, S.M. and Lynch, D.V. (2000) Compartmentation and topology of gluco-sylceramide synthesis. *Biochem. Soc. Trans.*, 28, 748–750.

Carpita, N.C. and Delmer, D.P. (1981) Concentration and metabolic turnover of UDP-glucose in developing cotton fibers. *J Biol Chem.*, 256, 308–315.

Carpita, N.C. and Gibeaut, D.M. (1993) Structural models of primary cell walls in flowering plants: consistency of molecular structure with the physical properties of the walls during growth. *Plant J.*, 3, 1–30.

Carpita, N. and Vergara, C. (1998) A recipe for cellulose. *Sci.*, 279, 672–673.

Carpita, N., Tierney, M. and Campbell, M. (2001) Molecular biology of the plant cell wall: search-ing for the genes that define structure, architecture and dynamics. *Plant Mol. Biol.*, 47, 1–5.

Charnock, S.J. and Davies, G.J. (1999) Structure of the nucleotide-diphospho-sugar transferase, SpsA from *Bacillus subtilis*, in native and nucleotide-complexed forms. *Biochem.*, 38, 6380–6385.

Charnock, S.J., Henrissat, B. and Davies, G.J. (2001) Three-dimensional structures of UDP-sugar glycosyltransferases illuminate the biosynthesis of plant polysaccharides. *Plant Physiol.*, 125, 527–531.

Coates, S.W., Gurney, T. Jr, Sommers, L.W., Yeh, M. and Hirschberg, C.B. (1980) Subcellular localization of sugar nucleotide synthetases. *J. Biol. Chem.*, 255, 9225–9229.

Conklin, P.L, Norris, S.R, Wheeler, G.L, Williams, E.H., Smirnoff, N. and Last, R.L. (1999) Ge-netic evidence for the role of GDP-mannose in plant ascorbic acid (vitamin C) biosynthesis. *Proc. Natl. Acad. Sci. USA*, 96, 4198–4203.

Coutinho, P.M. and Henrissat, B. (1999) Carbohydrate-Active Enzymes server at URL: http://afmb.cnrs-mrs.fr/CAZY/index.html

Cui, X., Shin, H., Song, D., Laosinchai, W., Amano, Y. and Brown, R.M. Jr (2001) A putative plant homolog of the yeast β-1,3-glucan synthase subunit FKS1 from cotton (*Gossypium hirsutum* L.) fibers. *Planta*, 213, 223–230.

Cutler, S. and Somerville, C. (1997) Cellulose synthesis: cloning *in silico*. *Curr. Biol.*, 7, R108–R111.

Delmer, D.P. (1977) The biosynthesis of cellulose and other plant cell wall polysaccharides. In *Recent Advances in Phytochemistry*, Vol. 11 (eds F.A. Loewus and V.C. Runeckles), Plenum Publishing Corporation, New York, pp. 45–77.

Delmer, D.P. (1987) Cellulose biosynthesis. *Annu. Rev. Plant Physiol.*, 38, 259–290.

Delmer, D.P. (1999) Cellulose biosynthesis: exciting times for a difficult field of study. *Annu. Rev. Plant Physiol. Plant Mol. Biol.*, 50, 245–276.

Delmer, D.P. and Amor, Y. (1995) Cellulose biosynthesis. *Plant Cell*, 7, 987–1000.

Delmer, D.P. and Stone, B.A. (1988) Biosynthesis of plant cell walls. In *The Biochemistry of Plants,* Vol 14 (ed. J. Preiss), Academic Press, San Francisco, pp. 373–420.

Delmer, D.P., Read, S.M. and Cooper, G. (1987) Identification of a protein receptor in cotton fibers for the herbicide 2,6-dichlorobenzonitrile. *Plant Physiol.*, 84, 415–420.

Delmer, D.P., Solomon, M. and Read, S.M. (1991) Direct photolabeling with [^{32}P]UDP-glucose for identification of a subunit of cotton fiber callose synthase. *Plant Physiol.* 95, 556–563.

Delmer, D.P., Volokita, M., Solomon, M., Fritz, U., Delphendahl, W. and Herth, W. (1993) A monoclonal antibody recognises a 65 kDa higher plant membrane polypeptide which undergoes cation-dependent association with callose synthase in vitro and co-localizes with sites of high callose deposition in vivo. *Protoplasma*, 176, 33–42.

Delmer, D.P., Pear, J.R., Andrawis, A. and Stalker, D.M. (1995) Genes encoding small GTP-binding proteins analogous to mammalian *rac* are preferentially expressed in developing cotton fibers. *Mol. Gen. Genet.*, 248, 43–51.

Desprez, T., Vernhettes, S., Fagard, M. *et al.* (2002) Resistance against herbicide isoxaben and cellulose deficiency caused by distinct mutations in same cellulose synthase isoform CESA6. *Plant Physiol.* 128, 482–490.

Dhugga, K.S. (2001) Building the wall: genes and enzyme complexes for polysaccharide synthases. *Curr. Op. Plant Biol.*, 4, 488–493.

Dhugga, K.S. and Ray, P.M. (1994) Purification of 1,3-β-D-glucan synthase activity from pea tissue. Two polypeptides of 55 kDa and 70 kDa copurify with enzyme activity. *Eur. J. Biochem.* 220, 943–953.

Ding, L. and Zhu, J.K. (1997) A role for arabinogalactan proteins in root epidermal cell expansion. *Planta,* 203, 289–294.

Doblin, M.S., De Melis, L., Newbigin, E., Bacic, A. and Read, S.M. (2001) Pollen tubes of *Nicotiana alata* express two genes from different β-glucan synthase families. *Plant Physiol.*, 125, 2040–2052.

Doblin, M.S., Eshed, R., Hogan, P. *et al.* (2003) Promoter-GUS fusion analyses of the Arabidopsis *CesA* and *CslD* gene families, manuscript in preparation.

Dörmann, P. and Benning, C. (1996) Functional expression of uridine 5′-diphospho-glucose 4-epimerase from *Arabidopsis thaliana* in *Saccharomyces cerevisiae* and *Escherichia coli*. *Arch. Biochem. Biophys.*, 327, 27–34.

Dörmann, P. and Benning, C. (1998) The role of UDP-glucose epimerase in carbohydrate metabolism of *Arabidopsis*. *Plant J.*, 13, 641–652.

Douglas, C.M., Foor, F., Marrinan, J.A. *et al.* (1994) The *Saccharomyces cerevisiae* Fks1 (Etg1) gene encodes an integral membrane protein which is a subunit of 1,3-beta-D-glucan synthase. *Proc. Natl. Acad. Sci. USA*, 91, 12907–12911.

Edwards, M.E., Bulpin, P.V., Dea, I.C.M. and Reid, J.S.G. (1989) Biosynthesis of legume-seed galactomannans *in vitro*. *Planta*, 178, 41–51.

Edwards, M., Scott, C., Gidley, M.J., Reid, J.S.G. (1992) Control of mannose/galactose ratio during galactomannan formation in developing legume seeds. *Planta*, 187, 67–74.

Edwards, M.E., Dickson, C.A., Chengappa, S., Sidebottom, C., Gidley, M.J. and Reid, J.S.G. (1999) Molecular characterisation of a membrane-bound galactosyltransferase of plant cell wall matrix polysaccharide biosynthesis. *Plant J.,* 19, 691–697.

El-Sherbeini, M. and Clemas, J.A. (1995) Cloning and characterization of *GNS1*: a *Saccharomyces cerevisiae* gene involved in synthesis of 1,3-β-glucan *in vitro*. *J. Bacteriol.*, 177, 3227–3234.

Fagard, M., Desnos, T., Desprez, T. *et al.* (2000) *PROCUSTE1* encodes a cellulose synthase required for normal cell elongation specifically in roots and dark-grown hypocotyls of Arabidopsis. *Plant Cell,* 12, 2409–2423.

Favery, B., Ryan, E., Foreman, J. *et al.* (2001) *KOJAK* encodes a cellulose synthase-like protein required for root hair cell morphogenesis in *Arabidopsis*. *Genes Dev.,* 15, 79–89.

Feingold, D.S. and Avigad, G. (1980) Sugar nucleotide transformation in plants. In *The Biochemistry of Plants: a comprehensive treatise,* Vol. 3 (eds P.K. Stumpf and E.E. Conn), Academic Press, New York, pp. 101–170.

Ferguson, C., Teeri, T.T., Siika-aho, M., Read, S.M. and Bacic, A. (1998) Location of cellulose and callose in pollen tubes and grains of *Nicotiana tabacum*. *Planta,* 206, 452–460.

Fincher, G.B. and Stone, B.A. (1981) Metabolism of noncellulosic polysaccharides. In *Encyclopedia of Plant Physiology, New Series, Plant Carbohydrates II* (eds W. Tanner and F.A. Loewus), Springer-Verlag, Berlin, pp. 68–132.

Frost, D.J., Read, S.M., Drake, R.R., Haley, B.E. and Wasserman, B.P. (1990) Identification of the UDPG-binding polypeptide of (1–3)-β-glucan synthase from a higher plant by photoaffinity labelling with 5-azido-UDP-glucose. *J. Biol. Chem.,* 265, 2162–2167.

Gastinel, L.N., Cambillaum, C. and Bournem, Y. (1999) Crystal structures of the bovine β-4-galactosyltransferase catalytic domain and its complex with uridine diphosphogalactose. *EMBO J.,* 18, 3546–3557.

Gibeaut, D.M. (2000) Nucleotide sugars and glycosyltransferases for synthesis of cell wall matrix polysaccharides. *Plant Physiol. Biochem.,* 38, 69–80.

Gibeaut, D.M. and Carpita, N.C. (1993) Synthesis of (1→3),(1→4)-β-D-glucan in the Golgi apparatus of maize coleoptiles. *Proc. Natl. Acad. Sci. USA,* 90, 3850–3854.

Gibeaut, D.M. and Carpita, N.C. (1994a) Biosynthesis of plant cell wall polysaccharides. *FASEB J.,* 8, 904–915.

Gibeaut, D.M. and Carpita, N.C. (1994b) Improved recovery of (1,3),(1,4)-β-D-glucan synthase activity from Golgi apparatus of *Zea mays* (L.) using differential flotation centrifugation. *Protoplasma,* 180, 92–97.

Gillmor, C.S., Poindexter, P., Lorieau, J., Palcic, M.M. and Somerville, C. (2002) α-Glucosidase I is required for cellulose synthesis and morphogenesis in *Arabidopsis*. *J. Cell Biol.,* 156, 1003–1013.

Girard, V. and Maclachlan, G. (1987) Modulation of pea membrane β-glucan synthase activity by calcium, polycation, endogenous protease, and protease inhibitor. *Plant Physiol.,* 85, 131–136.

Gordon, R. and Maclachlan, G. (1989) Incorporation of UDP-[^{14}C]glucose into xyloglucan by pea membranes. *Plant Physiol.,* 91, 373–378.

Grabenhorst, E. and Conradt, H.S. (1999) The cytoplasmic, transmembrane, and stem regions of glycosyltransferases specify their in vivo functional sublocalization and stability in the Golgi. *J. Biol Chem.,* 274, 36107–36116.

Ha, M.-A., Apperley, D.C., Evans, B.W. *et al.* (1998) Fine structure in cellulose microfibrils: NMR evidence from onion and quince. *Plant J.* 16, 183–190.

Haigler, C.H., Ivanova-Datcheva, M., Hogan, P.S. *et al.* (2001) Carbon partitioning to cellulose synthesis. *Plant Mol. Biol.,* 47, 29–51.

Hayashi, T., Read, S.M., Busell, J. *et al.* (1987) UDP-glucose:(1,3)-β-glucan synthases from mung bean and cotton. Differential effects of Ca^{2+} and Mg^{2+} on enzyme properties and on macromolecular structure of the glucan product. *Plant Physiol.* 83, 1054–1062.

Hazen, S., Scott-Craig, J.S. and Walton, J.D. (2002) Cellulose synthase-like (CSL) genes of rice. *Plant Physiol.,* 128, 336–340.

Henrissat, B., Coutinho, P.M. and Davies, G.J. (2001) A census of carbohydrate-active enzymes in the genome of *Arabidopsis thaliana*. *Plant Mol. Biol.,* 47, 55–72.

Henry, R.J. and Stone, B.A. (1982) Factors influencing β-glucan synthesis by particulate enzymes from suspension-cultured *Lolium multiflorum* endosperm cells. *Plant Physiol.,* 69, 632–636.

His, I., Driouich, A., Nicol, F., Jauneau, A. and Höfte, H. (2001) Altered pectin composition in primary cell walls of *korrigan*, a dwarf mutant of Arabidopsis deficient in a membrane-bound endo-1,4-β-glucanase. *Planta,* 212 348–358.

Hobbs, M.C., Delarge, M.H.P., Baydoun, E.A.-H. and Brett, C.T. (1991) Differential distribution of a glucuronyltransferase involved in glucuronoxylan synthesis within the Golgi apparatus of pea (*Pisum sativum* var. Alaska). *Biochem. J.*, 277, 653–658.

Holland, N., Holland, D., Helentjaris, T., Dhugga, K.S., Xoconostle-Cazares, B. and Delmer, D.P. (2000) A comprehensive analysis of the plant cellulose synthase (*CesA*) gene family. *Plant Physiol.*, 123, 1313–1323.

Hong, Z., Delauney, A.J. and Verma, D.P.S. (2001a) Cell plate specific callose synthase and its interaction with phragmoplastin. *Plant Cell*, 13, 755–768.

Hong, Z., Zhang, Z., Olson, J.M. and Verma, D.P.S. (2001b) A novel UDP-glucose transferase is a part of the callose synthase complex and interacts with phragmoplastin at the forming cell plate. *Plant Cell*, 13, 769–780.

Hu, W.J., Harding, S.A., Lung, J. *et al.* (1999) Repression of lignin biosynthesis promotes cellulose accumulation and growth in transgenic trees. *Nature Biotech.*, 17, 808–812.

Jacob, S.R. and Northcote, D.H. (1985) *In vitro* glucan synthesis by membranes of celery petioles: the role of the membrane in determining the type of linkage formed. *J. Cell Sci.*, S2, 1–11.

Jing, W. and De Angelis, P.L. (2000) Dissection of the two transferase activities of the *Pasteurella multocida* hyaluronan synthase: two active sites exist in one polypeptide. *Glycobiol.*, 10(9), 883–889.

Kauss, H. and Jeblick, W. (1986) Influence of free fatty acids, lysophosphatidylcholine, platelet-activating factor, acylcarnitine, and echinocandin B on 1,3-β-D-glucan synthase and callose synthesis. *Plant Physiol.*, 80, 7–13.

Keegstra, K. and Raikhel, N. (2001) Plant glycosyltransferases. *Curr. Op. Plant Biol.*, 4, 219–224.

Kimura, S., Laosinchai, W., Itoh, T., Cui, X., Linder, C.R. and Brown, R.M. Jr (1999a) Immuno-gold labeling of rosette terminal cellulose-synthesizing complexes in the vascular plant *Vigna angularis*. *Plant Cell*, 11, 2075–2085.

Kimura, S., Sakurai, N. and Itoh, T. (1999b) Different distribution of cellulose synthesizing complexes in Brittle and non-Brittle strains of barley. *Plant Cell Physiol.*, 40, 335–338.

Kobayashi, H., Fukuda, H. and Shibaoka, H. (1998) Interrelation between the spatial disposition of actin filaments and microtubules during the differentiation of tracheary elements in cultured *Zinnia* cells. *Protoplasma*, 143, 29–37.

Kohle, H., Jeblick, W., Poten, F., Blaschek, W. and Kauss, H. (1985) Chitosan-elicited callose synthesis in soybean cells is a Ca^{2+}-dependent process. *Plant Physiol.*, 77, 544–551.

Kokubo, A., Sakurai, N., Kuraishi, S. and Takeda, K. (1991) Culm brittleness of barley (*Hordeum vulgare* L.) mutants is caused by smaller number of cellulose molecules in cell wall. *Plant Physiol.*, 97, 509–514.

Kudlicka, K. and Brown, R.M. Jr (1997) Cellulose and callose biosynthesis in higher plants. *Plant Physiol.*, 115, 643–656.

Kudlicka, K., Brown, R.M. Jr, Li, L., Lee, J.H., Shin, H. and Kuga, S. (1995) β-Glucan synthesis in the cotton fiber. IV In vitro assembly of the cellulose I allomorph. *Plant Physiol.*, 107, 111–123.

Kurek, I., Kawagoe, Y., Jacob-Wilk, D., Doblin, M. and Delmer, D.P. (2002) Dimerization of cotton fiber cellulose synthase catalytic subunits occurs via oxidation of the zinc-binding domains. *Proc. Natl. Acad. Sci. USA*, 99, 11109–11114.

Lane, D.R., Wiedemeier, A., Peng, L. *et al.* (2001) Temperature-sensitive alleles of *RSW2* link the KORRIGAN endo-1,4,β-glucanase to cellulose synthesis and cytokinesis in Arabidopsis. *Plant Physiol.*, 126 278–288.

Li, L., Drake, R.R. Jr, Clement, S. and Brown, R.M. Jr (1993) β-glucan synthesis in the cotton fiber III. Identification of UDP-glucose-binding subunits of β-glucan synthases by photoaffinity labeling with [β-^{32}P]5′-N$_3$-UDP-glucose. *Plant Physiol.*, 101, 1149–1156.

Li, H., Bacic, A. and Read, S.M. (1997) Activation of pollen tube callose synthase by detergents – evidence for different mechanisms of action. *Plant Physiol.*, 114, 1255–1265.

Li, H., Bacic, A. and Read, S.M. (1999) Role of a callose synthase zymogen in regulating wall deposition in pollen tubes of *Nicotiana alata*. *Planta,* 208, 528–538.

Li, J., Wardak, A.Z., Burton, R.A., Hains, P., Stone, B.A. and Fincher, G.B. (2001) Linking the amino acid and nucleotide sequences of barley (1,3)-β-glucan synthases: a case study in functional genomics. In *Proceedings of the Ninth International Cell Wall Meeting,* 2–7 September, Toulouse, France, Abstract 87.

Lin, F.C. and Brown, R.M. Jr (1989) Purification of cellulose synthase from *Acetobacter xylinum*. In *Cellulose and Wood-Chemistry and Technology* (ed. C. Scheurch), John Wiley and Sons, New York, pp. 473–492.

Lin, F.C., Brown, R.M. Jr, Drake, R.R. Jr and Haley, B.E. (1990) Identification of the uridine 5′-diphosphoglucose (UDP-Glc) binding subunit of cellulose synthase in *Acetobacter xylinum* using the photoaffinity probe 5-azido-UDP-Glc. *J. Biol. Chem.*, 265, 4782–4784.

Lukowitz, W., Nickle, T.C., Meinke, D.W., Last, R.L., Conklin, P.L. and Somerville, C.R. (2001) Arabidopsis *cyt1* mutants are deficient in a mannose-1-phosphate guanylyltransferase and point to a requirement of N-linked glycosylation for cellulose biosynthesis. *Proc. Natl. Acad. Sci. USA*, 98, 2262–2267.

Lynch, M.A. and Staehelin, L.A. (1992) Domain-specific and cell-type specific localization of two types of cell wall matrix polysaccharides in the clover root tip. *J. Cell Biol.*, 118, 467–479.

McCormack, B.A., Gregory, A.C.E., Kerry, M.E., Smith, C. and Bolwell, G.P. (1997) Purification of an elicitor-induced glucan synthase (callose synthase) from suspension cultures of French bean (*Phaseolus vulgaris* L.): purification and immunolocation of a probable Mr-65 000 subunit of the enzyme. *Planta* 203, 196–203.

Mayer, U., Torres-Ruiz, R.A., Berleth, T., Miséra, S. and Jürgens, G. (1991) Mutations affecting body organization in the developing Arabidopsis embryo. *Nature*, 353, 402–407.

Meinert, M. and Delmer, D.P. (1977) Changes in biochemical composition of the cell wall of the cotton fibre during development. *Plant Physiol.*, 59, 1088–1097.

Mizuno, K. (1998) Cellulose synthase-tubulin interactions in Azuki bean. In *Proceedings of the Plant Polysaccharides Symposium*, University of California-Davis, California, U.S.A.

Mohnen, D. (1999) Biosynthesis of pectins and galactomannans. In *Carbohydrates and their Derivatives Including Tannins, Cellulose, and Related Lignins, Comprehensive Natural Products Chemistry*, Vol. 3 (ed. B.M. Pinto), Elsevier, Oxford, pp. 497–527.

Moore, P.J., Swords, K.M., Lynch, M.A. and Staehelin, L.A. (1991) Spatial organization of the assembly pathways of glycoproteins and complex polysaccharides in the Golgi apparatus of plants. *J. Cel. Biol.*, 112, 589–602.

Muñoz, P., Norambuena, L. and Orellana, A. (1996) Evidence for a UDP-glucose transporter in Golgi apparatus-derived vesicles from pea and its possible role in polysaccharide biosynthesis. *Plant Physiol.*, 112, 1585–1594.

Nam, J., Mysore, K.S., Zheng, C., Knue, M.K., Matthysse, A.G. and Gelvin, S.B. (1999) Identification of T-DNA tagged Arabidopsis mutants that are resistant to transformation by *Agrobacterium*. *Mol. Gen. Genet.*, 261, 429–438.

Nebenführ, A. and Staehelin, L.A. (2001) Mobile factories: Golgi dynamics in plant cells. *Trends Plant Sci.*, 6, 160–167.

Neckelmann, G. and Orellana, A. (1998) Metabolism of uridine 5′-diphosphate-glucose in Golgi vesicles from pea stems. *Plant Physiol.*, 117, 1007–1014.

Nickle, T.C. and Meinke, D.W. (1998) A cytokinesis-defective mutant of *Arabidopsis* (*cyt1*) characterized by embryonic lethality, incomplete cell walls, and excessive callose accumulation. *Plant J.,* 15, 321–332.

Nicol, F., His, I., Jauneau, A., Vernhettes, S., Canut, H. and Höfte, H. (1998) A plasma membrane-bound putative endo-1,4-β-D-glucanase is required for normal wall assembly and cell elongation in Arabidopsis. *EMBO J.,* 17, 5563–5576.

Nobles, D.R., Romanovicz, D.K. and Brown, R.M. Jr (2001) Cellulose in cyanobacteria. Origin of vascular plant cellulose synthase? *Plant Physiol.*, 127, 529–542.

Norambuena, L., Marchant, L., Berninsone, P., Hirschberg, C., Silva, H. and Orellana, A. (2002) Transport of UDP-galactose in plants: identification and functional characterization of AtUTR1, an *Arabidopsis thaliana* UDP-galactose/UDP-glucose transporter. *J. Biol. Chem.*, 277, 32923–32929.

Ohana, P., Benziman, M. and Delmer, D. P. (1993) Stimulation of callose synthesis in vivo correlates with changes in intracellular distribution of the callose synthase activator β-furfuryl-β-glucoside. *Plant Physiol.* 101, 187–191.

Orellana, A., Neckelmann, G. and Norambuena, L. (1997) Topography and function of Golgi uridine-5′-diphosphatase from pea stems. *Plant Physiol.*, 114, 99–107.

Østergaard, L., Petersen, M., Mattsson, O. and Mundy, J. (2002) An Arabidopsis callose synthase. *Plant Mol. Biol.*, 49, 559–566.

Paech, N.A., Fincher, G.B. and Koltunow, A.M. (1999) Functional characterisation of an *Arabidopsis* (1,3)-β-glucan synthase homologue. In *Proceedings of the Tenth International Conference on Arabidopsis Research*, University of Melbourne, Parkville, Victoria, Australia, Abstract 10–18.

Paulson, J.C. and Colley, K.J. (1989) Glycosyltransferases – structure, localization, and control of cell-specific glycosylation. *J. Biol. Chem.*, 264, 17614–17618.

Pear, J.R., Kawagoe Y., Schreckengost, W.E., Delmer, D.P. and Stalker, D.M. (1996) Higher plants contain homologs of the bacterial celA genes encoding the catalytic subunit of cellulose synthase. *Proc. Natl. Acad. Sci. USA*, 93, 12637–12642.

Peng, L., Hocart, C.H., Redmond, J.W. and Williamson, R.E. (2000) Fractionation of carbohydrates in *Arabidopsis* root cell walls shows that three radial swelling loci are specifically involved in cellulose production. *Planta*, 211, 406–414.

Peng, L., Xiang, F., Roberts, E. *et al.* (2001) The experimental herbicide CGA 325'615 inhibits synthesis of crystalline cellulose and causes accumulation of non-crystalline β-1,4-glucan associated with CesA protein. *Plant Physiol.*, 126, 981–992.

Peng, L., Kawagoe, Y., Hogan, P. and Delmer, D. (2002) Sitosterol-β-glucoside as primer for cellulose synthesis in plants. *Sci.*, 295, 147–150.

Perrin, R.M. (2001) How many cellulose synthases to make a plant? *Curr. Biol.*, 11, R213-R216.

Perrin, R.M., DeRocher, A.E., Bar-Peled, M. *et al.* (1999) Xyloglucan fucosyltransferase, an enzyme involved in plant cell wall biosynthesis. *Sci.*, 284, 1976–1979.

Piro, G., Dalessandro, G. and Northcote, D.H. (1993) Glucomannan synthesis in pea epicotyls: the mannose and glucose transferases. *Planta*, 190, 206–220.

Potikha, T.S. and Delmer, D.P. (1995) A mutant of *Arabidopsis thaliana* displaying altered patterns of cellulose deposition. *Plant J.*, 7, 453–460.

Potikha, T.S., Collins, C.C., Johnson, D.I., Delmer, D.P. and Levine, A. (1999) The involvement of hydrogen peroxide in the differentiation of secondary walls in cotton fibers. *Plant Physiol.*, 119, 849–858.

Puhlmann, J., Bucheli, E., Swain, M.J. *et al.* (1994) Generation of monoclonal antibodies against plant cell wall polysaccharides. I. Characterization of a monoclonal antibody to a terminal α(1→2)-linked fucosyl-containing epitope. *Plant Physiol.*, 104, 699–710.

Qadota, H., Python, C. P., Inoue, S. B. *et al.* (1996) Identification of yeast rho1p GTPase as a regulatory subunit of 1,3-ß-glucan synthase. *Sci.*, 272, 279–281.

Read, S.M. and Bacic, T. (2002) Prime time for cellulose. *Sci.*, 295, 59–60.

Read, S.M. and Delmer, D.P. (1990) Biochemistry and regulation of cellulose synthesis in higher plants. In *Biosynthesis and Biodegradation of Cellulose* (eds C.H. Haigler and P.J. Weimer), Marcel Dekker, New York, pp. 177–200.

Reid, J.S.G., Edwards, M., Gidley, M.J. and Clark, A.H. (1995) Enzyme specificity in galactomannan biosynthesis. *Planta*, 195, 489–495.

Reiter, W.-D. and Vanzin, G.F. (2001) Molecular genetics of nucleotide sugar interconversion pathways in plants. *Plant Mol. Biol.*, 47, 95–113.

Reiter, W-D., Chapple, C. and Somerville, C.R. (1997) Mutants of *Arabidopsis thaliana* with altered cell wall polysaccharide composition. *Plant J.*, 12, 335–345.

Richmond, T.A. and Somerville, C.R. (2000) The cellulose synthase superfamily. *Plant Physiol.*, 124, 495–498.

Richmond, T. and Somerville, C. (2001) Integrative approaches to determining Csl function. *Plant Mol. Biol.*, 47, 131–143.

Ridley, B.L., O'Neill, M.A. and Mohnen, D. (2001) Pectins: structure, biosynthesis, and oligo-galacturonide-related signaling. *Phytochem.*, 57, 929–967.

Rodkiewicz, B. (1970) Callose in cell walls during megasporogenesis in angiosperms. *Planta*, 93, 39–47.

Salinikov, V.V., Grimson, M.J., Delmer, D.P. and Haigler, C.H. (2001) Sucrose synthase localizes to cellulose synthesis sites in tracheary elements. *Phytochem.*, 57, 823–833.

Samuels, A.L., Giddings, T.H. Jr and Staehelin, L.A. (1995) Cytokinesis in tobacco BY-2 and root tip cells: a new model of cell plate formation in higher plants. *J. Cell Biol.*, 13, 1345–1357.

Sarria, R., Wagner, T.A., O'Neill, M.A. *et al.* (2001) Characterization of a family of *Arabidopsis* genes related to xyloglucan fucosyltransferase 1. *Plant Physiol.*, 127, 1595–1606.

Sato, S., Kato, T., Kakegawa, K. *et al.* (2001) Role of the putative membrane-bound endo-1,4-β-glucanase KORRIGAN in cell elongation and cellulose synthesis in *Arabidopsis thaliana*. *Plant Cell Physiol.*, 42, 251–263.

Saxena, I.M. and Brown, R.M. Jr (1995) Identification of a second cellulose synthase gene (acsAII) in *Acetobacter xylinum*. *J. Bacteriol.*, 177, 5276–5283.

Saxena, I.M., Lin, F.C. and Brown, R.M. Jr (1990) Cloning and sequencing of the cellulose syn-thase catalytic subunit gene of *Acetobacter xylinum*. *Plant Mol. Biol.*, 15, 673–683.

Saxena, I.M., Kudlicka, K., Okuda, K., Brown, R.M. Jr (1994) Characterization of genes in the cel-lulose-synthesizing operon (acs operon) of *Acetobacter xylinum*: implications for cellulose crystallization. *J. Bacteriol.*, 176, 5735–5752.

Saxena, I.M., Brown, R.M. Jr, Fevre, M., Geremia, R.A. and Henrissat, B. (1995) Multidomain architecture of β-glycosyl transferases: implications for mechanism of action. *J. Bacteriol.*, 177, 1419–1424.

Saxena, I.M., Brown, R.M. Jr and Dandekar, T. (2001) Structure-function characterization of cel-lulose synthase relationship to other glycosyltransferases. *Phytochem.*, 57, 1135–1148.

Scheible, W-R., Eshed, R., Richmond, T., Delmer, D. and Somerville, C. (2001) Modifications of cellulose synthase confer resistance to isoxaben and thiazolidinone herbicides in *Arabidop-sis Ixr1* mutants. *Proc. Natl. Acad. Sci. USA*, 98, 10079–10084.

Scherp, P., Grotha, R. and Kutschera, U. (2001) Occurrence and phylogenetic significance of cytokinesis-related callose in green algae, bryophytes and seed plants. *Plant Cell Rep.*, 20, 143–149.

Schiefelbein, J.W. and Somerville, C. (1990) Genetic control of root hair development in *Arabi-dopsis thaliana*. *Plant Cell*, 2, 235–243.

Schlüpmann, H., Bacic, A. and Read, S. (1993) A novel callose synthase from pollen tubes of *Nicotiana*. *Planta* 191, 470–481.

Schlüpmann, H., Bacic, A. and Read, S.M. (1994) UDP-glucose metabolism and callose synthesis in cultured pollen tubes of *Nicotiana alata* Link et Otto. *Plant Physiol.*, 105, 659–670.

Schwartz, N.B. (2000) Biosynthesis and regulation of expression of proteoglycans. *Frontiers Biosci.*, 5, D649–D655.

Seagull, R.W. (1990) The effects of microtubule and microfilament disrupting agents on cytoskel-etal arrays and wall deposition in developing cotton fibers. *Protoplasma*, 159, 44–59.

Seitz, B., Klos, C, Wurm, M. and Tenhaken, R. (2000) Matrix polysaccharide precursors in *Ara-bidopsis* cell walls are synthesized by alternate pathways with organ-specific expression patterns. *Plant J.*, 21, 537–546.

Shedletzky, E., Shmuel, M., Trainin, T., Kalman, S. and Delmer, D. (1992) Cell wall structure in cells adapted to growth on the cellulose-synthesis inhibitor 2.6-dichlorobenzonitrile. *Plant Physiol.*, 100 120–130.

Sherrier, D.J. and VandenBosch, K.A. (1994) Secretion of cell wall polysaccharides in *Vicia* root hairs. *Plant J.*, 5, 185–195.

Somerville, C. and Cutler, S. (1998) Use of genes encoding xylan synthase to modify plant cell wall composition. International patent application WO 098/55596 A1, Application no. US9811531.

Staehelin, L.A. and Moore, I. (1995) The plant Golgi apparatus: structure, functional organization and trafficking mechanisms. *Annu. Rev. Plant Physiol. Plant Mol. Biol.*, 46, 261–288.

Stasinopoulos, S.J., Fisher, P.R., Stone, B.A. and Stanisich, V.A. (1999) Detection of two loci involved in (1,3)-β-glucan (curdlan) biosynthesis by *Agrobacterium* sp. ATCC31749, and comparative sequence analysis of the putative curdlan synthase gene. *Glycobiol.*, 9, 31–41.

Stone, B.A. and Clarke, A.E. (1992) *Chemistry and biology of (1,3)-β-glucans*, La Trobe University Press, Melbourne.

Taylor, N.G., Scheible, W.-R., Cutler, S., Somerville, C.R. and Turner, S.R. (1999) The irregular xylem 3 locus of *Arabidopsis* encodes a cellulose synthase required for secondary cell wall synthesis. *Plant Cell*, 11, 769–779.

Taylor, N.G., Laurie, S. and Turner, S.R. (2000) Multiple cellulose synthase catalytic subunits are required for cellulose synthesis in Arabidopsis. *Plant Cell*, 12, 2529–2539.

Taylor, N.G., Howells, R.M., Huttly, A.K., Vickers, K. and Turner, S.R. (2003) Interactions among three distinct CesA proteins essential for cellulose synthesis. *Proc. Natl. Acad. Sci. USA*, 100, 1450–1455.

The Arabidopsis Genome Initiative (2000) Analysis of the genome sequence of the flowering plant *Arabidopsis thaliana*. *Nature*, 408, 796–815.

Turner, S.R. and Somerville, C.R. (1997) Collapsed xylem phenotype of *Arabidopsis* identifies mutants deficient in cellulose deposition in the secondary cell wall. *Plant Cell*, 9, 689–701.

Turner, A., Bacic, A., Harris, P.J. and Read, S.M. (1998) Membrane fractionation and enrichment of callose synthase from pollen tubes of *Nicotiana alata*. *Planta* 205, 380–388.

Ünligil, U.M., Zhou, S., Yuwaraj, S., Sarkar, M., Schachter, H. and Rini, J.M. (2000) X-ray crystal structure of rabbit N-acetylglucosaminyltransferase I: catalytic mechanism and a new protein superfamily. *EMBO J.*, 19(20), 5269–5280.

Vanzin, G.F., Madson, M., Carpita, N.C., Raikhel, N.V., Keegstra, K. and Reiter, W.-D. (2002) The *mur2* mutant of *Arabidopsis thaliana* lacks fucosylated xyloglucan because of a lesion in fucosyltransferase AtFUT1. *Proc. Natl. Acad. Sci. USA*, 99, 3340–3345.

Vergara, C.E. and Carpita, N.C. (2001) β-D-glycan synthases and the CesA gene family: lessons to be learned from the mixed-linkage (1→3),(1→4)β-D-glucan synthase. *Plant Mol. Biol.*, 47, 145–160.

Wagner, K.G. and Backer, A.I. (1992) Dynamics of nucleotides in plants studied on a cellular basis. *Int. Rev. Cytol.*, 134, 1–84.

Waldron, K.W. and Brett, C.T. (1987) Subcellular localization of a glucuronyltransferase involved in glucuronoxylan biosynthesis in pea (*Pisum sativum*) epicotyls. *Plant Sci.*, 49, 1–8.

Wang, X., Cnops, G., Vanderhaeghen, R., De Block, S., Van Montagu, M. and Van Lijsebettens, M. (2001) *AtCSLD3*, a cellulose synthase-like gene important for root hair growth in Arabidopsis. *Plant Physiol.*, 126, 575–586.

Wardak, A.Z., Jacobs, A.K., Anderson, M.A., Fincher, G.B. and Stone, B.A. (1999) The (1,3)-β-glucan synthase from suspension-cultured *Lolium multiflorum* (rye grass) endosperm. In *Proceedings of the 43rd Annual Australian Society for Biochemistry and Molecular Biology, 18th Annual Australian and New Zealand Society for Cell and Developmental Biology and 39th Annual Australian Society of Plant Physiologists Conferences*, Conrad Jupiters, Gold Coast, Queensland, Australia, Poster Mon-04.

Warnecke, D.C., Baltrusch, M., Buck, F., Wolter, F.P. and Heinz, E. (1997) UDP-glucose:sterol glucosyltransferase: cloning and functional expression in *Escherichia coli*. *Plant Mol. Biol.*, 35, 597–603.

Williamson, R.E., Burn, J.E., Birch, R. *et al.* (2001a) Morphology of *rsw1*, a cellulose-deficient mutant of *Arabidopsis thaliana*. *Protoplasma*, 215, 116–127.

Williamson, R.E., Burn, J.E. and Hocart, C.H. (2001b) Cellulose synthesis: mutational analysis and genomic perspectives using *Arabidopsis thaliana*. *Cell. Mol. Life Sci.*, 58, 1475–1490.

Winter, H., Huber, J.L. and Huber, S.C. (1998) Identification of sucrose synthase as an actin-bind-
    ing protein. *FEBS Lett.*, 430, 205–208.
Wolucka, B.A., Persiau, G., Van Doorsselaere, J. *et al.* (2001) Partial purification and identification
    of GDP-mannose 3´,5´-epimerase of *Arabidopsis thaliana*, a key enzyme of the plant vitamin
    C pathway. *Proc. Natl. Acad. Sci. USA*, 98, 14843–14848.
Wulff, C., Norambuena, L. and Orellana, A. (2000) GDP-fucose uptake into the Golgi apparatus
    during xyloglucan biosynthesis requires the activity of a transporter-like protein other than
    the UDP-glucose transporter. *Plant Physiol.*, 122, 867–877.
Yoshida, M., Itano, N., Yamada, Y. and Kimata, K. (2000) *In vitro* synthesis of hyaluronan by a
    single protein derived from mouse HAS1 gene and characterization of amino acid residues
    essential for the activity. *J. Biol. Chem.*, 255, 497–506.
Zablackis, E., York, W.S., Pauly, M. *et al.* (1996) Substitution of L-fucose by L-galactose in cell
    walls of Arabidopsis *mur1*. *Sci.*, 272, 1808–1810.
Zhang, G.F., and Staehelin, L.A. (1992) Functional compartmentation of the Golgi apparatus of
    plant cells. *Plant Physiol.*, 99, 1070–1083.
Zuo, J., Niu, Q-W., Nishizawa, N., Wu, Y., Kost, B. and Chua, N-H. (2000) KORRIGAN, an Arabi-
    dopsis endo-1,4-β-glucanase, localizes to the cell plate by polarized targeting and is essential
    for cytokinesis. *Plant Cell*, 12, 1137–1152.

7 WAKs: cell wall associated kinases

Jeff Riese, Josh Ney and Bruce D. Kohorn

7.1 Preface

The wall associated kinases, WAKs, physically link the plasma membrane to the carbohydrate and protein matrix that comprises the plant cell wall. WAKs are of importance as they have the potential to directly signal cellular events through their cytoplasmic kinase domain. There are five WAKs that vary only in their extracellular domain, and that collectively are expressed in most tissues. WAK mRNA and protein are present in vegetative meristems, junctions of organ types, and areas of cell expansion. They are also induced by pathogen infection and wounding. WAKs can be found bound to pectin in the cell wall, and at least one isoform is associated with a secreted glycine-rich protein (GRP). Disruption of WAK expression leads to the loss of cell expansion in a variety of cell types, indicating that WAKs play an important role in plant development. It remains to be discovered how WAKs, pectins and GRPs combine to regulate the patterns of cell expansion in coordination with the complex architecture and mechanical constraints of the plant cell wall.

7.2 Introduction

The plant extracellular matrix (ECM), also known as the cell wall, is highly structured, such that it has both strength and the ability to change in cell shape, size and volume during cell differentiation (Cosgrove, 1997; see also Chapter 8). Throughout the development of the plant, cells enlarge and modify the components of their ECM. At present we have a minimal understanding of how the ECM might communicate with the cytoplasm and interior compartments of the cell in response to cellular events. While there are a variety of proteins, small peptides and oligosaccharides which may interact with proteins on the cell surface, their role in cell wall-membrane communication is not clear (McCarthy and Chory, 2000; Torii, 2000). Although the identification of cell wall associated proteins is limited, there are few classes of proteins that have been found to associate with both the plasma membrane and the cell wall. This chapter focuses on the *Arabidopsis* wall associated kinases (WAKs), which are tightly associated with the cell wall, span the plasma membrane, and have a cytoplasmic kinase domain (Figure 7.1) (He *et al.*, 1996, 1999; Kohorn, 2000, 2001; Anderson *et al.*, 2001; Wagner and Kohorn, 2001).

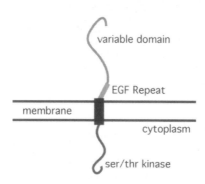

Figure 7.1   Cartoon of the topology of WAKs in the membrane and cell wall.

7.3   The cell wall and membrane

The cell wall and plasma membrane are likely to be in direct contact, as most electron micrographs show them to be appressed (Roberts, 1994). When cells are plasmolysed, lingering contacts between the plasma membrane and cell wall are readily evident (Figure 7.2a,b). Some of these contacts persist and extend from the collapsed plasma membrane to the cell wall and these have been termed Hechtian strands (Hecht, 1912; Roberts, 1994; Lang-Pauluzzi, 2000; see Figure 7.2c, thin arrow). Contacts can also be induced by growth of cultured cells in high salt, but the composition and role of these and Hechtian strands remains unknown. Some have termed these 'adhesion sites', suggesting some homology with similar sites in metazoan cells. However, the integrins, receptors and regulatory molecules that define metazoan adhesion sites appear not to be present in angiosperms (Carpita and Gibeaut, 1993; Canut *et al.*, 1998; Laval *et al.*, 1999). Despite these old and numerous observations, it is presently unknown how the cytoskeleton, cytoplasm and the cell wall communicate. The modulation of cell wall polysaccharides, such as pectins, must also be included in our understanding of how the wall is involved in developmental processes. The WAKs are of particular interest to us as they provide an avenue to explore these relationships, which may be essential to the regulation of cell expansion during development.

7.4   Cell wall contacts

Before exploring the WAKs in detail, it is important to briefly review other molecules that play a role in cell wall communication. One of these, *COBRA (COB)*, encodes a protein which is anchored to the extracellular surface of the plasma membrane of root

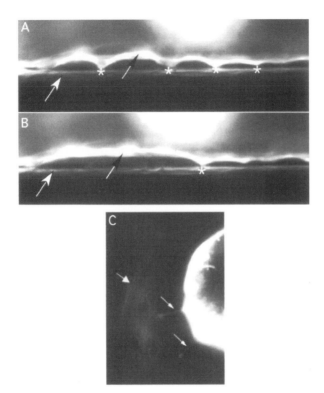

Figure 7.2 Membrane cell wall contacts. Onion skin cells were transformed with a green fluorescent protein (GFP), plasmolysed in 40% sucrose, and observed under UV light. (a) and (b) differ in extent of plasmolysis. White arrow points to cell wall, black arrow to plasma membrane, stars indicate contact sites. (c) Hechtian strands (long arrows) persist in fully plasmolysed cells bordered by the cell wall (short arrow).

cells by a glycosyl phosphatidylinositol (GPI) moiety (Schindelman *et al.*, 2001). COBRA is thought to regulate oriented cell expansion in root cells. Another gene, *LRX1*, is expressed in root hair cells and encodes a chimeric leucine-rich repeat/extensin protein (see Chapter 4), and is localized to the cell wall of the root hair. The interaction between the cell wall and the LRX1 protein is important for proper root hair development and expansion (Baumberger *et al.*, 2001). A family of secreted proteins called SCAs (stigma/stylar cysteine-rich adhesion) were identified that are required for lily pollen tube adhesion to the wall material of the style. The identification of the requirement of pectin for SCA activity is particularly intriguing in the context of plant cell adhesion to the ECM of adjacent cells (Mollet *et al.*, 2000).

Other proteins that associate with both the plasma membrane and the cell wall include the arabinogalactan-rich proteins (AGPs) (Oxley and Bacic, 1999; Svetek *et al.*, 1999; Majewska-Sawka and Nothnagel, 2000), some of which are found anchored to the membrane via a carboxy-terminal GPI. Cellulose synthases (Pear *et al.*,

1996) provide synthetic functions, but their location allows them to mediate signals between the structures they create and the cytoplasm that provides their polymer building blocks. Other proteins that are secreted by the cell and may serve indirectly in communication with the cytoplasm include the hydroxyproline-rich glycoproteins (Showalter, 1993), proline-rich proteins (Fowler *et al.*, 1999), glycine-rich proteins (GRPs) (Cheng *et al.*, 1996), xyloglucan endotransglycosylases (recently renamed XTHs, see Chapters 8 and 9) (Xu *et al.*, 1996; Vissenberg *et al.*, 2000), endoglucanases (Zuo *et al.*, 2000) and expansins (Cho and Cosgrove, 2000). These proteins somehow interact with each other and the ECM to influence cell growth and determine developmental processes in response to cellular and environmental signals.

7.5   The WAK family

Of the approximately 600 receptor-like proteins encoded by the *Arabidopsis* genome (*Arabidopsis* Genome Initiative, 2000), some likely use components of the ECM as their ligands. This chapter focuses on the family of *Arabidopsis* cell wall associated kinases (WAKs) which have an amino-terminus that is tightly linked to the cell wall and binds a cell wall glycine-rich protein (GRP). WAKs contain a transmembrane domain that separates the extracellular sequence from a cytoplasmic protein kinase. WAKs are distinguished from other proteins by the fact that they physically link the plasma membrane to cell wall and have the potential to signal cellular events through their kinase domain. Indeed, WAKs appear to be required for cell expansion and play some role in responses to pathogens (Lally *et al.*, 2001; Wagner and Kohorn, 2001).

7.6   A transmembrane protein with a cytoplasmic protein kinase and cell wall domain

WAKs were discovered when an antibody to a serine/threonine kinase domain reacted with a protein in the cell wall. This protein was detected using an immunogold labelled antiserum in the cell wall of plasmolysed cells, where the membrane has shrunk away from the wall (Figure 7.3). The same serum reacted with a 68-kDa protein that could only be released into a low speed supernatant if tissue was boiled in SDS and DTT. Subsequent analysis showed that the 68-kDa protein could also be detected by Western blots in the plasma membrane fractions if tissue was digested with cell wall digesting enzymes (He *et al.*, 1996). These experiments also demonstrated that the WAKs have a cytoplasmic kinase domain, and span the plasma membrane. Thus, WAKs indeed bind the cell wall, traverse the plasma membrane, and have a cytoplasmic carboxyl tail containing a protein kinase (Figure 7.1).

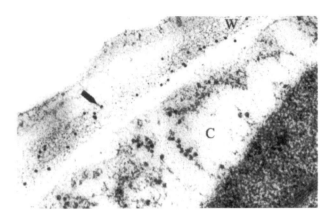

Figure 7.3 WAKs are cell wall proteins. Electron micrograph of leaf cells. Gold labelled antibodies (arrow) line the cell wall (W). The cytoplasm (C) contains more diffuse ribosome spheres and is separated from the wall by a clear space due to fixation.

7.7   WAKs are bound to pectin

It was initially noticed that WAKs could be released into a microsomal fraction after the enzymatic digestion of the cell wall with unpurified degradative enzyme (He *et al.*, 1996). Purified preparations of a number of enzymes specific for different cell wall components indicated that only pectinase treatments released WAK from the cell wall. The WAK protein that was released by pectinase reacted with WAK serum on Western blots, but also against antibodies raised to homogalacturonan (Knox *et al.*, 1990; Knox, 1997; Wagner and Kohorn, 2001, JIM5 and JIM7). Thus, it appears that this species of WAK has a covalently bound molecule that is likely to be a pectic fragment. However, the nature of this covalent bond and the composition of the pectic fragment remain to be determined. It is also not clear if the bond is induced by the extraction method, or what role this pectin binding plays in the physiology of the cell.

7.8   Genomic organization of WAKs

Western blots with WAK antiserum detect a 68-kDa band, and mapping studies using oligonucleotides specific to the kinase domain found only one locus on chromosome 1. Thus, it was thought that WAK was encoded by one gene. However, genomic cloning and the newly compiled *Arabidopsis* genome project define five tandemly arrayed WAK genes in a 30-kilobase cluster. Each WAK contains an extracellular domain, a transmembrane region, and similar kinase domain (He *et al.*, 1999). Each of the kinase domains is 87% identical, whereas the extracellular amino-termini are only 40–64% identical. All WAKs contain two distinct EGF-like cysteine repeats (Kohorn *et al.*, 1996; He *et al.*, 1999). The WAK serum used to define the location in the membrane and wall is directed to the conserved kinase region, and as such, the data are consistent with each of the 5 WAKs having a tight cell wall association. The recent *Arabidopsis* genome sequencing project has identified 16 other proteins (7 of which are also clustered) that also have the same three domains that characterize WAKs. These proteins are best denoted WLK, for WAK-like, since it is unknown whether these proteins indeed associate with the cell wall, and their kinase regions do not react with the WAK serum. The WLK 'extracellular domain' is only 20–30% similar to that of the WAKs, but does contain a variable number of degenerate EGF repeats. The expression patterns of the WLK family are not known.

7.9   EGF repeats

It appears that WAK and WLK represent the only receptor-like proteins in *Arabidopsis* that contain EGF repeats. Eukaryotic EGF repeats generally contain six cysteine residues with a conserved spacing that allows three disulfide bonds and a distinct shape (Davis *et al.*, 1987). EGF repeats are characterized as either binding or not binding calcium, and WAK has been found to contain both types. All characterized EGF repeats are involved in protein-protein interactions and some EGF repeats regulate ligand binding (Bork *et al.*, 1996). EGF repeats are well known in metazoans for their role in the dimerization of proteins, often receptors, but this has not been demonstrated in plants. In plants, other cysteine-repeat motifs, distinct from the EGF repeats, are required for the irreversible binding of proteins to the cell wall (Domingo *et al.*, 1999). In some fungi (Zhang *et al.*, 1998), cysteine-rich repeats mediate the binding to cellulose, but the WAK and WLK EGF repeats remain to be characterized.

7.10   WAK expression

The five WAK isoforms are expressed in various tissues, including vegetative meristems, junctions of organ types, and areas of cell expansion. *WAK1* and *WAK2* are the most ubiquitously expressed isoforms (He *et al.*, 1999; Wagner and Kohorn, 2001). The developmental and cell type differences in WAK isoform expression have been examined using promoter-GUS fusions (Figure 7.4). WAK expression is first detected in the embryo. The *WAK1* promoter is active in the cotyledons before the seed coat breaks and then activity decreases. WAK1-GUS was also found in the vasculature as the cotyledons opened. *WAK1* expression was detected at the junction between the cotyledons and the hypocotyl and remained at this junction throughout the growth of the shoot apical meristem (Wagner and Kohorn, 2001). The emerging leaf has very little *WAK1* expression but the level increases in older leaves, which show expression in the blade and petiole (Wagner and Kohorn, 2001). *WAK2* expres-

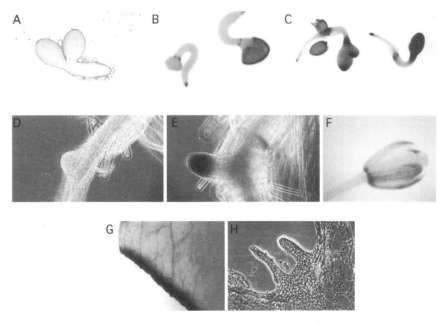

Figure 7.4 Expression of WAK genes. WAK promoters were fused to the GUS coding region and expressed (blue strain) in transformed *Arabidopsis* plants. (a) Embryo; (b) and (c) seedlings; (d) and (e) from same plant line; expression is only apparent after lateral root begins to emerge; (f) floral sepals; (g) wounded leaf margins; (h) vegetative meristems (thin cross-section).

sion was found in all layers of the apical meristem and leaf primordia (Anderson *et al.*, 2001). More specifically, *WAK2* expression was found in the primary root tip and at the junction of the cotyledon and hypocotyl. In older plants *WAK2* was found expressed at the base of flowers and at the tip of expanding sepals (Anderson *et al.*, 2001). *WAK2* expression continues throughout the cotyledon and entire leaf until cell expansion ceases. *WAK2* is expressed at the margins of leaves and at the leaf tip, whereas *WAK1* was found more often in the vasculature and to a lesser extent in the emerging leaves (Kohorn, 2001). At early stages in seed germination, the WAK1 and WAK2 promoters displayed distinct yet overlapping expression. Similar to the WAK1 promoter, WAK2 promoter-GUS stained throughout the shoot apical meristem and throughout the cotyledon. Also, the root/hypocotyl region of both WAK1 and WAK2 promoter-GUS seedlings showed GUS staining.

WAK3 was also detected by RNA gel blot analysis in the leaves and stems, but to a lesser extent than *WAK1* or *WAK2* (He *et al.*, 1999). WAK3 promoter-GUS staining was expressed in a spotty pattern, sometimes in an area the size of a single cell and sometimes in a larger pattern (Kohorn, 2001). *WAK4* and *WAK5* are not widely expressed. Northern blot analysis of RNA from leaves and stems shows that *WAK5* expression is barely detectable (He *et al.*, 1999). WAKs are expressed in the root but only in meristems and at the junction of the root and hypocotyls. WAKs are also not apparent in floral organs besides the base, sepals, and ovaries (Wagner and Kohorn, 2001). Significantly, expression of WAKs was not found in the elongation zone of the roots.

7.11 WAKs and cell expansion

The expression patterns of WAKs correlate with some, but not all, areas of cell expansion. In order to understand WAK function in cell expansion, dominant WAK mutations were generated. Antisense WAK2 was expressed under the control of an inducible promoter. Because the constructs contained the conserved kinase domain, the induced antisense WAK plants had reduced expression of all the WAK isoforms. These plants exhibited a loss of cell expansion in the tissues where WAKs are normally expressed (Lally *et al.*, 2001; Wagner and Kohorn, 2001). Scanning electron microscopy was used to determine if the reduced leaf size of emerging leaves of WAK antisense plants was due to reduced cell expansion, cell division, or a combination of both. The smaller leaves of the antisense WAK plants were found to have the same number of cells as wild-type plants, indicating that cell expansion, rather than cell division, was inhibited (Figure 7.5). WAK expression has also been reduced by RNA interference (RNAi), where a double stranded RNA copy of WAK RNA was expressed in transgenic plants. Using a constitutive promoter to express a region that is identical in all five WAK isoforms, it was possible to generate dwarf *Arabidopsis* plants that had reduced levels of cell expansion (Kohorn, unpublished). RNAi specific to *WAK1* or to *WAK2* expression also led to a dwarf phenotype, indicating that these two WAK isoforms have non-overlapping functions. Plant hormones, including gibberellins (GA; Phillips, 1998; Silverstone and Sun, 2000), brassinosteroids (Mussig and Altmann, 1999), auxins (Jones, 1998) and ethylene (Bleecker, 1999;

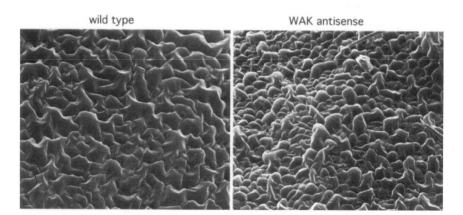

Figure 7.5 WAKs are required for cell expansion. SEM of epidermal leaf cells from wild-type and WAK antisense *Arabidopsis.*

Chang and Shockey, 1999) have all been shown to regulate cell expansion in some way. In all cases, a link between the events that occur just before expansion and those that occur during expansion has not been fully characterized. We therefore attempted to rescue the dwarf phenotype of WAK RNAi by treatment of the affected seedlings with hormones. Of all those tested, only GA was shown to reverse the effect of the WAK RNAi, indicating some link with this signalling pathway (Kohorn, unpublished). Future work will require an analysis of GA mutants, and the effect of GAs on WAK gene expression.

7.12 WAKs and pathogenesis

WAKs also appear to play some role in the response *Arabidopsis* has to pathogen infection. This is of particular interest since pathogens first contact plant cells at the cell wall. Systemic acquired resistance (SAR) is accompanied by increases in the levels of salicylic acid (SA) and the induction of pathogen-related proteins. WAK expression is also induced when *Arabidopsis* plants are infected by pathogens or stimulated by exogenous SA or its analogue, INA. This gene induction is dependent upon the transcriptional regulator NPR1/NIM1 (He *et al.*, 1998). Plants expressing an antisense WAK and a dominant negative allele of WAK1 were severely affected in growth when exposed to SA, and it was concluded that the induced expression of WAK1 is required during the pathogenesis response. In addition, ectopic expression of full length WAK1 or the kinase domain alone provides resistance to otherwise lethal SA levels (He *et al.*, 1998). Thus, there appears to be some role for a wall associated receptor kinase in the pathogen response, but the mechanism remains undefined.

7.13 WAK ligands

WAKs are covalently bound to pectin in the cell wall, and it appears that the extra-cellular domain also associates with at least one cell wall protein. At present, it is not clear if pectin or other proteins act as true ligands for the WAK receptors. The amino-terminal extra-cytoplasmic domain of WAK1 was used to screen an *Arabidopsis* library in the yeast two-hybrid assay to identify proteins that might bind with relatively high affinity. A recent report shows that WAK1 binds *in vitro* to one of a family of *Arabidopsis* glycine-rich proteins, GRPs (Park *et al.*, 2001). In pea, GRPs are secreted cell wall proteins that are often associated with the generation of highly elaborate cell walls in vascular tissue, and hence have been termed 'structural proteins' (Cheng *et al.*, 1996; see also Chapter 4). GRPs are characteristically rich in glycine residues and are represented by over 50 genes in *Arabidopsis*, some of which have signal sequences that direct them to the cell wall. Park *et al.* (2001) provide convincing evidence that GRP3, but not some of the other GRP family members, binds WAK1 but not WAK2 in yeast, and *in vitro*. Using an antibody to GRPs, a serum not specific for GRP3, they show that some GRPs are bound to a 100-kDa protein to form a larger 500-kDa complex in *Arabidopsis* extracts. An antibody generated to a WAK1 peptide reacts with this 100-kDa protein. However, this same antibody does not react with WAKs themselves. Since the 100-kDa protein is relatively large, and the 500-kDa complex it forms with GRPs is soluble in the detergent triton X-100, it is not likely to represent the wall associated kinases: WAKs are defined as 68-kDa triton X-100-insoluble, wall associated proteins. It is more likely that Park *et al.* (2001) have identified a new membrane protein that binds members of the GRP family. The results are more understandable when plant cell extracts are analysed by non-denaturing gels, which permit the visualization of triton X-100-insoluble particulate fractions. Here, WAKs can be found to be in a complex with a 14-kDa protein that cross-reacts with a GRP antiserum. This GRP is far less abundant than the one that associates with the unidentified 100-kDa membrane protein reported in Park *et al.* (2001). Thus, it is likely that there are several GRP receptors, one of which may well be WAK1 (Figure 7.6). The expression of GRP3 overlaps that of WAK1, as assayed by *in situ* hybridization and RT-PCR (Kohorn, unpublished). Currently, a definitive analysis of GRP-WAK interaction awaits the genetic dissection of their interaction *in vivo*. It will be of special interest to determine whether the WAK binding of GRP and pectin is correlated with WAK kinase activity, and if these interactions play a role in cell expansion (Figure 7.7).

7.14 WAK substrates

A complete understanding of WAK function also requires an analysis of the WAK kinase substrates. WAKs are predicted to be serine threonine kinases, and indeed when expressed in *E. coli* and assayed *in vitro*, they have protein kinase activity

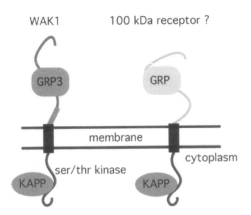

Figure 7.6 Model for the binding of GRP3 to WAK1, and of another GRP to an unidentified 100-kDa protein. KAPP is a cytoplasmic phosphatase.

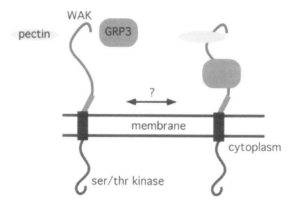

Figure 7.7 Model for the binding of pectin and GRP3 to WAKs. It is proposed that GRP and pectin binding may be coordinated and significant in WAK kinase signalling.

(Anderson *et al.*, 2001). However, WAKs have only been observed to phosphorylate other WAK kinase domains and not other substrates (Kohorn, unpublished). Site directed mutagenesis indicates that the conserved lysine at amino acid 298 is essential for this activity. The WAK kinase itself is found to be phosphorylated *in vivo* since anti phospho-threonine serum recognizes WAK protein in SDS/DTT and pectinase released extracts (Anderson *et al.*, 2001). However, it is not known if the *in vivo* phosphorylation is due to WAK kinase activity or to another kinase. The best indicator of WAK activity might be to identify a cytoplasmic substrate and to monitor its phosphorylation state, but such substrates have yet to be found. Recent two-hybrid analysis indicates that a transcriptional adaptor protein may bind to the WAK kinase domain (Kohorn, unpublished). This is of great interest, as this appears to be similar to examples found with metazoan serine threonine kinases which directly activate transcription factors and by-pass the extensive cascades of the tyrosine kinase pathways (Kohorn, 2000). A detailed analysis of the WAK associated transcription factor may be instructive in understanding WAK function.

7.15 Summary

The family of WAKs has caught our attention, as they represent one of the few characterized direct physical connections between the plasma membrane and cell wall, and they have the capacity to directly signal the cytoplasm through their kinase domains. Experiments demonstrate that these receptors indeed span the membrane, have a cytoplasmic active kinase, and can be covalently linked to pectin in the cell wall. The extracellular domain can also bind to a glycine-rich protein. Disruption of WAK expression by several methods leads to the inhibition of cell expansion, and the mechanism may well involve hormone stimulated pathways. The expression patterns of the WAK family are also consistent with a role of WAKs in cell expansion.

The challenge for the future is to understand how the binding to pectins and GRPs is related to the activation of yet to be defined cytoplasmic pathways, such that the cell wall can have a regulatory influence on the extent and timing of cell expansion.

Acknowledgements

We thank J. Kilmartin and B. Freddie for their assistance with this manuscript. This work was funded by the National Science Foundation and the Pew Charitable Trusts.

References

Anderson, C.M., Wagner, T.A., Perret, M., He, Z.H., He, D. and Kohorn, B.D. (2001) WAKs: cell wall-associated kinases linking the cytoplasm to the extracellular matrix. *Plant Mol. Biol.*, 47, 197–206.

Arabidopsis Genome Initiative (2000) Analysis of the flowering plant *Arabidopsis thaliana*. *Nature*, 408, 796–815.

Baumberger, N., Ringli, C. and Keller, B. (2001) The chimeric leucine-rich repeat/extensin cell wall protein LRX1 is required for root hair morphogenesis in *Arabidopsis thaliana*. *Genes Dev.*, 15, 1128–1139.

Bleecker, A.B. (1999) Ethylene perception and signaling: an evolutionary perspective. *Trends Plant Sci.*, 4, 269–274.

Bork, P., Downing, A.K., Kieffer, B. and Campbell, I.D. (1996) Structure and distribution of modules in extracellular proteins. *Q. Rev. Biophys.*, 29, 119–167.

Canut, H., Carrasco, A., Galaud, J.P. *et al.* (1998) High affinity RGD-binding sites at the plasma membrane of *Arabidopsis thaliana* links the cell wall. *Plant J.*, 16, 63–71.

Carpita, N. and Gibeaut, D. (1993) Structural models of primary cell walls in flowering plants: consistency of molecular structure with the physical properties of the walls during growth. *Plant J.*, 3, 1–30.

Chang, C. and Shockey, J.A. (1999) The ethylene-response pathway: signal perception to gene regulation. *Curr. Opin. Plant Biol.*, 2, 352–358.

Cheng, S.H., Keller, B. and Condit, C.M. (1996) Common occurrence of homologues of petunia glycine-rich protein-1 among plants. *Plant Mol. Biol.*, 31, 163–168.

Cho, H.T. and Cosgrove, D.J. (2000) From the cover: altered expression of expansin modulates leaf growth and pedicel abscission in *Arabidopsis thaliana*. *Proc. Natl. Acad. Sci. USA*, 97, 9783–9788.

Cosgrove, D.J. (1997) Assembly and enlargement of the primary cell wall in plants. *Annu. Rev. Cell Devel. Biol.*, 13, 171–201.

Davis, C.G., Goldstein, J.L., Sudhof, T.C., Anderson, R.G.W., Russell, D.W. and Brown, M.S. (1987) Acid-dependent ligand dissociation and recycling of LDL receptor mediated by growth factor homology region. *Nature*, 326, 760–765.

Domingo, C.Y., Sauri, A., Mansilla, E., Conejero, V. and Vera, P. (1999) Identification of a novel peptide motif that mediates cross-linking of proteins to cell walls. *Plant J.*, 20, 563–570.

Fowler, T., Bernhardt, C. and Tierney, M. (1999) Characterization and expression of four proline-rich cell wall protein genes in *Arabidopsis* encoding two distinct subsets of multiple domain proteins. *Plant Physiol.*, 121, 1081–1092.

He, Z.H., Fujiki, M. and Kohorn, B.D. (1996) A cell wall-associated, receptor-like kinase. *J. Biol. Chem.*, 271, 19789–19793.

He, Z.H., He, D. and Kohorn, B.D. (1998) Requirement for the induced expression of a cell wall associated receptor kinase for survival during the pathogen response. *Plant J.*, 14, 55–63.

He, Z.H., Cheeseman, I., He, D. and Kohorn, B.D. (1999) A cluster of five cell wall-associated receptor kinase genes, *Wak1–5*, are expressed in specific organs of *Arabidopsis*. *Plant Mol. Biol.*, 39, 1189–1196.

Hecht, K. (1912) Studien uber den Vorgang der Plasmolyse. *Beitr. Biol. Pflanz*, 11, 133–189.

Jones, A.M. (1998) Auxin transport: down and out and up again. *Sci.*, 282, 2201–2203.

Knox, J.P. (1997) The use of antibodies to study the architecture and developmental regulation of plant cell walls. *Int. Rev. Cytol.*, 171, 79–120.

Knox, J.P., Linstead, P.J., King, J., Cooper, C. and Roberts, K. (1990) Pectin esterification is spatially regulated both within cell walls and between developing tissues of root apices. *Planta*, 181, 512–521.

Kohorn, B.D. (2000) Plasma membrane-cell wall contacts. *Plant Physiol*, 124, 31–38.

Kohorn, B.D. (2001) WAKs: cell wall associated kinases. *Curr. Opin. Cell Biol.*, 13, 529–533.

Kohorn, B.D., He, Z.H., Fujiki, M. and Freddie, B. (1996) Elusin: a receptor-like kinase with an EGF domain in the cell wall. In *Protein Phosphorylation in Plants* (eds P.R. Shewry, N.G. Halford and R. Hooley), Clarendon Press, Oxford, pp. 297–305.

Lally, D., Ingmire, P., Tong, H. and He, Z.H. (2001) Antisense expression of a cell wall associated kinase WAK4 inhibits cell elongation and alters morphology. *Plant Cell*, 13, 1317–1332.

Lang-Pauluzzi, I. (2000) The behaviour of the plasma membrane during plasmolysis: a study by UV microscopy. *J. Microsc.*, 198, 188–198.

Laval, V., Chabannes, M., Carriere, M. *et al.* (1999) A family of *Arabidopsis* plasma membrane receptors presenting animal beta-integrin domains. *Biochim. Biophys. Acta*, 1435, 61–70.

Majewska-Sawka, A. and Nothnagel, E.A. (2000) The multiple roles of arabinogalactan proteins in plant development. *Plant Physiol.*, 122, 3–9.

McCarthy, D.R., and Chory, J. (2000) Conservation and innovation in plant signaling pathways. *Cell*, 103, 201–209.

Mollet, J.C., Park, S.Y., Nothnagel, E.A. and Lord, E.M. (2000) A lily stylar pectin is necessary for pollen tube adhesion to an in vitro stylar matrix. *Plant Cell*, 12, 1737–1749.

Mussig, C. and Altmann, T. (1999) Physiology and molecular mode of action of brassinosteroids. *Plant Physiol. Biochem.*, 37, 363–372.

Oxley, D. and Bacic, A. (1999) Structure of the glycosylphosphatidylinositol anchor of an arabinogalactan protein from *Pyrus communis* suspension-cultured cells. *Proc. Natl. Acad. Sci. USA*, 25, 14246–14251.

Park, A.E., Cho, S.K., Yun, U.J. *et al.* (2001) Interaction of the *Arabidopsis* receptor protein kinase Wak1 with a glycine-rich protein, AtGRP-3. *J. Biol. Chem.*, 276, 26688–26693.

Pear, J.R., Kawagoe, Y., Schreckengost, W.E., Delmer, D.P. and Stalker, D.M. (1996) Higher plants contain homologs of the bacterial *celA* genes encoding the catalytic subunit of cellulose synthase. *Proc. Natl. Acad. Sci. USA*, 25, 12637–12642.

Phillips, A.L. (1998) Gibberellins in Arabidopsis. *Plant Physiol. Biochem.*, 36, 115–124.

Roberts, K. (1994) The plant extracellular matrix: in a new expansive mood. *Curr. Opin. Cell Biol.*, 6, 688–694.

Schindelman, G., Morikami, A., Jung, J. *et al.* (2001) COBRA encodes a putative GPI-anchored protein, which is polarly localized and necessary for oriented cell expansion in *Arabidopsis*. *Genes Dev.*, 15, 1115–1127.

Showalter, A.M. (1993) Structure and function of plant cell wall proteins. *Plant Cell*, 5, 9–23.

Silverstone, A.L. and Sun, T.P. (2000) Gibberellins and the green revolution. *Trends Plant Sci.*, 5, 1–2.

Svetek, J., Yadav, M.P. and Nothnagel, E.A. (1999) Presence of a glycosylphosphatidylinositol lipid anchor on rose arabinogalactan proteins. *J. Biol. Chem.*, 274, 14724–14733.

Torii, K.U. (2000) Receptor kinase activation and signal transduction in plants: an emerging picture. *Curr. Opin. Plant Biol.*, 3, 361–367.

Vissenberg, K., Martinez-Vilchez, I.M., Verbelen, J.P., Miller, J.G. and Fry, S.C. (2000) In vivo colocalization of xyloglucan endotransglycosylase activity and its donor substrate in the elongation zone of *Arabidopsis* roots. *Plant Cell*, 12, 1229–1237.

Wagner, T.A., and Kohorn, B.D. (2001) Wall associated kinases, WAKs, are expressed throughout development and are required for cell expansion. *Plant Cell*, 13, 303–318.

Xu, W., Campbell, P., Vargheese, A.K. and Braam, J. (1996) The *Arabidopsis* XET-related gene family: environmental and hormonal regulation of expression. *Plant J.*, 9, 879–889.

Zhang, Y., Brown, R.J. and West, C. (1998) Two proteins of the *Dictyostelium* spore coat bind to cellulose in vitro. *Biochem.*, 37, 10766–10779.

Zuo, J., Niu, Q.W., Nishizawa, N., Wu, Y., Kost, B. and Chua, N.H. (2000) KORRIGAN, an *Arabidopsis* endo-1,4-β-glucanase, localizes to the cell plate by polarized targeting and is essential for cytokinesis. *Plant Cell*, 12, 1137–1152.

8   Expansion of the plant cell wall

Daniel J. Cosgrove

8.1   Introduction

The plant protoplast secretes to its surface a complex set of polysaccharides, structural proteins, enzymes and other materials that assemble into a strong, yet flexible, polymeric network, the cell wall, which protects the protoplast and enables it to attain large, irregular, elongated and stable shapes. The cell wall also permits plant cells to develop high turgor pressure, which is important for the mechanical stability of plant tissues and, moreover, it greatly influences the water relations and water economy of plants. Because of the high turgor pressure in growing cells (typically 0.3 to 1 MPa), the relatively thin plant cell wall is under high mechanical tension or stress – equivalent to 10 to 100 MPa tensile stress (150 to 1500 pounds per square inch, in Imperial units). Such high stresses are possible because the cellulose microfibrils, which act as strong, inextensible reinforcing elements in the wall, are linked together by matrix polymers, forming a strong network capable of supporting high tensile forces (Carpita and Gibeaut, 1993).

To grow, a plant cell must enlarge its wall in a way that avoids the danger of aneurism or other forms of mechanical failure due to weakening, thinning or fragmentation of the cell wall. Normally, the wall undergoes controlled, large-scale, irreversible expansion under the continuous mechanical load (tensile force) arising from cell turgor. This process, called 'wall yielding', is not merely a passive turgor-driven expansion of a pliant polymeric material – that is, it is not a simple visco-elastic creep, but instead it is tightly linked to the action of one or more wall loosening processes that modify the linkage of microfibrils to one another and to the wall matrix polymers (Cosgrove, 1997, 2000a). The loosening action enables a kind of *turgor-driven polymer creep* that results in stress relaxation and expansion of the wall. This has been termed *chemorheological creep*, to distinguish it from simple polymer viscoelasticity (Ray, 1987). At the same time, new wall polymers are deposited at the cell surface and soon become incorporated into the growing wall, thereby strengthening it and enabling it to expand for prolonged periods without risk of fragmentation.

This chapter is concerned with the physical and molecular aspects of *cell wall loosening*, which I define as a modification of the cell wall resulting in stress relaxation and irreversible expansion of the wall (see below). Walls may be loosened in small regions, resulting in localized growth, as occurs for example when a root epidermal cell initiates a root hair. Alternatively, cell wall loosening may be more dispersed, resulting in a pattern of diffuse growth over the whole cell surface. Wall expansion is also regulated on a time scale ranging from seconds to days, and the different time scales may involve distinctive mechanisms to control cell wall growth properties.

This chapter begins with a summary of the concept of wall stress relaxation (the physical aspect of wall loosening) and then considers some of the molecular under-pinnings of this process, which is so crucial to plant growth and morphogenesis. Other reviews may also be consulted for more focused summaries of expansins (Cosgrove, 2000b), xyloglucan endotransglucosylase-hydrolase (Rose *et al.*, 2002), and cell wall structure (Carpita and Gibeaut, 1993).

8.2   Wall stress relaxation, water uptake and cell enlargement

In physical terms, the key difference between the walls of growing and of non-grow-ing cells is that the former are able to undergo prolonged *stress relaxation* (Cosgrove, 1997). Stress relaxation refers to the reduction in wall stress without a concomitant change in dimensions of the wall. Imagine taking a wall specimen and pulling it very quickly until it attains a certain mechanical tension, or stress. Thereafter, you hold the wall dimensions constant. In a growing wall, the polymeric network continues to yield slowly to wall stress, thereby decreasing it, whereas in a non-growing wall this yielding and wall stress relaxation process is much more limited.

Wall stress relaxation is a key concept for understanding how growing plant cells take up water and thereby increase in size. Most of the volume increase in rapidly enlarging cells is due to water uptake (i.e. vacuole enlargement) and the origin of the sustained driving force for water uptake during cell growth can be understood as fol-lows: Because wall stress and cell turgor pressure are equal and opposite counterbal-ancing forces, stress relaxation of the wall results in reduced cell turgor pressure and water potential, thereby generating the requisite water potential difference across the plasma membrane needed for water uptake by the cell. Water uptake increases cell volume and stretches the cell wall, thereby restoring wall stress.

Under ordinary circumstances the process of wall stress relaxation is not evident because it is dynamically matched by water uptake, such that turgor stays constant and the wall expands. However, by preventing cells from taking up water, one can readily observe the underlying wall relaxation process (Cosgrove *et al.*, 1984). For example, if the growing region of a stem or leaf is carefully excised from the plant and held in a humid chamber (to prevent transpiration), but is otherwise isolated from a water supply, then its turgor pressure and water potential will gradually decline as a result of continued wall relaxation which is not, under these conditions, matched by water uptake (this assumes that excision was gentle enough not to disrupt the biochemical processes causing wall relaxation). Such wall relaxation can be moni-tored with a psychrometer to measure the decay in water potential or with a pressure probe to monitor the decay in turgor pressure (Figure 8.1). In ideal conditions, the reduction in turgor pressure approximates an exponential decay to a stable, positive

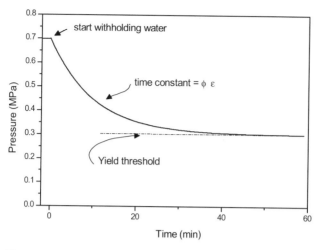

Figure 8.1 Time course for stress relaxation *in vivo*.

value of turgor pressure (Cosgrove, 1985), and from this decay one may estimate two important biophysical parameters that characterize growing cells, namely wall extensibility (ϕ) and the wall yield threshold (Y). In this context, wall extensibility is the sensitivity of growth rate to changes in turgor pressure (i.e. the local slope of the growth rate versus turgor curve) and the yield threshold is the minimum turgor pressure needed for growth. It is evident from a large body of experimental results that ϕ and Y are not purely physical properties of the cell wall, but depend upon action of biochemical wall loosening processes that act on the wall (Cosgrove, 1993). Although we would like to understand the molecular and structural nature of ϕ and Y, such an understanding still eludes us at this time, although we are beginning to get clues about their molecular underpinnings.

8.3 Alternative models of the plant cell wall

Wall relaxation is believed to be the result of a subtle modification of wall structure, e.g. a cutting or a movement of one or more of the load-bearing polymers in the cell wall. To understand the physical and molecular nature of this process, it is important to have a detailed understanding of how the polymers in the wall are linked together to form a strong network that can resist the mechanical forces generated by cell turgor, yet still undergo a controlled and limited process of yielding to these forces. This was one of the aims of the first detailed model of the plant cell wall published thirty years ago by Keegstra *et al.* (1973), but to this day some of the key features of cell wall architecture, important for our understanding of wall expansion, are unresolved.

The Keegstra *et al.* model of the plant cell wall proposed that cellulose microfibrils were coated with a layer of xyloglucans, which were covalently linked to xyloglucan chains on adjacent microfibrils via pectic polysaccharides and structural protein (Figure 8.2a). In this model, the xyloglucans, pectic polysaccharides and structural proteins formed one giant macromolecular complex that cross-linked microfibrils together.

The presence and structural significance of xyloglucan-pectin-protein cross-links is now uncertain (Talbott and Ray, 1992b), and alternative models of wall architecture have been proposed. Perhaps the most prevalent model of the cell wall today posits that hemicelluloses directly tether microfibrils together, forming a primary network of cellulose-xyloglucan, with pectic polysaccharides and structural proteins filling the spaces in between (Fry, 1989; Hayashi, 1989; Pauly *et al.*, 1999). Alternatively, neighbouring microfibrils may be held in place by lateral adhesion and viscous forces between matrix polysaccharides that bind to the surface of cellulose and tend to stick to each other (Talbott and Ray, 1992b). Yet another model, which

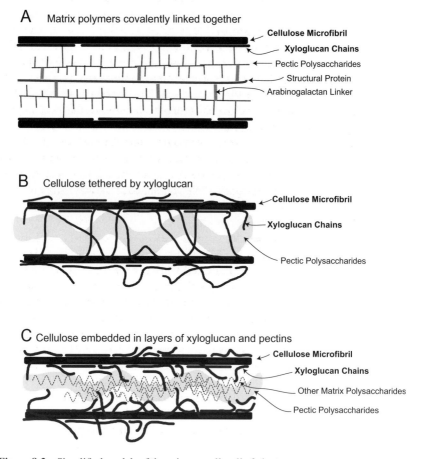

Figure 8.2 Simplified models of the primary cell wall of plants.

harkens back to the original model of Keegstra *et al.* (1973), proposes that pectic polysaccharides are covalently linked to xyloglucans that bind to the surface of cellulose. A schematic representation of these models is shown in Figure 8.2.

These models are not necessarily mutually exclusive and cell walls may have aspects of more than one of these models. Moreover, cell wall structure may change developmentally. For example, when growing cells mature, their walls become modified such that they are no longer extensible. Many changes in wall structure occur during this process. Phenolic cross-linking between polymers increases; pectins are de-esterified and form stiffer calcium-cross-linked gels; linkages between xyloglucans, pectins, and structural proteins may be formed. These changes may have the effect of reducing wall extensibility, but a quantitative assessment of this question remains to be made.

Physical measurements of cell wall stress/strain relationships (Probine and Preston, 1962) long ago made it clear that matrix components, rather than the cellulose microfibrils, determine the viscoelastic properties of plant cell walls, but unfortunately the measurements do not yet allow us to distinguish between the differing models of the wall. In support of these earlier conclusions, recent work on artificial composites made from cellulosic pellicles of *Acetobacter* have shown that matrix polysaccharides can greatly influence the mechanical properties of the pellicles (Whitney *et al.*, 1999; Chanliaud *et al.*, 2002). Many of these mechanical effects were attributed to changes in the structure of the cellulosic pellicle, e.g. local ordering and aggregation of microfibrils. Such effects are probably not relevant to plant cell walls, which control wall structure by other mechanisms, e.g. cellular control of the direction of microfibril synthesis and deposition to the wall surface. Nonetheless these artificial composites are useful for analysing interactions between cellulose and matrix polysaccharides in a controlled manner that cannot be achieved with natural plant cell walls, and they indicate that the matrix polysaccharides, especially xyloglucans, make these cellulosic composites much more pliant than pure cellulosic pellicles.

8.4   The meaning of wall loosening and wall extensibility

Another relevant issue concerns the meaning and measurement of 'wall extensibility' and 'wall loosening'. The problem is that these are rather general terms that have been assessed using different techniques which are difficult, if not impossible, to compare quantitatively or even in some cases qualitatively.

There are many ways in which wall structure may be modified so as to make it mechanically weaker, e.g. as measured by breaking strength, but without actually making the wall more extensible, that is, able to extend in the way that naturally occurs during cell enlargement. A biological example in point might be wall disassembly during fruit softening and abscission, where tissue-breaking strength is reduced, but the walls are not highly extensible (See Chapter 9). Such structural modifications, if they do not lead to a more extensible wall, are not considered to be wall-loosening processes, as I am using the term in this chapter.

When trying to assess the extensibility of a cell wall, one may get different answers, and therefore draw different conclusions, when one uses different techniques. This is hardly surprising, given that the wall is a complex structure and may be modified in many different ways. The effects of such modifications on wall extensibility are not always predictable. This topic was reviewed at some length a decade ago (Cosgrove, 1993), and here I will only summarize and update that assessment.

Wall extensibility, in its general sense, means the ability of the cell wall to extend irreversibly and thus inevitably includes the continuing action of wall-loosening processes, which may or may not make a significant change in wall structure. Thus, wall extensibility, in this sense, is not merely a property of wall structure at any given moment. One representation of wall extensibility is captured in the biophysical model of cell growth, the so-called 'Lockhart Model' where

$$GR = \phi (P - Y),$$

in which GR is growth rate (units: h^{-1}), ϕ is the specific wall extensibility (units: h^{-1} MPA^{-1}), P is turgor pressure and Y is the yield threshold (both in units of MPa).

In this equation wall extensibility in the general sense is captured by two terms: ϕ and Y, where an increase in ϕ or a decrease in Y results in a more extensible wall. I prefer to refer to ϕ and Y as 'wall yielding properties' to avoid the confusion between the specific meaning of ϕ (wall extensibility) and the more general notion of wall extensibility, which includes both ϕ and Y.

To measure these wall yielding properties, as defined in the Lockhart equation, requires living cells in which turgor is varied and the growth rate is measured. Alternatively, one can prevent the growing cells from taking up water and measure the consequent decay in turgor pressure (so-called *in-vivo* stress relaxation, as described above). With some assumptions, turgor is then expected to decay to the yield threshold Y with an exponential decay that depends on ϕ and also on the elastic modulus of the cell wall (Cosgrove, 1985). The values of ϕ and Y are likely to depend on wall structure as well as on the current activity of wall loosening processes. These *in-vivo* methods produce the most physiologically relevant assessments of wall yielding properties; those determined by both wall structure and the rate of various wall-loosening activities.

Another set of *in-vitro* techniques has been employed in attempts to measure wall extensibility in isolated (and sometimes heat-treated) cell wall specimens. These methods do not measure the physiological parameters ϕ and Y, and it is usually difficult to relate these measurements to ϕ and Y. They are probably best for assessing either a change in the structure of the cell wall or a change in the activity of wall-loosening enzymes strongly bound to the cell wall. Since the activity of these enzymes may depend on wall pH and other ephemeral conditions that are destroyed during preparation of the wall samples, it should be evident that the results obtained with isolated walls are some steps removed from the situation in the living, growing cells.

These *in vitro* techniques fall into two categories: for *stress/strain methods* a force is applied to the wall and the resulting extension is measured, whereas for *stress relaxation methods* a wall is extended to a predetermined force, then held constant in length and the subsequent decay in force is measured as the cell wall polymers move in response to the force. For stress/strain methods the amount, duration and rate of application of force are important variables; these methods also typically distinguish between elastic (reversible) and plastic (irreversible) extensibilities. For stress relaxation methods, one measures the rate of relaxation (typically on a log time scale) and the minimum relaxation time.

One of the complicating factors with these *in-vitro* assessments of wall extensibilities is that different techniques do not always give the same answer. For example, α-expansin-treated walls have enhanced stress relaxation in the time range of 1– >100 s (McQueen-Mason and Cosgrove, 1995), but stress/strain techniques (such as the Instron extensometry method; see Cleland, 1984) do not register a change in wall properties with α-expansin action (Yuan *et al.*, 2001). This is the case probably because the method applies a force to the wall for too brief a time to detect α-expansin's action. Thus by the Instron method α-expansin does not change wall extensibility, yet it does cause cell wall extension and stress relaxation. As a different example, a xyloglucan-degrading endoglucanase has been found to increase wall plastic extensibility, as measured by Instron technique, but not to enhance wall relaxation at times >1 s (Yuan *et al.*, 2001). Thus, these various techniques are best viewed as complementary ways to characterize how wall structure is changed by a certain treatment, but interpreting the results in terms of physiological wall extensibility is fraught with fundamental problems.

8.5   Time scales for changes in cell growth

The rate of cell expansion can be stimulated and inhibited quickly, in minutes or even seconds (Parks *et al.*, 2001), too fast to entail appreciable changes in either composition of the wall or its bulk physical properties. Inhibition of respiration, e.g. with cyanide or cold temperature, inhibits cell expansion within seconds (Ray and Ruesink, 1962) and some growth responses, e.g. to mechanical stimulation or to blue light, likewise occur within seconds. These characteristics of cell growth have been interpreted to mean that cell wall expansion requires continued action of one or more wall loosening processes. At the other end of the time scale, cell expansion persists for a period of many hours or days, and as the growth rate ceases the wall may change its composition substantially, eventually becoming permanently inextensible. This is thought to be due to cross-linking of the wall's polymers, perhaps in concert with changes in wall composition, both of which produce a wall more resistant to wall-loosening processes (Cosgrove and Li, 1993).

Relatively little is known for certain about the molecular nature of the growth cessation process, but more attention has been given to the process of wall loosening. The remainder of the chapter will address this issue.

8.6   Candidates for wall-loosening agents

Cell wall models are important for understanding cell enlargement because they present potential targets for the action of cell wall-loosening agents. This latter term refers to enzymes or other substances or processes that lead to wall stress relaxation and wall enlargement. For example, the model in Figure 8.2a suggests that enzymes that cut any of the polymers making up the xyloglucan-pectin-protein complex might act to loosen the bonding between microfibrils and thereby allow wall extension. On the other hand, the model in Figure 8.2b suggests xyloglucan as the only polymer worth considering as the target of wall loosening agents (cellulose microfibrils are considered to be stable, biochemically inert and immune to significant breakdown or loosening by plant enzymes).

Early biophysical studies hinted that there are multiple ways to loosen the cell wall (Cleland, 1971; Cosgrove, 1986), and so it is not to be expected that plants employ one and only one mechanism to enlarge their cell walls. Depending on the developmental state of the cell and the kind of growth stimuli that are acting on it, one or more of the following wall loosening agents may serve as a key control point for cell enlargement (the contribution of these agents to wall modification in non-growing cells is also discussed in Chapter 9):

- *Expansins* stimulate stress relaxation and wall enlargement *in vitro* and *in vivo*, and have a pH dependence that matches the 'acid growth' phenomenon.
- *Xyloglucan endotransglucosylase-hydrolases (XTHs)* comprise a family of enzymes that cut and, in some cases, ligate xyloglucans.
- *Endo-1,4-β-glucanases* hydrolyse the backbone of 1,4-β-glucans.
- The *hydroxyl radical* is a highly reactive molecule capable of breaking many chemical bonds, including glycosidic linkages that make up the backbone of wall polysaccharides.
- *Yieldin* is a wall-bound protein that changes the in-vitro yield threshold of walls in extensometer assays.

These five agents work in distinctive ways and should leave different 'signatures' of their action on the cell wall. By acting on different sites in the cell wall, different wall loosening processes may act synergistically. It is also likely that over longer times (many hours or days) the cell modulates its growth process via changes in wall composition, either by secreting new wall polymers with different assembly, bonding and cross-linking characteristics or by secreting enzymes that change the ability of wall polymers to bind to other wall polymers or to serve as substrates of wall-loosening enzymes.

Moreover, other wall enzymatic activities may change the interaction of wall polymers with each other, and thereby change the yielding properties of the cell wall. For example, pectin methyl esterase activity may result in a stronger pectin gel network, or glycosidase activities may strip the side chains from a branched glucan or xylan, thereby strengthening its binding to cellulosic surfaces or to other polysaccharides. These enzymes clearly are present in the wall and presumably modify wall polysaccharides, but it is not so certain whether these wall modifications actually change wall extensibility in a significant way. This issue requires critical examination.

8.7 Expansins

These wall-loosening proteins were first isolated and characterized as mediators of 'acid growth', which refers to the stimulation of cell elongation by low pH [see review in Cosgrove (2000b) and http://www.bio.psu.edu/expansins/]. When expansins are added back to heat-inactivated walls, they rapidly restore the wall's ability to extend in response to low pH. Expansins also enhance stress relaxation of the cell wall in a characteristic way, over a time range of ~1 to >100 s. These two rheological effects on cell walls are characteristic and diagnostic (up to now at least) for expansin proteins.

Today we recognize two multi-gene families of expansins, named α-expansins (*EXP*) and β-expansins (*EXPB*). In addition, there is a third group of related genes, named expansin-like (*EXPL*), that were discovered in the genomes of *Arabidopsis* and rice (see http://www.bio.psu.edu/expansins/). However, the function and potential wall-loosening activity of the EXPL proteins have not been established. These three protein families are rather divergent from each other, with ~25% amino acid identity between the EXP and EXPB families, ~20% between EXP and EXPL, and ~30% between EXPB and EXPL. Both rice and *Arabidopsis* also have a single gene, named expansin-related (*EXPR*), that is even more distant in sequence from α-expansins, but is of unknown functionality.

Most of the biochemical characterization of expansin action has been based on α-expansins, which act most effectively on walls of dicotyledons and monocotyledons outside the Poaceae (grasses). These have been named 'type I walls' by Carpita and Gibeaut (1993). α-Expansins may also cause extension of grass cell walls, but with lower effectiveness than in type I walls (McQueen-Mason *et al.*, 1992). No clear enzymatic activity has been found with α-expansins, although a very weak and cryptic endoglucanase activity has been observed in some α-expansins extracted from cucumber hypocotyls (D.J. Cosgrove and D.M. Durachko, unpublished data). α-expansins do not progressively weaken the wall in a time dependent manner, as one would expect of a hydrolytic enzyme, nor do they hydrolyse the major matrix polysaccharides of the wall. There is indirect data that they enable glucans to dissociate from each other. For example, paper is weakened by α-expansins and artificial cellulosic composites are rapidly loosened by α-expansin (McQueen-Mason and Cosgrove, 1995; Whitney *et al.*, 2000). Moreover, expansins improve the effectiveness of digestion of cellulose by cellulase (Baker *et al.*, 2000). This result is interpreted to mean that expansins can make the individual glucan chains more available to enzymatic attack, e.g. by lifting a surface glucan from the crystalline microfibril.

Sequence analysis indicates that expansins likely consist of two domains (Figure 8.3). Domain 1 has a series of highly conserved cysteines and short motifs; this domain has a distant but statistically significant sequence similarity to family-45 endoglucanases. Domain 2, at the carboxy terminus of the protein, has a series of conserved tryptophans and other aromatic residues, some of which may function in polysaccharide binding.

In *Arabidopsis*, rice and other plants there exist genes encoding a protein that is homologous to expansin domain 1. These were first identified as 12-kD proteins abundant in the vascular system of citrus tree struck with citrus blight (Ceccardi *et al.*, 1998). The function of these 'p12' proteins is unknown and assays for wall extension activity have not yielded positive results (D.M. Durachko and D.J. Cosgrove, unpublished data). Ludidi *et al.* (2002) have speculated that they have a signalling function, whereas Li *et al.* (2002) have called this group of proteins γ-expansins. I think it better to restrict the term 'expansin' to protein groups that either are homologous to the whole expansin protein (that is, contain both Domains 1 and 2) or that been shown to possess *bona fide* wall-loosening activity characteristic of expansins. For now, 'p12-like proteins' might be a better designation, until their function is definitively established.

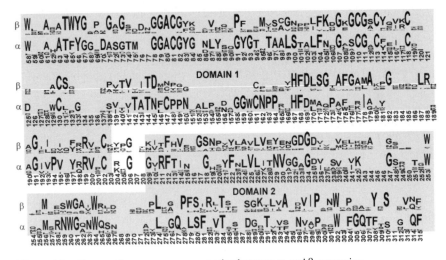

Figure 8.3   Comparative sequence conservation between α- and β-expansins.

Grass pollen produces a small protein that is homologous to expansin Domain 2. In rice and maize it is encoded by a small family of genes (D.J. Cosgrove, unpublished observations), but homologs are not present in the *Arabidopsis* genome and there is no evidence of homologs outside of the grass family. Thus, it appears that these small proteins may be an innovation restricted to grasses. These proteins are known by immunologists as group 2 grass pollen allergens (Ansari *et al.*, 1989; Dolecek *et al.*, 1993; De Marino *et al.*, 1999), but their biological function in the plant is unknown.

The wall-loosening activity of β-expansins has only been studied in a form of the protein that is abundantly produced by grass pollen (Cosgrove *et al.*, 1997). These pollen proteins were first isolated and partially characterized by immunologists because they are major pollen allergens. Immunologists classify allergens based on antibody cross-reactivity and classify these proteins as grass group 1 pollen allergens, with the allergens named in the following format: Zea m 1 for the allergen from *Zea mays* (maize), Lol p 1 for the homolog from *Lolium perenne* (ryegrass), Phl p 1 for the protein from *Phleum pratens*, etc.

The pollen allergen class of β-expansins appear to be exclusively expressed in pollen and their sequences are rather divergent from the other β-expansin genes, which are expressed in diverse vegetative tissues (Lee and Kende, 2001; Reidy *et al.*, 2001; Wu *et al.*, 2001a). Because of their divergent sequences, I suspect that their biochemical properties and actions on the cell wall may be different from that of the other β-expansins, but so far other β-expansins have not been tested. Thus, at this stage one should be wary of generalizations based only on studies of the pollen allergen class of this family.

The step-wise proteolysis of a β-expansin protein (CIM1) from soybean cell cultures has been studied (Downes *et al.*, 2001), but the significance of this processing for wall loosening has not been tested and remains an open issue.

Recent work has suggested that β-expansins are thiol proteases that may loosen the wall via proteolysis of wall structural protein, but this hypothesis is disputed. This idea originated from studies of recombinant Phlp1, a group-1 grass pollen allergen belonging to the β-expansin family. Recombinant Phlp1 was found to be highly unstable and was proteolytically degraded during purification (Grobe *et al.*, 1999). In contrast, Li and Cosgrove (2001) could not detect protease activity in group-1 pollen allergens and showed that C1 proteases (as well as other proteases) did not mimic expansin's wall-loosening effects. Subsequent work (Poppelmann *et al.*, 2002) discovered that expression of recombinant Phlp1 induced an endogenous protease in the yeast expression system, and this protease likely contaminated the protease assays of recombinant Phlp1. Thus the experimental basis for the idea that expansin has proteolytic activity seems to be an experimental artifact. Furthermore, sequence analysis gives substantially better support for homology with family-45 endoglucanases than with C1 proteases, and proteases have not been shown to cause wall loosening (Li and Cosgrove, 2001).

The expansin gene family is large and the biological functions proposed for expansins are diverse. Many α- and β-expansin genes have been implicated in the control of cell enlargement. Examples include control of growth in leaves (Keller and Cosgrove, 1995; Pien *et al.*, 2001; Reidy *et al.*, 2001), stems (Cho and Kende, 1997a, b; Lee and Kende, 2001), roots (Wu *et al.*, 2001a) and in specialized cells such as root hairs (Baluska *et al.*, 2000; Cho and Cosgrove, 2002), cotton fibres (Orford and Timmis, 1998; Ruan *et al.*, 2001; Harmer *et al.*, 2002), vascular cells (Im *et al.*, 2000), and incipient leaf primordia on the shoot apical meristem (Reinhardt *et al.*, 1998). Transcript levels of specific expansin genes are altered after treatment with growth hormones (Downes and Crowell, 1998; Hutchison *et al.*, 1999; Caderas *et al.*, 2000; Catala *et al.*, 2000; Chen *et al.*, 2001; Lee and Kende, 2001; Cho and Cosgrove, 2002; Mbeguie *et al.*, 2002), and expansin gene expression is also implicated in adaptive responses of maize roots to drought stress (Wu *et al.*, 2001b) and *Rumex* shoots to submergence (Baluska *et al.*, 2000; Vriezen *et al.*, 2000). Some α-expansin genes are up-regulated in ripening fruit, leading to the suggestion that expansins are involved in cell wall disassembly (Rose *et al.*, 1997; Brummell *et al.*, 1999b; Harrison *et al.*, 2001; Mbeguie *et al.*, 2002). This proposal has been partially supported by antisense experiments in tomato fruit (Brummell *et al.*, 1999a; Brummell *et al.*, 2002). As described above, the pollen allergen class of β-expansins is abundantly expressed in grass pollen and is thought to assist its penetration into the stigma and style.

In addition to these examples from plants, some microbial proteins contain expansin-like domains. These include a pathogenesis factor from the plant bacterial pathogen *Clavibacter michiganensis* (Laine *et al.*, 2001) and a protein from the fungus *Trichoderma* (Saloheimo *et al.*, 2002). Both of these microbial proteins contain other domains, and so are examples of modular protein construction. In plants there is no evidence that expansins are combined with other domains.

The exact biochemical mechanism of action of expansin and the identity of its target site of action are uncertain. One of the hallmarks of expansin action is an increase in the stress relaxation of isolated cell walls. This action requires the continued presence of expansin during the stress relaxation assay. For example, the pre-treatment of walls with α-expansin, followed by inactivation of expansin, did not change the mechanical properties of the wall (McQueen-Mason and Cosgrove, 1995). Similarly, walls treated with α-expansin did not become progressively weaker, as occurs when walls are treated with hydrolytic enzymes. Moreover, α-expansin does not hydrolyse the major structural polysaccharides in the cell wall, and it does not act as a xyloglucan endotransglycosylase. Thus, these results indicate that α-expansin does not change wall structure substantially. Similar studies need to be done with β-expansins, to determine whether they have similar properties.

Expansin may weaken the noncovalent adhesion of glucans to one another (e.g. xyloglucan to cellulose). Evidence for this is seen in its weakening effect on paper (a hydrogen-bonded network of glucans) and in the action of chaotropic agents such as 2-M urea, which can partially mimic and synergize the effect of expansins on cell walls (McQueen-Mason and Cosgrove, 1994). Additional support for this idea comes from a study of a fungal protein from *Trichoderma reesei*, also known as *Hypocrea jecorina* (Saloheimo *et al.*, 2002). The protein, called 'swollenin', contains an expansin-like domain linked to a cellulose-binding domain. Swollenin was expressed in yeast and crude extracts were incubated with cotton fibres. Swollenin caused local disruptions of the fibre structure after sonication. It also weakened filter paper and caused an apparent dispersion of *Valonia* cell wall structure, but without release of soluble sugars.

To assess the biological function of expansins, several approaches have been tried:

1. analysing gene expression patterns;
2. ectopic expression of specific expansin genes;
3. antisense reduction of expansin expression; and
4. identification and characterization of expansin knockout mutants.

The results of these studies have been reviewed recently (Cosgrove *et al.*, 2002) and are beyond the scope of this chapter. I can summarize their conclusions as follows: Many expansin genes are expressed in a spatial and temporal pattern correlated with cell growth, whereas some other expansin genes are expressed at times and locations that suggest a different wall loosening function, such as wall breakdown during fruit ripening, cell separation during abscission, or pollen tube penetration of the stigma. Altered expression of α-expansins has been shown to cause predicted changes in cell growth and cell wall properties in some cases, but not in others. Plants harbouring mutations in individual expansin genes may display a phenotype in some cases, whereas in other cases no obvious phenotype is seen under laboratory growth conditions. Taken together, these studies support the conclusion that expansins function not only in cell growth, but also in other situations where wall loosening is an important aspect of cell development.

8.8   Xyloglucan endotransglucosylase/hydrolases (XTHs)

Because xyloglucan is thought to be an important structural polysaccharide that binds to the surface of cellulose and perhaps directly links cellulose microfibrils together, enzymes that modify xyloglucan have figured prominently in the literature of the cell wall and cell wall loosening (Fry, 1989; Hayashi, 1989; Hoson, 1993). Most notably, xyloglucan modification has been implicated in auxin-induced wall loosening (Labavitch and Ray, 1974; Talbott and Ray, 1992a; Hoson *et al.*, 1993).

In plants there are at least three families of enzymes that may be important for this function, namely glucoside hydrolase families 5, 9, 16 (GH5, GH9 and GH16; see http://afmb.cnrs-mrs.fr/CAZY/). In *Arabidopsis* these enzyme families are large, with 13 members in GH5, 25 members in GH9, and 33 members in GH16. GH5 and GH9 enzymes will be discussed in the next section; here we will focus on GH16 enzymes.

Plant enzymes in GH16 have been described variously as xyloglucan endotransferases, hydrolases, endotransglycosylases and endo-xyloglucan transferases.

Enzymes in this family have been studied in greater detail than those in GH9, at least with respect to enzymes from plant sources.

The nomenclature for GH16 enzymes from plants was recently modified to reduce the inconsistent names used in the past for this class of enzyme (Rose *et al.*, 2002). To summarize the revised nomenclature, XTH is used for enzymes that cut xyloglucan somewhere in the middle of the backbone and transfer one of the fragments (the one with the new reducing end) either to another xyloglucan (called endotransglucosylase, or XET, activity) or to water (called endohydrolase, or XEH, activity). Some *XTH* genes encode enzymes with strict XET activity, others with mostly XEH activity, and some with hybrid activities.

Strict XET enzymes form a covalent intermediate with the cut xyloglucan and this property has been usefully employed for the isolation of XETs (Sulova and Farkas, 1998; Sulova *et al.*, 1998; Steele and Fry, 1999). Several clever techniques have been devised for assaying XET activity *in vitro* using acceptor xyloglucan oligosaccharides (XGOs) modified with a radioactive or fluorescent tag (reviewed in detail by Rose *et al.*, 2002). Moreover, fluorescent XGOs have been used for histological localization of XET activity (Vissenberg *et al.*, 2000, 2001, 2003).

By cutting xyloglucan, and in some cases rejoining with new xyloglucan chains, it has been proposed that XTHs may catalyse a type of wall loosening that leads to cell wall extension (Fry *et al.*, 1992; Nishitani and Tominaga, 1992). However, such loosening has not actually been demonstrated to result from XTH activity, in the manner that expansins have been shown to act. If XTH activity led directly to wall extension, this should be readily demonstrated, e.g. by assays similar to those used to assess expansin activity (above). This has been tested at least once with an XET preparation, but with negative results (McQueen-Mason *et al.*, 1993). Since the pH optimum for XTH enzymes is typically ~5.5, they are not good candidates for mediators of acid growth, where the optimum is typically <pH 4.5.

A recent study lends some support for a role of XEH in wall loosening (Kaku *et al.*, 2002). Treatment of azuki bean epidermal strips with an XEH for 48 h at 37°C (a very long incubation indeed!) caused wall loosening, as assessed by mechanical tensile tests that measured both total wall extensibility (amount of extension per unit force) and subsequent relaxation of isolated wall strips. The extensibility increased substantially in XEH-treated walls, and there was a very modest reduction in the minimum relaxation time. The ability of this XEH to cause wall extension in creep assays was not tested, and in view of the very long incubations required to change the wall extensibility, it probably has weak, if any, wall extension activity. But this is a point that requires direct testing. It is worth noting that the cell walls in this study were treated in such a way (methanol boiling) that endogenous expansins would probably have retained their activity. It is thus possible that the wall loosening induced by XEH was due to the combined effects of the added XEH and the endogenous expansin.

These effects of a plant XEH may be compared with those found for a fungal endoglucanase (a GH Family 12 enzyme) with the ability to cut xyloglucan (Yuan *et al.*, 2001). This enzyme, called Cel12A, caused a similar increase in wall extensibility and a decrease in the minimum relaxation time in isolated wall strips, as described above for XEH, but after only 2 h incubation at room temperature (the enzyme concentrations were similar to those used in the XEH study above). Thus, Cel12A is apparently many times more potent in causing changes in wall properties than is the XEH from azuki bean. Cel12A also caused prolonged cell wall extension (wall creep) under a constant load, with a lag time that depended on enzyme concentration. Thus, there is some reason to expect that the azuki bean XEH would also have wall extension activity, if applied in sufficient concentration. However, not all wall hydrolases that increase wall mechanical extensibility lead to faster wall extension, so the presumed wall extension activity of this XEH needs to be tested directly. Also, it is of interest to know whether XEH at physiological concentrations can cause these effects on wall extensibility in the absence of activity expansin, and for this the experimental protocol would need to be altered to inactivate endogenous expansins in hot water (boiling methanol is ineffective).

In addition to their possible wall-loosening functions, XTHs with XET activity may help integrate newly synthesized xyloglucans into the existing wall structure. Integration of xyloglucans into walls of cultured rose cells has been demonstrated by use of isotopically-labelled xyloglucans (Thompson and Fry, 2001). This action may result in a less-extensible cell wall because of greater cross-linking of the cellulose, but this possibility has not actually been tested.

A recent study (Takeda *et al.*, 2002) supports the concept that xyloglucan secretion and integration into the cell wall may significantly affect wall growth properties. This study found that incubation of pea epicotyl segments in xyloglucans of high molecular weight resulted in slowed growth and less extensible walls, measured by a dynamic stress/strain analyser cycling a load at 0.05 Hz. In contrast, addition of xyloglucan oligosaccharide had the opposite effect. These results were interpreted to mean that plant cell growth could be controlled by the secretion and integration of large xyloglucans into the wall (thereby restricting growth) or secretion of small xyloglucans to the wall (thereby promoting growth). Integration was envisioned to be the result of physical binding of xyloglucans to the nascent cellulose microfibrils and XTH-aided integration of xyloglucans into the load-bearing part of the cell wall.

In addition to a potential role in wall loosening, the roles of XTH enzymes may also include wall restructuring during growth and fruit ripening, and integration of newly synthesized xyloglucans into the existing cell wall. *XTH* gene expression is often high in cells during and after cell enlargement, and also in cells undergoing cell wall breakdown, such as softening fruit and in aerenchyma (see review by Rose *et al.*, 2002). By use of rhodamine-labelled xyloglucan oligosaccharides, a recent study also found high XET activity in the secondary walls (i.e. non-expanding walls) of xylem and phloem fibres in poplar, leading to the proposal that XTH creates and reinforces the connection between primary and secondary cell walls (Bourquin *et al.*, 2002).

In summary, XTH enzymes modify xyloglucans in ways that hypothetically may either loosen or reinforce (stiffen) the cell wall, but the current evidence for either of these roles is indirect. Although it is possible that different enzymes within the family have distinctive roles, *in-vitro* tests of 10 different XTH enzymes did not support this idea (Steele *et al.*, 2001; Vissenberg *et al.*, 2003).

8.9   Endo-1,4-β-ᴅ-glucanases

These enzymes hydrolyse 1,4-β-ᴅ-glucan chains, typically at random points on the glucan backbone. Potential cell wall substrates for plant endo-1,4-β-ᴅ-glucanases include cellulose, xyloglucan, and the mixed-linked β-glucan specific to grass cell walls. The crystalline regions of cellulose microfibrils are generally considered to be inaccessible to attack by plant endoglucanases, but disordered regions of the microfibril or stray individual 1,4–β-ᴅ-glucans may be targets of these enzymes. Although xyloglucans were previously considered to be likely substrates of these enzymes, this point is not well established and some plant endo-1,4-β-ᴅ-glucanases exhibit relatively poor activity against xyloglucans (Nakamura and Hayashi, 1993), whereas others have appreciable xyloglucanase activity (Woolley *et al.*, 2001).

Some members of the XTH family (in GH16), described in the previous section, have endoglucanase, as well as transglucosylase activity. In this section we will focus on two other families of possible endo-1,4-β-ᴅ-glucanases, namely GH5 and GH9.

The *Arabidopsis* genome contains 13 genes classified as members of GH5 (see http://afmb.cnrs-mrs.fr/CAZY/). Apparently, plant enzymes within GH5 are only known from their sequence. It is possible that some plant enzymes that were extracted from plant tissues and characterized biochemically are members of GH5, but without protein sequence identification this issue is unclear. GH5 enzymes have been characterized from microbial sources and include not only endo-1,4-β-ᴅ-glucanases, but also endo-xylanases, endo-mannanases and hydrolases with other substrate preferences. Thus, plant GH5 genes may encode endo-1,4-β-ᴅ-glucanases, but it is possible that they encode other enzymatic activities. Their biological functions and potential role in wall loosening have not been explored, but they are potential candidates for a new class of wall-loosening enzymes.

Members of GH9 are encoded by 25 *Arabidopsis* genes and numerous entries from other plant species. Although the GH9 enzyme activities encoded by the *Arabidopsis* genes have not been determined, those from other plant species have been partially characterized. Perhaps the best characterized plant endo-1,4-β-ᴅ-glucanase from growing plant cells is that from poplar cells, studied by Hayashi's group. This enzyme has relatively weak activity against xyloglucan (Nakamura and Hayashi, 1993), but is active against carboxymethyl cellulose, an artificial substrate commonly used to assess 'cellulase' activity, and against phospho-swollen cellulose. Ohmiya *et al.* (2000) showed a correlation between expression of mRNAs for endo-1,4-β-ᴅ-glucanase in poplar cell cultures and accumulation of cello-oligosaccharides in the culture medium. On the basis of this and other evidence, they proposed that poplar endo-1,4-β-ᴅ-glucanases act on the non-crystalline regions of cellulose, e.g. the region hypothesized to be involved in physical entrapment of xyloglucan within cellulose microfibrils. While these sites may indeed be the targets of plant endo-1,4-β-ᴅ-glucanases, this specific point is difficult to prove definitively, and other targets of these enzymes have not been excluded.

Support for a role of GH9 endo-1,4-β-ᴅ-glucanases in cell growth comes from analysis of promoter activities for some of these genes. The promoter of one of the

Arabidopsis genes, named *cel1*, is active in elongation zones of tobacco roots and shoots (Shani *et al.*, 1997). The expression of two GH9 endoglucanase genes from poplar are not so neatly associated with elongation, but are more closely associated with cellulose synthesis, which is only partially correlated with cell elongation (Ohmiya *et al.*, 2000).

Further evidence for involvement in cell growth comes from transgenic plants where expression of endo-1,4-β-D-glucanase genes was altered. In transgenic poplar with suppressed expression of the endogenous endo-1,4-β-D-glucanase genes, leaf growth (surface area enlargement) was reduced by ~30% (Levy *et al.*, 2002). Transgenic expression of the poplar endo-1,4-β-D-glucanase genes in *Arabidopsis* resulted in plants with longer hypocotyls (~15% longer), larger rosettes, and more leaves (Park *et al.*, 2003). Moreover, cell walls from the transgenic *Arabidopsis* had increased plastic extensibility (~27% increase, as measured by applying force to methanol-killed leaf strips). Analogous growth results were reported when an *Arabidopsis* endoglucanase gene (*cel1*) was expressed in poplar (Shani *et al.*, 1999).

Park *et al.* (2003) interpreted their results (including previously published data) to mean that poplar endo-1,4-β-D-glucanases trim the disordered glucan chains from cellulose microfibrils, thereby freeing xyloglucans that are intercalated within the disordered domains of the microfibrils, resulting in reduced cross-linking of microfibrils by xyloglucan. Although this is reasonable, the supporting evidence is circumstantial and other possible explanations for the growth effects of endo-1,4-β-D-glucanases can be imagined. What is missing from these studies is direct evidence that these endo-1,4-β-D-glucanases have wall extension activity in and of themselves.

Some of the plant enzymes within the GH9 family have a membrane anchor domain and are localized to membrane vesicles, rather than the cell wall (see review by Molhoj *et al.*, 2002). The *Arabidopsis* KORRIGAN mutant, which is defective in one of these membrane-anchored endoglucanases, has growth defects that are interpreted to be due to a defect in cellulose synthesis, leading to poorly formed microfibrils. This in turns leads to pleiotropic effects: abnormal cytokinesis, misshapen cells, changes in pectin composition of the wall, and a dwarf phenotype (Nicol *et al.*, 1998; Zuo *et al.*, 2000; Lane *et al.*, 2001; Sato *et al.*, 2001). The exact role of this endoglucanase in cellulose synthesis is not yet clear. Molhoj *et al.* (2002) consider that it might cleave a sterol-cellodextrin precursor in the β-1,4-glucan synthesis pathway or it might function in the assembly of glucan chains in cellulose microfibrils.

The plant GH9 family also includes genes each of which encodes a presumptive endo-1,4-β-D-glucanase with a C-terminal extension that resembles a carbohydrate-binding domain (e.g. GenBank Accession No. AAD08699 by C. Catala, and A.B. Bennett). Carbohydrate-binding domains are commonly found in microbial cellulases, xylanases and other hydrolytic enzymes that degrade plant cell walls, but they are not common to analogous enzymes from plants. The putative carbohydrate-binding domain in the plant GH9 enzymes consists of ~125 residues with many conserved aromatic residues (tryptophan, phenylalanine, tyrosine). Aromatic residues have been implicated in carbohydrate-protein binding. The role of this C-terminal extension has not been established by experimental evidence, but may modify the activity of this sub-class of plant GH9 enzymes.

In addition to these studies of endo-1,4-β-D-glucanases of known sequence, there is a large body of earlier work implicating metabolism of matrix glucans (e.g. xyloglucan in dicotyledons, mixed-linked glucan in grasses) in auxin-induced cell elongation (see reviews by Masuda, 1990; Hoson, 1993), at least over a period of hours to days after start of auxin treatment. These effects on matrix glucan metabolism are probably too slow to account for the rapid induction of cell elongation by auxin, but there is no compelling reason to believe that auxin stimulates cell growth by only a single molecular mechanism. Indeed, recent studies have proposed that auxin-induced cell elongation is mediated by hydroxyl radicals, which are the next candidate wall loosening agents to be considered.

8.10 Non-enzymatic scission of wall polysaccharides by hydroxyl radicals

The hydroxyl radical ($\cdot$OH) is a highly reactive oxygen species capable of attacking proteins, lipids, nucleic acids, carbohydrates – in short, most constituents of the cell. While considerable attention has been paid to the role of $\cdot$OH radicals and other reactive oxygen species in oxidative stress and in defence responses against pathogens (e.g. the hypersensitive response), some recent studies have raised the possibility that $\cdot$OH radicals may serve a role in cell wall loosening by cleaving cell wall polysaccharides non-enzymatically (Fry, 1998; Schweikert et al., 2000).

In vitro, hydroxyl radicals may be generated via the non-enzymatic Fenton reaction, which can be formed with a mixture of H_2O_2, ascorbate and Cu^{2+} or Fe^{3+}. When produced by this means, hydroxyl radicals were shown to be able to cleave xyloglucan and pectins (Fry, 1998; Fry et al., 2001; Miller and Fry, 2001; Tabbi et al., 2001) and also to induce extension of isolated cell walls clamped under tension in an extensometer (Schopfer, 2001). Hydroxyl radicals may also be generated enzymatically by peroxidase supplied with O_2 and NADH, likewise leading to polysaccharide degradation (Chen and Schopfer, 1999; Schweikert et al., 2000, 2002).

These in-vitro systems for generating $\cdot$OH radicals are rather artificial, so naturally the relevance of these results to the situation in living, growing cells needs to be addressed. One approach is to look for evidence of substantial $\cdot$OH-induced cleavage of wall polysaccharides in growing tissues (e.g. enough to account for a significant fraction of the wall loosening in these cells). Results of this sort have not yet been published, but Fry's group has characterized unique products of $\cdot$OH attack of wall polysaccharides and found evidence for their accumulation during the softening process of ripening pears (Fry et al., 2001). We await a parallel testing made on growing cell walls.

Taking another approach, Schopfer et al. (2002) have shown that free radical scavengers, which competitively react with $\cdot$OH as well as other radicals, are able to suppress auxin-induced growth, that auxin enhances superoxide production by oat coleoptiles slightly (25–30%), and that artificially generated ·OH (using H_2O_2, ascorbate, Fe^{2+}) can induce elongation of living coleoptiles. These and other results led the authors to propose the following hypothesis: auxin stimulates production of superoxide radicals in the cell wall via a plasma membrane NADH oxidase. Superoxide is converted to H_2O_2, which in turn is converted to ·OH by wall-bound peroxidase. The ·OH then cleaves wall polysaccharides, causing cell wall extension. While this is an appealing hypothesis, there are some experimental observations that present problems with it. For example, the effect of auxin on superoxide production does not correspond well with the timing and magnitude of auxin-stimulated growth (see Figure 2a and Table 1 in Schopfer *et al.*, 2002). Nevertheless, this is an intriguing hypothesis that warrants further examination.

In a related study, Joo *et al.* (2001) observed changes in reactive oxygen species in maize roots during gravitropism. Whether the generation of reactive oxygen species is sufficient in amount and in localization to account for cell wall loosening in these responses (auxin, gravitropism) is an issue that requires further attention. Also, it is unclear whether non-enzymatic degradation of wall polysaccharides by the ·OH radical has sufficient selectivity for cell wall loosening.

8.11 Yieldin

As described in the introduction, growing plant cell walls must be under substantial tensile stress in order to extend. Wall extension commonly exhibits a yield threshold stress, below which walls do not extend. Although its molecular nature is not well understood, one way to think about the yield threshold is to consider that it depends on the size of the shearing units within the cell wall – the smaller the size, the lower the threshold. The yield threshold is apparently under control of the cell and can rapidly shift up or down in response to some stimuli (Cramer and Bowman, 1991; Nakahori *et al.*, 1991; Cramer *et al.*, 1998; Hsiao *et al.*, 1998; Kitamura *et al.*, 1998).

A yield threshold can also be observed *in vitro*, when isolated cell walls are allowed to creep under the action of an external force (Okamoto and Okamoto, 1994). Whether this type of yield threshold is due to the same phenomenon as that observed in growing cells has not been established, but is a possibility. Okamoto and co-workers characterized and later identified a pH-dependent wall protein from cowpea hypocotyls that was able to reduce the yield threshold measured in an *in-vitro* assay (Okamoto and Okamoto, 1994, 1995; Okamoto-Nakazato *et al.*, 2000a). In this work the walls cells were from cowpea hypocotyls that were bored out (central core removed) and subsequently stored for 2–3 weeks in 50% glycerin at –15°C. The hypocotyls were then clamped in an extensometer in acidic buffer, a graded series of tensile forces were applied to the specimen and the corresponding extension rates were measured. When the extension rate was plotted against the applied force, a clear yield threshold was detected. To demonstrate protein activity, the glycerinated hypocotyls were incubated in 1 mg/mL of purified protein at 10°C, then assayed as described above (Okamoto-Nakazato *et al.*, 2000a). After incubation for 8–48 h, the yield threshold shifted to lower values. Shorter incubations had no effect on wall extension.

Cloning of a cDNA for this protein, named 'yieldin', revealed it to be homologous (70% protein sequence similarity) to acidic class III endochitinases (Okamoto-Nakazato *et al.*, 2000b). This is a surprising result because polymers of N-acetyl glucosamine – the presumptive substrate of endochitinases – are not known to make up the plant cell wall. The mechanism by which this protein changes the yield threshold is a mystery.

Yieldin belongs to glycoside hydrolase family 18 (GH18; http://afmb.cnrs-mrs.fr/CAZY/) and all of the characterized enzymes in this family are endo-β-N-acetylglucosaminidases, which hydrolyse the 1,4-β-linkages of N-acetyl-D-glucosamine polymers of chitin (or of glycoproteins, in rare cases). Recombinant yieldin, synthesized in *E. coli*, had low but significant endochitinase activity (Okamoto-Nakazato *et al.*, 2000b). Whether it is able to cut N-acetyl-D-glucosamine residues within glycosyl chains attached to wall proteins has not been tested. Such glycosyl chains may be the target of yieldin, but how this action might change the yield threshold is unclear.

In the *Arabidopsis* genome there are ten GH18 genes, which are divided into two deeply divided subfamilies: 9 genes in one subfamily, and a single gene (GenBank Access No. AAA32768) in a second family which is the same family that contains the cowpea yieldin along with two other cowpea chitinase genes (Figure 8.4; unpublished observations of the author). It has not been determined whether these other two cowpea acidic endochitinases have yieldin activity. There is less than 10% protein identity between the subfamily of 9 GH18 genes in *Arabidopsis* and the single *Arabidopsis* ortholog of yieldin. Thus, these two GH18 subfamilies are very

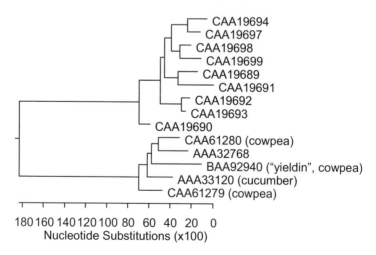

Figure 8.4   Phylogenetic tree based on protein sequences for GH18 (endo-β-N-acetylglucosaminidases) from *Arabidopsis* and cowpea.

divergent. It may be that the subfamily containing the 9 *Arabidopsis* chitinases is involved in defence against fungi, whereas the smaller, divergent subfamily containing yieldin has another function, e.g. wall loosening. This is a rather different situation compared with expansins, XTHs and endo-1,4-β-D-glucanases, which are relatively large multigene families.

In cowpea seedlings, immunolocalization with a polyclonal antibody indicates that yieldin protein is located principally in the apical hook, i.e. in the cells prior to rapid cell elongation, and is not detected at all in the roots. This pattern of expression is also rather different from that found for expansins, XTH and endo-1,4-β-D-glucanases, and it does not suggest a general function in cell elongation, but something specialized to the apical hook. Samac *et al.* (1990) cloned and partially characterized the *Arabidopsis* ortholog of the cowpea yieldin. They were not able to detect expression of this gene by Northern analysis, but did find that the promoter was active when introduced into *Arabidopsis* leaves with the biolistics particle gun. They inferred that expression of the gene was induced by wounding caused by entry of the tungsten micro-particles. Stable transformation of *Arabidopsis* with this construct indicated a complicated pattern of expression that was not specific to growing regions and that was clearly induced around necrotic lesions caused by pathogen invasion (Samac and Shah, 1991). In cucumber an orthologous gene was found to be induced by salicylic acid and by infection by the tobacco necrosis virus (Metraux *et al.*, 1989). Whether these orthologous genes in *Arabidopsis* and cucumber have yieldin activity has not been determined, but it is noteworthy that *Arabidopsis* has only a single apparent ortholog of yieldin and its expression, as judged from available the Stanford Micro-Array Database (http://genome-www5.stanford.edu/MicroArray/SMD/), is not linked with growth nor is it up-regulated by growth hormones such as auxin and gibberellin.

Finally, the localization of yieldin protein in cowpea seedlings and the prolonged incubation required to see yieldin action (see above) are difficult to reconcile with the physiological results showing rapid changes in yield threshold hypocotyls and in roots (Okamoto *et al.*, 1990; Nakahori *et al.*, 1991). Thus, there remain numerous unanswered questions about the physiological relevance of yieldin action. Even so, understanding the mechanism by which this protein causes a change in the yield threshold of cell walls may yield valuable insights into wall structure and the mechanisms of wall expansion.

8.12 Summary

In the past decade we have moved from a situation where wall loosening activities were very vaguely defined at the protein level and not a single gene for a wall loosening enzyme was identified, to the current situation today, in which we can point to hundreds of genes with potential wall-loosening roles, but a dearth of information on the biochemical and biophysical actions of these enzymes on the cell wall. The challenge for the next decade will to be integrate the wealth of recent data on genomics and gene expression profiling with biochemical and biophysical investigations of how plant cell wall expansion is controlled by the growing cell.

References

Ansari, A.A., Ponniah, S. and Marsh, D.G. (1989) Complete amino acid sequence of a *Lolium perenne* (perennial rye grass) pollen allergen, Lol p II. *J. Biol. Chem.*, 264, 11181–11185.

Baker, J.O., King, M.R., Adney, W.S. *et al.* (2000) Investigation of the cell-wall loosening protein expansin as a possible additive in the enzymatic saccharification of lignocellulosic biomass. *Appl. Biochem. Biotechnol.*, 84–86, 217–223.

Baluska, F., Salaj, J., Mathur, J. *et al.* (2000) Root hair formation: F-Actin-dependent tip growth is initiated by local assembly of profilin-supported F-Actin meshworks accumulated within expansin-enriched bulges. *Dev. Biol.*, 227, 618–632.

Bourquin, V., Nishikubo, N., Abe, H. *et al.* (2002) Xyloglucan endotransglycosylases have a function during the formation of secondary cell walls of vascular tissues. *Plant Cell*, 14, 3073–3088.

Brummell, D.A., Harpster, M.H., Civello, P.M., Palys, J.M., Bennett, A.B. and Dunsmuir, P. (1999a) Modification of expansin protein abundance in tomato fruit alters softening and cell wall polymer metabolism during ripening. *Plant Cell*, 11, 2203–2216.

Brummell, D.A., Harpster, M.H. and Dunsmuir, P. (1999b) Differential expression of expansin gene family members during growth and ripening of tomato fruit. *Plant Mol. Biol.*, 39, 161–169.

Brummell, D.A., Howie, W.J., Ma, C. and Dunsmuir, P. (2002) Postharvest fruit quality of transgenic tomatoes suppressed in expression of a ripening-related expansin. *Postharvest Biol. Technol.*, 25, 209–220.

Caderas, D., Muster, M., Vogler, H. *et al.* (2000) Limited correlation between expansin gene expression and elongation growth rate. *Plant Physiol.*, 123, 1399–1414.

Carpita, N.C. and Gibeaut, D.M. (1993) Structural models of primary cell walls in flowering plants: consistency of molecular structure with the physical properties of the walls during growth. *Plant J.*, 3, 1–30.

Catala, C., Rose, J.K.C. and Bennett, A.B. (2000) Auxin-regulated genes encoding cell wall-modifying proteins are expressed during early tomato fruit growth. *Plant Physiol.*, 122, 527–534.

Ceccardi, T.L., Barthe, G.A. and Derrick, K.S. (1998) A novel protein associated with citrus blight has sequence similarities to expansin. *Plant Mol. Biol.*, 38, 775–783.

Chanliaud, E., Burrows, K.M., Jeronimidis, G. and Gidley, M.J. (2002) Mechanical properties of primary plant cell wall analogues. *Planta*, 215, 989–996.

Chen, S.X. and Schopfer, P. (1999) Hydroxyl-radical production in physiological reactions. A novel function of peroxidase. *Eur. J. Biochem.*, 260, 726–735.

Chen, F., Dahal, P. and Bradford, K.J. (2001) Two tomato expansin genes show divergent expression and localization in embryos during seed development and germination. *Plant Physiol.*, 127, 928–936.

Cho, H.T. and Cosgrove, D.J. (2002) Regulation of root hair initiation and expansin gene expression in Arabidopsis. *Plant Cell*, 14, 3237–3253.

Cho, H.T. and Kende, H. (1997a) Expansins and internodal growth of deepwater rice. *Plant Physiol.*, 113, 1145–1151.

Cho, H.T. and Kende, H. (1997b) Expression of expansin genes is correlated with growth in deepwater rice. *Plant Cell*, 9, 1661–1671.

Cleland, R.E. (1971) Cell wall extension. *Annu. Rev. Plant Physiol.*, 22, 197–222.

Cleland, R.E. (1984) The Instron technique as a measure of immediate-past wall extensibility. *Planta*, 160, 514–520.

Cosgrove, D.J. (1985) Cell wall yield properties of growing tissues. Evaluation by in vivo stress relaxation. *Plant Physiol.*, 78, 347–356.

Cosgrove, D.J. (1986) Biophysical control of plant cell growth. *Annu. Rev. Plant Physiol.*, 37, 377–405.

Cosgrove, D.J. (1993) Wall extensibility: its nature, measurement, and relationship to plant cell growth. *New Phytol.*, 124, 1–23.

Cosgrove, D.J. (1997) Relaxation in a high-stress environment: the molecular bases of extensible cell walls and cell enlargement. *Plant Cell*, 9, 1031–1041.

Cosgrove, D.J. (2000a) Expansive growth of plant cell walls. *Plant Physiol. Biochem.*, 38, 1–16.

Cosgrove, D.J. (2000b) Loosening of plant cell walls by expansins. *Nature*, 407, 321–326.

Cosgrove, D.J. and Li, Z.C. (1993) Role of expansin in cell enlargement of oat coleoptiles (analysis of developmental gradients and photocontrol). *Plant Physiol.*, 103, 1321–1328.

Cosgrove, D.J., Van Volkenburgh, E. and Cleland, R.E. (1984) Stress relaxation of cell walls and the yield threshold for growth: demonstration and measurement by micro-pressure probe and psychrometer techniques. *Planta*, 162, 46–52.

Cosgrove, D.J., Bedinger, P. and Durachko, D.M. (1997) Group I allergens of grass pollen as cell wall-loosening agents. *Proc. Natl. Acad. Sci. USA*, 94, 6559–6564.

Cosgrove, D.J., Li, L.C., Cho, H.T., Hoffmann-Benning, S., Moore, R.C. and Blecker, D. (2002) The growing world of expansins. *Plant Cell Physiol.*, 43, 1436–1444.

Cramer, G.R. and Bowman, D.C. (1991) Kinetics of maize leaf elongation. I. Increased yield threshold limits short-term, steady-state elongation rates after exposure to salinity. *J. Exper. Bot.*, 42, 1417–1426.

Cramer, G.R., Krishnan, K., Abrams, S.R., Munns, R. and Cramer, G.R. (1998) Kinetics of maize leaf elongation. IV. Effects of (+)- and (−)-abscisic acid. Is coordination of leaf and root growth mediated by abscisic acid? Opinion. *J. Exper. Bot.*, 319, 195–198.

De Marino, S., Morelli, M.A., Fraternali, F. *et al.* (1999) An immunoglobulin-like fold in a major plant allergen: the solution structure of Phl p 2 from timothy grass pollen. *Structure Fold Des.*, 7, 943–952.

Dolecek, C., Vrtala, S., Laffer, S. *et al.* (1993) Molecular characterization of Phl p II, a major timothy grass (*Phleum pratense*) pollen allergen. *FEBS Lett.*, 335, 299–304.

Downes, B.P. and Crowell, D.N. (1998) Cytokinin regulates the expression of a soybean β-expansin gene by a post-transcriptional mechanism. *Plant Mol. Biol.*, 37, 437–444.

Downes, B.P., Steinbaker, C.R. and Crowell, D.N. (2001) Expression and processing of a hormonally regulated beta-expansin from soybean. *Plant Physiol.*, 126, 244–252.

Fry, S.C. (1989) The structure and functions of xyloglucan. *J. Exper. Bot.*, 40, 1–12.

Fry, S.C. (1998) Oxidative scission of plant cell wall polysaccharides by ascorbate-induced hydroxyl radicals. *Biochem. J.*, 332, 507–515.

Fry, S.C., Smith, R.C., Renwick, K.F., Martin, D.J., Hodge, S.K. and Matthews, K.J. (1992) Xyloglucan endotransglycosylase, a new wall-loosening enzyme activity from plants. *Biochem. J.*, 282, 821–828.

Fry, S.C., Dumville, J.C. and Miller, J.G. (2001) Fingerprinting of polysaccharides attacked by hydroxyl radicals in vitro and in the cell walls of ripening pear fruit. *Biochem. J.*, 357, 729–737.

Grobe, K., Becker, W.M. and Petersen, A. (1999) Grass group I allergens (β-expansins) are novel, papain-related proteinases. *Eur. J. Biochem.*, 263, 33–40.

Harmer, S.E., Orford, S.J. and Timmis, J.N. (2002) Characterisation of six alpha-expansin genes in *Gossypium hirsutum* (upland cotton). *Mol. Genet. Genomics*, 268, 1–9.

Harrison, E.P., McQueen-Mason, S.J. and Manning, K. (2001) Expression of six expansin genes in relation to extension activity in developing strawberry fruit. *J. Exper. Bot.*, 52, 1437–1446.

Hayashi, T. (1989) Xyloglucans in the primary cell wall. *Annu. Rev. Plant Physiol. Plant Mol. Biol.*, 40, 139–168.

Hoson, T. (1993) Regulation of polysaccharide breakdown during auxin-induced cell wall loosening. *J. Plant Res.*, 103, 369–381.

Hoson, T., Sone, Y., Misaki, A. and Masuda, Y. (1993) Role of xyloglucan breakdown in epidermal cell walls for auxin-induced elongation of azuki bean epicotyl segments. *Physiol. Plant.*, 87, 142–147.

Hsiao, T.C., Frensch, J. and Rojas-Lara, B.A. (1998) The pressure-jump technique shows maize leaf growth to be enhanced by increases in turgor only when water status is not too high. *Plant Cell Environ.*, 21, 1–42.

Hutchison, K.W., Singer, P.B., McInnis, S., Diaz-Sala, C. and Greenwood, M.S. (1999) Expansins are conserved in conifers and expressed in hypocotyls in response to exogenous auxin. *Plant Physiol.*, 120, 827–832.

Im, K.H., Cosgrove, D.J. and Jones, A.M. (2000) Subcellular localization of expansin mRNA in xylem cells. *Plant Physiol.*, 123, 463–470.

Joo, J.H., Bae, Y.S. and Lee, J.S. (2001) Role of auxin-induced reactive oxygen species in root gravitropism. *Plant Physiol.*, 126, 1055–1060.

Kaku, T., Tabuchi, A., Wakabayashi, K., Kamisaka, S. and Hoson, T. (2002) Action of xyloglucan hydrolase within the native cell wall architecture and its effect on cell wall extensibility in azuki bean epicotyls. *Plant Cell Physiol.*, 43, 21–26.

Keegstra, K., Talmadge, K.W., Bauer, W.D. and Albersheim, P. (1973) The structure of plant cell walls. III. A model of the walls of suspension-cultured sycamore cells based on the interconnections of the macromolecular components. *Plant Physiol.*, 51, 188–196.

Keller, E. and Cosgrove, D.J. (1995) Expansins in growing tomato leaves. *Plant J.*, 8, 795–802.

Kitamura, S., Mizuno, A. and Katou, K. (1998) IAA-Dependent adjustment of the in vivo wall-yielding properties of hypocotyl segments of *Vigna unguiculata* during adaptive growth recovery from osmotic stress. *Plant Cell Physiol.*, 39, 627–631.

Labavitch, J.M. and Ray, P.M. (1974) Relationship between promotion of xyloglucan metabolism and induction of elongation by IAA. *Plant Physiol.*, 54, 499–502.

Laine, M., Haapalainen, M., Wahlroos, T. *et al.* (2001) The cellulase encoded by the native plasmid of *Clavibacter michiganensis* subsp. sepedonicus plays a role in virulence and contains an expansin-like domain. *Physiol. Mol. Plant Pathol.*, 57, 221–233.

Lane, D.R., Wiedemeier, A., Peng, L. *et al.* (2001) Temperature-sensitive alleles of RSW2 link the KORRIGAN endo-1,4-beta- glucanase to cellulose synthesis and cytokinesis in Arabidopsis. *Plant Physiol.*, 126, 278–288.

Lee, Y. and Kende, H. (2001) Expression of beta-expansins is correlated with internodal elongation in deepwater rice. *Plant Physiol.*, 127, 645–654.

Levy, I., Shani, Z. and Shoseyov, O. (2002) Modification of polysaccharides and plant cell wall by endo-1,4-beta-glucanase and cellulose-binding domains. *Biomol. Engng*, 19, 17–30.

Li, L.C. and Cosgrove, D.J. (2001) Grass group I pollen allergens (beta-expansins) lack proteinase activity and do not cause wall loosening via proteolysis. *Eur. J. Biochem.*, 268, 4217–4226.

Li, Y., Darley, C.P., Ongaro, V. *et al.* (2002) Plant expansins are a complex multigene family with an ancient evolutionary origin. *Plant Physiol.*, 128, 854–864.

Ludidi, N.N., Heazlewood, J.L., Seoighe, C., Irving, H.R. and Gehring, C.A. (2002) Expansin-like molecules: novel functions derived from common domains. *J. Mol. Evol.*, 54, 587–594.

Masuda, Y. (1990) Auxin-induced cell elongation and cell wall changes. *Bot. Mag. Tokyo*, 103, 345–370.

Mbeguie, A.M., Gouble, B., Gomez, R.M., Audergon, J.M., Albagnac, G. and Fils-Lycaon, B. (2002) Two expansin cDNAs from *Prunus armeniaca* expressed during fruit ripening are differently regulated by ethylene. *Plant Physiol. Biochem.*, 40, 445–452.

McQueen-Mason, S. and Cosgrove, D.J. (1994) Disruption of hydrogen bonding between plant cell wall polymers by proteins that induce wall extension. *Proc. Natl. Acad. Sci. USA*, 91, 6574–6578.

McQueen-Mason, S.J. and Cosgrove, D.J. (1995) Expansin mode of action on cell walls. Analysis of wall hydrolysis, stress relaxation, and binding. *Plant Physiol.*, 107, 87–100.

McQueen-Mason, S., Durachko, D.M. and Cosgrove, D.J. (1992) Two endogenous proteins that induce cell wall expansion in plants. *Plant Cell*, 4, 1425–1433.

McQueen-Mason, S.J., Fry, S.C., Durachko, D.M. and Cosgrove, D.J. (1993) The relationship between xyloglucan endotransglycosylase and in-vitro cell wall extension in cucumber hypocotyls. *Planta*, 190, 327–331.

Metraux, J.P., Burkhart, W., Moyer, M. *et al*. (1989) Isolation of a complementary DNA encoding a chitinase with structural homology to a bifunctional lysozyme/chitinase. *Proc. Natl. Acad. Sci. USA*, 86, 896–900.

Miller, J.G. and Fry, S.C. (2001) Characteristics of xyloglucan after attack by hydroxyl radicals. *Carbohydr. Res.*, 332, 389–403.

Molhoj, M., Pagant, S. and Hofte, H. (2002) Towards understanding the role of membrane-bound endo-beta-1,4-glucanases in cellulose biosynthesis. *Plant Cell Physiol.*, 43, 1399–1406.

Nakahori, K., Katou, K. and Okamoto, H. (1991) Auxin changes both the extensibility and the yield threshold of the cell wall of *Vigna* hypocotyls. *Plant Cell Physiol.*, 32, 121–129.

Nakamura, S. and Hayashi, T. (1993) Purification and properties of an extracellular endo-1,4-β-glucanase from suspension-cultured poplar cells. *Plant Cell Physiol.*, 34, 1009–1013.

Nicol, F., His, I., Jauneau, A., Vernhettes, S., Canut, H. and Hofte, H. (1998) A plasma membrane-bound putative endo-1,4-β-D-glucanase is required for normal wall assembly and cell elongation in Arabidopsis. *EMBO J.*, 17, 5563–5576.

Nishitani, K. and Tominaga, T. (1992) Endo-xyloglucan transferase, a novel class of glycosyltransferase that catalyzes transfer of a segment of xyloglucan molecule to another xyloglucan molecule. *J. Biol. Chem.*, 267, 21058–21064.

Ohmiya, Y., Samejima, M., Shiroishi, M. *et al*. (2000) Evidence that endo-1,4-beta-glucanases act on cellulose in suspension-cultured poplar cells. *Plant J.*, 24, 147–158.

Okamoto, H. and Okamoto, A. (1994) The pH-dependent yield threshold of the cell wall in a glycerinated hollow cylinder (in vitro system) of cowpea. *Plant Cell Environ.*, 17, 979–983.

Okamoto, A. and Okamoto, H. (1995) Two proteins regulate the cell wall extensibility and the yield threshold in glycerinated hollow cylinders of cowpea hypocotyl. *Plant Cell Environ.*, 18, 827–830.

Okamoto, H., Miwa, C., Masuda, T., Nakahori, K. and Katou, K. (1990) Effects of auxin and anoxia on the cell wall yield threshold determined by negative pressure jumps in segments of cowpea hypocotyl. *Plant Cell Physiol.*, 31, 783–788.

Okamoto-Nakazato, A., Nakamura, T. and Okamoto, H. (2000a) The isolation of wall-bound proteins regulating yield threshold tension in glycerinated hollow cylinders of cowpea hypocotyl. *Plant Cell Environ.*, 23, 145–154.

Okamoto-Nakazato, A., Takahashi, K., Kido, N., Owaribe, K. and Katou, K. (2000b) Molecular cloning of yieldins regulating the yield threshold of cowpea cell walls: cDNA cloning and characterization of recombinant yieldin. *Plant Cell Environ.*, 23, 155–164.

Orford, S.J. and Timmis, J.N. (1998) Specific expression of an expansin gene during elongation of cotton fibres. *Biochim. Biophys. Acta*, 139(8), 342–346.

Park, Y.T., Tominaga, R., Sugiyama, J. *et al*. (2003) Enhancement of growth by expression of poplar cellulase in *Arabidopsis thaliana*. *Plant J.*, 33, 1087-1097.

Parks, B.M., Folta, K.M. and Spalding, E.P. (2001) Photocontrol of stem growth. *Curr. Opin. Plant Biol.*, 4, 436–440.

Pauly, M., Albersheim, P., Darvill, A. and York, W.S. (1999) Molecular domains of the cellulose/xyloglucan network in the cell walls of higher plants. *Plant J.*, 20, 629–639.

Pien, S., Wyrzykowska, J., McQueen-Mason, S., Smart, C. and Fleming, A. (2001) From the cover: local expression of expansin induces the entire process of leaf development and modifies leaf shape. *Proc. Natl. Acad. Sci. USA*, 98, 11812–11817.

Poppelmann, M., Becker, W.M. and Petersen, A. (2002) Combination of zymography and immuno-detection to analyze proteins in complex culture supernatants. *Electrophoresis*, 23, 993–997.

Probine, M.C. and Preston, R.D. (1962) Cell growth and the structure and mechanical properties of the wall in internodal cells of *Nitella opaca*. *J. Exper. Bot.*, 13, 111–127.

Ray, P.M. (1987) Principles of Plant Cell Growth. In *Physiology of Cell Expansion During Plant Growth (Symposium in Plant Physiology, Penn State Univ)* (eds D.J. Cosgrove and D.J. Knievel), American Society of Plant Physiologists, Rockville, pp. 1–17.

Ray, P.M. and Ruesink, A.W. (1962) Kinetic experiments on the nature of the growth mechanism in oat coleoptile cells. *Dev. Biol.,* 4, 377–397.

Reidy, B., McQueen-Mason, S., Nosberger, J. and Fleming, A. (2001) Differential expression of alpha- and beta-expansin genes in the elongating leaf of *Festuca pratensis. Plant Mol. Biol.,* 46, 491–504.

Reinhardt, D., Wittwer, F., Mandel, T. and Kuhlemeier, C. (1998) Localized upregulation of a new expansin gene predicts the site of leaf formation in the tomato meristem. *Plant Cell,* 10, 1427–1437.

Rose, J.K.C., Lee, H.H. and Bennett, A.B. (1997) Expression of a divergent expansin gene is fruit-specific and ripening-regulated. *Proc. Natl. Acad. Sci. USA,* 94, 5955–5960.

Rose, J.K.C., Braam, J., Fry, S.C. and Nishitani, K. (2002) The XTH family of enzymes involved in xyloglucan endotransglucosylation and endohydrolysis: current perspectives and a new unifying nomenclature. *Plant Cell Physiol.,* 43, 1421–1435.

Ruan, Y., Llewellyn, D. and Furbank, R. (2001) The control of single-celled cotton fiber elongation by developmentally reversible gating of plasmodesmata and coordinated expression of sucrose and K$^+$ transporters and expansin. *Plant Cell,* 13, 47–60.

Saloheimo, M., Paloheimo, M., Hakola, S. *et al.* (2002) Swollenin, a *Trichoderma reesei* protein with sequence similarity to the plant expansins, exhibits disruption activity on cellulosic materials. *Eur. J. Biochem.,* 269, 4202–4211.

Samac, D.A. and Shah, D.M. (1991) Developmental and pathogen-induced activation of the Arabidopsis acidic chitinase promoter. *Plant Cell,* 3, 1063–1072.

Samac, D.A., Hironaka, C.M., Yallaly, P.E. and Shah, D.M. (1990) Isolation and characterization of the genes encoding basic and acidic chitinase in *Arabidopsis thaliana. Plant Physiol.,* 93, 907–914.

Sato, S., Kato, T., Kakegawa, K. *et al.* (2001) Role of the putative membrane-bound endo-1,4-beta-glucanase KORRIGAN in cell elongation and cellulose synthesis in *Arabidopsis thaliana. Plant Cell Physiol.,* 42, 251–263.

Schopfer, P. (2001) Hydroxyl radical-induced cell-wall loosening in vitro and in vivo: implications for the control of elongation growth. *Plant J.,* 28, 679–688.

Schopfer, P., Liszkay, A., Bechtold, M., Frahry, G. and Wagner, A. (2002) Evidence that hydroxyl radicals mediate auxin-induced extension growth. *Planta,* 214, 821–828.

Schweikert, C., Liszkay, A. and Schopfer, P. (2000) Scission of polysaccharides by peroxidase-generated hydroxyl radicals. *Phytochem.,* 53, 565–570.

Schweikert, C., Liszkay, A. and Schopfer, P. (2002) Polysaccharide degradation by Fenton reaction – or peroxidase-generated hydroxyl radicals in isolated plant cell walls. *Phytochem.,* 61, 31–35.

Shani, Z., Dekel, M., Tsabary, G. and Shoseyov, O. (1997) Cloning and characterization of elongation specific endo-1,4-beta-glucanase (cel1) from *Arabidopsis thaliana. Plant Mol. Biol.,* 34, 837–842.

Shani, Z., Dekel, M., Tzbary, G. *et al.* (1999) Expression of *Arabidopsis thaliana* endo-1,4-β-glucanase *(cel1)* in transgenic plants. In *Plant Biotechnology and In Vitro Biology in the 21st Century* (ed. A. Altman, M. Ziv and S. Izhar), Kluwer Academic Publishers, Dordrecht, the Netherlands, pp. 209–211.

Steele, N.M. and Fry, S.C. (1999) Purification of xyloglucan endotransglycosylases (XETs): a generally applicable and simple method based on reversible formation of an enzyme-substrate complex. *Biochem. J.,* 340, 207–211.

Steele, N.M., Sulova, Z., Campbell, P., Braam, J., Farkas, V. and Fry, S.C. (2001) Ten isoenzymes of xyloglucan endotransglycosylase from plant cell walls select and cleave the donor substrate stochastically. *Biochem. J.,* 355, 671–679.

Sulova, Z. and Farkas, V. (1998) A method for purification of XET based on affinity sorption of XET:xyloglucan complex on cellulose (Abstract). *8th International Cell Wall Meeting* 7.41.

Sulova, Z., Takacova, M., Steele, N.M., Fry, S.C. and Farkas, V. (1998) Xyloglucan endotransglycosylase: evidence for the existence of a relatively stable glycosyl-enzyme intermediate. *Biochem. J.*, 330(3), 1475–1480.

Tabbi, G., Fry, S.C. and Bonomo, R.P. (2001) ESR study of the non-enzymic scission of xyloglucan by an ascorbate- H_2O_2-copper system: the involvement of the hydroxyl radical and the degradation of ascorbate. *J. Inorg. Biochem.*, 84, 179–187.

Takeda, T., Furuta, Y., Awano, T., Mizuno, K., Mitsuishi, Y. and Hayashi, T. (2002) Suppression and acceleration of cell elongation by integration of xyloglucans in pea stem segments. *Proc. Natl. Acad. Sci. USA*, 99, 9055–9060.

Talbott, L.D. and Ray, P.M. (1992a) Changes in molecular size of previously deposited and newly synthesized pea cell wall matrix polysaccharides. *Plant Physiol.*, 98, 369–379.

Talbott, L.D. and Ray, P.M. (1992b) Molecular size and separability features of pea cell wall polysaccharides. Implications for models of primary wall structure. *Plant Physiol.*, 92, 357–368.

Thompson, J.E. and Fry, S.C. (2001) Restructuring of wall-bound xyloglucan by transglycosylation in living plant cells. *Plant J.*, 26, 23–34.

Vissenberg, K., Martinez-Vilchez, I.M., Verbelen, J.P., Miller, J.G. and Fry, S.C. (2000) In vivo colocalization of xyloglucan endotransglycosylase activity and its donor substrate in the elongation zone of Arabidopsis roots. *Plant Cell*, 12, 1229–1238.

Vissenberg, K., Fry, S.C. and Verbelen, J.P. (2001) Root hair initiation is coupled to a highly localized increase of xyloglucan endotransglycosylase action in Arabidopsis roots. *Plant Physiol.*, 127, 1125–1135.

Vissenberg, K., Van, S., V, Fry, S.C. and Verbelen, J.P. (2003) Xyloglucan endotransglucosylase action is high in the root elongation zone and in the trichoblasts of all vascular plants from *Selaginella* to *Zea mays*. *J. Exper. Bot.*, 54, 335–344.

Vriezen, W.H., De Graaf, B., Mariani, C. and Vosenek, L.A.C.J. (2000) Submergence induces expansin gene expression in flooding tolerant *Rumex palustris* and not in flooding intolerant *R. acetosa. Planta*, 210, 956–963.

Whitney, S.E., Gothard, M.G., Mitchell, J.T. and Gidley, M.J. (1999) Roles of cellulose and xyloglucan in determining the mechanical properties of primary plant cell walls. *Plant Physiol.*, 121, 657–664.

Whitney, S.E., Gidley, M.J. and McQueen-Mason, S.J. (2000) Probing expansin action using cellulose/hemicellulose composites. *Plant J.*, 22, 327–334.

Woolley, L.C., James, D.J. and Manning, K. (2001) Purification and properties of an endo-β-1,4-glucanase from strawberry and down-regulation of the corresponding gene, *cell. Planta*, 214, 11–21.

Wu, Y., Meeley, R.B. and Cosgrove, D.J. (2001a) Analysis and expression of the alpha-expansin and beta-expansin gene families in maize. *Plant Physiol.*, 126, 222–232.

Wu, Y., Thorne, E.T., Sharp, R.E. and Cosgrove, D.J. (2001b) Modification of expansin transcript levels in the maize primary root at low water potentials. *Plant Physiol.*, 126, 1471–1479.

Yuan, S., Wu, Y. and Cosgrove, D.J. (2001) A fungal endoglucanase with plant cell wall extension activity. *Plant Physiol.*, 127, 324–333.

Zuo, J., Niu, Q.W., Nishizawa, N., Wu, Y., Kost, B. and Chua, N.H. (2000) KORRIGAN, an *Arabidopsis* endo-1,4-beta-glucanase, localizes to the cell plate by polarized targeting and is essential for cytokinesis. *Plant Cell*, 12, 1137–1152.

9 Cell wall disassembly

Jocelyn K.C. Rose, Carmen Catalá, Zinnia H. Gonzalez-Carranza and Jeremy A. Roberts

9.1 Introduction

As with the walls of man-made structures, the 'lifetime' of plant walls involves the fabrication of the constituent materials and transport to the site of construction, the assembly of the building blocks into the framework of the final structure, architectural remodeling when needed, and finally demolition. Previous chapters in this volume have outlined some of the molecular processes involved in wall biosynthesis (Chapter 6), and assembly and reorganization (Chapter 8), but the aim of this chapter is to provide a current perspective on the mechanisms of plant cell wall disassembly. As reviewed in Chapter 8, cell expansion also involves localized wall loosening; however, a clear distinction is made between the transient wall restructuring during cell elongation, and wall disassembly as defined here. During cell growth, wall loosening is reversible, wall tensile strength is typically maintained and extensive coupled wall biosynthesis is essential. In contrast, wall disassembly during processes such as fruit softening and organ abscission involves net wall depolymerization and, while there are some parallels with wall modification in expanding cells (Rose and Bennett, 1999), in general the changes in the wall architecture in these instances are irreversible.

There are many examples where cell wall depolymerization is an essential step in a more complex developmental program and, in this context, the term 'disassembly' is useful as it has connotations of coordinated, regulated deconstruction, rather than non-specific degradation. Again, by analogy with building demolition, the use of a range of precision tools targeted to specific walls of a building scheduled for destruction, and load-bearing elements within those walls, will likely have a more predictable and incremental outcome than a single blast of high explosive.

Particular emphasis is placed here on describing wall disassembly during fruit ripening and organ abscission and dehiscence, since these processes have been the most extensively studied in this regard, reflecting their agricultural and commercial importance. Other somewhat less characterized examples are also reviewed, such as aerenchyma formation and radicle emergence in germinating seeds, which suggest some apparent mechanistic similarities. Finally, an overview is given drawing together some common themes that have emerged from studies of wall disassembly in different systems, including some of the questions that have been raised in the face of the increasing apparent complexity and that remain to be addressed.

9.2 Fruit softening

The ripening of most fleshy fruits involves a significant, and often dramatic, combination of textural changes that are collectively referred to as 'softening', but that reflect multiple sensory attributes (Redgwell and Fischer, 2001). While factors such as cellular turgor and morphology (Lin and Pitt, 1986; Shackel *et al.*, 1991; Harker *et al.*, 1997) contribute to overall fruit texture, the loss of fruit firmness is believed to result principally from cell wall disassembly (Tucker, 1993; Wakabayashi, 2000) and a reduction in cell-cell adhesion following dissolution of the pectinaceous middle lamella (Diehl and Hamann, 1980; Harker *et al.*, 1997). The relative contributions of these factors to the textural changes in the fruit are difficult to define and appear to vary between species, and between fruit of the same species at different ripening stages (Jackman and Stanley, 1992; De Belie *et al.*, 2000). Similarly, the rate, extent and sensory attributes of softening differ considerably among fruits from divergent species and additional parameters may also need to be considered in specific cases; for example, starch degradation has been suggested to contribute an important textural component in ripening banana fruit (Tucker, 1993).

The variation in softening that is exhibited by different fruits is mirrored in the range of structural changes in the cell wall and middle lamella that are apparent through microscopy and immunolocalization studies (Ben-Arie *et al.*, 1979; Crookes and Grierson, 1983; Hallett *et al.*, 1992; Redgwell *et al.*, 1997b; Sutherland *et al.*, 1999). Moreover, fruit cell walls from various species undergo substantially different degrees of swelling both *in vitro* and *in vivo* during ripening, which can be correlated with wall modification (Redgwell *et al.*, 1997b) and the ionic conditions in the apoplast (MacDougall *et al.*, 2001). It is therefore not surprising that reports over the last 30 years of ripening-related structural and compositional changes in the cell wall, and the expression and activities of wall modifying proteins, in a range of ripening fruits show great qualitative and quantitative differences. There are also many suggestions of variability in aspects of cell wall modification among different cultivars of the same species (e.g. Blumer *et al.*, 2000). An evaluation of cell wall disassembly in ripening fruit should therefore be approached from the perspective of underlying variability, and although many basic biochemical mechanisms of softening may be conserved, considerable variation appears to exist at the quantitative, temporal and regulatory levels. Thus, while tomato (*Lycopersicon esculentum*) represents by far the best-studied system in terms of evaluating the molecular processes that underlie cell wall disassembly in softening fruit (e.g. see Table 9.1), the results from such studies should be used with caution to elucidate softening in other fruit species.

A number of recent reviews have described aspects of fruit ripening-related wall metabolism (Brownleader *et al.*, 1999; Rose *et al.*, 2000b; Wakabayashi, 2000; Brummell and Harpster, 2001; Giovannoni, 2001; Huber *et al.*, 2001; Redgwell and Fischer, 2001) and detailed current interpretations of cell wall polysaccharide structures and interactions are described elsewhere in this volume (e.g. Chapter 1).

Therefore, rather than provide an exhaustive review of the primary literature, the goal of this section is to highlight key papers and reviews, to summarize the major directions that are currently being pursued in elucidating the complexities of ripening-related wall disassembly, and to suggest some critical questions that remain to be resolved. In many cases, tomato is used to illustrate examples of ripening-related wall disassembly, reflecting the extensive literature and use of tomato as a model system to study fruit ripening (Giovannoni, 2001).

9.2.1   Pectins and pectinases

Fruit tissue is composed mainly of parenchymatous cells that show little lignification and generally have a wall polysaccharide composition similar to that of type I primary walls (see Chapter 1; Carpita and Gibeaut, 1993), with approximately equal amounts of cellulose and hemicelluloses and a pectin content of typically 40–60% (Redgwell and Fischer, 2001). Pectin solubilization and depolymerization are among the most pronounced and widely reported changes in the walls of ripening fleshy fruits, and these changes are usually closely correlated with softening and with cell wall swelling (Huber, 1983; Fischer and Bennett, 1991; Huber and O'Donoghue, 1993; Redgwell et al., 1997b). Experimental analysis of the changes in the molecular masses of pectins during ripening typically involves fractionating them into several classes based on different solvents that are used to extract them from the wall. A typical sequential extraction generates water-, chelator- (e.g. CDTA or EGTA), and sodium carbonate-soluble fractions (Rose et al., 1998; Brummell and Harpster, 2001). These subsets are generally described as corresponding to pectins that are freely soluble in the apoplast, ionically associated with the wall, or linked into the wall by covalent bonds, respectively. During ripening, enzymatic and probably also non-enzymatic processes result in the solubilization of pectins into different fractions, thus sodium carbonate-extractable pectin in an immature fruit may be extracted from ripe fruit in the water- or chelator-soluble fraction. A common observation is that ripening-related increases in water-soluble polyuronides are paralleled by equivalent decreases in the amounts of pectins in the wall-associated fractions (Rose et al., 1998; Wakabayashi et al., 2000). However, the proportion of water-soluble polyuronides in ripe fruit varies substantially among different species, from 10% in grapefruit (Hwang et al., 1990) to more than 85% in avocado (Wakabayashi et al., 2000).

The bases of pectin solubilization and depolymerization have not been fully resolved, but probably result from a number of mechanisms that may vary between fruits. Indeed, the extent of changes in pectin molecular weight in vivo during ripening, and the degree to which they are associated with the wall, are difficult to assess experimentally (Redgwell and Fischer, 2001) and are affected by factors such as the methods of extraction and size estimation (Huber, 1983, 1991; Huber and O'Donoghue, 1993; Brummell and Labavitch, 1997).

9.2.1.1   Polyuronide hydrolysis and polygalacturonase

The enzyme polygalacturonase (PG) hydrolyses the α-1,4-D-galacturonan backbone of pectic polysaccharides and PG activity has long been known to increase substantially in many species of ripening fruit, concomitant with polyuronide depolymerization (Hobson, 1962; Gross and Wallner, 1979; Huber, 1983; Redgwell and Fischer, 2001). As with many cell wall modifying enzymes, PG occurs as a family of genes and the corresponding proteins can act as either endo-, or exo-hydrolases (Hadfield and Bennett, 1998); however, the endo-acting enzymes are more likely to contribute significantly to pectin depolymerization in ripening fruit. Ripening-related PG has been studied in detail in tomato (reviewed in Giovannoni *et al.*, 1991; Brummell and Harpster, 2001; Redgwell and Fischer, 2001) where the abundance of PG mRNA can increase more than 2,000 fold, reaching levels that comprise more than 1% of total poly(A)+ mRNA (DellaPenna *et al.*, 1986; Fischer and Bennett, 1991). In contrast, in the ripening-impaired tomato mutants *rin, nor* and *Nr,* which exhibit delayed and/or reduced softening, PG mRNA levels and the rate of transcription are substantially lower (DellaPenna *et al.*, 1987, 1989). The abundance and prevalence of PG in many ripening fruit, close temporal association with polyuronide depolymerization in some species, and correlation with loss of fruit firmness, originally led to the attractive hypothesis that PG-mediated pectin depolymerization represented the enzymatic basis of softening.

This model has been tested directly using reverse genetics in two different approaches to alter PG expression in transgenic tomato fruit. Firstly, PG expression was suppressed in wild-type tomato by approximately 99% using a constitutive antisense PG transgene (Sheehy *et al.*, 1988; Smith *et al.*, 1988; Table 9.1). The transgenic fruit exhibited substantially reduced polyuronide depolymerization; however, softening was essentially unaffected (Sheehy *et al.*, 1988; Smith *et al.*, 1988). Subsequent analyses indicated a marginal effect of the transgene on fruit firmness during postharvest storage (Kramer *et al.*, 1992), pectin mobility (Fenwick *et al.*, 1996) and delayed depolymerization of some pectin fractions (Carrington *et al.*, 1993; Brummell and Labavitch, 1997). Pectin solubility may also be reduced in the transgenic fruit (Carrington *et al.*, 1993), although another report suggested no major difference compared with wild-type fruit (Brummell and Labavitch, 1997). Conclusions from all these experiments are complicated by the fact that while PG activity is effectively suppressed in the antisense lines, 0.5–1% of wild-type activity levels remains (Smith *et al.*, 1990; Kramer *et al.*, 1992). This may be sufficient to catalyse significant pectin depolymerization and alter the degree of association with the wall, especially given the high levels of PG activity in wild-type tomato.

Secondly, in a related study (DellaPenna *et al.*, 1990), PG was expressed ectopically in the non-softening *rin* tomato mutant (Giovannoni *et al.*, 1989), which normally shows minimal ripening-related PG activity, pectin depolymerization and solubilization (Seymour *et al.*, 1987a). Polyuronide depolymerization in the transgenic *rin* expressing high levels of PG occurred to the same extent as in wild-type fruit (DellaPenna *et al.*, 1990), but the fruit did not soften (Giovannoni *et al.*, 1989;

Table 9.1 Summary of experimental results derived from transgenic tomato lines in which fruit ripening-related genes have been suppressed.

Protein family	Gene name	Effect of suppression on fruit cell walls and firmness	Reference
Polygalacturonase (PG)	*Le-PG2*	Reduced polyuronide depolymerization No major change in softening Slight increase in firmness during storage Delayed polyuronide depolymerization	Sheehy *et al.* (1988); Smith *et al.* (1988); Kramer *et al.* (1992); Brummell and Labavitch (1997)
Polygalacturonase β-subunit		Increased polyuronide solubilization Increased softening	Watson *et al.* (1994); Chun and Huber (2000)
Pectin methylesterase (PME)	*PE2*	Decreased pectin depolymerization and amount of chelator-soluble pectins. No effect on softening	Tieman *et al.* (1992); Hall *et al.* (1993)
β-galactosidase	*TBG1*	No detectable effect on wall composition or fruit firmness	Carey *et al.* (2001)
	TBG3	Increase in wall galactosyl content. No major effect on softening. Some increase in postharvest quality	de Silva and Verhoeyen (1998)
	TBG4	Some fruit up to 40% firmer than wild-type. Little difference in wall galactosyl content	Smith *et al.* (2002)
Endo-β-1,4-glucanase ('cellulase'/EGase)	*Cel1* *Cel2*	No effect on fruit firmness. No analysis of cell wall composition performed	Lashbrook *et al.* (1998); Brummell *et al.* (1999a)
Expansin	*LeExp1*	Small increase in fruit firmness. Secondary effects on matrix glycans Change in postharvest characteristics	Brummell *et al.* (1999b); Brummell *et al.* (2002)
Xyloglucan endotransglucosylase-hydrolase (XTH); formerly termed XET	*TXET B1*	No apparent change in fruit texture No reported cell wall analysis	Referred to in: de Silva *et al.* (1994); Brummell and Harpster (2001)

DellaPenna *et al.*, 1990). More recently still, a transposon-tagged tomato line was reported with an insertion in PG, inactivating the gene and massively reducing PG expression, while the fruit were described as exhibiting normal softening (Cooley and Yoder, 1998).

Taken together, these studies demonstrate that in ripening tomato fruit, while PG activity plays a significant role in polyuronide depolymerization, and particularly in the hydrolysis and solubilization of a pectin fraction that is covalently bound to the wall (Brummell and Labavitch, 1997), it is not the major factor responsible for softening. Furthermore, there is considerable qualitative variability in the patterns of polyuronide depolymerization and solubilization between fruits and in many cases a poor correlation with PG activity is observed. For example, avocado polyuronides appear to undergo a considerably greater downshift in molecular mass than those in tomato, despite high PG activity levels in both fruits (Huber and O'Donoghue, 1993). It should also be noted however that variability in the methods of polysaccharide extraction and analysis that are used by different research groups might exacerbate or mask differences in wall composition between extracts.

Other analyses of the PG-antisense fruit suggest that PG may influence different aspects of fruit quality, including shelf-life, the onset of postharvest microbial infection (Schuch *et al.*, 1991; Kramer *et al.*, 1992; Langley *et al.*, 1994) and tomato paste and juice characteristics during processing (Errington *et al.*, 1998; Porretta *et al.*, 1998; Lang and Dörnenburg, 2000).

Tomato fruit PG is generally described as comprising three isoforms: PG1, PG2A and PG2B (reviewed in Fischer and Bennett, 1991; Brummell and Harpster, 2001). PG2A and PG2B are differentially glycosylated proteins derived from the gene that was used for the antisense experiments described above (Sheehy *et al.*, 1988; Smith *et al.*, 1988), and represent the catalytic component of PG. PG1 is a heterodimer of either of the PG2 isoforms tightly bound to a protein that has been termed the β-subunit (Zheng *et al.*, 1992). The β-subunit has no pectinolytic activity itself, but down-regulation of β-subunit expression in transgenic tomato by antisense resulted in fruits with increased levels of chelator-soluble pectins and an overall increase in polyuronide depolymerization (Watson *et al.*, 1994; Table 9.1). Other studies also reported an increase in polyuronide extractability in green fruit, prior to the expression of PG2 and therefore preceding the formation of the PG1 heterodimer (Chun and Huber, 1997). This observation suggests that the β-subunit can affect pectin solubility by a mechanism that is independent of PG activity. The same authors have also proposed, based on ultrastructural studies of PG-treated fruit tissues, that the β-subunit reduces PG2-mediated pectin hydrolysis *in muro* (Chun and Huber, 2000). Thus the β-subunit appears to play both a PG2-dependent and -independent role in pectin metabolism. While it has been proposed that the function of the β-subunit is to alter and restrict the biochemical activity or localization of PG2 *in vivo* (DellaPenna *et al.*, 1996), there are also suggestions that the PG1 dimer is an artifact generated during protein extraction and does not occur *in vivo* (Moore and Bennett, 1994). A recent report described the purification of a PG from avocado fruit and analysis of its activity against native substrates (Wakabayashi and Huber, 2001). However, although evidence was obtained that the PG had a restricted capacity to solubilize polyuronides from the walls of pre-ripe fruit, no evidence was found for the presence of β-subunit-like proteins. The authors suggest that this may explain, at least in part, the observation that polyuronide depolymerization during avocado ripening is more extensive than that in tomato.

PG provides a clear illustration of the considerable quantitative and qualitative differences in the expression of cell wall modifying enzymes among different fruit species. PG activity is exceptionally high in tomato and avocado (Huber and O'Donoghue, 1993) and yet has been reported as undetectable in many fruits, including muskmelon (McCollum et al., 1989) and banana (Wade et al., 1992). However, the more recent detection of PG activity and/or gene expression in both melon (Hadfield et al., 1998) and banana (Neelam et al., 2000) suggests that factors may impede experimental detection of activity in vitro. From a qualitative standpoint, the pattern of PG gene expression and regulation also shows considerable variability in different fruits (Redgwell and Fischer, 2001) and the expression of a single predominant ripening-specific PG, as is seen in tomato, is certainly not the case in all fruits (Redgwell and Fischer, 2001).

In conclusion, the role of PG in fruit softening is still open to debate. Undoubtedly the enzyme(s) catalyses substantial depolymerization and solubilization of a subset of wall polyuronides in many ripening fruits, but there is apparent restriction of PG action by a range of possible factors (see also section 9.2.1.7) and the relationship between PG, pectin depolymerization and solubilization and specific textural changes is considerably more complex than originally conceived. Only two examples have been reported to date that demonstrate a genetic linkage between a softening phenotype and PG. The *melting flesh* locus in peach cosegregates closely with an endo-PG gene (Lester et al., 1994, 1996) and the S gene in pepper that controls a soft flesh phenotype has also been shown to exhibit complete linkage to PG (Rao and Paran, 2003). Interestingly, PG activity assayed in vitro is not especially high in either of these species, being 50 times (peach) and up to 164 times (bell pepper) less than levels in tomato (Pressey and Avants, 1978; Jen and Robinson, 1984). It is likely that the relative contributions of PG to softening and other textural changes in different fruit species will be easier to assess once genetic mapping projects progress and PG levels can be manipulated in transgenic fruits other than tomato. One such example was recently reported, where over-expression of PG in transgenic apples had a variety of effects on plant phenotype and cell-cell adhesion (Atkinson et al., 2002).

9.2.1.2   Pectin deesterification: pectin methylesterase and pectin acetylesterase
Pectins are believed to be synthesized and deposited into the wall in a highly methylesterified form during wall assembly and then to undergo enzyme-mediated demethylation (Kauss and Hassid, 1967; Lau et al., 1985). Pectin methylation and associated properties have been examined in some detail in tomato fruit (Roy et al., 1992; MacDougall et al., 1996; Blumer et al., 2000) and quantitative studies indicate that the degree of methyl esterification of galacturonosyl residues in polyuronides declines from levels as high as 90% at a pre-ripe stage, to 35% in ripe fruit (Har-

riman, 1990). Similar values were observed in ripening avocado (Wakabayashi *et al.*, 2000). Demethylation of the pectin galacturonosyl residues would result in the generation of carboxylate ions, which can then bind cations such as calcium, of which there can be relatively high concentrations in the fruit apoplast (Almeida and Huber, 1999). Blocks of adjacent unesterified galacturonic acid residues are believed to contribute to a so-called 'egg-box' structure, where homogalacturonan chains associate non-covalently at specific junction zones through interactions between the free carboxyl groups of the demethylesterified galacturonosyl residues and calcium ions, resulting in the formation of a pectate gel (Grant *et al.*, 1973; Jarvis, 1984; Carpita and Gibeaut, 1993; Pérez *et al.*, 2000). Changes in the degree of methylation could thus influence pectin charge density and cation-binding capacity, apparent molecular size, aggregation, solubility, gelation properties, wall-association and ultimately, function in the wall and middle lamella. The physicochemical properties of pectins is too extensive a topic to be described in detail here, but some recent excellent reviews provide descriptions of the structure and multi-functional role of the pectin network and examine some of these characteristics (Pérez *et al.*, 2000; Ridley, *et al.*, 2001; Willats *et al.*, 2001a).

In addition to the direct effect on structure and associated characteristics, pectin methylesterification also has significance in terms of facilitating hydrolysis of the pectin backbone by PG and pectate lyases. Highly methylated polyuronides represent poor substrates for PG (Rexová-Benkova and Markovič, 1976; Yoshioka *et al.*, 1992) and so pectin demethylation is an important prelude to PG-mediated polyuronide depolymerization and solubilization, and represents another potential mechanism for regulating the activity of PG (Koch and Nevins, 1989; Wakabayashi, 2000; Wakabayashi *et al.*, 2000) and perhaps other pectinases.

Pectin demethylation is catalysed by pectin methylesterases (PME), which are also referred to in the literature as pectin esterases (PEs). PMEs are expressed ubiquitously in plants and sizeable PME gene families have been detected in several species (Micheli, 2001). Remarkably, the *Arabidopsis* PME gene family is predicted to have a up to 79 members (Arabidopsis Genome Initiative, 2000), and although the expression of only a small proportion of these genes has been reported (Micheli *et al.*, 1998), it is clear that PMEs are associated with many developmental events (reviewed in Micheli, 2001). One of the most widely proposed roles for PME is the demethylation of homogalacturonan in ripening fruit, and ripening-related PME activity and gene expression have been described in a range of fruits (Fischer and Bennett, 1991; Wegrzyn and MacRae, 1992; Glover and Brady, 1994; Nairn *et al.*, 1998). However, as is the case with PG, the most extensive studies of PME activity, gene and protein expression and function have been in tomato.

Tomato PME activity is detectable throughout fruit development, peaking at an early ripening stage (Tucker *et al.*, 1982; Fischer and Bennett, 1991), and corresponds to the cumulative actions of at least three fruit-specific PME isozymes (Gaffe *et al.*, 1994). In contrast to the large number of predicted *Arabidopsis* PME genes, only a few have been cloned from tomato (listed in Brummell and Harpster, 2001). To gain insight into the contribution of PME action to fruit softening, two separate research groups used an antisense approach to down-regulate the expression of the tomato *PME2* gene, which encodes the major fruit-specific PME isoform (Tieman *et al.*, 1992; Hall *et al.*, 1993; see Table 9.1). The transgenic fruits softened at the same rate as wild-type fruit during ripening although they showed substantially more tissue disintegration during postharvest storage than untransformed tomatoes (Tieman and Handa, 1994). Cell wall analysis revealed an increase in the degree of methylesterification of the pectin from the transgenic fruit, and Tieman *et al.* (1992) reported both a reduction in ripening-related pectin depolymerization and a decrease in the amount of chelator-soluble polyuronides. The reduction in pectin depolymerization probably reflects the reduced susceptibility of the more methylesterified pectins to hydrolysis by PG, while the loss of tissue integrity during storage has been attributed, at least in part, to a reduction in the cation-binding potential of the walls of the transgenic fruit, and specifically to the amount of bound calcium (Tieman and Handa, 1994). Loss of wall-bound calcium has previously been proposed to enhance PG activity (Buescher and Hobson, 1982).

Based on the studies in tomato, it can be concluded that PME is not solely responsible for fruit softening, but is likely to play an important role in influencing pectin disassembly through enhancing the susceptibility of pectins to attack by other pectinases, altering the charge density on demethylesterified pectic polymers and changing the ionic and pH conditions in the apoplast. These factors are superimposed on many other variables that need to be considered when evaluating the importance and regulation of pectin metabolism in wall disassembly (see also section 9.2.1.7). To conclude, although considerable progress has been made in understanding the biochemical modes of action of plant PMEs (Micheli, 2001; Willats *et al.*, 2001b), and the first three-dimensional structure of a plant PME has recently been reported (Johansson *et al.*, 2002), there are many details of plant PME function and the significance of PME action that remain unexplored.

Even less is known about another class of pectin esterases from plants: pectin acetylesterases (PAEs). Pectins from some sources, such as sugar beet, are known to be highly acetylated at the C-2 and/or C-3 position of the galacturonosyl residues and the presence of the acetyl groups has a significant effect on the gelation properties of the pectin and the action of PMEs (Oosterveld *et al.*, 2000). Thus, pectin deacetylation could represent an important component of pectin metabolism. Only a few plant PAEs have been experimentally characterized (Breton *et al.*, 1996; Stratilova *et al.*, 1998; Vercauteren *et al.*, 2002), although the *Arabidopsis thaliana* genome contains 13 putative PAE gene family members. The potential role of PAEs in fruit softening has not yet been addressed, but PAE-like sequences are expressed in grape berries (Terrier *et al.*, 2001) and are present in EST collections from ripening tomato and citrus (J. Rose, unpublished data), suggesting that PAE expression may be a common feature of ripening in many fruit species.

9.2.1.3   *Pectin depolymerization and pectate lyases*

Another class of plant pectinases, and one that has been less studied that PGs and PMEs in the context of fruit ripening or other developmental processes, are pectate lyases (PLs), or pectate transeliminases. These enzymes catalyse the cleavage of unesterified α-1→4-galacturonosyl linkages by a β-elimination reaction, in contrast with the hydrolytic mechanism of PGs. Microbial PLs were first identified 40 years ago and have been studied in some detail in terms of their activity, expression, structure and role in plant wall degradation during pathogenesis (Collmer and Keen, 1986; Annis and Goodwin, 1997; Jedrzejas, 2000; Hoondal et al., 2002). In contrast, relatively few examples of plant PLs were identified prior to the sequencing of the *Arabidopsis* genome, which was recently reported to encode a family of 27 PLs, or PL-like proteins (Marín-Rodríguez, et al., 2002). To date, plant PLs have been associated with a number of developmental events that involve extensive wall disassembly, including pollen tube penetration of stylar tissue, tracheary element differentiation, laticifer growth and fruit ripening (reviewed in Marín-Rodríguez, et al., 2002).

Ripening-related PL genes have been identified in banana (Dominguez-Puigjaner et al., 1997; Pua et al., 2001), grape (Nunan et al., 2001; Ishimaru and Kobayashi, 2002) and strawberry (Medina-Escobar et al., 1997; Jiménez-Bermúdez et al., 2002) and although PL activity was reported to be absent in tomato fruit (Besford and Hobson, 1972), PL-like ESTs have been identified in several tomato tissues, and appear to be expressed at relatively high levels in ripening fruit (Marín-Rodríguez, et al., 2002; http://www.tigr.org/tdb/tgi/lgi/index.html). To date, the contribution of PLs to pectin degradation and fruit softening remains relatively unexplored; however, suppression of a strawberry pectate lyase gene resulted in significantly firmer fruit and a reduction in both ripening-related wall swelling *in vitro* and the amount of chelator-soluble pectins (Jiménez-Bermúdez et al., 2002). Thus the role of PLs in fruit ripening represents an important question and will no doubt be the subject of considerable study in the near future.

9.2.1.4   Pectin side chain modification: galactanases/β-galactosidases and arabinosidases

Another factor that potentially contributes to pectin mobilization during ripening is the loss of arabinan and galactan side chains from RG-I and, accordingly, a decline in the amount of neutral sugars is a common feature of cell wall composition in most ripening fruits. The loss of galactose is particularly prevalent, with decreases of up to 70% during ripening (Gross and Sams, 1984; Redgwell et al., 1997a), and studies in several fruits suggest that, while galactose content decreases in several wall fractions during ripening, it is lost predominantly from pectic polysaccharides (e.g. Gross, 1984; Seymour et al., 1990; Rose et al., 1998). However, it has also been suggested that loss of pectin neutral sugar side chains has no significant effect on the physiochemical properties of the wall and there appears not to be a consistent correlation between galactose loss and extent or type of softening (Redgwell and Percy, 1992; Redgwell and Fischer, 2001).

To date, endo-β-1,4-galactanase activity has not been reported in fruit tissue, while the enzyme β-galactosidase (for a discussion of nomenclature see Smith and Gross, 2000), which acts as an exo-hydrolase, removing terminal non-reducing β-D-galactosyl residues, has been well documented in many fruit species (e.g. Pressey, 1983; Ranwala *et al.*, 1992; Lazan *et al.*, 1995; Barnavon *et al.*, 2000; Tateishi *et al.*, 2001). β-galactosidases are involved in multiple biochemical pathways and can act on a spectrum of galactose-containing substrates in addition to wall polysaccharides, including galactolipids and glycoproteins. Not surprisingly then, β-galactosidases are present as gene families and the *Arabidopsis* genome is predicted to have 18 putative β-galactosidases (as of November 2002 at http://www.arabidopsis.org). Although few of these have been studied experimentally, it is likely that they exhibit differences in substrate specificity, and spatial and temporal expression patterns. The recent identification of two genes from strawberry fruits that appear to encode β-galactosidases with lectin-like domains (Trainotti *et al.*, 2001), which act as carbohydrate-binding modules in animals, also suggests functional divergence among members of this gene family. It should be noted however, that while the recombinant strawberry proteins exhibited β-galactosidases activity against an artificial substrate, no activity against a wall polysaccharide was reported (Trainotti *et al.*, 2001). β-galactosidases from several fruits have been shown to release galactose from pectins, although activity is typically also seen against other galactose-containing hemicellulosic polymers (Ranwala *et al.*, 1992; Ross *et al.*, 1993, 1994; Carey *et al.*, 1995; Kitagawa *et al.*, 1995).

At least seven β-galactosidase genes are expressed in tomato at different stages of fruit development (Smith and Gross, 2000). Similarly, studies of several of the corresponding isozymes show a range of expression patterns, with some present at high levels in green fruits and others detectable only during ripening (Pressey, 1983; Carey *et al.*, 1995; Carrington and Pressey, 1996). The net result is that total β-galactosidase activity remains at high levels throughout tomato fruit development and does not change substantially during ripening (Wallner and Walker, 1975; Pharr *et al.*, 1976; Carey *et al.*, 1995), despite considerable fluctuations in the abundance of individual isozymes (reviewed in Brummell and Harpster, 2001).

The potential contributions of three tomato β-galactosidases to fruit softening and wall disassembly has been assessed in transgenic plants (Table 9.1). Reduction of the expression of TBG1 mRNA to 10% of wild-type levels had no observed effect on softening, β-galactosidase activity or wall galactose content, despite the demonstration that the corresponding recombinant protein could release galactose from tomato wall galactans *in vitro* (Carey *et al.*, 2001). In contrast, antisense inhibition of *TBG3* substantially reduced exo-galactanase activity and increased wall galactose content, although no effect was observed on the rate or extent of softening, other than under postharvest storage conditions (de Silva and Verhoeyen, 1998). More recently, Smith *et al.* (2002) reported the antisense suppression of *TBG4*, which resulted in a substantial reduction of exo-galactanase activity when assayed against lupin galactan, but had no significant effect on β-galactosidase activity when measured using an artificial glycoside. Levels of free galactose in pre-ripe fruit were approximately half those in wild-type tomatoes, but no difference was seen in ripe fruits and no effects were seen on wall galactosyl content. Interestingly, the antisense tomatoes were up to 40% firmer than wild-type fruits at the ripe stage, confirming a role for *TBG4* in softening, although the *in vivo* substrate has still not been determined (Smith *et al.*, 2002).

The importance of pectin galactan hydrolysis for fruit softening is still unclear. Smith *et al.* (2002) proposed that the galactans might regulate wall porosity and influence the activities of other wall modifying enzymes. Interestingly, it has also been shown that exogenous galactose application can induce ripening in immature fruits (Gross, 1985; Kim *et al.*, 1987) by inducing the activity of ACC synthase, which in turn regulates ethylene biosynthesis.

In addition to galactans, the backbone of rhamnogalacturonan I is also substituted with arabinans (see Chapter 1); however, while most fruit walls show a loss of arabinose during ripening (Gross and Wallner, 1979; Gross and Sams, 1984), compared with galactan metabolism and the action of galactanases/β-galactosidases, relatively little has been reported about the degradation of arabinans in fruit, or ripening-related arabinarase/α-*L*-arabinofuranosidases activities. The α-*L*-arabinofuranosidase family of enzymes are exo-acting hydrolases that remove terminal non-reducing arabinosyl residues from a range of arabinose-containing pectic and hemicellulosic polysaccharides and other glycoconjugates, as described in a recent comprehensive review (Saha, 2000). Much of the existing research to date has focused on the activities, regulation and properties of microbial arabinofuranosidases, since they have potential industrial value in the bioconversion of hemicelluloses, such as arabinoxylans, to fermentable sugars. However, several plant arabinofuranosidases have been isolated from leaves, seeds and cell cultures (Saha, 2000) and fruit ripening-related increases in arabinofuranosidase activity have been reported in Japanese pear (Tateishi *et al.*, 1996), carambola (Chin *et al.*, 1999) and tomato (Sozzi *et al.*, 2002a,b). Three tomato arabinofuranosidase isozymes have recently been identified in fruit extracts and shown to exhibit differential activity against tomato hemicelluloses and carbonate-soluble pectins (Sozzi *et al.*, 2002b), but the corresponding genes have not yet been reported and their contributions to fruit softening have yet to be explored in transgenic fruits.

9.2.1.5 *Rhamnogalacturonase*

Rhamnogalacturonase A (RGase A), a hydrolase that cleaves galacturonosyl-1,2-rhamnosyl glycosidic bonds, that was first identified in *Aspergillus aculeatus* (Schols *et al.*, 1990), has been proposed as another enzyme that may act in concert with PG and other pectinases to degrade the pectin network (Redgwell and Fischer, 2001). RGase A has been detected in apple, grape and tomato fruits (Gross *et al.*, 1995), but no plant RGase protein or gene sequence has been published and the *Arabidopsis* genes that appear to be most closely related to the *Aspergillus* RGase A are annotated as putative PGs in the database (J. Rose, unpublished data).

9.2.1.6 *Regulation of pectin disassembly in ripening fruit*

An important observation that has emerged from several studies in tomato is that pectin depolymerization *in vivo* does not occur to the same extent as can occur *in vitro*, where extensive pectin degradation occurs within a few hours in tomato fruit extracts (Seymour *et al.*, 1987b; Huber and O'Donoghue, 1993; Brummell and La-bavitch, 1997). This implies that pectin breakdown is substantially limited *in muro*, and a number of mechanisms have been proposed that may restrict the activity and/or mobility of pectinases in the walls and middle lamella *in vivo*.

Firstly, the pH and ionic conditions in the apoplast may play a critical role in regulating pectinase activity, either by affecting the enzyme directly or by altering the wall porosity. Almeida and Huber (1999) demonstrated that in ripening tomato fruit the apoplastic pH drops from greater than 6.7 in pre-ripe fruit to 4.4 during ripening. Such a change might provide a critical regulatory mechanism for the activation of cell wall modifying proteins, as Chun and Huber (1998) reported that tomato PG2 only appears to catalyse pectin depolymerization at pH values of less than 6.0. Consequently, a picture emerges in which endo-PG activity may be severely limited by the fruit apoplastic pH (Huber *et al.*, 2001), and since it has also been shown that PME action can also be strongly pH-dependent (Denès *et al.*, 2000), it is likely that other pectinases show similar pH-related control.

The concentrations of apoplastic ions, such as potassium and calcium, may also be critical factors. Calcium is believed to modulate the structure and properties of the pectin network (Jarvis, 1984; Carpita and Gibeaut, 1993; Pérez *et al.*, 2000; Ryden *et al.*, 2000) and can also accumulate in the apoplast to levels that strongly inhibit the autolytic release of pectins from cell wall extracts (Rushing and Huber, 1987; Almedia and Huber, 1999). Furthermore, the apoplastic calcium and potassium concentrations have been shown to alter PG action: at low pH, potassium promotes PG-mediated pectin depolymerization while calcium substantially suppresses PG activity (Chun and Huber, 1998). Salts may also play a role in promoting the non-covalent association of pectin macromolecular complexes (Fishman *et al.*, 1989).

A second factor that may be of crucial importance in determining the contribution that ripening-related pectin metabolism has to changes in fruit texture is the spatial distribution and frequency of sites of pectin disassembly within each cell wall. The observations that there is variability in wall structure and/or composition at specific locations within the fruit wall (Orfila and Knox, 2000), and that PG and PME expression can occur in distinct microdomains, with colocalized change in pectin structure (Steele *et al.*, 1997; Morvan *et al.*, 1998) suggest that wall disassembly may involve a far greater degree of spatial control than has previously been suspected. Steele *et al.* (1997) suggested that compartmentation of pectinases such as PME within distinct domains in the wall contributes to pectin modification being far more restricted *in vivo* than is observed once walls and pectinases are co-extracted and incubated *in vitro*. Wall microheterogeneity will become easier to resolve with continued advances in wall immunocytochemistry (see Chapter 3).

Thirdly, the identification of multiple classes of pectin modifying enzymes raises the obvious potential for synergistic activity among pectinases. It is well established that PME acting to demethylesterify pectins can enhance the subsequent activity of PG, and the same may apply for pectate lyases and other pectolytic enzymes. It is possible to imagine many similar such scenarios in which the actions of specific classes of pectinases facilitate the activities of others. Pectin disassembly may therefore be

viewed as a consequence of the net activity of numerous enzymes that act coopera-
tively to alter, and are themselves affected by, the structure and physicochemical
properties of the pectic network. Furthermore, this 'system' is located within, and
profoundly affected by, a complex and changing apoplastic environment with substantial pH and ionic gradients in time and/or space.

It also appears that some of the fruit textural disorders, such as mealiness/
wooliness that occur during postharvest treatments or prolonged storage, may reflect an abnormal expression of pectinases, particularly PG and PME, and altered
apoplastic conditions (Dawson *et al.*, 1992; Zhou *et al.*, 2000a,b).

The complexity and multiplicity of factors that affect pectin structure and properties in softening fruit make experimental elucidation extremely challenging. Studies
to date of single genes and proteins, such as PG and PME, have provided valuable
insights into specific aspects of pectin turnover, but at the same time have revealed
the importance of a broader perspective. The application of genome-scale analysis
(Chapter 10) or the characterization of pleiotropic mutants with altered pectin biochemistry and metabolism are two additional approaches that should prove useful.
For example, a tomato mutant *Cnr* (*colourless non-ripening*), that shows reduced
cell-cell adhesion and wall swelling in ripening fruit, has abnormal deposition of
$(1\rightarrow5)$-α-arabinan in the pericarp at the pre-ripe stage (Orfila *et al.*, 2001). The
mutant also has a reduced calcium-binding capacity, possibly as a consequence of
suppressed homogalacturonan (HGA) de-esterification, and lower levels of expression of both PG (Thompson *et al.*, 1999) and an isoform of PME (cited in Orfila *et
al.*, 2001, 2002). Further analysis of cell walls from *Cnr* suggests other changes, such
as the extractability of HGA and other pectin fractions, and increased susceptibility
of HGA to PG-mediated hydrolysis (Orfila *et al.*, 2002). The genetic basis of this
mutation is not yet known, but pleiotropic mutants such as *Cnr* should prove to be
very helpful in understanding the structural relationship between different components of the pectin network, and the biochemical mechanism and regulation of its
assembly and disassembly.

9.2.2 *Cellulose and cellulose-interacting proteins*

Cellulose microfibrils are aggregates of β-$(1\rightarrow4)$-glucan chains that are aligned
in parallel and that form strong self-associations; a consequence of inter- and intra-
chain bonding that is predicted to result principally from van der Waals forces, with
hydrogen bonding contributing additional cohesive energy (Cousins and Brown,
1995). The packing of the β-$(1\rightarrow4)$-glucan chains is such that crystalline structures
are formed, resulting in increased strength, resistance to hydrolysis and insolubility.
Many aspects of microfibril structure and synthesis have been extremely difficult
to resolve, and while these are described only briefly here, several recent publica-
tions provide extensive background information (O'Sullivan, 1997; Brown, 1999;
Delmer, 1999; Williamson, 2001; Zugenmaier, 2001; Viëtor *et al.*, 2002).

Historically, there has been considerable debate about whether cellulose microfibrils are composed of single or multiple crystals (crystallites). It is generally agreed

that large highly crystalline algal cellulose fibrils are comprised of single crystalline units (Revol, 1982; Koyama *et al.*, 1997); however, there is still no such consensus for plant primary cell wall microfibrils and recent studies have reached differing conclusions, proposing multiple subunits (Ha *et al.*, 1998) or single crystallites (Davies and Harris, 2003). The situation is further complicated by variation in microfibril and crystalline diameter between species.

One of the least understood aspects of cellulose molecular ordering and microfibril organization is the relative amounts and distribution of ordered crystalline and non-crystalline (also termed 'paracrystalline', or 'amorphous') cellulose and the consequences of the loss of crystallinity for microfibril properties. Early studies using X-ray diffraction suggested that more than 50% of primary wall cellulose is non-crystalline (Frey-Wyssling, 1954), whereas more recent reports based on nuclear magnetic resonance (NMR) spectroscopy indicate that cellulose from primary walls of several monocotyledons and dicotyledons is essentially all crystalline (Newman *et al.*, 1996; Smith *et al.*, 1998). The small crystallite size of higher plant cellulose means that up to two thirds of the glucan chains may be on the crystallite surface and many of these are thought to be disordered. The surface chains show a different conformation from the interior chains, and one that would promote interactions with external molecules (Viëtor *et al.*, 2002), and so there is considerable interest in determining the organization, formation and interactions of the surface chains with other cell wall polymers. This is experimentally problematical to evaluate since plant walls cannot yet be synthesized *in vitro* and techniques to isolate cellulose and cellulose-associated matrix glycans are likely to disrupt and alter many of the interactions that are present *in vivo*.

However, a number of studies of cellulose-matrix glycan interactions have been made using artificial composites that are generated by growing cultures of the cellulose-synthesizing bacterium *Acetobacter xylinum* in the presence of different plant cell wall polysaccharides (Whitney *et al.*, 1995, 1998, 1999; Chanliaud and Gidley, 1999; Hackney *et al.*, 1999; Astley *et al.*, 2001; Tokoh *et al.*, 2002). These composites are similar in many ways to the equivalent structure in plant primary walls and studies indicate that assembly of the microfibrils in the presence of hemicelluloses, such as xyloglucan, xylan and mannan, results in a reduction in microfibril crystallinity, cellulose isoform composition and tensile properties. The interaction between cellulose and the hemicelluloses is likely a combination of the adsorption of the hemicelluloses onto the surface of the crystallites and their entrapment within the microfibril structure during synthesis.

The difficulties in evaluating the structural organization and heterogeneity of cellulose microfibrils, and their interactions with other wall components, ensure that it is equally difficult to determine whether cellulose undergoes modification and degradation during processes such as fruit ripening. Cellulose is typically extracted from plant cell walls using solvents such as high concentrations of alkali that first remove associated matrix glycans (Brummell and Harpster, 2001); however, these are conditions under which hydrogen bonds would be disrupted and disordered surface glucans might also be solubilized. Thus, while studies to date often, but not always, report that the cellulose that is solubilized from fruits shows little change in total amount (Gross and Wallner, 1979; Gross *et al.* 1986; Kojima *et al.*, 1994; Nunan *et al.*, 1998) or molecular mass (Maclachlan and Brady, 1994) during ripening, it should be noted that this strategy specifically evaluates crystalline cellulose. Significant and subtle disassembly of the cellulose-hemicellulose framework and degradation of amorphous cellulose would probably be overlooked.

Less destructive evaluations of the molecular ordering and polymer mobility in ripening fruit using ^{13}C solid state NMR spectroscopy suggest that there may be considerable variability in the extent of cellulose metabolism in different species. For example, while analyses of cell walls from ripening strawberry (Koh *et al.*, 1997) and kiwifruit (Newman and Redgwell, 2002) indicated that the degree of cellulose crystallinity does not change during ripening, Stewart *et al.* (2001) reported extensive cellulose degradation and loss of crystallinity in raspberry. Studies of cellulose in ripening avocado using X-ray diffraction suggested an increase in the degree of crystallinity, which was interpreted to reflect a loss of amorphous cellulose (O'Donoghue *et al.*, 1994), although similar evaluations in peach revealed no change (Sterling, 1961).

Ultrastructural studies of fruit cell walls also suggest differences in the degree of cellulose microfibril disruption between species. The microfibrillar component in the walls of avocado (Platt-Aloia *et al.*, 1980), apple and pear (Ben Arie *et al.*, 1979) fruits undergoes substantial ripening-related disruption, while immunolocalization studies of kiwifruit suggested that cellulose remained stable during ripening (Sutherland *et al.*, 1999).

Given the current limited knowledge of microfibril architecture, it is difficult at this point to come to any firm conclusions regarding the extent and nature of cellulose microfibril degradation during ripening. In particular, the dynamics of the interaction between the glucan chains on the microfibril surface and associating hemicelluloses is likely to be an important factor in determining the structural properties of the hemicellulose-cellulose framework, but one that is hard to elucidate. This is likely to remain the case until additional analytical tools and experimental approaches become available. The identification of enzymes and cellulose-interacting proteins that affect inter- and intramicrofibrillar associations, or that alter cellulose-hemicellulose interactions, may be one means of better understanding cellulose metabolism *in vivo*.

9.2.2.1 C_x cellulases/Endo-β-1,4-glucanases

Endo-1,4-β-D-glucanases (EGases) comprise a class of enzymes that can hydrolyse the 1,4-β-D linkages between two unsubstituted glucose moieties, and so potential substrates in plant cell walls include cellulose, xyloglucan and glucomannan. EGases are also referred to as 'cellulases' since many, such as those that are secreted by microbes, can hydrolyse crystalline cellulose (Levy *et al.*, 2002). However, some EGases, including most EGases from plants that have been studied to date, appear to have little activity against crystalline cellulose, but are instead active against the artificial soluble cellulose derivative carboxymethylcellulose (C_x cellulose, or CMC), and so these are also termed 'C_x cellulases' or 'CMCases'.

CMCase activity has long been associated with processes involving wall disassembly and typically increases during fruit ripening and softening (Huber, 1983; Fischer and Bennett, 1991; Brummell et al., 1994; Redgwell and Fischer, 2001), although there are dramatic differences in activity levels between extracts from various fruit species. For example, CMCase activity in ripe avocado fruit (Awad and Young, 1979; Christoffersen et al., 1984; Tucker and Laties, 1984) is more than 700 times great than that reported for tomato (cited in Brummell and Harpster, 2001).

Early studies with purified CMCases associated with avocado fruit ripening and bean organ abscission lead to the cloning of the first plant 'CMCase/endo-1,4-β-glucanase/cellulase' genes, at which point it was revealed that the plant CMCases, that had been described for some time in the literature as endo-1,4-β-glucanases (EGases), did indeed share substantial sequence similarity to the catalytic domain of EGases from cellulolytic bacteria (Brummell et al., 1994; del Campillo, 1999). Recent classification of glycosyl hydrolases into 12 families based on sequence similarity, has now placed all plant EGases within family 9, together with a subset of homologous microbial EGases (Henrissat and Bairoch, 1993, 1996).

As with the various classes of pectinases described above, plant EGases are encoded by large multigene families, such as the 27 members that are predicted in Arabidopsis (Henrissat et al., 2001) and not surprisingly, there are numerous examples of differential patterns of EGase gene expression and regulation in various plant species. Ripening-related EGase genes have been described in avocado (Cass et al., 1990), peach (Bonghi et al., 1998), strawberry (Llop-Tous et al., 1999; Trainotti et al., 1999a,b), pepper (Harpster et al., 1997) and tomato (Lashbrook et al., 1994; Kalaitzis et al., 1999), and EGase expression has now been suppressed in transgenic fruits from three different species. The first example was in tomato, where Cel1 and Cel2, which are expressed in early or late ripening stages, respectively, were down-regulated using an antisense approach (Lashbrook et al., 1998; Brummell et al., 1999a; Table 9.1). In both cases, no effect was seen on fruit softening and no cell wall analysis was described, although floral abscission-related phenotypes were reported (see section 9.3.2). A ripening-related EGase gene from strawberry, also named Cel1, was strongly down-regulated resulting in some transgenic lines having <0.5% of the wild-type levels of Cel1 mRNA; however, as with tomato, no significant effect on fruit firmness was detectable (Woolley et al., 2001). More recently still, the function of a ripening-related EGase from pepper (Capsicum annum), CaCel1, was investigated using two complementary approaches: down-regulating expression of the native CaCel1 gene in transgenic pepper, and over-expressing the same gene in transgenic tomato fruit (Harpster et al., 2002a,b). In both cases, the presence of the transgene did not influence softening and the cell wall composition of the suppressed pepper lines was indistinguishable from that of the wild-type fruit. The only observable difference from either set of transgenic plants was a small decrease in the amount of an unidentified matrix polysaccharide, that was bound tightly to cellulose, in the tomato fruit that had up to 20-fold higher levels of CMCase activity following overexpression of CaCel1 (Harpster et al., 2002b). Collectively, these experiments suggest that EGases alone do not make a substantial contribution to wall disassembly in many ripening fruit, although it would be interesting to examine the effects of suppressing EGases in avocado fruit where CMCase activity is exceptionally high, and where EGases may play a more significant role in fruit textural changes.

Even though plant CMCases/EGases have been studied for more than thirty years, remarkably, their *in vivo* substrate(s) still remains to be identified. A number of EGase proteins have been isolated following extraction from different plant tissues (Wong *et al.*, 1977; Maclachlan and Wong, 1979; Hayashi *et al.*, 1984; Hatfield and Nevins, 1986; Durbin and Lewis, 1988; Maclachlan, 1988; Truelsen and Wyndaele, 1991; O'Donohue and Huber, 1992; Nakamura and Hayashi, 1993; Sanwal, 1999; Ohmiya *et al.*, 2000; Woolley *et al.*, 2001; Harpster *et al.*, 2002a); however, assays of their hydrolytic activities against a range of glucan substrates have failed to reveal a common pattern. For example, a purified avocado fruit EGase hydrolysed $(1 \rightarrow 3),(1 \rightarrow 4)\beta$-glucans and showed no activity against native cellulose or xyloglucan (Hatfield and Nevins, 1986; O'Donohue and Huber, 1992), while conversely, a strawberry fruit EGase showed relatively high activity against xyloglucan, but minimal hydrolysis of $(1 \rightarrow 3),(1 \rightarrow 4)\beta$-glucan (Woolley *et al.*, 2001). Similarly, the most consistently observed feature of EGase activities appears to be a lack of activity against crystalline cellulose, although exceptions have been reported, such as a cellulase from *Catharanthus roseus* stems (Sanwal, 1999).

Given that EGases typically have a relatively high degree of sequence similarity, it is perhaps surprising that there is such divergence in their apparent enzymatic activities. It should be noted however that in many cases only a few substrates were tested, assay conditions varied substantially and the activities were rarely assayed using native substrates. Another important consideration is that some potential substrates such as glucomannans, appear never to have been tested, and the EGases may also act on a substrate that is difficult to isolate. One candidate is the disordered β-1,4-glucan chains in the paracrystalline regions of the cellulose microfibrils, and indeed the *in vitro* assays generally suggest that EGases act preferentially on unsubstituted linear glucans. For example, several EGases have been reported to show relatively high hydrolytic activity against linear $(1 \rightarrow 3),(1 \rightarrow 4)\beta$-glucans, which have consecutive $(1 \rightarrow 4)\beta$-linkages. However, these mixed linkage glucans are not generally considered to be major components of type I walls, which would represent the substrate for all the EGase proteins isolated to date. Studies with a tobacco EGase also indicated that activity was higher against native xyloglucan, which is predicted to have long contiguous stretches of unsubstituted glucan backbone, than against amyloid xyloglucan (Truelsen and Wyndaele, 1991). On the other hand, some EGases show similar activities against glucans with different degrees of side-chain substitution. Interestingly, Harpster *et al.* (2002b) reported that a pepper fruit EGase appears to act on an unidentified non-xyloglucan matrix glycan that was extracted in a wall fraction that was tightly bound to cellulose, and non-crystalline cellulose might be extracted under such conditions. Lastly, O'Donoghue *et al.* (1994) showed that incubation of avocado EGase with unripe fruit tissue caused a decrease in the proportion of non-crystalline cellulose and ultimately a loss of microfibril cohesiveness. They further proposed that a major role for avocado EGase is to degrade the peripheral paracrystalline regions of cellulose microfibrils.

It is therefore tempting to speculate that EGases degrade the non-crystalline glucan chains on the microfibril periphery, thus influencing microfibril organization and the integrity of the cellulose-hemicellulose framework. However, there has been insufficient detailed study to draw any conclusions and it might well be that divergent EGases from different sources act on different substrates, or multiple substrates. To complicate matters further, structurally divergent EGases have recently been identified, including a class of membrane-bound EGases (Reiter, 2002) and EGases that have a putative cellulose-binding domain (CBD) (Catalá and Bennett, 1998; Trainotti et al., 1999b). The CBDs are commonly found in microbial EGases where they facilitate high affinity binding to crystalline cellulose (Levy et al., 2002) and the identification of the first examples of EGases with putative CBDs in ripening strawberry and tomato fruit is intriguing. The role, substrate specificity and diversity of plant EGases and their importance in wall disassembly are questions that have been raised in the literature for some time, and yet no approach has emerged that clearly promises to provide new insight into these enzymes. It may be that some of the emerging techniques and tools for wall analysis, such as those described in Chapters 2 and 3, when used in conjunction with transgenic plants with suppressed or upregulated EGase expression, will provide the necessary resolution to detect some of the more subtle changes in wall architecture that may reflect the substrates of plant EGases in vivo.

9.2.2.2   Expansins

Expansins are proteins that promote wall loosening but which appear to lack hydrolytic activity (see Chapter 8; Cosgrove, 2000a,b). The mechanism of expansin action has still not been resolved and, as with EGases, their site(s) of action in muro are obscure. However, several pieces of evidence suggest that they weaken glucan-glucan interactions by disrupting hydrogen bonds and that they act in vivo at the cellulose-hemicellulose interface (McQueen-Mason and Cosgrove, 1994, 1995; Cosgrove, 2000a,b). Moreover, Whitney et al. (2000) examined the activity of expansin on a variety of Acetobacter xylinum cellulose-hemicellulose composites (see section 9.2.2) and saw the greatest effect on the tensile properties of a cellulose-xyloglucan matrix. The exact nature of the glucan interaction that is affected is unresolved, but given the apparent association between expansins and the cellulose microfibril environment, it seems appropriate to include expansins in this section of cellulose-interacting proteins.

As the name suggests, expansins were first proposed as proteins that function to promote wall stress relaxation during turgor-driven expansion. However, Rose et al. (1997) identified an expansin gene, LeExp1, that is specifically and abundantly expressed at the onset of ripening in tomato fruit, in addition to ripening-related orthologs in melon and strawberry. Cell expansion does not occur in tomato fruit during ripening and turgor pressure declines (Shackel et al., 1991), which suggests that expansins may also contribute to wall disassembly in the absence of cell expansion. Subsequently, ripening-related expansins have been identified in a number of fruits including strawberry (Civello et al., 1999; Harrison et al., 2001), apricot (Mbeguie-A-Mbeguie et al., 2002), peach (Hayama et al., 2000) and pear (Hiwasa et al., 2003).

Expression of LeExp1 mRNA and protein is closely correlated with the degree of softening in wild-type tomato and non-softening tomato mutants (Rose et al., 1997, 2000a), suggesting that expansins play a role in wall disassembly in the absence of cell expansion, thus influencing fruit softening. This hypothesis was tested (Table 9.1) in transgenic tomato plants that had either suppressed or elevated levels of LeExp1 (Brummell et al., 1999b). In the suppressed lines, fruit softening still occurred, although a small increase in firmness was observed, particularly in the late ripening stage when fruits had substantially softened. Conversely, in the LeExp1 over-expressing lines, fruits softened more rapidly, particularly at the early ripening stages. As with the PG antisense plants (see section 9.2.1.1), a change in the postharvest and processing characteristics of the LeExp1-suppressed fruits, such as paste viscosity, was noted (Brummell et al., 2002). Cell wall analysis of the LeExp1-suppressed fruits showed reduced depolymerization of CDTA-soluble pectins compared with control fruits, but no change in the molecular size of other wall polymers such as xyloglucans. The only other detected change in wall composition in either set of transgenic fruits was an increased depolymerization of xyloglucan in the LeExp1 over-expressing fruits. These data further support the idea that expansins may alter the interaction between xyloglucan and cellulose microfibrils, either directly or indirectly. Brummell et al. (1999b) suggested that expansins may also influence the accessibility of pectinases to their substrates in the later ripening stages; however, as pointed out by Redgwell and Fischer (2001), this seems unlikely, given that the walls already exhibit substantial swelling and dissolution by this time and so pectins should be readily accessible. Expansins may indeed be involved in mediating ripening-related wall swelling, although the amount of extractable expansin protein from different fruits (Rose et al., 2000a) appears not to correlate with the extent of wall swelling observed in vitro and in vivo (Redgwell et al., 1997b). For example, pear does not undergo wall swelling and yet has the greatest relative amount of detectable expansin protein of any fruits examined to date (Rose et al., 2000a; Rose, unpublished data). It is of course possible that the relative amounts of immunoreactive expansin do not reflect the extent or significance of expansin action in vivo although, of the fruit extracts tested, pear also had the highest expansin activity (Rose et al., 2000a).

The cumulative studies to date suggest that expansins are involved in cell wall disassembly during fruit ripening and influence changes in fruit texture although, as with several other transgenic tomato lines (Table 9.1), in the absence of a complete gene knockout, their quantitative contribution to softening cannot be determined. The low levels of residual LeExp1 (approximately 3%; Brummell et al., 1999b) may be sufficient to mediate a significant proportion of normal expansin action. Important future goals will clearly be to elucidate the site of action of expansins, to evaluate better their effects on wall architecture, and particularly different domains of the cellulose-hemicellulose framework (Pauly *et al.*, 1999; Whitney *et al.*, 1999), and to address their possible synergistic role in the context of other cell wall modifying enzymes such as EGases and hemicellulases (Rose and Bennett, 1999; Rose *et al.*, 2000a).

9.2.3 Hemicelluloses and hemicellulases

The primary walls of fruits from dicotyledons contain three basic classes of hemicelluloses, whose relative abundance varies between species: xyloglucans, glucuronoarabinoxylans and related polymers (xylans and arabinoxylans), and ga-lactoglucomannans and related polymers (mannans and glucomannans). Although the absolute amount of hemicellulose in the wall typically shows less of a ripen-ing-related change than is observed for pectins, hemicellulose depolymerization is a common feature of fruit softening, and is believed to contribute significantly to changes in texture (Huber, 1984; Gross *et al.*, 1986; Tong and Gross, 1988; Fischer and Bennett, 1991; Maclachlan and Brady, 1994; Rose *et al.*, 1998; Brummell and Harpster, 2001).

9.2.3.1 Xyloglucan and xyloglucanases

Xyloglucan is frequently ascribed a critical structural role in type I walls (see Chapter 1) since current models describe it as enmeshed with, and hydrogen bonded to, the cel-lulose microfibril surface, with individual xyloglucans tethering adjacent microfibrils (Pauly *et al.*, 1999; Rose and Bennett, 1999). The resulting cellulose-xyloglucan net-work is thought to be the major load-bearing structure in type I walls, where its physi-cal properties both provide strength and allow for controlled extensibility (Whitney *et al.*, 1999). The xyloglucans may be divided into at least three structural domains: xy-loglucan cross-links that are susceptible to enzymatic cleavage, a xyloglucan fraction that is tightly bound to the microfibril surface and a third component that is trapped within the microfibril periphery (Pauly *et al.*, 1999). This latter fraction is resistant to experimental extraction; however, the first two domains are typically analysed in cell wall analyses of ripening fruits following extraction with strong alkali to disrupt the hydrogen bonding (Brummell and Harpster, 2001). Given the apparent structural importance of xyloglucan, there is considerable interest in understanding the nature and mechanism of xyloglucan metabolism in softening fruits.

Xyloglucan depolymerization in ripening fruit
Xyloglucans are the predominant hemicelluloses in the cell walls of many fruits, including tomato (Sakurai and Nevins, 1993; Maclachlan and Brady, 1994), avocado (O'Donoghue and Huber, 1992), melon (McCollum *et al.*, 1989; Rose *et al.*, 1998) and kiwifruit (Redgwell *et al.*, 1988). While some papers report limited xyloglucan degradation during ripening in certain fruits, most analyses conclude that xyloglu-can undergoes depolymerization, particularly in the early softening stages (Wakaba-yashi, 2000; Redgwell and Fischer, 2001). For example, in the Charentais variety of melon, which exhibits particularly rapid softening, and so represents a useful model

system to examine the relative timing of cell wall disassembly, one of the first detectable changes is depolymerization of a xyloglucan fraction that is apparently tightly bound to the microfibrils (Rose *et al.*, 1998). Similarly, early depolymerization of xyloglucan has been observed in tomato (Maclachlan and Brady, 1994), avocado (Sakurai and Nevins, 1997), and kiwifruit (MacRae and Redgwell, 1992), although it is not known from which domain(s) this xyloglucan originates *in vivo*, or whether inter-microfibrillar xyloglucan chains are preferentially cleaved.

These observations suggest the existence of enzymes that depolymerize xyloglucan and for some time it was thought that EGases were perhaps primarily responsible for xyloglucan hydrolysis. However, as described in section 9.2.2.1, biochemical and transgenic studies with EGases from different fruits now indicate that the situation is probably more complex, and that some other xyloglucanolytic activity may be involved.

Xyloglucan endotransglucosylase-hydrolase (XTH)

While xyloglucan depolymerization was for many years usually attributed to the action of EGases and other undefined hydrolases, an additional candidate emerged following the identification of a class of enzymes from growing plant tissues that catalyse the endo-cleavage and religation of xyloglucans in a transglycosylation reaction. These transglycosylases were originally described by two separate research groups and termed either xyloglucan endotransglycosylases (XETs; Smith and Fry, 1991) or endoxyloglucan transferases (EXTs; Nishitani and Tominaga, 1992). In related studies a 'xyloglucanase', or 'xyloglucan-specific endo-β-1,4-glucanase', that also exhibits endotransglycosylase activity under certain conditions *in vitro*, was isolated from germinating nasturtium seeds, where it catalyses the depolymerization of the xyloglucan seed storage reserves (Farkaš *et al.*, 1992; Fanutti *et al.*, 1993). The genes encoding these proteins were later shown to have substantial sequence homology to each other, but little similarity to the class of plant EGases described above in section 9.2.3.1.1. Subsequently, a number of groups have identified large gene families encoding these proteins, and these are typically divided into three or four distinct phylogenetic subgroups (reviewed in Nishitani, 1997; Campbell and Braam, 1999; Rose *et al.*, 2002a). However, unfortunately a number of contradictory and confusing nomenclatures have also been used in the literature to describe essentially the same class of proteins, and in some cases different names have been assigned to the same gene. To remedy this situation, a new unifying nomenclature has recently been proposed where a member of this class of genes/proteins is now referred to as a xyloglucan endotransglucosylase-hydrolase (XTH), and these proteins are now described as having either xyloglucan endotransglucosylase (XET) or xyloglucan endohydrolase (XEH) activities (Rose *et al.*, 2002a). All XTHs described to date specifically use xyloglucan as a substrate and show no activity against other glycans.

Some XTH isoforms appear to catalyse only the transglucosylation of xyloglucan, while others preferentially function as xyloglucan-specific endohydrolases, and so the consequence of XTH action probably depends largely on the nature of the xyloglucan substrates. This is illustrated in the schematic diagram of the cellulose-xyloglucan network in Figure 9.1. Panel A shows two adjacent cellulose microfibrils that are cross-linked by a load-bearing xyloglucan chain (black line), a second xyloglucan polymer that is hydrogen bonded to the microfibril surface but does not cross-link the microfibrils, a freely mobile xyloglucan oligosaccharide (grey line) and an XTH protein. Panels B–D summarize some of the reactions that are thought to be catalysed by XTHs, involving transglucosylation between two xyloglucan polymers, transglucosylation between a xyloglucan polymer and a xyloglucan oligomer, and xyloglucan endohydrolysis, respectively. When considering the mechanisms by which XTHs might contribute to the restructuring and disassembly of the cellulose-xyloglucan network, the reactions depicted here represent only some of the permutations, and several papers and reviews present additional and more detailed models (Nishitani, 1997; Campbell and Braam, 1999; Thompson and Fry, 2001; Rose *et al.*, 2002a). However, to summarize, XTHs probably play a role in cell expansion by both rearranging load-bearing xyloglucan cross-links and incorporating newly synthesized xyloglucan chains into the expanding wall (e.g. Figure 9.1 panels A and B). This might provide a means of allowing incremental wall

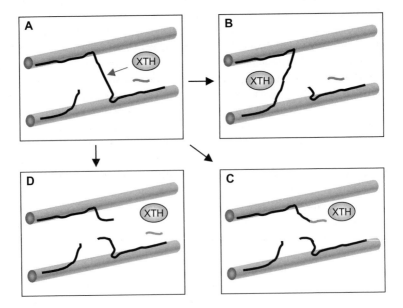

Figure 9.1 Schematic model of XTH action against the cellulose-xyloglucan matrix. Panel A shows two cellulose microfibrils that are cross-linked by a xyloglucan polymer (black line), while another xyloglucan polymer is shown bound to the microfibril surface. A xyloglucan oligosaccharide (grey line) and an XTH protein are also shown and the arrow indicates the site of XTH-mediated cleavage of the xyloglucan background. Panels B and C shows XET activity involving polymer-polymer, or polymer-oligomer transglycosylation, respectively. Panel D represents XEH activity where water is transferred (hydrolysis) onto the newly generated reducing end of the cleaved xyloglucan chain.

loosening with no new decrease in xyloglucan molecular size. In contrast, they can also catalyse the transfer of the reducing end of a cleaved xyloglucan chain to either a xyloglucan oligosaccharide, or a water molecule in a hydrolytic reaction (Figure 9.1 panels C and D, respectively), resulting in xyloglucan depolymerization. Support for this latter mechanism has emerged over the last 10 years or so, through association of XTH gene expression and activity with a number of developmental processes where xyloglucan depolymerization is a major feature, including fruit ripening.

Ripening-related XTH activities or gene expression have been reported in kiwifruit (Redgwell and Fry, 1993; Schröder *et al.*, 1998), tomato (Maclachlan and Brady, 1994; Arrowsmith and deSilva, 1995), grape (Nunan *et al.*, 2001) and persimmon (Cutillas-Iturralde *et al.*, 1994), although one report suggested that in apple, XTH activity correlated with cell division but did not increase during ripening (Percy *et al.*, 1996). However, determination of XTH activity is complicated since both XET and XEH activities may be catalysed by the same proteins, different assays are involved and the apparent activities may be significantly influenced by the abundance of xyloglucan oligosaccharides, the type of xyloglucan and other experimental conditions (Maclachlan and Brady, 1994; Schröder *et al.*, 1998; Catalá *et al.*, 2001). It was concluded from a study of xyloglucanase, xyloglucan endotransglucosylase and CMCase activities in protein extracts from ripening tomato fruit, that a spectrum of enzymes that cleave β-1,4-glucans is present, and that all the activities are lower in the non-softening *rin* mutant (Maclachlan and Brady, 1992, 1994). However, no proteins were purified or sequenced and it is difficult to predict how XTHs specifically contribute to xyloglucan disassembly and fruit softening *in vivo*, although Faik *et al.* (1998) proposed that XET activity alone is sufficient to account for all the xyloglucanase activity in expanding tomato fruits, when assayed using seed storage xyloglucan. Two review articles have referred to an experiment in which the expression of a ripening-related tomato *XTH* gene (*LeXETB1*) was down-regulated in transgenic fruit and no differences were reported with regard to fruit texture and softening characteristics or postharvest and processing characteristics (de Silva *et al.*, 1994; Brummell and Harpster, 2001). However, no detailed data have yet been published describing the consequences of down-regulating XTH expression in transgenic fruits and so the significance of XTHs for softening or restructuring of cell wall architecture in ripening fruit remains unresolved.

9.2.3.2 *Mannans and mannanases*

Compared with xyloglucan, relatively little is known about the metabolism of mannans and related hemicelluloses in ripening fruits. Galactoglucomannans and/or glucomannans, which are also hydrogen bonded to cellulose (Tong and Gross, 1988; Fischer and Bennett, 1991), appear to undergo minimal depolymerization or solubilization from the wall in ripening kiwifruit (Redgwell *et al.*, 1991) and tomato (Sakurai and Nevins, 1993; Carrington *et al.*, 2002), although depolymerization of a mannose-rich polymer has been detected in ripening melons (Rose *et al.*, 1998). Evaluation of the metabolism of this class of hemicelluloses may be complicated by the suggestion that glucomannan biosynthesis continues during ripening (Tong and Gross, 1988; Greve and Labavitch, 1991) and so any turnover of mannan-based polymers may be masked by coincident synthesis.

A role for mannan degradation in cell wall disassembly during fruit ripening is also supported by the detection of endo-β-mannanase activity, genes and/or proteins during ripening in several species of fruit, including tomato (Pressey, 1989; Sozzi *et al.*, 1996; Bewley *et al.*, 2000; Carrington *et al.*, 2002), watermelon, peach, melon and nectarine (Bourgault *et al.*, 2001). Moreover, endo-β-mannanase activity is substantially reduced in a number of non-ripening tomato mutants (Bewley *et al.*, 2000). Interestingly, in tomato fruit, endo-β-mannanase mRNA and protein are present in the early ripening stages, prior to any detectable activity, suggesting post-translational regulation of activity (Bewley *et al.*, 2000). The reason for this inactivity is not currently known, but the causes are distinct from a recent report that described a permanently inactive isoform of endo-β-mannanase in one specific tomato cultivar, that resulted from a mutation that caused a truncation at the C-terminus of the protein (Banik *et al.*, 2001; Bourgault and Bewley, 2002). The cultivar harbouring this inactive mutant isoform exhibited no endo-β-mannanase activity and yet softened essentially normally, suggesting that endo-β-mannanase is not required for softening. Other potential substrates for mannanases are N-linked glycoproteins; however, no attempts have been described in the literature to date to identify the *in vivo* substrate(s) and to assess the contribution of endo-β-mannanases to wall disassembly using transgenic plants.

9.2.3.3   Xylans and xylanases

Xylans and the related hemicelluloses arabinoxylan and glucuronoarabinoxylan (GAX) represent a major hemicellulosic component of the cell walls of monocotyledons, but also typically comprise about 5% of the primary cell wall in dicotyledons (Carpita and Gibeaut, 1993). However, this varies between species and xylan is relatively abundant in Spanish pear (Martin-Cabrejas *et al.*, 1994) and guava (Marcelin *et al.*, 1993) and has been detected in pear (Labavitch and Greve, 1983), while the walls of pineapple are enriched in GAX (Smith and Harris, 1995). Xylans are bound to cellulose and have been extracted in a complex with pectin from tomato (Seymour *et al.*, 1990) and pear (Martin-Cabrejas *et al.*, 1994), and covalent linkages between xylan and xyloglucan have been suggested to exist in olive fruit (Coimbra *et al.*, 1995). Thus, xylan metabolism might have multiple effects on wall organization and the architecture of different polysaccharide complexes; however, the extent of xylan depolymerization has not been determined for many fruits, and one study of kiwifruits revealed no change in molecular size (Redgwell *et al.*, 1991).

Despite the lack of evidence for xylan degradation during ripening, endoxylanase activity has been detected in a wide range of fruits, including avocado (Ronen *et al.*, 1991), papaya (Paull and Chen, 1983), Japanese pear (Yamaki and Kakiuchi, 1979) and tomato (Barka *et al.*, 2000). As with other cell wall hydrolases, there is considerable variability between fruits, such that activity increases dramatically during ripening in some species (Yamaki and Kakiuchi, 1979; Paull and Chen, 1983), and yet is undetectable in others (Ahmed and Labavitch, 1980). Currently, no ripening-related xylanase genes have been described, although it seems inevitable that xylanase gene families will be identified, with individual members expressed at different stages during fruit development, as is the case for the other families of cell wall-related genes described in the previous sections. Indeed, the *Arabidopsis* genome is predicted to have at least 12 xylanases (Henrissat *et al.*, 2001).

9.2.4 Scission of cell wall polysaccharides by reactive oxygen species (ROS)

The preceding sections include examples of proteins that mediate cell wall disassembly by interacting directly with polysaccharides and disrupting intracellular covalent glycosidic bonds or intermolecular hydrogen bonds. An alternative mechanism for polysaccharide depolymerization has recently received renewed attention, in which reactive oxygen species (ROS) are proposed to cause scission of cell wall polymers. Two different sources of ROS have been proposed in this regard. In earlier studies Miller (1986) described the peroxide-mediated degradation of pectin, polygalacturonic acid and cell walls from tomato and cucumber fruits, and Fry (1998) subsequently obtained evidence that highly reactive hydroxyl radicals ($\cdot$OH) might be generated in a 'non-enzymatic' Fenton reaction system, derived from ascorbate and Cu^{2+}. It was further demonstrated that $\cdot$OH generated in this way under physiological conditions could induce scission of xyloglucan and pectins *in vitro* (Fry, 1998; Fry *et al.*, 2001), and that the side chains of xyloglucan are more susceptible to $\cdot$OH attack than the xyloglucan backbone (Miller and Fry, 2001). An association between this mechanism and cell wall degradation during fruit ripening was also provided with the observation that cell walls from ripening pear fruit showed evidence of $\cdot$OH-induced breakage of glycosidic bonds *in vivo* (Fry *et al.*, 2001).

An alternative 'enzymatic' route for the generation of wall polysaccharide depolymerizing $\cdot$OH has been proposed in the form of peroxidase-mediated reactions (Chen and Schopfer, 1999; Schweikert *et al.*, 2000) and peroxidase-derived $\cdot$OH have been shown to degrade cell walls from maize and soybean seedlings *in vitro* (Schweikert *et al.*, 2002).

At this point, while ROS represent an interesting additional factor to consider when evaluating mechanisms of transient wall loosening during cell expansion, or wall degradation accompanying processes such as fruit ripening (Fry, 1998; Fry *et al.*, 2001; Schopfer, 2001; Schopfer *et al.*, 2002), it has not yet been definitively proven that any type of ROS-mediated cell wall polysaccharide scission occurs *in vivo*. Indeed, historically, peroxides have been proposed to induce cross-linking, rather than degradation, of cell wall components (Fry, 1986; Hatfield *et al.*, 1999). Moreover, application of a partially purified wall-bound tomato fruit peroxidase to tomato fruit tissues has recently been reported to result in an increase in tissue stiffness (Andrews *et al.*, 2002), although it is important to note that this experiment was performed *in vitro* using freeze-thawed tissue and so many influential factors that occur *in vivo* were likely to be absent.

For this to represent an effective means of controlled wall modification, the sites of ROS action would have to be carefully controlled to avoid damage to the plasma membrane and important apoplastic enzymes. Accordingly, numerous wall-bound peroxidases have been identified in the literature (Bolwell and Wojtaszek, 1997) and it is conceivable that they may deliver ·OH in a highly localized fashion such that specific polysaccharides, and even glycosidic linkages, would be attacked. Another element to this argument is that the ROS would need a sufficiently active lifetime before being scavenged by apoplastic antioxidants (Vanacker *et al.*, 1998). Fruit ripening is typically described as an oxidation-related phenomenon (Brennan and Frenkel, 1977), and while antioxidant systems are upregulated during ripening (Jimenez *et al.*, 2002), gene expression studies indicate that fruit experience, or at least respond to counter the onset of, oxidative stress (Aharoni and O'Connell, 2002; Aharoni *et al.*, 2002). Thus, there may be a shift in the antioxidant potential in the apoplast during ripening that would consequently alter the stability of the ·OH population. Such hypotheses remain to be tested, but a role for ROS in wall disassembly during fruit ripening and other systems remains an intriguing possibility.

9.2.5 Summary of wall disassembly during fruit ripening

The overarching theme that emerges from studies of cell wall disassembly in ripening fruit is the bewildering complexity of wall-related changes, enzymatic and non-enzymatic processes that regulate those changes, and the numerous additional factors that appear to influence the biochemical properties of the wall modifying proteins and the wall polymers themselves.

To complicate matters further, cell wall analyses typically provide static snapshots of wall composition and architecture, which can be severely limiting since wall structure is inherently dynamic. It has been proposed that cell wall synthesis occurs during ripening in some fruit (Tong and Gross, 1988; Greve and Labavitch, 1991; Huysamer *et al.*, 1997a,b), which might obscure some changes in wall composition and architecture, although the nature of the wall components that are synthesized can also change depending on the experimental approach that is used to monitor the fruit (Redgwell, 1996; Redgwell and Fischer, 2001). Currently then, ripening-related wall biosynthesis *in vivo* is still a relatively unexplored phenomenon.

It is clear that disassembly of the cell wall during ripening is associated with a huge variety of wall modifying proteins, and that this apparent diversity will increase as additional classes of proteins are identified that disrupt intra- and inter-molecular associations in the wall. A few types of activities that have only recently been identified are suggested in the previous sections, but novel genes/proteins will doubtless emerge as new assay strategies are developed. For example, while xyloglucan-specific transglycosylases are now widely studied (see section 9.2.3.1.2), related enzymes that act on other polysaccharides may also exist, although preliminary attempts to find an activity from fruits that could utilize pectin oligosaccharides in a transglycosylation reaction failed to identify such an activity (Garciaromera and Fry, 1994). Another class of wall component that has received relatively little attention in the context of fruit ripening is the potential degradation of structural proteins by ripening-related proteases. The amount of covalently bound protein in fruit is generally small; however, proteins such as extensin may contribute to the overall fruit textural characteristics.

Lastly, while there are many key features that are common to different species of ripening fruits, there are also dramatic qualitative and quantitative differences between species in terms of wall composition and the suites of proteins that interact with them. The significance of these differences is difficult to discern at this time, but rapid progress is being made in elucidating the numerous ways in which genetic divergence and evolutionary selection have given rise to 'fruit softening' in all its forms.

9.3   Abscission and dehiscence

Cell wall disassembly plays a key role in the shedding of organs such as leaves, flowers and fruit and in the dehiscence of anthers and pods. In these instances, wall breakdown is highly restricted (see Figure 9.2) and may be limited to only one or two layers of cells (Roberts *et al.*, 2002). Invariably the cells that comprise the abscission zone (AZ) or dehiscence zone (DZ) can be identified prior to the induction of wall dissolution and this observation has led to the hypothesis that they may constitute a specific target tissue.

A major site of degradation during both abscission and dehiscence is the middle lamella (Sexton and Roberts, 1982; Meakin and Roberts, 1990a), with a decline in the staining intensity of this pectin-rich component of the wall preceding cell separation. The degree of disassembly of the primary and secondary cell wall may vary from species to species, but in *Sambucus nigra*, where the leaflet AZ is composed

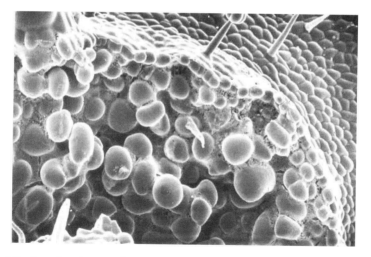

Figure 9.2   Scanning electron micrograph (SEM) of a tomato leaf petiole abscission zone after exposure to ethylene, showing cell separation.

of many layers, separated cells are released almost as protoplasts (Osborne and Sargent, 1976).

One of the first visible signs that the abscission process has been induced is the dilation of the Golgi cisternae within the AZ cells. This observation, coupled with the accumulation of vesicles within the cytoplasm of AZ cells, suggests that the process of wall breakdown is driven by the secretion of hydrolytic enzymes into the wall matrix. Degradation often begins in regions surrounding plasmodesmatal connections and it has been proposed that enzyme secretion may initially be focused at these sites (Sexton and Roberts, 1982).

9.3.1   Signals that regulate abscission and dehiscence

Both abscission and dehiscence normally take place at predictable times during the life cycle of a plant although premature shedding can be induced as a consequence of stress (Roberts et al., 2000). These observations indicate that the processes are highly coordinated and imply that regulation by a signaling cascade follows initiation by a series of environmental and developmental triggers (Taylor and Whitelaw, 2001). A role for ethylene in promoting the onset of leaf, flower and fruit abscission is now well established and there is good evidence that the plant growth regulator auxin (IAA) dictates the sensitivity of the AZ cells to ethylene (Roberts et al., 2002).

Recently a role for jasmonic acid (JA) in the regulation of anther dehiscence has been demonstrated by the cloning and characterization of the defective genes in non-dehiscing *Arabidopsis thaliana* mutants (Sanders et al., 2000; Ishiguro et al., 2001). There is no evidence that JA deficient mutants exhibit delayed pod dehiscence; however, in *Brassica napus* both IAA and ethylene have been implicated in regulating the timing of pod shatter (Chauvaux et al., 1997).

9.3.2   Biochemical and molecular events associated with wall disassembly

The activities or expression of a spectrum of cell wall modifying proteins including endo-1,4-β-D-glucanases, polygalacturonases, and expansins have been shown to increase specifically in AZ or DZ cells prior to wall dissolution. These increases seem to be the result of *de novo* protein synthesis arising as a consequence of an up-regulation of abscission- or dehiscence-related genes (Gonzalez-Carranza et al., 1998; Roberts et al., 2000; Roberts et al., 2002).

Endo-β-1,4-glucanases (EGases)

Although the initial signs of wall breakdown are restricted to the middle lamella, the first hydrolytic enzyme proposed to contribute to wall breakdown during abscission was 'cellulase' (EGase, or CMCase; see section 9.2.2.1). Increases in the activity of this enzyme have been documented during the abscission of a spectrum of tissues including *Phaseolus vulgaris* leaves (Lewis and Varner, 1970; Durbin and Lewis, 1988), *S. nigra* leaflets (Webb et al., 1993), and pepper flowers and leaves (Ferrarese et al., 1995; Trainotti et al., 1998). Activity has been shown to be elevated specifically in the AZ tissues after exposure to ethylene, while treatment with IAA prevents this increase. Further evidence to implicate endo-β-1,4-glucanases in the wall loosening process is the discovery that in a non-abscinding mutant of *Lupinus angustifolius* no increase in EGase activity could be detected in the leaflet abscission zones after incubation in ethylene (Henderson *et al.*, 2001b).

A rise in EGase activity has also been correlated with dehiscence in *Brassica napus* pods (Meakin and Roberts, 1990b; Child *et al.*, 1998) and again, application of auxin was shown to delay both the pod dehiscence process and the rise in enzyme activity (Chauvaux *et al.*, 1997). A recent comparison of shatter susceptible and resistant varieties of soybean has revealed that EGase activity increases in the DZ of the former but not the latter (Agrawal *et al.*, 2002).

EGases comprise large gene families in plants (see section 9.2.2.1) and in tomato at least eight different EGase genes (*Cel1* to *Cel8*) have been identified (Lashbrook *et al.*, 1994; del Campillo and Bennett, 1996; Brummell *et al.*, 1997a; Catalá *et al.*, 1997; Catalá and Bennett, 1998). Of these, *Cel1, Cel2* and *Cel5* have been associated with flower abscission (del Campillo and Bennett, 1996; Gonzalez-Bosch *et al.*, 1997; Kalaitzis *et al.*, 1999). Down-regulation of *Cel1* and *Cel2* using an antisense RNA strategy resulted in an increase in the required force to bring about flower shedding, although the fruit ripening process remained unaffected, suggesting that different EGase isozymes contribute to wall loosening at the two different sites (Lashbrook *et al.*, 1998; Brummell *et al.*, 1999). There are up to 27 EGase gene family members in *Arabidopsis* (del Campillo, 1999; Henrissat *et al.*, 2001), although it is unknown which of these might contribute to abscission. A preliminary study of leaf abscission in *B. napus* has led to the isolation of 6 distinct EGases, suggesting that more than one isoform may contribute to the cell wall degradation in the abscission zone (Gonzalez-Carranza, unpublished). Similarly, a number of isoforms of the enzyme have been isolated during dehiscence of *B. napus* pods (Roberts, unpublished).

Although there is strong circumstantial evidence that EGases contribute to the wall loosening process during abscission and dehiscence, their precise role in bringing about cell separation is unclear. Whether they hydrolyse critical load-bearing bonds in the wall, or facilitate the action of other proteins such as PGs or expansins (see below), remains open to question.

Polygalacturonases

An increase in the activity of the pectin-degrading enzyme polygalacturonase (PG) has been frequently observed during abscission. Extensive characterization of abscission-related PGs has been reported from peach leaves and fruit (Bonghi *et al.*, 1992), oil palm fruits (Henderson *et al.*, 2001a), *S. nigra* leaflets (Taylor *et al.*, 1993) and tomato leaves and flowers (Tucker *et al.*, 1984; Taylor *et al.*, 1990). Whilst a correlation between elevated PG activity and cell separation does not prove that the enzyme plays a role in wall breakdown, biochemical analyses of wall material extracted from the AZ of *S. nigra* leaflets has revealed that after incubation in ethylene the size of the soluble polyuronides is much more heterodisperse with smaller fragments becoming increasingly prevalent in extracts from the cell wall matrix.

This observation is consistent with the activity of an endo-acting PG, and the enzyme that can be extracted from *S. nigra* AZ tissue has been shown to have these properties (Taylor *et al.*, 1993). PG activity has been reported to increase prior to pod shatter in *B. napus* and this enzyme also has the characteristics of being able to cleave its substrate in an endo-fashion with only a small quantity of free galacturonic acid being released during the process (Petersen *et al.*, 1996).

Like EGases, PGs comprise a large gene family (see also section 9.2.1.1) and it has been demonstrated that in *S. nigra* at least two isoforms increase in activity during abscission (Taylor *et al.*, 1993). In tomato, three abscission-related PGs (*TAPG1, TAPG2* and *TAPG4*) have been identified (Kalaitzis *et al.*, 1997). Both *TAPG2* and *TAPG4* can be detected in flower pedicel AZ tissue prior to cell separation, while the expression of *TAPG1* increases during the shedding process. Fusion of the *β-glucuronidase* (*GUS*) reporter gene to the promoters of *TAPG1* or *TAPG4* has revealed that expression of these genes is restricted to the AZ of leaf petioles, flower and fruit pedicels, petal corolla and stigma. Expression was increased in the presence of ethylene and inhibited by IAA. Expression of *TAPG4* was largely limited to the vascular tissue and it has been suggested that the enzyme might release pectic fragments from the stele to coordinate the wall loosening process (Thompson and Osborne, 1994; Hong and Tucker, 2000).

Recently, an abscission-related PG has been isolated from *B. napus* (*PGAZBRAN*) and used to isolate its ortholog in *Arabidopsis* (*PGAZAT*) (Gonzalez-Carranza *et al.*, 2002). *PGAZBRAN* expression occurs during leaf and floral abscission and the mRNA accumulates prior to ethylene-promoted leaf shedding. When the promoter of *PGAZAT* was fused to *GUS* or green fluorescent protein (*GFP*), expression was detected in a two-cell layer zone located at the base of anther filaments, petals and sepals (Figure 9.3). Not all cells showed evidence of expression, suggesting that wall breakdown might be localized to reduce the likelihood of the developing silique becoming severed from the pedicel (Gonzalez-Carranza *et al.*, 2002).

Analysis of gene products isolated from *B. napus* pods has led to the identification of a PG that is upregulated specifically prior to dehiscence (Jenkins *et al.*, 1996). Fusion of the promoter of this gene to GUS revealed that expression was restricted specifically to the DZ cells of the pod and anther, and the junction between the funiculus and the seed (Jenkins *et al.*, 1999). This observation suggests that the process that brings about wall dissolution at these three sites may be similar, although this does not preclude the regulatory mechanisms being different. Further evidence was obtained that the site of PG expression is critical for wall dissolution by the demonstration that, if the DZ cells are ablated using the targeted expression of a ribonuclease, then plants are sterile due to a failure of the anthers to dehisce (Jenkins *et al.*, 1999). Recently an endo-PG has been shown to be up-regulated in the DZ of soybean pods (Christiansen *et al.*, 2002) and, interestingly, *Arabidopsis* transformants carrying approximately 1.2 kb of the promoter of this gene fused to GUS showed expression not only in the pod DZ, but also in the floral AZ cells.

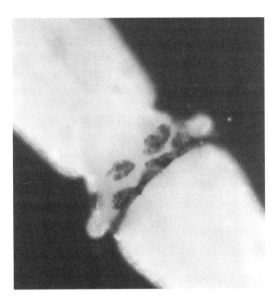

Figure 9.3   Expression of *PGAZAT::GUS* in the floral abscission zones of *Arabidopsis thaliana*.

Expansins

A role for expansins in wall loosening associated with growth and fruit softening is now well documented (Rose *et al.*, 1997; 2000a; Cosgrove, 2000a,b); however, it is still not yet clear whether these proteins contribute to shedding processes such as abscission and dehiscence.

In *Arabidopsis*, the expression of the expansin gene *AtEXP10* was found to be up-regulated at the base of the leaf petiole, and an increase in force was required to separate the petiole from the body of the plant when the gene was down-regulated (Cho and Cosgrove, 2000). However, these data are difficult to interpret as *Arabidopsis* petioles do not naturally abscind and expression of *ATEXP10* was not observed during shedding of the floral parts. A more convincing demonstration that expansin proteins may play a role in abscission is the observation that expansin activity increases in the leaflet AZ of *S. nigra* after exposure to ethylene (McQueen-Mason, unpublished). Two expansin genes have now been cloned from this species that show up-regulated expression during leaflet abscission (E. Belfield unpublished).

9.3.3   Strategies to study cell wall dissolution during abscission and dehiscence

Use of knockout mutants

Cell wall degradation is a complex process and, as with wall metabolism in softening fruit (see section 9.2), the shedding of plant organs certainly involves the action of a cocktail of wall modifying proteins. Moreover, many of the enzymes that have been implicated are themselves members of large gene families and this situation enhances the complexity. In the *Arabidopsis* genome it is possible to identify over 50 putative PGs (Roberts *et al.*, 2000), and although there is some evidence that sites of expression may overlap, it is likely that each of these acts at a specific location. One approach that can be used to ascertain the contribution of each of these isoforms is to examine the spatial and temporal wall loosening that takes place in plants where selected genes have been down-regulated. Lines where a T-DNA or transposon has inserted into a gene of interest can be obtained from a range of sources (see also Chapter 10) and it can be a relatively straightforward task to isolate knockout (KO) plants and compare them with negative segregants. A detailed analysis of a *PGAZAT* KO line has revealed that these mutants show a delayed time course of abscission in the presence of ethylene, further implicating the enzyme in wall dissolution (K. Elliott, unpublished). As cell separation still takes place, albeit at a slower rate, it is likely that either an additional PG is active (Sander *et al.*, 2001), and perhaps induced as a consequence of the down-regulation of *PGAZAT*, or that PG-mediated wall disassembly is not essential. No ultrastructural analysis has yet been undertaken to determine whether wall loosening is different in the mutant compared to the wild-type line.

Wall analysis and substrate specificity

There is convincing evidence that the AZ and DZ comprise different types of cells and that *de novo* gene expression is necessary to bring about wall loosening in these tissues. One question that has yet to be addressed is whether the cell wall matrix at these sites is also different to facilitate rapid disassembly. The AZ and DZ from most plant species do not constitute a sufficient number of cell layers for a biochemical analysis. However, NMR analysis has recently been used to show that the wall matrix in oil palm fruit, which has an unusually large AZ, contains elevated amounts of unmethylated pectins at the site of shedding (Henderson *et al.*, 2001a). The heterogeneity of wall architecture in the AZ and DZ represents a promising future area of study as new techniques develop for examining cell walls on a smaller scale.

The availability of KO lines that have a phenotypic impact on wall loosening provides an excellent opportunity for experiments to be carried out to investigate substrate specificity of potential wall loosening enzymes. For instance, the *Arabidopsis* PGs associated with DZ and AZ show close amino acid sequence homology and it might be predicted that they can attack similar cell wall domains. By driving *DZPG* expression at the site of abscission in *PGAZT* KO plants it should be possible to ascertain whether this hypothesis is correct, and whether other members of the PG family can also bring about cell separation of the AZ cell walls. As we discover more about the spatial and temporal expression patterns of abscission-related genes and isolate appropriate KO lines, the substrate specificity of other putative wall loosening proteins can be investigated.

9.4 Other examples of cell wall disassembly

Given the extensive literature and detailed experimental studies that relate to fruit ripening and organ abscission, as summarized in the previous sections, it can be informative to survey other examples of cell wall disassembly that have been less well characterized, but which in some cases are now being examined in great detail, and identify conserved or unique features. A few examples are given here.

Seed germination

The embryos of most seeds are surrounded by rigid endosperm tissue which acts as a physical restraint, and at the onset of germination and seedling growth, the radicle must penetrate this layer (Bewley, 1997). For this to occur, the region of the endosperm opposite the radicle tip, called the endosperm cap, undergoes weakening, and this is believed to result primarily from cell wall disassembly (Groot and Karssen, 1987). The cell walls of the endosperm cap during germination have been shown to undergo hydrolysis (Watkins *et al.*, 1985; Groot *et al.*, 1988; Sánchez *et al.*, 1990; Dutta *et al.*, 1994) and a broad range of cell wall hydrolases are expressed in germinating seeds, including EGase, PME, PG, arabinosidase, mannosidase, galactosidase, endo-β-1,3-glucanase and endo-β-mannanase (Groot *et al.*, 1988; Leviatov, *et al.*, 1995; Downie *et al.*, 1998; Sitrit *et al.*, 1999; Bradford *et al.*, 2000; Nonogaki *et al.*, 2000; Wu *et al.*, 2001). In some species, such as tomato, the cell walls are enriched in mannan-related hemicelluloses and much attention has been paid to endo-β-mannanase (Groot *et al.*, 1988; Nomaguchi *et al.*, 1995). However, although endo-β-mannanase activity typically increases during radicle emergence (Nonogaki and Morohashi, 1996; Nonogaki *et al.*, 2000), there are also instances where emergence fails to occur, despite high activity levels, and so endo-β-mannanase is not believed to be sufficient for germination in every case (Toorop *et al.*, 1996; Dahal *et al.*, 1997; Still and Bradford, 1997). More recently, the identification of new classes of wall-modifying proteins in ripening fruit has driven the search for similar activities and genes in germinating seeds. For example, an expansin (*LeEXP4*) and an XTH (*LeXET4*) have now been associated with weakening of the endosperm cap during tomato germination (Chen and Bradford, 2000; Chen *et al.*, 2002) and Figure 9.4A shows that *LeEXP4* mRNA is specifically expressed in the micropylar region (Chen and Bradford, 2000), while Figure 9.4B presents similar specific expression patterns for *LeXET4* mRNA and also shows that the gene was expressed while the seed was being imbibed in water, but prior to radicle emergence (Chen *et al.*, 2002).

Some species of plants also accumulate substantial quantities of cell wall polysaccharides as their major seed storage reserve and this material is then rapidly depolymerized and mobilized upon germination. These polymers include xyloglucan, galactomannans and galactans, and while details of this system are not described here, an excellent review by Buckeridge *et al.* (2000) provides an in-depth description of the enzymes that are involved and their regulation. While these polysaccharides do not play a structural role, there are numerous similarities

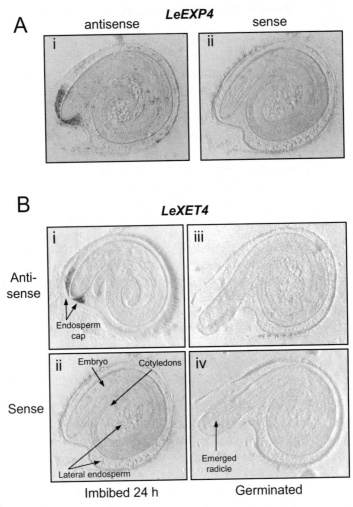

Figure 9.4 Tissue prints of germinating tomato seedlings to evaluate the spatial expression of an expansin and an XTH (formerly XET) gene. (a) Localization of *LeXET4*. Tomato seeds were imbibed for 24 h, bisected and the cut surfaces printed onto two membranes. The membranes were hybridized with either an antisense (i), or a sense (ii) *LeXXP4* probe. This part of the figure is reprinted from Chen and Bradford (2000) and is copyrighted by the American Society of Plant Biologists and reprinted with kind permission. (b) Localization of *LeXET4* expression. Tomato seeds were imbibed for 24 h (i and ii), or 60 h (iii and iv) and printed onto two membranes as described for A. The membranes were hybridized with an antisense (i and iii) or sense (ii and iv) *LeXET4* probe. This part of the figure is reprinted from Chen *et al.* (2002) and is copyrighted by the Society for Experimental Biology, and reprinted with kind permission.

between the mechanisms of their mobilization and the disassembly of primary cell wall polymers.

It appears that, as with fruit softening, a battery of wall modifying proteins are expressed in geminating seeds with an exquisite degree of temporal and spatial regulation, and it would not be surprising to see additional parallels emerge with the existing studies that have been made in fruit.

Tracheary element formation

The formation of xylem vessels of higher plants and their constituent tracheary elements is a striking example of cell differentiation. Briefly, cells from the procambium and vascular cambium undergo a highly orchestrated series of developmental changes, concluding in the formation of a series of dead hollow tubes through which water and mineral transport occur (Fukuda, 1996; McCann, 1997; Roberts and McCann, 2000). Among the many processes that are involved, primary cell wall disassembly is believed to be an integral component, and a number of recent detailed studies of gene expression during tracheary element differentiation support this idea. Such studies been made possible using an elegant experimental system in which isolated mesophyll cells from the leaves of Zinnia elegans undergo highly synchronized differentiation into tracheary elements without cell division, when grown in culture using a specific hormonal regime (Fukuda and Komamine, 1980). Both cDNA microarray analysis (Demura et al., 2002) and large-scale cDNA–AFLP analysis (Milioni et al., 2001, 2002) have been used to examine the expression patterns of thousands of genes during differentiation. These include members of many of the cell wall-related gene families that have already been discussed, including numerous pectinases, hemicellulases and expansins. Again, the results point to the coordinated modification of multiple wall components by suites of hydrolases and other proteins, rather than one key wall degrading enzyme. Despite the high degree of synchronization of the system, it is not yet possible to discern a particular pattern for the published data, such that a chronology of wall modification can be established. For example, it would be useful to be able to ask questions such as whether pectins are hydrolyzed prior to hemicelluloses, or whether expansins act before the majority of hydrolases. Such a degree of resolution may not be possible, or equally, wall disassembly may not involve such sequential steps.

Aerenchyma formation

Plants that adapt to grow in water-logged environments can develop tissue called aerenchyma, which is an extensive interconnected gas-filled space surrounded by intact cortical cells, that provides a direct connection between the oxygen-deprived roots and the aerial environment. The aerenchymatous tissue can take a number of forms that involve cell separation, differential expansion and also localized cell death (reviewed in Jackson and Armstrong, 1999) and the extensive cell wall disassembly that is involved (Drew et al., 1979; Saab, 1999) can produce some dramatic changes in root morphology (see Figure 9.5). Relatively little is known about the nature of the wall reorganization, although CMCase activity is induced during flooding or ethyl-

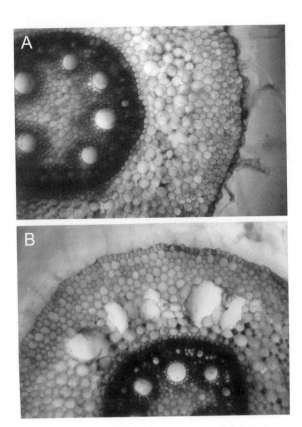

Figure 9.5 Induction of aerenchyma in the cortex of the maize primary root by flooding. Sections were taken from roots of seedlings that were kept aerated (A) or flooded (B) for 72 hours. This photograph is taken from Saab and Sachs (1996) and is copyrighted by the American Society of Plant Biologists and reprinted with kind permission.

ene or hypoxia treatments (Kawase, 1979; He *et al.*, 1994; Huang *et al.*, 1997), and an XTH gene was reported to be induced by flooding in maize root mesocotyl tissue that subsequently developed into aerenchyma (Saab and Sachs, 1996). Moreover, XTH gene expression correlated with either induction or repression or aerenchyma formation. A more recent immunolocalization study described changes in deesterified and esterified pectins early in aerenchyma formation (Gunawaredena *et al.*, 2001) and the authors reached the conclusion that wall disassembly involving numerous wall modifying proteins was an integral part of early aerenchyma formation and subsequent cell death.

Pith autolysis
When many herbaceous dicotyledons are exposed to drought stress, the pith in the stem undergoes autolysis (Aloni and Pressman, 1981; Pressman *et al.*, 1983; Carr

and Jaffe, 1995; Carr *et al.*, 1995), where the cell walls are completely degraded and the protoplasts are lysed. PG and CMCase activities both increase in the stem during induced drought stress, coincident with a large increase in apoplastic galacturonic acid, and it seems likely that both hemicelluloses and pectins undergo rapid hydrolysis (Huberman *et al.*, 1993). Little more is known about the other factors that contribute to the wall disassembly, beyond these preliminary observations.

These examples have been selected to illustrate the range of information that has been obtained in different systems, from detailed studies of thousands of genes, to a limited number of studies of the activities of one or two wall hydrolases. However, there are many other examples of wall disassembly associated with diverse developmental programs, or following interactions with pathogens or symbionts. These include the release of root border cells (Stephenson and Hawes, 1994; Wen *et al.*, 1999), the intrusive growth of laticifers (Mahlberg, 1993; Pilatzke-Wunderlich and Nessler, 2001; Serpe *et al.*, 2002) and the formation of symbiotic associations with mycorrhizal fungi (RejonPalomares *et al.*, 1996; Perotto *et al.*, 1997; van Buuren *et al.*, 1999) and nodulating bacteria (Munoz *et al.* 1998; Parniske, 2000; Lievens *et al.*, 2002). Of course, extensive wall disassembly is also a hallmark of colonization by microbial pathogens, but this will not be discussed here, since arguably this falls into a different conceptual class of wall degradation, where normal plant health and viability is compromised. It is however worth noting that infection of plant roots by nematodes also involves wall disassembly and, while an array of cell wall hydrolases are synthesized and secreted by nematodes to facilitate infection (Smant *et al.*, 1998; Popeijus *et al.*, 2000; Jaubert *et al.*, 2002), there is also evidence that the nematodes can induces the expression of endogenous cell wall hydrolases in the plant host (Goellner *et al.*, 2001; Vercauteren *et al.*, 2002). Thus, the nematodes appear to mediate wall disassembly not only by secreting exogenous wall modifying proteins, but also by hijacking the plant's endogenous mechanism of wall loosening. This ingenious combination may be a widespread phenomenon but has yet to be reported in interactions with organisms other than nematodes.

9.5 Conclusions, questions and future directions

Wall disassembly is becoming increasingly complicated and current models of processes such as wall deconstruction during fruit softening are vastly more complex than earlier notions of tissue disintegration resulting from the activities of one, or perhaps two polysaccharide hydrolases. The application of new analytical tools and approaches continues to uncover novel mechanisms of wall degradation and numerous new layers of regulation, that may seem to obscure, rather than clarify, the picture. For example, the completion of the first two plant genome sequences (*Arabidopsis thaliana* and rice) and the huge collections of EST sequences from several other plant species (see Chapter 10) indicate that cell wall hydrolase gene families typically contain dozens of members, as described throughout this chapter. It is unlikely that prior to the *Arabidopsis* genome sequence being published, anyone would have predicted that a plant might have more than 50 PG genes (Roberts *et al.*, 2000) together with correspondingly large families of EGases, PMEs, expansins and XTHs, and virtually any other class of wall modifying protein.

There is insufficient space in this chapter to discuss in detail the numerous facets of wall disassembly that remain enigmatic; however, a series of questions are posed that perhaps represent some of the more challenging and immediate goals within this field.

To what extent is cell wall disassembly a synergistic process?

It now seems that each structural component of primary wall architecture, or complex of components such as the cellulose-xyloglucan network (Rose and Bennett, 1999), is targeted by multiple classes of wall modifying proteins. There is therefore the potential for synergistic action, where one class of proteins enhances the activity of another. There are many ways in which this might occur, especially given the structural specificity of many wall hydrolases and so, for example, exo-acting glycosidases or esterases acting on a polysaccharide might expose the polymer backbone and allow more substantial and rapid depolymerization by endo-acting hydrolases. This has been proposed to be the case for PME, which demethylesterifies polyuronides, thereby generating a suitable substrate for PG and perhaps pectate lyase (see section 9.2.1.2). The disruption of non-covalent inter-polysaccharide associations may also be an important component of synergistic wall disassembly, and in this regard, expansin action may alter the accessibility of a range of enzymes to their substrates (see section 9.2.2.2). It has been known for some time that microbes utilize a complex synergistic system of hydrolases that can include more than 20 enzymes acting on the same substrate (Warren, 1996; Amano and Kanda, 2002; Murashima *et al.*, 2002; Watson *et al.*, 2002). These systems are likely to extend beyond hydrolases and recent data support the involvement of pectate lyases (Tamaru and Doi, 2001) and expansin-like proteins (Laine *et al.*, 2000; Saloheimo *et al.*, 2002). Even though the purpose of microbial-mediated degradation is somewhat different from endogenous wall disassembly, and is generally designed to degrade the wall polymers for use as a carbon source, often saprophytically, it seems highly likely that complex synergistic systems of wall modifying proteins also exist in plants.

Are there many wall modifying proteins that remain to be identified?

Given the range of covalent and non-covalent associations between structural components of the cell wall, the number of biochemically characterized classes of wall modifying proteins certainly does not reflect the entire complement. A recent analysis of secreted proteins from *Arabidopsis* revealed a significant number which could not be assigned a function based on sequence homology (Chivasa *et al.*, 2002) and similar results have been obtained from studies of cell wall-related proteins in tomato fruit (J. Rose, unpublished data).

The identification of new wall modifying proteins is limited to a great extent by the development of new biochemical assays and substrates. In many cases it is extremely difficult to extract native polysaccharides without modifying their structure and it is even more challenging to extract polysaccharide complexes without disrupting associations or creating artifacts. For example, expansins were originally identified following the development of a biophysical assay using an extensometer (McQueen-Mason *et al.*, 1992), rather than measuring the chemical changes in a purified cell wall polymer, and expansin action is currently only detectable by this technique. There may be many other proteins that require similar novel experimental assays for their detection.

The identification of new wall modifying proteins will likely come from a number of sources, including the development of new assays, database mining to identify homologs of known wall modifying proteins or predicted proteins with structural domains that indicate an association with the wall (e.g. Catalá and Bennett, 1998; Trainotti *et al.*, 2001), proteomic analysis of cell wall protein populations, and through forward genetics using mutants with cell wall-related phenotypes (see Chapter 10).

Is wall disassembly typically restricted in vivo and by which mechanisms?
The best-studied example of cell wall disassembly to date is in ripening tomato fruits, where it appears that pectin depolymerization is substantially restricted *in vivo* and rapid degradation occurs following extraction *in vitro* (see section 9.2.1.7). A number of factors have been proposed that may limit the amount of wall modification by PG and other wall hydrolases. These include ionic and pH conditions in the apoplast, immobilization of wall modifying proteins through binding to wall components, limited access to substrates as a consequence of conformational, structural and steric effects, synthesis and secretion of enzymes in an inactive form that require apoplastic post-translational proteolytic processing for activation, and the involvement of proteins that modulate enzymatic activities, such as the interaction between tomato PG2 and the β-subunit (see section 9.2.1.1). It is not known whether there are other examples of polysaccharide modification in fruits, or wall disassembly in other tissues such as those that are described in sections 9.3 and 9.4, that are similarly restricted.

A number of cell wall hydrolase inhibitor proteins are known to be synthesized by plants that can inhibit isoforms of PG (Leckie *et al.*, 1999; Esquerré-Tugayé *et al.*, 2000; Stotz *et al.*, 2000), xylanase (McLauchlan *et al.*, 1999), xyloglucanase (Qin *et al.*, 2003), or pectin lyase (Bugbee, 1993) from microbial pathogens, although they have not been shown to inhibit the equivalent endogenous plant enzymes. Conversely, a class of proteins termed glucanase inhibitor proteins (GIPs) has recently been identified in oomycetes, that can specifically inhibit plant endo-β-1,3-glucanases (Ham *et al.*, 1997; Rose *et al.*, 2002b). It is likely that additional classes of inhibitor proteins will be identified that act to limit wall degrading activities, although it remains to be seen whether these occur outside the arena of plant-pathogen interactions.

To what extent does wall disassembly occur in specific domains within tissues or individual walls?

Studies in tomato fruit indicate that pectin disassembly occurs in domains within the fruit pericarp, rather than uniformly over a broad area, and that this may reflect heterogeneous wall architecture (Steele *et al.,* 1997; Orfila and Knox, 2000). Similarly, Morvan *et al.* (1998) used immunoanalysis to localize PME and acidic pectins to microdomains within specific walls of flax hypocotyls. As techniques develop that improve the resolution of wall architecture (see Chapters 2 and 3), it will be interesting to see the extent to which the disassembly of other classes of polymers occurs in a highly localized manner, and whether wall turnover in fruit and other tissues occurs in microdomains within a single wall.

Are the mechanisms of cell wall disassembly conserved in different developmental programs and different plant species, and how are they related to wall loosening during cell expansion?

The detailed studies of wall disassembly in ripening fruit, and particularly in tomato, have driven much of what is currently known about this process. While it is tempting to apply mechanistic models that are developed from observations in tomato fruit to other fruit species, and indeed to other examples of wall disassembly such as organ abscission, far less is known about other systems. Thus, it is not yet understood whether 'wall disassembly' is a highly conserved process, with a small amount of variability in terms of the constituent molecular changes and their regulation between different species and tissues, or whether there are many fundamental differences. Similarly, while there are a number of parallels between wall loosening during cell expansion and terminal wall degradation (Rose and Bennett, 1999), it might be that these are two extremes of a generic 'wall modification program', that are distinguished by a few differences such as coincident wall biosynthesis, and that there is a simple shift in the equilibrium between net biosynthesis and degradation. On the other hand, it may be that the apparent similarities are superficial and that many fundamental differences will emerge once wall metabolism becomes better resolved. Ongoing initiatives to profile global gene expression during different developmental processes should help answer such questions by providing holistic insights into multiple metabolic pathways and signaling cascades.

Acknowledgements

The authors would like to thank Drs Don Huber, Ken Gross, Dave Smith, Maria Harrison, John Labavitch and Derek Bewley for providing pre-publication manuscripts, data, or valuable discussion, and Drs Kent Bradford and Imad Saab for Figures 9.4 and 9.5, respectively. The American Society of Plant Biologists and the Society for Experimental Biology are also acknowledged for permission to reprint the copyrighted photographs in Figures 9.4 and 9.5. J.R. is funded in this area by the

National Science Foundation (grant IBN-009109) and the United States Department of Agriculture (grants 2001-52100-1137 and 2002-01406).

References

Agrawal, A.P., Basarkar, P.W., Salimath, P.M. and Patil, S.A. (2002) Role of cell wall-degrading enzymes in pod-shattering process of soybean, *Glycine max* (L.). *Curr. Sci.*, 82, 58–61.

Aharoni, A. and O'Connell, A.P. (2002) Gene expression analysis of strawberry achene and receptacle maturation using DNA microarrays. *J. Exp. Bot.*, 53, 2073–2087.

Aharoni, A., Keizer, L.C.P., van den Broeck, H.C. *et al.* (2002) Novel insight into vascular, stress, and auxin-dependent and -independent gene expression programs in strawberry, a non-climacteric fruit. *Plant Physiol.*, 129, 1019–1031.

Ahmed, A.E.R. and Labavitch, J.M. (1980) Cell wall metabolism in ripening fruit. II. Changes in carbohydrate degrading enzymes in ripening 'Bartlett' pears. *Plant Physiol.*, 65, 1014–1016.

Almeida, D.P.F. and Huber, D.J. (1999) Apoplastic pH and inorganic ion levels in tomato fruit: a potential means for regulating cell wall metabolism during ripening. *Physiol. Plant.*, 105, 506–512.

Aloni, B. and Pressman, E. (1981) Stem pith autolysis in tomato plants: the effect of water stress and the role of abscisic acid. *Physiol. Plant.*, 51, 39–44.

Amano, Y. and Kanda, T. (2002) New insights into cellulose degradation by cellulases and related enzymes. *Trends Glycosci. Glycotechnol.*, 14, 27–34.

Andrews, J., Adams, S.J., Burton, K.S. and Edmondson, R.N. (2002) Partial purification of tomato fruit peroxidase and its effect on the mechanical properties of tomato fruit skin. *J. Exp. Bot.*, 53, 2393–2399.

Annis, S.L. and Goodwin, P.H. (1997) Recent advances in the molecular genetics of plant cell wall-degrading enzymes produced by plant pathogenic fungi. *Eur. J. Plant Pathol.*, 103, 1–14.

Arabidopsis Genome Initiative (2000) Analysis of the genome sequence of the flowering plant *Arabidopsis thaliana*. *Nature*, 408, 796–815.

Arrowsmith, D.A. and deSilva, J. (1995) Characterisation of two tomato fruit-expressed cDNAs encoding xyloglucan *endo*-transglycosylase. *Plant Mol. Biol.*, 28, 391–403.

Astley, O.M., Chanliaud, E., Donald, A.M. and Gidley, M.J. (2001) Structure of *Acetobacter* cellulose composites in the hydrated state. *Int. J. Biol. Macromol.*, 29, 193–202.

Atkinson, R.G., Schroder, R., Hallett, I.C., Cohen, D. and MacRae, E.A. (2002) Overexpression of polygalacturonase in transgenic apple trees leads to a range of novel phenotypes involving changes in cell adhesion. *Plant Physiol.*, 129, 122–133.

Awad, M. and Young, R.E. (1979) Postharvest variation in cellulase, polygalacturonase and pectin methylesterase in avocado (*Persea americana* Mill, cv *Fuerte*) fruits in relation to respiration and ethylene production. *Plant Physiol.*, 64, 306–308.

Banik., M., Bourgault, R. and Bewley, J.D. (2001) Endo-β-mannanase is present in an inactive form in ripening tomato fruits of the cultivar Walter. *J. Exp. Bot.*, 52, 105–111.

Barka, E.A., Kalantari, S., Makhlouf, J. and Arul, J. (2000) Impact of UV-C irradiation on the cell wall degrading enzymes during ripening of tomato (*Lycopersicon esculentum* L.) fruit. *J. Ag. Food Chem.*, 48, 667–671.

Barnavon, L., Doco, T., Terrier, N., Ageorges, A., Romieu, C. and Pellerin, P. (2000) Analysis of cell wall neutral sugar composition, β-galactosidase activity and a related cDNA clone throughout development of *Vitis vinifera* grape berries. *Plant Physiol. Biochem.*, 38, 289–300.

Ben-Arie, R., Kislev, N. and Frenkel, C. (1979) Ultrastructural changes in the cell walls of ripening apple and pear fruit. *Plant Physiol.*, 64, 197–202.

Besford, R.T., and Hobson, G.E. (1972) Pectic enzymes associated with softening of tomato fruit. *Phytochem.*, 11, 873–881.

Bewley, J.D. (1997) Seed germination and dormancy. *Plant Cell*, 9, 1055–1066.

Bewley, J.D., Banik, M., Bourgault, R., Feurtado, J.A., Toorop, P. and Hilhorst, H.W.M. (2000) Endo-β-mannanase activity increases in the skin and outer pericarp of tomato fruits during ripening. *J. Exp. Bot.*, 51, 529–538.

Blumer, J.M., Clay, R.P., Bergmann, C.W., Albersheim, P. and Darvill, A.G. (2000) Characterization of changes in pectin methylesterase expression and pectin esterification during tomato fruit ripening. *Can. J. Bot.*, 78, 607–618.

Bolwell, G.P. and Wojtaszek, P. (1997) Mechanisms for the generation of reactive oxygen species in plant defence – a broad perspective. *Physiol. Mol. Plant Path.*, 51, 347–366.

Bonghi, C., Rascio, N., Ramina, A., and Casadoro, G. (1992) Cellulase and polygalacturonase involvement in the abscission of leaf and fruit explants of peach. *Plant Mol. Biol.*, 20, 839–848.

Bonghi, C., Ferrarese, L., Ruperti, B., Tonutti, P. and Ramina, A. (1998) Endo-β-1,4-glucanases are involved in peach fruit growth and ripening, and are regulated by ethylene. *Physiol. Plant.*, 102, 346–352.

Bourgault, R. and Bewley, J.D. (2002) Variation in its C-terminal amino acids determines whether endo-β-mannanase is active or inactive in ripening tomato fruits of different cultivars. *Plant Physiol.*, 130, 1254–1262.

Bourgault, R., Bewley, J.D., Alberici, A. and Decker, D. (2001) Endo-β-mannanase activity in tomato and other ripening fruits. *Hort. Sci.*, 36, 72–75.

Bradford, K.J., Chen, F., Cooley, M.B. *et al.* (2000) Gene expression prior to radicle emergence in imbibed tomato seeds. In *Advances and Applications in Seed Biology* (eds M. Black, K.J. Bradford and J. Vazquez-Ramos), CAB International, Wallingford, UK. pp. 231–251.

Breton, C., Bordenave, M., Richard, L. *et al.* (1996) PCR cloning and expression analysis of a cDNA encoding a pectinacetylesterase from *Vigna radiata* L. *FEBS Lett.*, 388, 139–142.

Brown, R.M. (1999) Cellulose structure and biosynthesis. *Pure Appl. Chem.*, 71, 767–775.

Brownleader, M.D., Jackson, P., Mobasheri, A. *et al.* (1999) Molecular aspects of cell wall modification during fruit ripening. *Crit. Rev. Food Sci. Nutr.*, 39, 149–164.

Brummell, D.A. and Harpster, M.H. (2001) Cell wall metabolism in fruit softening and quality and its manipulation in transgenic plants. *Plant Mol. Biol.*, 47, 311–340.

Brummell, D.A., Lashbrook, C.C. and Bennett, A.B. (1994) Plant endo-1,4-β-D-glucanases: structure, properties and physiological function. In *Enzymatic Conversion of Biomass for Fuels Production* (eds M.E. Himmel, J.O. Baker and P.R. Overend). ACS Symposium Series 566, ACS Washington DC, pp. 100–129.

Brummell, D.A. and Labavitch, J.M. (1997) Effect of antisense suppression of endopolygalacturonase activity on polyuronide molecular weight in ripening tomato fruit and in fruit homogenates. *Plant Physiol.*, 115, 717–725.

Brummell, D.A., Bird, C.R., Schuch, W. and Bennett, A.B. (1997) An endo-1,4-β-glucanase expressed at high levels in rapidly expanding tissues. *Plant Mol. Biol.*, 33, 87–95.

Brummell, D.A., Hall, B.D. and Bennett, A.B. (1999a) Antisense suppression of tomato endo-1,4-β-glucanase *Cel2* mRNA accumulation increases the force required to break fruit abscission zones but does not affect fruit softening. *Plant Mol. Biol.*, 40, 615–622.

Brummell, D.A., Harpster, M.H., Civello, P.M., Palys, J.M., Bennett, A.B. and Dunsmuir, P. (1999b) Modification of expansin protein abundance in tomato fruit alters softening and cell wall polymer metabolism during ripening. *Plant Cell*, 11, 2203–2216.

Brummell, D.A., Howie, W.J., Ma, C. and Dunsmuir, P. (2002) Postharvest fruit quality of transgenic tomatoes suppressed in expression of a ripening-related expansin. *Postharvest Biol. Technol.*, 25, 209–220.

Buckeridge, M.S., Pessoa dos Santos, H. and Tiné, M.A.S. (2000) Mobilisation of storage cell wall polysaccharides in seeds. *Plant Physiol. Biochem.*, 38, 141–156.

Buescher, R.W. and Hobson, G.E. (1982) Role of calcium and chelating agents in regulating the degradation of tomato fruit tissue by polygalacturonase. *J. Food Biochem.*, 6, 147–160.

Bugbee, W.M. (1993) A pectin lyase inhibitor protein from cell walls of sugar beet. *Phytopathol.*, 83, 63–68.

Campbell, P. and Braam, J. (1999) Xyloglucan endotransglycosylases: diversity of genes, enzymes and potential wall-modifying functions. *Trends Plant Sci.*, 4, 361–366.

Carey, A.T., Holt, K., Picard, S. *et al.* (1995) Tomato exo-β(1→4)-D-galactanase: isolation, changes during ripening in normal and mutant tomato fruit, and characterization of a related cDNA clone. *Plant Physiol.*, 108, 1099–1107.

Carey, A.T., Smith, D.L., Harrison, E. *et al.* (2001) Down-regulation of a ripening-related β-galactosidase gene (*TBG1*) in transgenic tomato fruits. *J. Exp. Bot.* 52, 663–668.

Carr, S.M. and Jaffe, M.J. (1995) Autolysis in herbaceous, dicotyledonous plants: experimental manipulation of pith autolysis in several cultivated species. *Ann. Bot.*, 75, 587–592.

Carr, S.M., Seifert, M., Delabaere, B. and Jaffe, M.J. (1995) Pith autolysis in herbaceous dicotyledonous plants: a physiological ecological study of pith autolysis under native conditions with special attention to the wild plant *Impatiens capensis* Meerb. *Ann. Bot.*, 76, 177–189.

Carrington, C.M.S. and Pressey, R. (1996) β-galactosidase II activity in relation to changes in cell wall galactosyl composition during tomato fruit ripening. *J. Am. Soc. Hort. Sci.*, 121, 132–136.

Carrington, C.M.S., Greve, L.C. and Labavitch, J.M. (1993) Cell wall metabolism in ripening fruit: V. Effect of the antisense polygalacturonase gene on cell wall changes accompanying ripening in transgenic tomatoes. *Plant Physiol.*, 103, 429–434.

Carrington, C.M.S., Vendrell, M. and Dominguez-Puigjaner, E. (2002) Characterisation of an *endo*-(1,4)-β-mannanase (*LeMAN4*) expressed in ripening tomato fruit. *Plant Sci.*, 163, 599–606.

Carpita, N.C. and Gibeaut, D.M. (1993) Structural models of primary cell walls in flowering plants: consistency of molecular structures with the physical properties of the wall during growth. *Plant J.*, 3, 1–30.

Cass, L.G., Kirven, K.A. and Christoffersen, R.E. (1990) Isolation and characterization of a cellulase gene family member expressed during avocado fruit ripening. *Mol. Gen. Genet.*, 223, 76–86.

Catalá, C. and Bennett, A.B. (1998) Cloning and sequence analysis of *TomCel8*, a new plant endo-β-1,4-D-glucanase gene encoding a protein with a putative carbohydrate binding domain (accession no. AF098292). *Plant Physiol.*, 118, 1535.

Catalá, C., Rose, J.K.C. and Bennett A.B. (1997) Auxin regulation and spatial localization of an endo-1,4-β-D-glucanase and a xyloglucan endotransglycosylase in expanding tomato hypocotyls. *Plant J.*, 12, 417–426.

Catalá, C., Rose, J.K.C., York, W.S., Albersheim, P., Darvill, A.G. and Bennett, A.B. (2001) Characterization of a tomato xyloglucan endotransglycosylase gene that is down-regulated by auxin in etiolated hypocotyls. *Plant Physiol.*, 127, 1180–1192.

Chanliaud, E. and Gidley, M.J. (1999) *In vitro* synthesis and properties of pectin/*Acetobacter xylinum* cellulose composites. *Plant J.*, 20, 25–35.

Chauvaux, N., Child, R., John, K. *et al.* (1997) The role of auxin in cell separation in the dehiscence zone of oilseed rape pods. *J. Exp. Bot.*, 48, 1423–1429.

Chen, F. and Bradford, K.J. (2000) Expression of an expansin is associated with endosperm weakening during tomato seed germination. *Plant Physiol.*, 124, 1265–1274.

Chen, S.-X. and Schopfer, P. (1999) Hydroxyl-radical production in physiological reactions. A novel function of peroxidase. *Eur. J. Biochem.*, 260, 726–735.

Chen, F., Nonogaki, H. and Bradford, K.J. (2002) A gibberellin-regulated xyloglucan endotransglycosylase gene is expressed in the endosperm cap during tomato seed germination. *J. Exp. Bot.*, 53, 215–223.

Child, R., Chauvaux, N., John, K., Ulvskov, P. and van Onckelen, H.A. (1998) Ethylene biosynthesis in oilseed rape pods in relation to pod shatter. *J. Exper. Bot.*, 49, 829–838.

Chin, L.H., Ali, Z.M. and Lazan, H. (1999) Cell wall modifications, degrading enzymes and softening of carambola fruit during ripening. *J. Exp. Bot.*, 50, 767–775.

Chivasa, S., Ndimba, B.K., Simon, W.J. *et al.* (2002) Proteomic analysis of the *Arabidopsis thaliana* cell wall. *Electrophoresis*, 23, 1754–1765.

Cho, H.-T. and Cosgrove, D.J. (2000) Altered expression of expansin modulates leaf growth and pedicel abscission in *Arabidopsis thaliana*. *Proc. Natl. Acad. Sci. USA,* 97, 9783–9788.

Christiansen, L.C., Dal Degan, F., Ulvskov, P. and Borkhardt, B. (2002) Examination of the dehiscence zone in soybean pods and isolation of a dehiscence-related endopolygalacturonase gene. *Plant Cell Environ.,* 25, 479–490.

Christoffersen, R.E., Tucker, M.L. and Laties, G.G. (1984) Cellulase gene expression in ripening avocado fruit: the accumulation of cellulase mRNA and protein as demonstrated by cDNA hybridization and immunodetection. *Plant Mol. Biol.*, 3, 385–391.

Chun, J.P. and Huber, D.J. (1997) Polygalacturonase isozyme 2 binding and catalysis in cell walls from tomato fruit: pH and β-subunit effects. *Physiol. Plant.,* 101, 283–290.

Chun, J.P. and Huber, D.J. (1998) Polygalacturonase-mediated solubilization and depolymerization of pectic polymers in tomato fruit cell walls. *Plant Physiol.*, 117, 1293–1299.

Chun, J.P. and Huber, D.J. (2000) Reduced levels of β-subunit protein influence tomato fruit firmness, cell wall ultrastructure, and PG-2 mediated pectin hydrolysis in excised pericarp tissue. *J. Plant Physiol.*, 157, 153–160.

Civello, P.M., Powell, A.L.T., Sabehat, A. and Bennett, A.B. (1999) An expansin gene expressed in ripening strawberry fruit. *Plant Physiol.*, 121, 1273–1279.

Coimbra, M.A., Rigby, N.M., Selvendran, R.R. and Waldron, K.W. (1995) Investigation of the occurrence of xylan-xyloglucan complexes in the cell walls of olive pulps (*Olea euopaea*). *Carbohydr. Polym.*, 27, 277–284.

Collmer, A. and Keen, N.T. (1986) The role of pectin enzymes in plant pathogenesis. *Annu. Rev. Phytopathol.*, 24, 383–409.

Cooley, M. and Yoder, J.I. (1998) Insertional inactivation of the tomato polygalacturonase gene. *Plant Mol. Biol.*, 38, 521–530.

Cosgrove, D.J. (2000a) New genes and new biological roles for expansins. *Curr. Opin. Plant Biol.*, 3, 73–78.

Cosgrove, D.J. (2000b) Loosening of plant cell walls by expansins. *Nature*, 407, 321–326.

Cousins, S.K. and Brown, R.M. (1995) Cellulose I microfibril assembly: computational molecular mechanics energy analysis favors bonding by van der Waals forces as the initial step in crystallization. *Polymer,* 36, 3885–3888.

Crookes, P.R. and Grierson, D. (1983) Ultrastructure of tomato fruit ripening and the role of polygalacturonase isoenzymes in cell wall degradation. *Plant Physiol.*, 72, 1088–1093.

Cutillas-Iturralde, A., Zarra, I., Fry, S.C. and Lorences, E.P. (1994) Implication of persimmon fruit hemicellulose metabolism in the softening process. Importance of xyloglucan endotransglycosylase. *Physiol. Plant.*, 91, 169–176.

Dahal, P., Nevins, D.J. and Bradford, K.J. (1997) Relationship of endo-β-mannanase activity and cell wall hydrolysis in tomato endosperm to germination rates. *Plant Physiol.*, 113, 1243–1252.

Davies, L.M. and Harris, P.J. (2003) Atomic force of microscopy of microfibrils in primary cell walls. *Planta*, 217, 283–289.

Dawson, D.M., Melton, L.D. and Watkins, C.B. (1992) Cell wall changes in nectarines (*Prunus persica*): solubilization and depolymerization of pectic and neutral polymers during ripening and in mealy fruit. *Plant Physiol.*, 100, 1203–1210.

De Belie, N., Hallett, I.C., Harker, F.R. and De Baerdemaeker, J.D. (2000) Influence of ripening and turgor on the tensile properties of pears: a microscopic study of cellular and tissue changes. *J. Amer. Soc. Hort. Sci.*, 125, 350–356.

de Silva, J., Arrowsmith, D., Hellyer, A., Whiteman, S. and Robinson, S. (1994) Xyloglucan endotransglycosylase and plant growth. *J. Exp. Bot.*, 45, 1693–1701.

de Silva, J. and Verhoeven, M.E. (1998) Production and characterisation of antisense-exogalactanase tomatoes. In *Report of the Demonstration Programme on Food Safety Evaluation of*

Genetically Modified Foods as a Basis for Market Introduction. Netherlands Ministry of Economic Affairs, The Hague, pp. 99–106.

del Campillo, E. (1999) Multiple endo-1,4-β-D-glucanase (cellulase) genes in Arabidopsis. *Curr. Topics Dev. Biol.*, 46, 39–61.

del Campillo, E. and Bennett, A.B. (1996) Pedicel breakstrength and cellulase gene expression during tomato flower abscission. *Plant Physiol.*, 111, 813–820.

DellaPenna, D., Alexander, D.C. and Bennett, A.B. (1986) Molecular cloning of tomato fruit polygalacturonase: analysis of polygalacturonase mRNA levels during ripening. *Proc. Natl. Acad. Sci. USA*, 83, 6420–6424.

DellaPenna, D., Kates, D.S. and Bennett, A.B. (1987) Polygalacturonase gene expression in Rutgers, *rin, nor* and *Nr* tomato fruits. *Plant Physiol.*, 85, 502–507.

DellaPenna, D., Lincoln, J.E., Fischer, R.L. and Bennett, A.B. (1989) Transcriptional analysis of polygalacturonase and other ripening-associated genes in Rutgers, *rin, nor* and *Nr* tomato fruit. *Plant Physiol.*, 90, 1372–1377.

DellaPenna, D., Lashbrook, C.C., Toenjes, K., Giovannoni, J.J., Fischer, R.L. and Bennett, A.B. (1990) Polygalacturonase isozymes and pectin depolymerization in transgenic *rin* tomato fruit. *Plant Physiol.*, 94, 1882–1886.

DellaPenna, D., Watson, C., Liu, J. and Schuchman, D. (1996) The β-subunit of tomato fruit polygalacturonase isoenzyme 1 defines a new class of plant cell wall proteins involved in pectin metabolism: AroGPs (Aromatic amino acid rich glycoproteins). In: Visser, J, Voragen, AGJ (eds) *Pectins and Pectinases.* Elsevier Science, Amsterdam, pp. 247–262.

Delmer, D.P. (1999) Cellulose biosynthesis: exciting times for a difficult field of study. *Annu. Rev. Plant Physiol. Plant Mol. Biol.*, 50, 245–276.

Demura, T., Tashiro, G., Horiguchi, G. *et al.* (2002) Visualization by comprehensive microarray analysis of gene expression programs during transdifferentiation of mesophyll cells into xylem cells. *Proc. Natl. Acad. Sci, USA*, 99, 15794–1579.

Denès, J.-M., Baron, A., Renard, C.M.G.C., Péan, C. and Drilleau, F.-F. (2000) Different action patterns for apple pectin methylesterase at pH 7.0 and 4.5. *Carbohydr. Res.*, 327, 385–393.

Diehl, K.C. and Hamann, D.D. (1980) Relationships between sensory profile parameters and fundamental mechanical parameters for raw potatoes, melons and apples. *J. Text. Studies*, 10, 401–420.

Dominguez-Puigjaner, E., Llop, I., Vendrell, M. and Prat, S. (1997) A cDNA clone highly expressed in ripe banana fruits shows homology to pectate lyases. *Plant Physiol.*, 114, 1882–1886.

Downie, B., Dirk, L.M.A., Hadfield, K.A., Wilkins, T.A., Bennett, A.B. and Bradford, K.J. (1998) A gel diffusion assay for quantification of pectin methylesterase activity. *Anal. Biochem.*, 264, 149–157.

Drew, M.C., Jackson, M.B. and Giffard, S. (1979) Ethylene promoted adventitious rooting and development of cortical air spaces in *Zea mays* L. *Planta*, 147, 83–88.

Durbin, M. and Lewis, L. (1988) Cellulases in *Phaseolus vulgaris. Meth. Enzymol.*, 160, 342–351.

Dutta, S., Bradford, K.J. and Nevins, D.J. (1994) Cell-wall autohydrolysis in isolated endosperms of lettuce (*Lactuca sativa* L.). *Plant Physiol.*, 104, 623–628.

Errington, N., Tucker, G.A. and Mitchell, J.R. (1998) Effect of genetic down-regulation of polygalacturonase and pectin esterase on rheology and composition of tomato juice. *J. Sci. Food Agric.*, 76, 515–519.

Esquerré-Tugayé, M.-T., Boudart, G. and Dumas, B. (2000) Cell wall degrading enzymes, inhibitory proteins, and oligosaccharides participate in the molecular dialogue between plants and pathogens. *Plant Physiol. Biochem.*, 38, 157–163.

Faik, A., Desveaux, D. and Maclachlan, G. (1998) Enzymic activities responsible for xyloglucan depolymerization in extracts of developing tomato fruits. *Phytochem.*, 49, 365–376.

Fanutti, C., Gidley, M.J. and Reid, J.S.G. (1993) Action of a pure xyloglucan endo-transglycosylase (formerly called xyloglucan-specific endo-(1)-β-D-glucanase) from the cotyledons of germinated nasturtium seeds. *Plant J.*, 3, 691–700.

Farkaš, V., Sulová, Z., Stratilova, E., Hanna, R. and Maclachlan, G. (1992) Cleavage of xyloglucan by nasturtium seed xyloglucanase and transglycosylation to xyloglucan subunit oligosaccharides. *Arch. Biochem. Biophys.,* 298, 365–370.

Fenwick, K.M., Jarvis, M.C., Apperley, D.C., Seymour, G.B. and Bird, C.R. (1996) Polymer mobility in cell walls of transgenic tomatoes with reduced polygalacturonase activity. *Phytochem.,* 42, 310–307.

Ferrarese, L., Trainotti, L., Moretto, P., Polverino de Laureto, P., Rascio, N. and Casadoro, G. (1995) Differential ethylene-inducible expression of cellulases in pepper plants. *Plant Mol. Biol.,* 29, 735–747.

Fischer, R.L. and Bennett, A.B. (1991) Role of cell wall hydrolases in fruit ripening. *Annu. Rev. Plant Physiol. Plant Mol. Biol.,* 42, 675–703.

Fishman, M.L., Gross, K.C., Gillespie, D.T. and Sondey, S.M. (1989) Macromolecular components of tomato fruit pectin. *Arch. Biochem. Biophys.,* 274, 179–191.

Frey-Wyssling, A. (1954) The fine structure of cellulose microfibrils. *Science,* 119, 80–82.

Fry, S.C. (1986) Cross-linking of matrix polymers in the growing cell wall of angiosperms. *Annu. Rev. Plant Physiol.,* 37, 165–1286.

Fry, S.C. (1998) Oxidative scission of plant cell wall polysaccharides by ascorbate-induced hydroxyl radicals. *Biochem. J.,* 332, 507–515.

Fry, S.C., Dumville, J.C. and Miller, J.G. (2001) Fingerprinting of polysaccharides attacked by hydroxyl radicals in vitro and in the cell walls of ripening pear fruit. *Biochem. J.,* 357, 729–737.

Fukuda, H. (1996) Xylogenesis: initiation, progression and cell death. *Annu. Rev. Plant Physiol. Plant Mol. Biol.,* 47, 299–325.

Fukuda, H. and Komamine, A. (1980) Establishment of an experimental system for the tracheary element differentiation for single cells isolated from the mesophyll of *Zinnia elegans*. *Plant Physiol.,* 52, 57–60.

Gaffe, J., Tieman, D.M. and Handa, A.K. (1994) Pectin methylesterase isoforms in tomato (*Lycopercison esculentum*) tissues. *Plant Physiol.,* 105, 199–203.

Garciaromera, I. and Fry, S.C. (1994) Absence of transglycosylation with oligogalacturonides in plant cells. *Phytochem.,* 35, 67–72.

Giovannoni, J.J. (2001) Molecular biology of fruit maturation and ripening. *Annu. Rev. Plant Physiol. Plant Mol. Biol.,* 52, 725–749.

Giovannoni, J.J., DellaPenna, D., Bennett, A.B. and Fischer, R.L. (1989) Expression of a chimeric polygalacturonase gene in transgenic *rin* (*ripening inhibitor*) tomato fruit results in polyuronide degradation but not fruit softening. *Plant Cell,* 1, 53–63.

Giovannoni, J.J., DellaPenna, D. and Bennett, A.B. (1991) Polygalacturonase and tomato fruit ripening. *Hortic. Rev.,* 13, 67–103.

Glover, H. and Brady, C. (1994) Purification of three pectin esterases from ripe peach fruit. *Phytochem.,* 37, 949–955.

Goellner, M., Wang, X.H. and Davis, E.L. (2001) Endo-beta-1,4-glucanase expression in compatible plant-nematode interactions. *Plant Cell,* 13, 2241–2255.

Gonzalez-Bosch, C., Brummell, D.A. and Bennett, A.B. (1997) Differential expression of two endo-1,4-beta-glucanase genes in pericarp and locules of wild-type and mutant tomato fruit. *Plant Physiol.,* 111, 1313–1319.

Gonzalez-Carranza, Z.H., Lozoya-Gloria, E. and Roberts, J.A. (1998) Recent developments in abscission: shedding light on the shedding process. *Trends Plant Sci.,* 3, 10–13.

Gonzalez-Carranza, Z.H., Whitelaw, C.A., Swarup, R. and Roberts, J.A. (2002) Temporal and spatial expression of a polygalacturonase during leaf and flower abscission in oilseed rape and Arabidopsis. *Plant Physiol.,* 128, 534–543.

Grant, G.T., Morris, E.R., Rees, D.A., Smith, P.J.C. and Thom, D. (1973) Biological interactions between polysaccharides and divalent cations: the egg box model. *FEBS Lett.,* 32, 195–198.

Greve, L.C. and Labavitch, J.M. (1991) Cell wall metabolism in ripening fruit. *Plant Physiol.,* 97, 1456–1461.

Groot, S.P.C., Kieliszewska-Rokicka, B., Vermeer, E. and Karssen, C.M. (1987) Gibberellin-induced hydrolysis of endosperm cell walls in gibberellin-deficient tomato seed prior to radicle protrusion. *Planta*, 174, 500–504.

Groot, S.P.C. and Karssen, C.M. (1988) Gibberellins regulate seed germination in tomato by endosperm weakening: a study with gibberellin-deficient mutants. *Planta*, 171, 525–531.

Gross, K.C. (1984) Fractionation and partial characterization of cell walls from normal and non-ripening mutant tomato fruit. *Physiol. Plant*, 62, 25–32.

Gross, K.C. (1985) Promotion of ethylene evolution and ripening of tomato fruit by galactose. *Plant Physiol.*, 79, 306–307.

Gross, K.C. and Sams, C.E. (1984) Changes in cell wall neutral sugar composition during fruit ripening: a species survey. *Phytochem.*, 23, 2547–2461.

Gross, K.C. and Wallner, S.J. (1979) Degradation of cell wall polysaccharides during tomato fruit ripening. *Plant Physiol.*, 63, 117–120.

Gross, K.C., Watada, K.E., Kang, M.S., Kim, S.D., Kim, K.S. and Lee, S.W. (1986) Biochemical changes associated with the ripening of hot pepper fruit. *Physiol. Plant.*, 66, 31–36.

Gross, K.C., Starrett, D.A. and Chen, H.L. (1995) Rhamnogalacturonase, β-galactosidase and α-galactosidase: potential role in fruit softening. *Acta Hort.*, 398, 121–130.

Gunawaredena, A.H.L.A.N., Pearce, D.M.E., Jackson, M.B., Hawes, C.R. and Evans, D.E. (2001) Rapid changes in cell wall pectic polysaccharide are closely associated with early stages of aerenchyma formation, a spatially localized form of programmed cell death in roots of maize (*Zea mays* L.) promoted by ethylene. *Plant Cell Environ.*, 24, 1369–1375.

Ha, M.-A., Apperley, D.C., Evans, B.W. *et al.* (1998) Fine structure in cellulose microfibrils: NMR evidence from onion and quince. *Plant J.*, 16, 183–190.

Hackney, J.M., Atalla, R.H. and VanderHart, D.L. (1999) Modification of crystallinity and crystalline structure of *Acetobacter xylinum* cellulose in the presence of water-soluble β-1,4-linked polysaccharides: ^{13}C-NMR evidence. *Int. J. Macromol.*, 16, 215–218.

Hadfield, K.A. and Bennett, A.B. (1998) Polygalacturonases: many genes in search of a function. *Plant Physiol.*, 117, 337–343.

Hadfield, K.A., Rose, J.K.C., Yaver, D.S., Berka, R.M. and Bennett, A.B. (1998) Polygalacturonase gene expression in ripe melon fruit supports a role for polygalacturonase in ripening-associated pectin disassembly. *Plant Physiol.*, 117, 363–373.

Hall, L.N., Tucker, G.A., Smith, C.J.S. *et al.* (1993) Antisense inhibition of pectin esterase gene expression in transgenic tomatoes. *Plant J.*, 3, 121–129.

Hallett, I.C., MacRae, E.A. and Wegrzyn, T.F. (1992) Changes in kiwifruit cell wall ultrastructure and cell packing during postharvest ripening. *Int. J. Plant Sci.*, 153, 49–60.

Ham, K.-S., Wu, S.-C., Darvill, A.G. and Albersheim, P. (1997) Fungal pathogens secrete an inhibitor protein that distinguishes isoforms of plant pathogenesis-related endo-β-1,3-glucanases. *Plant J.*, 12, 169–179.

Harker, F.R., Redgwell, R.J., Hallett, I.C., Murray, S.H. and Carter, G. (1997) Texture of fresh fruit. *Hort. Rev.*, 20, 121–224.

Harpster, M.H., Lee, K.Y. and Dunsmuir, P. (1997) Isolation and characterization of a gene encoding endo-β-1,4-glucanase from pepper (*Capsicum annum* L.). *Plant Mol. Biol.*, 33, 47–59.

Harpster, M.A., Brummell, D.A. and Dunsmuir, P. (2002a) Suppression of a ripening-related endo-1,4-β-glucanase in transgenic pepper fruit does not prevent depolymerization of cell wall polysaccharides during ripening. *Plant Mol. Biol.*, 50, 345–355.

Harpster, M.A., Dawson, D.M., Nevins, D.J., Dunsmuir, P. and Brummell, D.A. (2002b) Constitutive overexpression of a ripening-related pepper endo-1,4-β-glucanase in transgenic tomato fruit does not increase xyloglucan depolymerization or fruit softening. *Plant Mol. Biol.*, 50, 357–369.

Harriman, R.W. (1990) Molecular cloning and regulation of expression of pectin methylesterase in ripening tomato. Ph.D. thesis, Purdue University, West Lafayette, Ind.

Harrison, E.P., McQueen-Mason, S.J. and Manning, K. (2001) Expression of six expansin genes in relation to extension activity in developing strawberry fruit. *J. Exp. Bot.*, 52, 1437–1446.

Hatfield, R. and Nevins, D.J. (1986) Characterization of the hydrolytic activity of avocado cellulase. *Plant Cell Physiol.*, 27, 541–552.

Hatfield, R.D., Ralph, J. and Grabber, J.H. (1999) Cell wall cross-linking by ferulates and diferulates in grasses. *J. Sci. Food Agric.*, 79, 403–407.

Hayama, H., Shimada, T., Haji, T., Ito, A., Kashimura, Y. and Yoshioka, H. (2000) Molecular cloning of a ripening-related expansin cDNA in peach: evidence for no relationship between expansin accumulation and change in fruit firmness during storage. *J. Plant Physiol.*, 157, 567–573.

Hayashi., T., Wong, Y.-S. and Maclachlan, G. (1984) Pea xyloglucan and cellulose. II. Hydrolysis by pea endo-1,4-β-glucanase. *Plant Physiol.*, 75, 605–610.

He, C.-J., Drew, M.C. and Morgan, P.W. (1994) Induction of enzymes associated with lysigenous aerenchyma formation in roots of *Zea mays* during hypoxia or nitrogen starvation. *Plant Physiol.*, 105, 861–865.

Henderson, J., Davies, H.A., Heyes, S.J. and Osborne, D.J. (2001a) The study of a monocotyledon abscission zone using microscopic, chemical, enzymatic and solid state ^{13}C CP/MAS NMR analyses. *Phytochem.*, 56, 131–39.

Henderson, J., Lyne, L. and Osborne, D. (2001b) Failed expression of an *endo*-β-1,4-glucanhydrolase (cellulase) in a non-abscinding mutant of *Lupinus angustifolius* cv. Danja. *Phytochem.*, 58, 1025–1034.

Henrissat, B. and Bairoch, A. (1993) New families in the classification of glycosyl hydrolases based on amino acid sequence similarities. *Biochem. J.*, 293, 781–788.

Henrissat, B. and Bairoch, A. (1996) Updating the sequence-based classification of glycosyl hydrolases. *Biochem. J.*, 316, 695–696.

Henrissat, B., Coutinho, P.M. and Davies, G.J. (2001) A census of carbohydrate-active enzymes in the genome of *Arabidopsis thaliana*. *Plant Mol. Biol.*, 47, 55–72.

Hiwasa, K., Rose, J.K.C., Nakano, R., Inaba, A. and Kubo, Y. (2003) Differential expression of seven expansin genes during growth and ripening in pear fruit. *Physiol. Plant.* 117, 564–572.

Hobson, G.E. (1962) Determination of polygalacturonase in fruits. *Nature*, 195, 804–805.

Hong, S.B. and Tucker, M.L. (2000) Molecular characterization of a tomato polygalacturonase gene abundantly expressed in the upper third of pistils from open and unopened flowers. *Plant Gene Reporter,* 19, 680–683.

Hoondal, G.S., Tiwari, R.P., Tewari, R., Dahiya, N. and Beg, Q.K. (2002) Microbial alkaline pectinases and their industrial applications: a review. *Appl. Microbiol. Biotech.*, 59, 409–418.

Huang, B., Johnson, J.W., Box, J.E. and NeSmith, D.S. (1997) Root characteristics and hormone activity of wheat in response to hypoxia and ethylene. *Crop Sci.*, 37, 812–818.

Huber, D.J. (1983) The role of cell wall hydrolases in fruit softening. *Hort. Rev.*, 5, 169–219.

Huber, D.J. (1984) Strawberry fruit softening: the potential roles of polyuronides and hemicelluloses. *J. Food Sci.*, 49, 1310–1315.

Huber, D.J. (1991) Acidified phenol alters tomato cell wall pectin solubility and calcium content. *Phytochem.*, 30, 2523–2527.

Huber, D.J. and O'Donoghue, E.M. (1993) Polyuronides in avocado (*Persea americana*) and tomato (*Lycopersicon esculentum*) fruits exhibit markedly different patterns of molecular weight downshifts during ripening. *Plant Physiol.*, 102, 473–480.

Huber, D.J., Karakurt, Y. and Jeong, J. (2001) Pectin degradation in ripening and wounded fruits. *Brazilian J. Plant Physiol.*, 13, 224–241.

Huberman, M., Pressman, E. and Jaffe, M. (1993) Pith autolysis in plants: IV. The activity of polygalacturonase and cellulase during drought stress induced pith autolysis. *Plant Cell Physiol.*, 34, 795–801.

Huysamer, M., Greve, L.C. and Labavitch, J.M. (1997a) Cell wall metabolism in ripening fruit VIII. Cell wall composition and synthetic capacity of two regions of the outer pericarp of mature green and red ripe cv. Jackpot tomatoes. *Physiol. Plant.*, 101, 314–322.

Huysamer, M., Greve, L.C. and Labavitch, J.M. (1997b) Cell wall metabolism in ripening fruit IX. Synthesis of pectic and hemicellulosic cell wall polymers in the outer pericarp of mature green tomatoes (cv. XMT-22). *Plant Physiol.,* 114, 1523–1531.

Hwang, Y.-S., Huber, D.J. and Albrigo, G. 1990) Comparison of cell wall components in normal and disordered juice vesicles of grapefruit. *J. Am. Soc. Hort. Sci.,* 115, 281–287.

Ishiguro S., Kawai-Oda A., Ueda J., Nishida I. and Okada K. (2001) The *DEFECTIVE IN AN-THER DEHISCENCE1* gene encodes a novel phospholipase a1 catalyzing the initial step of jasmonic acid biosynthesis, which synchronizes pollen maturation, anther dehiscence, and flower opening in Arabidopsis. *Plant Cell,* 13, 2191–2209.

Ishimaru, M. and Kobayashi, S. (2002) Expression of a xyloglucan endotransglycosylase gene is closely related to grape berry softening. *Plant Sci.,* 162, 621–628.

Jackman, R.L. and Stanley, D.W. (1992) Area- and perimeter-dependent properties and failure of mature-green and red-ripe tomato pericarp tissues. *J. Tex. Stud.,* 23, 461–474.

Jackson, M.B. and Armstrong, W. (1999) Formation of aerenchyma and the processes of plant ventilation in relation to soil flooding and submergence. *Plant Biol.,* 1, 274–287.

Jarvis, M.C. (1984) Structure and properties of pectin gels in plant cell walls. *Plant Cell Environ.,* 7, 153–164.

Jaubert, S., Laffaire, J.-B., Abad, P. and Rosso, M.-N. (2002) A polygalacturonase of animal origin isolated from the root-knot nematode *Meloidogyne incognita. FEBS Lett.,* 522, 109–112.

Jedrzejas, M.J. (2000) Structural and functional comparisons of polysaccharide-degrading enzymes. *Crit. Rev. Biochem. Mol. Biol.,* 35, 221–251.

Jen, J.J. and Robinson, M.L. (1984) Pectolytic enzymes in sweet bell peppers (*Capsicum annum* L.). *J. Food Sci.,* 49, 1085–1087.

Jenkins, E., Paul, W., Coupe, S.A., Bell, S.J., Davies, E.C. and Roberts, J.A. (1996) Characterization of an mRNA encoding a polygalacturonase expressed during pod development in oilseed rape (*Brassica napus* L). *J. Exp. Bot.,* 47, 111–115.

Jenkins, E., Paul, W., Craze, M., Whitelaw, C., Weigand, A. and Roberts, J.A. (1999) Dehiscence-related expression of an *Arabidopsis thaliana* gene encoding a polygalacturonase in transgenic plants of *Brassica napus. Plant Cell Environ.,* 22, 159–168.

Jimenez, A., Creissen, G., Kular, B. *et al.* (2002) Changes in oxidative processes and components of the antioxidant system during tomato fruit ripening. *Planta,* 214, 751–858.

Jiménez-Bermúdez, S., Redondo-Nevado, J., Muñoz-Blanco, J. *et al.* (2002) Manipulation of strawberry fruit softening by antisense expression of a pectate lyase gene. *Plant Physiol.,* 128, 751–759.

Johansson, K., El-Ahmad, M., Friemann, R., Jörnvall, H., Markovič, O. and Eklund, H. (2002) Crystal structure of plant pectin methylesterase. *FEBS Lett.,* 514, 243–249.

Kalaitzis, P., Solomos, T. and Tucker, M.L. (1997) Three different polygalacturonases are expressed in tomato leaf and flower abscission, each with a different temporal expression pattern. *Plant Physiol.,* 113, 1303–1308.

Kalaitzis, P., Hong, S.-B., Solomos, T. and Tucker, M.L. (1999) Molecular characterization of a tomato endo-β-1,4-glucanase gene expressed in mature pistils, abscission zones and fruit. *Plant Cell Physiol.,* 40, 905–908.

Kauss, H. and Hassid, W.Z. (1967) Enzymic introduction of the methyl ester groups of pectin. *J. Biol. Chem.,* 242, 3449–3453.

Kawase, M. (1979) Role of cellulase in aerenchyma development in sunflower. *Amer. J. Bot.,* 66, 183–190.

Kim, J., Gross, K.C. and Solomos, T. (1987) Characterization of the stimulation of ethylene production by galactose in tomato (*Lycopersicon esculentum* Mill.) fruit. *Physiol. Plant,* 85, 804–807.

Kitagawa, Y., Kanayama, Y. and Yamaki, S. (1995) Isolation of β-galactosidase fractions from Japanese pear: activity against native cell wall polysaccharides. *Physiol. Plant.,* 93, 545–550.

Koch, J.L. and Nevins, D.J. (1989) Tomato fruit cell wall. I. Use of purified tomato polygalacturonase and pectin methylesterase to identify developmental changes in pectins. *Plant Physiol.*, 91, 816–822.

Koh, T.H., Melton, L.D. and Newman, R.H. (1997) Solid state ^{13}C NMR characterization of cell walls of ripening strawberries. *Can. J. Bot.*, 75, 1957–1964.

Kojima, K., Sakurai, N. and Kuraishi, S. (1994) Fruit softening in banana: correlation among stress-relaxation parameters, cell wall components and starch during ripening. *Physiol. Plant.*, 90, 772–778.

Koyama, M., Sugiyama, J. and Itoh, I. (1997) Systematic survey on crystalline features of algal celluloses. *Cellulose*, 4, 147–160.

Kramer, M., Sanders, R., Bolkan, H., Waters, C., Sheehy, R.E. and Hiatt, W.R. (1992) Postharvest evaluation of transgenic tomatoes with reduced levels of polygalacturonase: processing, firmness and disease resistance. *Postharvest Biol. Technol.*, 1, 241–255.

Labavitch, J.M. and Greve, L.C. (1983) Cell wall metabolism in ripening fruit. III. Purification of an endo-β-1,4-xylanase that degrades a structural polysaccharide of pear fruit cell walls. *Plant Physiol.*, 72, 68–673.

Laine, M.J., Haapalainen, M., Wahlroos, T. *et al.* (2000) The cellulase encoded by the native plasmid of *Clavibacter michiganensis ssp sepedonicus* plays a role in virulence and contains an expansin-like domain. *Physiol. Mol. Plant Pathol.*, 57, 221–233.

Lang, C. and Dörnenburg, H. (2000) Perspectives in the biological function and the technological application of polygalacturonases. *Appl. Microbiol. Biotechnol.*, 53, 366–375.

Langley, K.R., Martin, A., Stenning, R. *et al.* (1994) Mechanical and optical assessment of the ripening of tomato fruit with reduced polygalacturonase activity. *J. Sci. Food Agric.*, 66, 547–554.

Lashbrook, C.C., Gonzalez-Bosch, C. and Bennett, A.B. (1994) Two divergent endo-β-1,4-glucanase genes exhibit overlapping expression in ripening fruits and abscising flowers. *Plant Cell*, 6, 1485–1493.

Lashbrook, C.C., Giovannoni, J.J., Hall, B.D., Fischer, R.L. and Bennett, A.B. (1998) Transgenic analysis of tomato endo-β-1,4-glucanase gene function. Role of *cel1* in floral abscission. *Plant J.*, 13, 303–310.

Lau, J.M., McNeil, M., Darvill, A.G. and Albersheim, P. (1985) Structure of backbone of rhamnogalacturonan I, a pectic polysaccharide in the primary walls of plants. *Carbohydr. Res.*, 137, 111–125.

Lazan, H., Selemat, M.K. and Ali, Z.M. (1995) β-galactosidase, polygalacturonase and pectinesterase in differential softening and cell wall modification during papaya fruit ripening. *Physiol. Plant.*, 95, 106–112.

Leckie, F., Mattie, B., Capodicasa, C. *et al.* (1999) The specificity of polygalacturonase inhibiting protein (PGIP): a single amino acid substitution in the solvent-exposed β-strand/-turn region of the leucine-rich repeats (LRRs) confers a new recognition capability. *EMBO J.*, 18, 2352–2363.

Lester, D.R., Speirs, J., Orr, G. and Brady, C.J. (1994) Peach (*Prunus persica*) endopolygalacturonase cDNA isolation and mRNA analysis in melting and non-melting peach cultivars. *Plant Physiol.*, 105, 225–231.

Lester, D.R, Sherman, W.B. and Atwell, B.J. (1996) Endopolygalacturonase and the melting flesh (M) locus in peach. *J. Amer. Soc. Hort. Sci.,* 121, 231–234.

Leviatov, S., Shoseyov, O. and Wolf, S. (1995) Involvement of endomannanase in the control of tomato seed germination under low temperature conditions. *Ann. Bot.*, 76, 1–6.

Levy, I, Shani, Z. and Shoseyov, O. (2002) Modification of polysaccharides and plant cell wall by endo-1,4-β-glucanase and cellulose-binding domain. *Biomol. Eng.*, 19, 17–30.

Lewis, L., and Varner, J. (1970) Synthesis of cellulase during abscission of *Phaseolus vulgaris* leaf explants. *Plant Physiology*, 46, 194–199.

Lievens, S., Goormachtig, S., Herman, S. and Holsters, M. (2002) Patterns of pectin methyl-esterase transcripts in developing stem nodules of *Sesbania rostrata*. *Mol. Plant-Microbe Interact.*, 15, 164–168.

Lin, Y.-T. and Pitt, R.E. (1986) Rheology of apple and potato tissue as affected by cell turgor pressure. *J. Text. Stud.*, 17, 291–313.

Llop-Tous, I., Dominuez-Puigjaner, E., Palomer, X. and Vendrell, M. (1999) Characterization of two divergent endo-β-1,4-glucanase cDNA clones highly expressed in nonclimacteric strawberry fruit. *Plant Physiol.*, 119, 1415–1421.

MacDougall, A.J., Needs, P.W., Rigby, N.M. and Ring, S.G. (1996) Calcium gelation of pectic polysaccharides isolated from unripe tomatoes. *Carbohydr. Res.*, 923, 235–249.

MacDougall, A.J., Rigby, N.M., Ryden, P., Tibbits, C.W. and Ring, S.G. (2001) Swelling behavior of the tomato cell wall network. *Biomacromol.*, 2, 450–455.

Maclachlan, G. (1988) β-glucanases from *Pisum sativum*. *Meth. Enzymol.*, 160, 382–391.

Maclachlan, G. and Brady, C.J. (1992) Multiple forms of 1,4-β-glucanase in ripening tomato fruits include a xyloglucanase activatable by xyloglucan oligosaccharides. *Aust. J. Plant Physiol.*, 19, 137–146.

Maclachlan, G. and Brady, C.J. (1994) Endo-1,4-β-glucanase, xyloglucanase and xyloglucan endo-transglycosylase activities versus potential substrates in ripening tomatoes. *Plant Physiol.*, 105, 965–974.

Maclachlan, G. and Wong, Y.-S. (1979) Two pea cellulases display the same catalytic mechanism despite major differences in physical properties. *Adv. Chem. Ser.*, 181, 347–360.

MacRae, E. and Redgwell, R.J. (1992) Softening in kiwifruit. *Postharvest News Info.*, 3, 49N-52N.

Mahlberg, P.G. (1993) Laticifers: an historical perspective. *Bot. Rev.*, 59, 1–23.

Marcelin, O. Williams, P. and Brillouet, J.-M. (1993) Isolation and characterization of the two main cell-wall types from Guava (*Psidium guajava* L.). *Carbohydr. Res.*, 240, 233–243.

Marín-Rodríguez, M.C., Orchard, J. and Seymour, G.B. (2002) Pectate lyases, cell wall degradation and fruit softening. *J. Exp. Bot.*, 53, 2115–2119.

Martin-Cabrejas, M.A., Waldron, K.W. and Selvendran, R.R. (1994) Cell wall changes in Spanish pear during ripening. *J. Plant Physiol.*, 144, 541–548.

Mbeguie-A-Mbeguie, D., Gouble, B., Gomez, R.M., Audergon, J.M., Albagnac, G. and Fils-Lycaon, B. (2002) Two expansin cDNAs from *Prunus armeniaca* expressed during fruit ripening are differently regulated by ethylene. *Plant Physiol. Biochem.*, 40, 445–452.

McCann, M. (1997) Tracheary element formation: building up to a dead end. *Trends Plant Sci.*, 2, 333–338.

McCollum, T.G., Huber, D.J. and Cantcliffe, D.J. (1989) Modification of polyuronides and hemicelluloses during muskmelon fruit softening. *Physiol. Plant.*, 76, 303–308.

McLauchlan, W.R., Garcia-Coneas, M.T., Williamson, G., Rza, M., Ravenstein, P. and Maat, J. (1999) A novel class of protein from wheat which inhibits xylanases. *Biochem. J.*, 338, 441–446.

McQueen-Mason, S. and Cosgrove, D.J. (1994) Disruption of hydrogen bonding between plant cell wall polymers by proteins that induce wall extension. *Proc. Natl. Acad. Sci. USA*, 91, 6574–6578.

McQueen-Mason, S. and Cosgrove, D.J. (1995) Expansin mode of action on cell walls. Analysis of wall hydrolysis, stress relaxation, and binding. *Plant Physiol.*, 107, 87–100.

McQueen-Mason, S., Durachko, D.M. and Cosgrove, D.J. (1992) Two endogenous proteins that induce cell wall extension in plants. *Plant Cell*, 4, 1425–1433.

Meakin, P.J. and Roberts, J.A. (1990a) Dehiscence of fruit in oilseed rape (*Brassica napus* L.). I. Anatomy of pod dehiscence. *J. Exp. Bot.*, 41, 995–1002.

Meakin, P.J. and Roberts, J.A. (1990b) Dehiscence of fruit in oilseed rape (*Brassica napus* L.). II. The role of cell wall degrading enzymes and ethylene. *J. Exp. Bot.*, 41, 1003–1011.

Medina-Escobar, N., Cardenas, J., Moyano, E., Caballero, J.L. and Muñoz-Blanco, J. (1997) Cloning, molecular characterization and expression of a strawberry ripening-specific cDNA with sequence homology to pectate lyase from higher plants. *Plant Mol. Biol.*, 34, 867–877.

Micheli, F. (2001) Pectin methylesterases: cell wall enzymes with important roles in plant physiology. *Trends Plant Sci.*, 6, 414–429.

Micheli, F., Holliger, C., Goldberg, R. and Richard, L. (1998) Characterization of the pectin methylesterase-like gene *AtPME3*: a new member of a gene family comprising at least 12 genes in *Arabidopsis thaliana*. *Gene*, 220, 13–20.

Miller, A.R. (1986) Oxidation of cell wall polysaccharides by hydrogen peroxide: a potential mechanism for cell wall breakdown in plants. *Biochem. Biophys. Res. Comm.*, 141, 238–244.

Miller, J.G. and Fry, S.C. (2001) Characteristics of xyloglucan after attack by hydroxyl radicals. *Carbohydr. Res.*, 332, 389–403.

Milioni, D., Sado, P.-E.R., Stacey, N.J., Domingo, D.C., Roberts, K. and McCann, M. (2001) Differential expression of cell-wall-related genes during the formation of tracheary elements in the *Zinnia* mesophyll cell system. *Plant Mol. Biol.*, 47, 221–238.

Milioni, D., Sado, P.-E.R., Stacey, N.J., Roberts, K. and McCann, M. (2002) Early gene expression associated with the commitment and differentiation of a plant treachery element is revealed by cDNA-amplified fragment length polymorphism analysis. *Plant Cell*, 14, 2813–2824.

Moore, T. and Bennett, A.B. (1994) Tomato fruit polygalacturonase isoenzyme 1. Characterization of the β-subunit and its state of assembly *in vivo*. *Plant Physiol.*, 106, 1461–1469.

Morvan, O., Quentin, M., Jauneau, A., Mareck, A. and Morvan, C. (1998) Immunogold localization of pectin methylesterase in the cortical tissues of flax hypocotyl. *Protoplasma*, 202, 175–184.

Munoz, J.A., Coronado, C., Perez-Hormaeche, J., Kondorosi, A., Ratet, P. and Palomares, A.J. (1998) *MsPG3*, a *Medicago sativa* polygalacturonase gene expressed during the alfalfa *Rhizobium meliloti* interaction. *Proc. Natl. Acad., Sci. USA*, 95, 987–9692.

Murashima, K., Kosugi, A. and Doi, R.H. (2002) Synergistic effects on crystalline cellulose degradation between cellulosomal cellulases from *Clostridium cellulovorans*. *J. Bacteriol.*, 184, 5088–5095.

Nairn, C.J., Lewandowski, D.J. and Burns, J.K. (1998) Genetics and expression of two pectinesterase genes in Valencia orange. *Physiol. Plant.*, 103, 226–235.

Nakamura, S. and Hayashi, T. (1993) Purification and properties of an extracellular endo-1,4-β-glucanase from suspension-cultured poplar cells. *Plant Cell Physiol.*, 34, 1009–1013.

Neelam, P., Sanjay, M. and Sanwal, G.G. (2000) Purification and characterization of polygalacturonase from banana fruit. *Phytochem.*, 54, 147–152.

Newman, R.H. and Redgwell, R. (2002) Cell wall changes in ripening kiwifruit: ^{13}C solid state NMR characterisation of relatively rigid cell wall polymers. *Carbohydr. Polymers*, 49, 121–129.

Newman, R.H., Davies, L.M. and Harris, P. J. (1996) Solid-state ^{13}C nuclear magnetic resonance characterization of cellulose in the cell walls of *Arabidopsis thaliana* leaves. *Plant Physiol.*, 111, 475–485.

Nishitani, K. (1997) The role of endoxyloglucan transferase in the organization of plant cell walls. *Int. Rev. Cytol.*, 173, 157–206.

Nishitani, K. and Tominaga, R. (1992) Endo-xyloglucan transferase, a novel class of glycosyltransferase that catalyzes transfer of a segment of xyloglucan molecule to another xyloglucan molecule. *J. Biol. Chem.*, 267, 21058–21064.

Nomaguchi, M., Nonogaki, H. and Morohashi, Y. (1995) Development of galactomannan-hydrolyzing activity in the micropylar endosperm tip of tomato seed prior to germination. *Physiol. Plant.*, 94, 105–109.

Nonogaki, H. and Morohashi, Y. (1996) A specific endo-β-mannanase develops exclusively in the micropylar endosperm of tomato seeds prior to radicle emergence. *Plant Physiol.*, 110, 555–559.

Nonogaki, H., Gee, O.H. and Bradford, K.J. (2000) A germination-specific endo-β-mannanase gene is expressed in the micropylar endosperm cap of tomato seeds. *Plant Physiol.*, 123, 1235–1245.

Nunan, K.J., Sims, I.M., Bacic, A., Robinson, S.P. and Fincher, G.B. (1998) Changes in cell wall composition during ripening of grape berries. *Plant Physiol.*, 118, 783–792.

Nunan, K.J., Davies, C., Robinson, S.P. and Fincher, G.B. (2001) Expression patterns of cell wall-modifying enzymes during grape berry development. *Planta*, 214, 257–264.

O'Donoghue, E.M. and Huber, D.J. (1992) Modification of matrix polysaccharides during avocado (*Persea americana*) fruit ripening: an assessment of the role of Cx-cellulase. *Physiol. Plant.*, 86, 33–42.

O'Donoghue, E.M., Huber, D.J., Timpa, J.D., Erdos, G.W. and Brecht, J.K. (1994) Influence of avocado (*Persea americana*) Cx-cellulase on the structural features of avocado cellulose. *Planta*, 194, 573–584.

Ohmiya, Y., Samjima, M., Shiroishi, M. *et al.* (2000) Evidence that endo-1,4-β-glucanases act on cellulose in suspension-cultured poplar cells. *Plant J.*, 24, 147–158.

Oosterveld, A., Beldman, G., Searle-van-Leeuwen, M.J.F. and Voragen, A.G.J. (2000) Effect of enzymatic deacetylation on gelation of sugar beet pectin in the presence of calcium. *Carbohydr. Polymers*, 43, 249–256.

Orfila, C. and Knox, J.P. (2000) Spatial regulation of pectic polysaccharides in relation to pit fields in cell walls of tomato fruit pericarp. *Plant Physiol.*, 122, 775–781.

Orfila, C., Seymour, G.B., Willats, W.G.T. *et al.* (2001) Altered middle lamella homogalacturonan and disrupted deposition of (1→5)-α-arabinan in the pericarp of *Cnr*, a ripening mutant of tomato. *Plant Physiol.*, 126, 210–221.

Orfila, C. Huisman, M.M.H., Willats, W.G.T. *et al.* (2002) Altered cell wall disassembly during ripening of *Cnr* tomato fruit: implications for cell adhesion and fruit softening. *Planta*, 215, 440–447.

Osborne, D.J. and Sargent, J.A. (1976) The positional differentiation of ethylene-responsive cells in rachis abscission zones in leaves of *Sambucus nigra* and their growth and ultrastructural changes at senescence and separation. *Planta*, 130, 203–210.

O'Sullivan, A.C. (1997) Cellulose; the structure slowly unravels. *Cellulose*, 4, 173–207.

Parniske, M. (2000) Intracellular accommodation of microbes by plants: a common developmental program for symbiosis and disease? *Curr Opin. Plant Biol.*, 3, 320–328.

Paull, R.E. and Chen, N.J. (1983) Postharvest variation in cell wall-degrading enzymes of papaya (*Caria papaya* L.) during fruit ripening. *Plant Physiol.*, 72, 382–385.

Pauly, M., Albersheim, P., Darvill, A.G. and York, W.S. (1999) Molecular domains of the cellulose/xyloglucan network in the cell walls of higher plants. *Plant J.*, 20, 629–639.

Percy, A.E., O'Brien, I.E.W., Jameson, P.E., Melton, L.D., MacRae, E.A. and Redgwell, R.J. (1996) Xyloglucan endotransglycosylase activity during fruit development and ripening of apple and kiwifruit. *Physiol. Plant.*, 96, 43–50.

Pérez, S., Mazeau, K. and Hervé du Penhoat, C. (2000) The three dimensional structures of the pectic polysaccharides. *Plant Physiol. Biochem.*, 38, 37–55.

Perotto, S., Coisson, J.D., Perugini, I., Cometti, V. and Bonfante, P. (1997) Production of pectin-degrading enzymes by ericoid mycorrhizal fungi. *New Phytol.*, 135, 151–162.

Peterson, M., Sander, L., Child, R., van Onckelen, H., Ulvskov, P. and Borkhardt, B. (1996) Isolation and characterization of a pod dehiscence zone-specific polygalacturonase from *Brassica napus*. *Plant Mol. Biol.*, 31, 517–527.

Pilatzke-Wunderlich, I. and Nessler, C.L. (2001) Expression and activity of cell-wall-degrading enzymes in the latex of opium poppy, *Papaver somniferum* L. *Plant Mol. Biol.*, 45, 567–576.

Platt-Aloia, K.A., Thomas, W.W. and Young, R.E. (1980) Ultrastructural changes in the cell walls of ripening avocados: transmission, scanning and freeze-fracture microscopy. *Bot. Gaz.*, 141, 366–373.

Popeijus, H., Overmars, H., Jones, J. *et al.* (2000) Enzymology – degradation of plant cell walls by a nematode. *Nature*, 406, 36–37.

Porretta, S., Poli, G. and Minuti, E. (1998) Tomato pulp quality from transgenic fruits with reduced polygalacturonase (PG). *Food Chem.*, 62, 283–290.

Pressey, R. (1983) β-galactosidases in ripening tomatoes. *Plant Physiol.*, 71, 32–135.

Pressey, R. (1989) Endo-β-mannanase in ripening tomatoes. *Phytochem.*, 28, 3277–3280.

Pressey, R. and Avants, J.K. (1978) Difference in polygalacturonase composition of clingstone and freestone peaches. *J. Food Sci.*, 43, 1415–1423.

Pressman, E., Hunerman, M., Aloni, B. and Jaffe, M.J. (1983) Thigmomorphogenesis: the effect of mechanical perturbation and ethrel on stem pith autolysis in tomato [*Lycopersicon esculentum* (Mill.)] plants. *Ann. Bot.*, 52, 93–100.

Pua, E.-C., Ong, C.K., Liu, P. and Liu, J.-Z. (2001) Isolation and expression of two pectate lyase genes during fruit ripening of banana (*Musa acuminata*). *Physiol. Plant.*, 113, 92–99.

Qin, Q., Bergmann, C., Rose, J.K.C., Saladie, M., Kumar Kolli, V.S., Albersheim, P., Darvill, A.G. and York, W.S. (2003) Characterization of a tomato protein that inhibits a xyloglucan-specific endoglucanase. *Plant J.*, 34, 327–338.

Rao, G.U. and Paran, I. (2003) Polygalacturonase: a candidate gene for the soft flesh and deciduous fruit mutation in *Capsicum. Plant Mol. Biol.*, 51, 135–141.

Ranwala, A.P., Suematsu, C. and Masuda, H. (1992) The role of β-galactosidases in the modification of cell wall components during muskmelon fruit ripening. *Plant Physiol.*, 100, 1318–1325.

Redgwell, R.J. (1996) Cell wall synthesis in kiwifruit following postharvest ethylene treatment. *Phytochem.*, 41, 407–413.

Redgwell, R.J. and Fischer, M. (2001) Fruit texture, cell wall metabolism and consumer perceptions. In *Fruit Quality and its Biological Basis.* Knee, M. (ed). CRC Press LLC, Boca Raton, FL.

Redgwell, R.J. and Fry, S.C. (1993) Xyloglucan endotransglycosylase activity increases during kiwifruit (*Actinidia deliciosa*) ripening, *Plant Physiol.*, 103, 407–413.

Redgwell, R.J. and Percy, A.E. (1992) Cell wall changes during on-vine softening of kiwifruit. *New Zealand J. Crop Hort. Sci.*, 20, 453–456.

Redgwell, R.J., Melton, L.D. and Brasch, D.J. (1988) Cell-wall polysaccharides of kiwifruit (*Actinidia deliciosa*): chemical features in different tissue zones of the fruit at harvest. *Carbohydr. Res.*, 182, 241–258.

Redgwell, R.J., Melton, L.D. and Brasch, D.J. (1991) Cell-wall polysaccharides of kiwifruit (*Actinidia deliciosa*): effect of ripening on the structural features of cell-wall materials. *Carbohydr. Res.*, 209, 191–202.

Redgwell, R.J., Fischer, M., Kendal, E. and MacRae, E. (1997a) Galactose loss and fruit ripening: high-molecular-weight arabinogalactans in the pectic polysaccharides of fruit cell walls. *Planta*, 203, 174–181.

Redgwell, R.J., MacRae, E., Hallett, I., Fischer, M., Perry, J. and Harker, R. (1997b) *In vivo* and *in vitro* swelling of cell walls during fruit ripening. *Planta*, 203, 162–173.

Reiter, W.-D. (2002) Biosynthesis and properties of the plant cell wall. *Curr. Opin. Plant Biol.*, 5, 536–542.

RejonPalomares, A., GarciaGarrido, J.M., Ocampo, J.A. and GarciaRomera, I. (1996) Presence of xyloglucan-hydrolyzing glucanases (xyloglucanases) in arbuscular mycorrhizal symbiosis. *Symbiosis*, 21, 249–261.

Revol, J.F. (1982) On the cross sectional shape of cellulose crystallites in *Valonia ventricosa. Carbohydr. Polym.*, 2, 123–134.

Rexová-Benkova, L. and Markovič, O. (1976) Pectic enzymes. *Adv. Carbohydr. Chem. Biochem.*, 33, 323–385.

Ridley, B.L., O'Neill, M.A. and Mohnen, D. (2001) Pectins: structure, biosynthesis and oligo-galacturonide-related signaling. *Phytochem.*, 57, 929–967.

Roberts, K. and McCann, M. (2000) Xylogenesis: the birth of a corpse. *Curr. Opin. Plant Biol.*, 3, 517–522.

Roberts, J.A., Whitelaw, C.A., Gonzalez-Carranza, Z.H. and McManus, M. (2000) Cell separation processes in plants – models, mechanisms and manipulation. *Annals. Bot.*, 86, 223–235.

Roberts, J.A., Elliott, K.A. and Gonzalez-Carranza, Z.H. (2002) Abscission, dehiscence and other cell separation processes. *Ann. Rev. Plant Biol.*, 53, 131–158.

Ronen, R., Zauberman, G., Akerman, M., Weksler, A., Rot, I. and Fuchs, Y. (1991) Xylanase and xylosidase activities in avocado fruit. *Plant Physiol.*, 95, 961–964.

Rose, J.K.C. and Bennett, A.B. (1999) Cooperative disassembly of the cellulose-xyloglucan network of plant cell walls: parallels between cell expansion and fruit ripening. *Trends Plant Sci.*, 4, 176–183.

Rose, J.K.C., Lee, H.H. and Bennett, A.B. (1997) Expression of a divergent expansin gene is fruit-specific and ripening-regulated. *Proc. Natl. Acad., Sci. USA*, 94, 5955–5960.

Rose, J.K.C., Hadfield, K.A., Labavitch, J.M. and Bennett, A.B. (1998) Temporal sequence of cell wall disassembly in rapidly ripening melon fruit. *Plant Physiology*, 117, 345–361.

Rose, J.K.C., Cosgrove, D.J., Albersheim, P., Darvill, A.G. and Bennett, A.B. (2000a) Detection of expansin proteins and activity during tomato fruit ontogeny. *Plant Physiol.*, 123, 1583–1592.

Rose, J.K.C., O'Neill, M.A., Albersheim, P. and Darvill A. (2000b) Functions of the plant primary cell wall. In: *Oligosaccharides in Chemistry and Biology. Vol. II Biology of Saccharides.* B. Ernst, G. Hart, P. Sinay (eds). Wiley/VCH. Weinheim, Germany.

Rose, J.K.C., Braam, J., Fry, S.C. and Nishitani, K. (2002a) The XTH family of enzymes involved in xyloglucan endotransglucosylation and endohydrolysis: current perspectives and a new unifying nomenclature. *Plant Cell Physiol.*, 43, 1421–1453.

Rose, J.K.C., Ham, K.-S., Darvill, A.G. and Albersheim, P. (2002b) Molecular cloning and characterization of glucanase inhibitor proteins: coevolution of a counterdefense mechanism by plant pathogens. *Plant Cell*, 14, 1329–1345.

Ross, G.S., Redgwell, R.J. and MacRae, E.A. (1993) Kiwifruit β-galactosidase: isolation and activity against specific cell wall polysaccharides. *Planta*, 189, 499–506.

Ross, G.S., Wegrzyn, T., MacRae, E.A. and Redgwell, R.J. (1994) Apple beta-galactosidase: activity against cell wall polysaccharides and characterization of a related cDNA clone. *Plant Physiol.*, 106, 521–528.

Roy, S., Vian, B. and Roland, J.-C. (1992) Immuncytochemical study of the deesterification pattern during cell wall autolysis in the ripening of cherry tomato. *Plant Physiol. Biochem.*, 30, 139–146.

Rushing, J.W. and Huber, D.J. (1987) Effects of NaCl, pH and Ca^{2+} on autolysis of isolated tomato fruit cell walls. *Physiol. Plant.*, 70, 78–84.

Ryden, P., MacDougall, A.J., Tibbits, C.W. and Ring, S.G. (2000) Hydration of pectic polysaccharides. *Biopolymers*, 54, 398–405.

Saab, I.N. (1999) Involvement of the cell wall in responses to water deficit and flooding. In *Plant Responses to Environmental Stresses. From Phytohormones to Genome Reorganization.* H.R. Lerner (ed). Marcel Dekker, Inc., New York, pp. 413–429.

Saab, I.N. and Sachs, M.S. (1996) A flooding-induced xyloglucan endo-transglycosylase homolog in maize is responsive to ethylene and associated with aerenchyma. *Plant Physiol.*, 112, 385–391.

Saha, B.C. (2000) Alpha-L-arabinofuranosidases: biochemistry, molecular biology and application in biotechnology. *Biotech. Adv.*, 18, 403–423.

Sakurai, N. and Nevins, D.J. (1993) Changes in physical properties and cell wall polysaccharides of tomato (*Lycopersicon esculentum*) pericarp tissue. *Physiol. Plant.*, 89, 681–686.

Sakurai, N. and Nevins, D.J. (1997) Relationship between fruit softening and wall polysaccharides in avocado (*Persea americana* Mill) mesocarp tissues. *Plant Cell Physiol.*, 38, 603–610.

Saloheimo, M., Paloheimo, M., Hakola, S. *et al.* (2002) Swollenin, a *Trichoderma reesei* protein with sequence similarity to the plant expansins, exhibits disruption activity on cellulosic materials. *Eur. J. Biochem.*, 269, 4202–4211.

Sánchez, R.A., Sunell, L., Labavitch, J.M. and Bonner, B.A. (1990) Changes in the endosperm cell walls of two *Datura* species before radicle protrusion. *Plant Physiol.*, 93, 89–97.

Sander, L., Child, R., Ulskov, P., Albrechtsen, M. and Borkjardt, B. (2001) Analysis of a dehiscence zone endo-polygalacturonase in oilseed rape (*Brassica napus*) and *Arabidopsis thaliana*: evidence for roles in cell separation in dehiscence and abscission zones, and in stylar tissues during pollen tube growth. *Plant Mol. Biol.*, 46, 469–479.

Sanders, P.M., Lee, P.Y., Biesgen, C. *et al.* (2000) The *Arabidopsis DELAYED DEHISCENCE1* gene encodes an enzyme in the jasmonic acid synthesis pathway. *Plant Cell*, 12, 1041–1061.

Sanwal, G.G. (1999) Purification and characterization of a cellulase from *Catharanthus roseus* stems. *Phytochem.*, 52, 7–13.

Sexton, R. and Roberts, J.A. (1982) Cell biology of abscission. *Annu. Rev. Plant Physiol.*, 33, 133–162.

Schols, H.A., Geraeds, C.J.M., Searle-van-Leeuwen, M.F., Kormelink, F.J.M. and Voragen, A.G.J. (1990) Rhamnogalacturonase: a novel enzyme that degrades the hairy regions of pectins. *Carbohydr. Res.*, 206, 105–115.

Schuch, W., Kanczler, J., Robertson, D. *et al.* (1991) Fruit quality characteristics of transgenic tomato fruit with altered polygalacturonase activity. *Hort. Sci.*, 26, 1517–1520.

Schröder, R., Atkinson, R.G., Langenkämper, G. and Redgwell, R.J. (1998) Biochemical and molecular characterisation of xyloglucan endotransglycosylase from ripe kiwifruit. *Planta*, 204, 242–251.

Schopfer, P. (2001) Hydroxyl radical induced cell wall loosening *in vitro* and *in vivo*: implications for the control of elongation growth. *Plant J.*, 28, 679–688.

Schopfer, P., Liszkay, A., Bechtold, M., Frahry, G. and Wagner, A. (2002) Evidence that hydroxyl radicals mediate auxin-induced extension growth. *Planta*, 214, 821–828.

Schweikert, C., Liszkay, A. and Schopfer, P. (2000) Scission of polysaccharides by peroxidase-generated hydroxyl radicals. *Phytochem.*, 53, 565–570.

Schweikert, C., Liszkay, A. and Schopfer, P. (2002) Polysaccharide degradation by Fenton reaction or peroxidase-generated hydroxyl radicals in isolated plant cell walls. *Phytochem.*, 61, 31–35.

Serpe, M.D., Muir, A.J. and Driouich, A. (2002) Immunolocalization of beta-D-glucans, pectins, and arabinogalactan-proteins during intrusive growth and elongation of nonarticulated laticifers in *Asclepias speciosa* Torr. *Planta*, 215, 357–370.

Sexton, R. and Roberts, J.A. (1982) Cell biology of abscission. *Annu. Rev. Plant Physiol.*, 33, 133–162.

Seymour, G.B., Harding, S.E., Taylor, A.J., Hobson, G.E. and Tucker, G.A. (1987a) Polyuronide solubilization during ripening of normal and mutant tomato fruit. *Phytochem.*, 26, 1871–1875.

Seymour, G.B., Lasslett, Y. and Tucker, G.A. (1987b) Differential effects of pectolytic enzymes on tomato polyuronide *in vivo* and *in vitro*. *Phytochem.*, 26, 3137–3139.

Seymour, G.B., Colquhoun, I.J., DuPont, M.S., Parsley, K.R. and Selvendran, R.R. (1990) Composition and structural features of cell wall polysaccharides from tomato fruits. *Phytochem.*, 29, 725–731.

Shackel, K.A., Greve, C., Labavitch, J.M. and Ahmadi, H. (1991) Cell turgor associated with ripening in tomato pericarp tissue. *Plant Physiol.*, 97, 814–816.

Sheehy, R.E., Kramer, M. and Hiatt, W.R. (1988) Reduction of polygalacturonase activity in tomato fruits by antisense RNA. *Proc. Natl. Acad. Sci. USA*, 85, 8805–8809.

Sitrit, Y., Hadfield, K.A., Bennett, A.B., Bradford, K.J. and Downie, B. (1999) Expression of a polygalacturonase associated with tomato seed germination. *Plant Physiol.*, 121, 419–428.

Smant, G., Stokkermans, J.P.W.G., Yan, Y.T. *et al.* (1998) Endogenous cellulases in animals: isolation of beta-1,4-endoglucanase genes from two species of plant-parasitic cyst nematodes. *Proc. Natl. Acad. Sci. USA*, 95, 4906–4911.

Smith, R.C. and Fry, S.C. (1991) Endotransglycosylation of xyloglucans in plant cell suspension cultures. *Biochem. J.*, 279, 529–535.

Smith, D.L. and Gross, K.C. (2000) A family of at least seven β-galactosidase genes is expressed during tomato fruit development. *Plant Physiol.*, 123, 1173–1183.

Smith, B.G. and Harris, P.J. (1995) Polysaccharide composition of unlignified cell walls of pineapple (*Ananas comosus* [L] Merr.) fruit. *Plant Physiol.*, 107, 1399–1409.

Smith, C.J.S., Watson, C.F., Ray, J. *et al.* (1988) Antisense RNA inhibition of polygalacturonase gene expression in transgenic tomatoes. *Nature*, 334, 724–726.

Smith, C.J.S., Watson, C.F., Morris, P.C. *et al.* (1990) Inheritance and effect on ripening of antisense polygalacturonase genes in transgenic tomatoes. *Plant Mol. Biol.*, 14, 369–379.

Smith, B.G., Harris, P.J., Melton, L.D. and Newman, R.H.Y. (1998) Crystalline cellulose in hydrated primary cell walls of three monocotyledons and one dicotyledon. *Plant Cell Physiol.*, 39, 711–720.

Smith , D.L., Abbott, J.A. and Gross, K.C. (2002) Down-regulation of tomato beta-galactosidase 4 results in decreased fruit softening. *Plant Physiol.*, 129, 1755–1762.

Sozzi, G.O., Cascone, O. and Fraschina, A.A. (1996) Effect of a high temperature stress on endo-β-mannanase and α- and β-galactosidase activities during tomato fruit ripening. *Postharvest Biol. Technol.*, 9, 49–63.

Sozzi, G.O., Fraschna, A.A., Navarro, A.A., Cascone, O., Greve, L.C. and Labavitch, J.M. (2002a) α-L-Arabinofuranosidase activity during development and ripening of normal and ACC synthase antisense tomato fruit. *Hort. Sci.*, 37, 564–566.

Sozzi, G.O., Greve, L.C., Prody, G.A. and Labavitch, J.M. (2002b) Gibberellic acid, synthetic auxins and ethylene differentially modulate α-L-arabinofuranosidase activities in antisense 1-aminocyclopropane-1-carboxylic acid synthase tomato pericarp discs. *Plant Physiol.*, 129, 1330–1340.

Steele, N.M., McCann, M.C. and Roberts, K. (1997) Pectin modification in cell walls of ripening tomatoes occurs in distinct domains. *Plant Physiol.*, 114, 373–381.

Stephenson, M.B. and Hawes, M.C. (1994) Correlation of pectin methylesterase activity in root caps of pea with root border cell separation. *Plant Physiol.*, 106, 739–745.

Sterling, C. (1961) Physical state of cellulose during ripening of peach. *J. Food Sci.*, 26, 95–98.

Stewart, D., Iannetta, P.P. and Davies, H.V. (2001) Ripening-related changes in raspberry cell wall composition and structure. *Phytochem.*, 56, 423–428.

Still, D.W. and Bradford, K.J. (1997) Endo-β-mannanase activity from individual tomato endosperm caps and radicle tips in relation to germination rates. *Plant Physiol.*, 113, 21–29.

Stotz, H.U., Bishop, J.G., Bergmann, C.W. *et al.* (2000) Identification of target amino acids that affect interactions of fungal polygalacturonases and their plant inhibitors. *Mol. Physiol. Plant Pathol.*, 56, 117–130.

Stratilova, E., Markovič, O., Dzurova, M., Malovikova, A., Capek, P. and Omelkova, J. (1998) The pectolytic enzymes of carrots. *Biologia*, 53, 731–738.

Sutherland, P., Hallett, I., Redgwell, R., Benhamou, N. and MacRae, E. (1999) Localization of cell wall polysaccharides during kiwifruit (*Actinidia deliciosa*) ripening. *Int. J. Plant Sci.*, 160, 1099–1109.

Tamaru, Y. and Doi, R. (2001) Pectate lyase A, an enzymatic subunit of the *Clostridium cellulovorans* cellulosome. *Proc. Natl. Acad. Sci. USA*, 98, 4125–4129.

Tateishi, A., Kanayama, Y. and Yamaki, S. (1996) α-L-arabinofuranosidase from cell walls of Japanese pear fruits. *Phytochem.*, 42, 295–299.

Tateishi, A., Inoue, H., Shiba, H. and Yamaki, S. (2001) Molecular cloning of beta-galactosidase from Japanese pear (*Pyrus pyrifolia*) and its gene expression with fruit ripening. *Plant Cell Physiol.*, 42, 492–498.

Taylor, J.E. and Whitelaw, C.A. (2001) Signals in abscission. *New Phytol.,* 151, 323–339.

Taylor, J.E., Tucker, G.A., Lasslett, Y. *et al.* (1990) Polygalacturonase expression during leaf abscission of normal and transgenic tomato plants. *Planta,* 183, 133–138.

Taylor, J.E., Webb, S.T.J., Coupe, S.A., Tucker, G.A. and Roberts, J.A. (1993) Changes in polygalacturonase activity and solubility of polyuronides during ethylene-stimulated leaf abscission in *Sambucus nigra. J. Exp. Bot.,* 44, 93–98.

Terrier, N., Ageorges, A., Abbal, P. and Romieu, C. (2001) Generation of ESTs from grape berry at various developmental stages. *J. Plant Physiol.,* 158, 1575–1583.

Thompson, J.E. and Fry, S.C. (2001) Restructuring of wall-bound xyloglucan by transglycosylation in living plant cells. *Plant J.,* 26, 23–34.

Thompson, D.S. and Osborne, D.J. (1994) A role for the stele in intertissue signalling in the initiation of abscission in bean leaves (*Phaseolus vulgaris* L). *Plant Physiol.,* 105, 341–347.

Thompson, A.J., Tor, M., Barry, C.S. *et al.* (1999) Molecular and genetic characterization of a novel pleiotropic tomato-ripening mutant. *Plant Physiol.,* 120, 383–389.

Tieman, D.M. and Handa, A.K. (1994) Reduction in pectin methylesterase activity modifies tissue integrity and cation levels in ripening tomato (*Lycopersicon esculentum* Mill.) fruits. *Plant Physiol.,* 106, 429–436.

Tieman, D.M., Harriman, R.W., Ramamohan, G. and Handa, A.K. (1992) An antisense pectin methylesterase gene alters pectin chemistry and soluble solids in tomato fruit. *Plant Cell,* 4, 667–679.

Tokoh, C., Takabe, K., Sugiyama, J. and Fujita, M. (2002) Cellulose synthesized by *Acetobacter xylinum* in the presence of plant cell wall polysaccharides. *Cellulose,* 9, 65–74.

Tong, C.B.S. and Gross, K.C. (1988) Glycosyl-linkage composition of tomato fruit cell wall hemicellulosic fractions during ripening. *Physiol. Plant.,* 74, 365–370.

Toorop, P.E., van Aelst, A.C. and Hilhorst, W.M. (1996) Endo-β-mannanase isoforms are present in the endosperm and embryo of tomato seeds but are not essentially linked to the completion of germination. *Planta,* 200, 153–158.

Trainotti, L., Ferrarese, L., Poznaski, E. and Della Vecchia, F. (1998) Endo-β-1,4-glucanase activity is involved in the abscission of pepper flowers. *J. Plant Physiol.,* 152, 70–77.

Trainotti, L., Ferrareses, L., Dalla Vecchia, F., Rascio, N. and Casadoro, G. (1999a) Two different endo-β-1,4-glucanases contribute to the softening of strawberry fruit. *J. Plant Physiol.,* 154, 355–362.

Trainotti, L., Spolaore, S., Pavanelo, A., Baldan, B. and Casadoro, G. (1999b) A novel E-type endo-β-1,4-glucanase with a putative cellulose-binding domain is highly expressed in ripening strawberry fruits. *Plant Mol. Biol.,* 40, 323–332.

Trainotti, L., Spinello, R., Piovam, A., Spolaore, S. and Casadoro, G. (2001) β-galactosidases with a lectin-like domain are expressed in strawberry. *J. Exp. Bot.,* 52, 1635–1645.

Truelsen, T.A. and Wyndaele, R. (1991) Cellulase in tobacco callus: regulation and purification. *J. Plant Physiol.,* 139, 129–134.

Tucker, G.A. (1993) *Biochemistry of Fruit Ripening* (G. Seymour, J.E. Taylor, and G.A. Tucker, eds), Chapman and Hall, London.

Tucker, M.L. and Laties, G.G. (1984) Interrelationship of gene expression, polysome prevalence and respiration during ripening of ethylene and/or cyanide-treated avocado fruit. *Plant Physiol.,* 74, 307–315.

Tucker, G.A., Robertson, N.G. and Grierson, D. (1982) Purification and changes in activities of tomato pectinesterase enzymes. *J. Sci. Food Agric.,* 33, 396–400.

Tucker, G.A., Schindler, C.B. and Roberts, J.A. (1984) Flower abscission in mutant tomato plants. *Planta,* 160, 164–167.

van Buuren, M.L., Maldonado-Mendoza, I.E., Trieu, A.T., Blaylock, L.A. and Harrison, M.J. (1999) Novel genes induced during an arbuscular mycorrhizal symbiosis formed between *Medicago truncatula* and *Glomus versiforme. Mol. Plant-Microbe Interact.,* 12, 171–181.

Vanacker, H., Harbinson, J., Ruisch, J., Carver, T.L.W. and Foyer, C.H. (1998) Antioxidant defenses in the apoplast. *Protoplasma,* 205, 129–140.

Vercauteren, I., de Almeida Engler, J., De Groodt, R. and Gheysen, G. (2002) An *Arabidopsis thaliana* pectin acetylesterase gene is upregulated in nematode feeding sites induced by root-knot and cyst nematodes. *Mol. Plant-Microbe Interact.*, 15, 404–407.

Viëtor, R.J., Newman, R.H., Ha, M.-A., Apperley, D.C. and Jarvis, M.C. (2002) Conformational features of crystal-surface cellulose from higher plants. *Plant J.*, 30, 721–731.

Wade, N.L., Kavanagh, E.E., Hockley, D.G. and Bradu, C.J. (1992) Relationship between softening and polyuronides in ripening banana fruit. *J. Sci. Food Agric.*, 60, 61–68.

Wakabayashi, K. (2000) Changes in cell wall polysaccharides during fruit ripening. *J. Plant Res.*, 113, 231–237.

Wakabayashi, K., and Huber, D.J. (2001) Purification and catalytic properties of polygalacturonase isoforms from ripe avocado (*Persea americana*) fruit mesocarp. *Physiol. Plant.*, 113, 210–216.

Wakabayashi, K., Chun, J.-P. and Huber, D.J. (2000) Extensive solubilization and depolymerization of cell wall polysaccharides during avocado (*Persea americana*) ripening involves concerted action of polygalacturonase and pectinmethylesterase. *Physiol. Plant.*, 108, 345–352.

Wallner, S.J. and Walker, J.E. (1975) Glycosidases in cell wall-degrading extracts of ripening tomato fruits. *Plant Physiol.*, 55, 94–98.

Warren, R.A.J. (1996) Microbial hydrolysis of polysaccharides. *Annu. Rev. Microbiol.*, 50, 183–212.

Watkins, J.T., Cantliffe, D.J., Huber, D.J. and Nell, T.A. (1985) Gibberellic acid stimulated degradation of endosperm in pepper. *J. Am. Soc. Hortic. Sci.*, 110, 61–65.

Watson, C.F., Zheng, L. and DellaPenna, D. (1994) Reduction of tomato polygalacturonase beta subunit expression affects pectin solubilization and degradation during ripening. *Plant Cell*, 6, 1623–1634.

Watson, D.L., Wilson, D.B. and Walker, L.P. (2002) Synergism in binary mixtures of *Thermobifida fusca* cellulases Cel6B, Cel9A, and Cel5A on BMCC and Avicel. *Appl. Biochem. Biotechnol.*, 101, 97–111.

Webb, S.T.J., Taylor, J.E., Coupe, S.A., Ferrarese, L. and Roberts, J.A. (1993) Purification of β-1,4 glucanase from ethylene-treated abscission zones of *Sambucus nigra*. *Plant Cell Environ.*, 16, 329–333.

Wegrzyn, T.F. and MacRae, E.A. (1992) Pectinesterase, polygalacturonase and beta-galactosidase during softening of ethylene-treated kiwifruit. *Hort. Sci.*, 27, 900–902.

Wen, F.S., Zhu, Y.M. and Hawes, M.C. (1999) Effect of pectin methylesterase gene expression on pea root development. *Plant Cell*, 11, 1129–1140.

Whitney, S.E.C., Brigham, J.E., Darke, A.H., Reid, J.S.G. and Gidley, M.J. (1995) *In vitro* assembly of cellulose/xyloglucan networks: ultrastructural and molecular aspects. *Plant J.*, 8, 491–504.

Whitney, S.E.C., Brigham, J.E., Darke, A.H., Reid, J.S.G. and Gidley, M.J. (1998) Structural aspects of the interaction of mannan-based polysaccharides with bacterial cellulose. *Carbohydr. Res.*, 307, 299–309.

Whitney, S.E.C., Gothard, M.G.E., Mitchell, J.T. and Gidley, M.J. (1999) Roles of cellulose and xyloglucan in determining the mechanical properties of primary plant cell walls. *Plant Physiol.*, 121, 657–663.

Whitney, S.E.C., Gidley, M.J. and McQueen-Mason, S.J. (2000) Probing expansin action using hemicellulose/cellulose composites. *Plant J.*, 22, 327–334.

Willats, W.G.T., McCartney, L., Mackie, W. and Knox, J.P. (2001a) Pectin: cell biology and prospects for functional analysis. *Plant Mol. Biol.*, 47, 9–27.

Willats, W.G.T., Orfila, C., Limberg, G. *et al.* (2001b) Modulation of the degree and pattern of methyl-esterification of pectic homogalacturonan in plant cell walls – implications for pectin methyl esterase action, matrix properties, and cell adhesion. *J. Biol. Chem.*, 276, 19404–19413.

Wong, Y., Fincher, G.B. and Maclachlan, G.A. (1977) Kinetic properties and substrate specificities of two cellulases from auxin-treated pea epicotyls. *J. Biol. Chem.,* 252, 1402–1407.

Woolley, L.C., James, D.J. and Manning, K. (2001) Purification and properties of an endo-β-1,4-glucanase from strawberry and down-regulation of the corresponding gene, *cell. Planta,* 214, 11–21.

Wu, C.-T., Leubner-Metzger, G., Meins Jr., F. and Bradford, K.J. (2001) Class I β-1,3-glucanase and chitinase are expressed specifically in the micropylar endosperm of tomato seeds prior to radicle emergence. *Plant Physiol.,* 126, 1299–1313.

Yamaki, S. and Kakiuchi, N. (1979) Changes in hemicellulose-degrading enzymes during development and ripening of Japanese pear fruit. *Plant Cell Physiol.,* 20, 301–309.

Yoshioka, H., Aoba, K. and Kashimua, Y. (1992) Molecular weight and degree of methoxylation in cell wall polyuronide during softening in pear and apple fruit. *J. Am. Soc. Hort. Sci.,* 117, 600–606.

Zhou, H.-W., Ben Arie, R. and Lurie, S. (2000a) Pectin esterase, polygalacturonase and gel formation in peach fruit fractions. *Phytochem.,* 55, 191–195.

Zhou, H.-W., Lurie, S., Lers, A., Khatchitski, A., Sonego, L. and Ben Arie, R. (2000b) Delayed storage and controlled atmosphere storage of nectarines: two strategies to prevent woolliness. *Postharv. Biol. Technol.,* 18, 133–141.

Zheng, L.S., Heupel, R.V.C. and DellaPenna, D. (1992) The beta-subunit of tomato fruit polygalacturonase isoenzyme-1. Isolation, characterization and identification of unique structural features. *Plant Cell,* 4, 1147–1156.

Zugenmaier, P. (2001) Conformation and packing of various crystalline cellulose fibers. *Prog. Polym. Sci.,* 26, 1341–1417.

10 Plant cell walls in the post-genomic era

Wolf-Rüdiger Scheible, Sajid Bashir and Jocelyn K.C. Rose

10.1 Introduction

The cell wall is one of the most complex and least understood plant structures. It has been estimated that among the ~27,000 *Arabidopsis thaliana* genes, some 15% are dedicated to cell wall biogenesis and modification during plant development (Carpita *et al.*, 2001). Those genes expected to be involved in cell wall synthesis, composition, modification and degradation include:

1. genes necessary for donor substrate generation, i.e. the nucleotide-sugar interconversion pathways;
2. polysaccharide synthesis (glucan synthases and glycosyltransferases);
3. genes in the secretory/protein targeting pathways;
4. genes for polymer assembly and architecture;
5. genes for dynamic rearrangement of the cell wall during growth and differentiation (e.g. expansins, xyloglucan transglucosylase hydrolases, endoglucanases, yieldins); and
6. genes involved in wall disassembly and catabolism (e.g. hydrolases, esterases and lyases).

In addition it can be assumed that there is a large number of genes necessary for the correct regulation of these processes during growth and development.

Prior to the development of extensive sequence databases and advanced molecular biology and bioinformatic techniques, the identification of some classes of cell wall related plant genes and enzymes has proved difficult. For example, biochemical approaches to identify enzymes that synthesize wall components have met with limited success, leading to the identification of only a few genes from *Arabidopsis*, pea or fenugreek (Edwards *et al.*, 1999; Perrin *et al.*, 1999; Faik *et al.*, 2000). The inability to purify enough protein to permit the cloning of the corresponding gene, as well as loss of enzymatic activity upon extraction, absence of acceptor substrates and cofactors, and disassembly of multi-enzyme complexes are all possible contributing factors (Kawagoe and Delmer, 1997).

Due to the availability of increasing amounts of genome sequence and precise genetic and physical maps, classical genetic approaches have recently been more successful in identifying cell wall related genes. For example, several *Arabidopsis* mutants deficient in cellulose were cloned using map-based approaches, and were shown to have defects in, e.g., cellulose synthase genes (Arioli *et al.*, 1998; Taylor *et al.*, 1999, 2000; Fagard *et al.*, 2000a).

The path towards the identification and functional characterization of many more, and eventually all, genes required to make a cell wall however is now paved by the public availability of the entire genome sequences for *Arabidopsis thaliana* (The Arabidopsis Genome Initiative, 2000) and rice (Goff *et al.*, 2002; Yu *et al.*, 2002), by increasing numbers of expressed sequence tag (EST) sequences from many other plant species, and by the development of new reverse genetics tools and genomics platform technologies for gene discovery and functional analysis. In this chapter, we summarize the principal genomics- and proteomics-related technologies and available resources, highlight additional emerging approaches for high-throughput and broad scale wall-related studies, and outline what this genome-scale information from *Arabidopsis*, and other plants, might provide and enable that has not been possible to date.

10.2   Genome annotation and identification of cell wall related genes and proteins

The publicly available genome sequences of *Arabidopsis* and rice provide unprecedented insights and opportunities for cell wall researchers. For the first time, there is access to a complete set of plant genes required to make a cell wall, and the opportunity to learn what a plant 'can and cannot do'. *Arabidopsis* has long been an important genetic plant model organism for various reasons (Meinke *et al.*, 1998), and for physiological and biochemical studies. *Arabidopsis* also has a cell wall that is representative of higher plants with type I walls (see Chapter 1; Carpita and Gibeaut, 1993; Zablackis *et al.*, 1995), and can therefore be used as a cell wall model for most dicotyledons. Similarly, rice may provide opportunities for studies of the type II walls that are typical of monocotyledons.

To obtain access to the information and extract the biological knowledge buried in a genome, it is crucial to correctly annotate and identify the genes (for an in-depth discussion of plant genome annotation see Aubourg and Rouzé, 2001). Genome annotation can be divided into structural and functional annotation. Structural annotation includes determining sites and regions involved in gene and genome functionality (splice sites, introns, exons, untranslated regions, transcriptional and translational start and control regions, promoters, matrix attachment points, etc.). The first structural annotation of the *Arabidopsis* genome was performed with a variety of computer programs, such as NetPlantGene, GenScan or Grail (The Arabidopsis Genome Initiative, 2000). Since each genome has its unique features, the programs, which also differ in their performances, were often not entirely correct with their automatic predictions, and different predictions resulted for the same gene. Therefore, new more refined software is still being developed and regular annotation updates that include help from experts with specialized knowledge of certain genes or gene families (see http://www.tigr.org/tdb/e2k1/ath1/ath1.shtml; http://mips.gsf.de/proj/

thal/db/about/externalanno.html). The existence of large sets of EST sequences from *Arabidopsis* (~175,000) and rice (~110,000) helps structural genome annotation by giving precise information about intron/exon borders. However, since EST sequences are typically rather short (~500bp), they usually do not cover entire open reading frames. In addition, the sum of the *Arabidopsis* EST sequences represents only a subset of the genes present in the genome. The construction of an entire set of full-length cDNAs (the ORFeome), ideally encompassing all sequences, from the CAP site to the poly A site, is therefore a focus for various genome projects, including *Arabidopsis* (Seki *et al.*, 2001, 2002; http://signal.salk.edu/SSP/index.html), and is an important goal if a true 'post-genome' perspective can be achieved. Structural annotation greatly benefits from this growing experimental resource.

Functional annotation initially assigns a basic cellular or biochemical function to a DNA sequence, or its predicted protein product, based on DNA or protein similarity (BLAST and Smith-Waterman algorithms), multiple sequence alignments (hidden Markov models), 3D-structure similarity and pattern or domain searches. Clearly, the quality of functional annotation is greatly dependent on the quality of structural annotation: accurate and comprehensive annotation of genome data is therefore crucial. Systematic analysis of the *Arabidopsis* genome, with computational sequence and structure similarity searches, has also been widely used in recent years to identify cell wall related genes and gene families in *Arabidopsis*, rice and other plants. For example, Richmond and Somerville (2000) identified 41 genes of the *Arabidopsis* cellulose synthase superfamily using the sequence of one *Arabidopsis* cellulose synthase, *RSW1* (Arioli *et al.*, 1998), and a cotton cellulose synthase protein sequence (Pear *et al.*, 1996). All the genes appear to encode processive glucosyltransferases, integral membrane proteins that share the common amino acid motif D, D, D, QxxRW, which has been proposed to define the nucleotide sugar-binding domain and the catalytic site of these enzymes (Saxena *et al.*, 1995). Based on predicted protein sequences and intron-exon structure, the members of the superfamily can be further subdivided in seven distinct families. There is already convincing genetic and biochemical evidence that at least five of the ten members of the *CesA* family encode cellulose synthases involved in primary and secondary cellulose synthesis (Arioli *et al.*, 1998; Taylor *et al.*, 1999, 2000; Fagard *et al.*, 2000a, b; Scheible *et al.*, 2001; Burn *et al.*, 2002). The enzymatic functions of the members of the additional six gene families, preliminarily termed 'cellulose-synthase-like' genes (*CslA*, *B*, 'C, *D*, *E* and *G*) are still unclear (as described in Chapter 6), but it has been speculated that these genes encode enzymes that catalyse the synthesis of other non-cellulosic cell wall polysaccharides such as xyloglucan (Cutler and Somerville, 1997; Richmond and Somerville, 2000, 2001). This has yet to be confirmed, but *Arabidopsis* plants with a mutation in *CslA9* display increased resistance to *Agrobacterium tumefaciens*, which binds to the plant cell wall during the infection process (Nam *et al.*, 1999), and *Arabidopsis CslD3* (KOJAK) mutants have root hair defects because of abnormal cell walls (Favery *et al.*, 2001; Wang *et al.*, 2001), supporting a wall-related function for these genes.

It has been known for many years that substantial differences exist between the cell wall compositions and polymer structures of dicotyledons and commelinoid monocotyledons, and even among the latter group, there are large differences between the poales (Poaceae, which include grasses and the cereals) and others (for review see Carpita, 1996). For example, substituted xyloglucans are the major cross-linking glycans in dicotyledons and non-commelinoid monocotyledons, whereas in the cell walls of the commelinoid line of monocotyledons, which include bromeliads, palms, gingers, cypresses and grasses, the major cross-linking glycan is glucurono-arabinoxylan (Carpita and McCann, 2000). In addition, poales contain a third type of cross-linking glycan called 'mixed-linkage' glucan; β-D-glucans that contain 1→3 as well as 1→4 linkages. Because of such substantial difference between the cell walls of dicotyledons and commelinoid monocotyledons, it is likely that there are numerous cell wall-related genes in certain monocotyledons that are either not expressed, or not present in the *Arabidopsis* genome. In this regard, the genome sequence of rice serves as an excellent comparison, and knowledge of the differences and similarities between the complement of genes in *Arabidopsis* and rice can yield valuable information to cell wall researchers. As an example, an analysis of approximately 50% of the rice genome by Hazen *et al.* (2002) has already led to the identification of 37 *Csl* genes in rice. Besides striking similarities between the *Csl* genes in rice and *Arabidopsis*, there are also differences, which might reflect the cross-linking glycan composition of dicotyledons and grasses. Rice contains a distinct group of *Csl* genes (named *CslF*), the products of which are related but nonetheless distinct from, and much shorter than, the CslD and CesA proteins. The rice *CslB* family also appears to constitute a distinct family that has diverged from the *Arabidopsis CslB* genes, and rice seems to be lacking the *CslG* family. Members of this family are widespread in dicotyledons but have not been found so far in any monocotyledon (Richmond and Somerville, 2001). In addition, rice has at least five members in the *CslE* family, whereas *Arabidopsis* contains just one.

Besides the family of cellulose synthase and cellulose synthase-like genes, many other surprisingly large families of cell wall related genes have been identified in *Arabidopsis*. For example, twelve sequences are related to β-1,3-glucan-synthases from yeast and fungi. Hong *et al.* (2001) characterized one of these *Arabidopsis* genes, demonstrating localization of the protein to the growing cell plate, an interaction with phragmoplastin, and enhanced callose synthase activity in transgenic tobacco cells. While there is no genetic evidence that these proteins are callose synthases, and *in vitro* enzymatic activity has not been demonstrated, it seems likely that this gene family, called *AtGsl* or *CalS*, encodes plant callose synthases. Other identified cell wall related gene families include 33 genes encoding xyloglucan endotransglucosylase-hydrolases (Yokohama and Nishitani, 2001; Rose *et al.*, 2002), 13 sequences encoding xyloglucan fucosyltransferases (Sarria *et al.*, 2001), 38 expansins and expansin-like genes (Li *et al.*, 2002), and 48 genes encoding arabinogalactan protein (AGP) protein backbones (Schultz *et al.*, 2002). Also, more than 170 genes appear to be devoted to pectin degradation in *Arabidopsis*, including 66 potential polygalacturonases, 58 potential pectin methylesterases, 28 pectin lyases,

13 potential pectin acetylesterases and seven potential rhamnogalacturonan lyases (Henrissat *et al.*, 2001).

The surprising size of the cell-wall related and other *Arabidopsis* gene families directly leads to the question 'why are there so many members?'. A logical explanation is that, although they likely perform the same or a similar biochemical function, the family members have specific biological roles and are expressed and required in different organs, cell types or intracellular compartments at different times during development, in response to different stimuli. In addition, the encoded proteins, many of which are enzymes, may have different substrate specificities and affinities. Hence, large gene families might reflect the need for differential expression of their members. As a rule, each gene is expressed in the specific cells and under the specific conditions in which its product makes a contribution to the fitness of an organism (Brown and Botstein, 1999).

It is also possible that two or more structurally related family members have broadly overlapping or identical biological roles (for examples see Leung *et al.*, 1997; Liljegren *et al.*, 2000; Pelaz *et al.*, 2000), where the gene products are functionally redundant. This redundancy could be interpreted, at least in some cases, as a need for biological duplication, since the loss of a single gene copy through mutation might be severely disadvantageous for the plant.

The National Science Foundation's '2010 project' has the primary goal of assigning at least some degree of biological function to most of the ~27,000 *Arabidopsis* genes/gene products by the year 2010 (Somerville and Dangl, 2000). This can be regarded as the second step towards functional annotation of the *Arabidopsis* genome, and may eventually permit the development of a 'virtual plant' – a computer model that will use information about each gene product to simulate the growth and development of a plant under many environmental conditions (Somerville and Dangl, 2000). Elucidating the biological function of every given gene – determining when, where and under what conditions these genes and their gene products are important for the fitness of the plant – will require extensive experimental investigation, such as the detection and analysis of mutant phenotypes using reverse genetics approaches (see below), as well as analysis of gene and protein expression and localization, and will represent a major step towards an integrated understanding of plant growth and development.

10.3 Assigning gene functions using reverse genetics and the tools of functional genomics

10.3.1 Overview of reverse genetics

Reverse genetics, in contrast to forward genetics (see section 10.4), is a strategy that leads from a gene of interest, usually via a corresponding loss-of-function mutant, to the associated mutant phenotypes and finally to the assignment of a biological role to the gene under study (Figure 10.1). Reverse genetics allows for testing of specific hypotheses, linking a certain candidate or suspect gene(s) with a biological pathway, process or phenotype. A candidate gene in a genome might be of interest to a researcher because its DNA or deduced protein sequence is similar to experimentally characterized ('known') proteins with defined biochemical functions, or because its expression pattern, revealed for instance in the course of a transcript or protein profiling experiment (see sections 10.5 and 10.6), specifically matches the expectations or the expression of other genes that are implicated in the biological process. With the recent expansion of sequence databanks and the availability of complete plant genomes, as well as increasingly efficient methods to produce and identify reduction- or loss-of-function mutants in plants, reverse genetics has become a very attractive alternative to phenotypic screens for functional analysis and will play an essential role in the process of assigning functions to a large number of plant genes.

10.3.2 DNA-insertion mutagenesis and identification of tagged mutants

DNA-insertion mutagenesis, or gene tagging, has become a well-established method to produce plant mutants. Two main approaches exist for DNA-insertion mutagenesis: namely, tagging with T-DNA from *Agrobacterium tumefaciens* (Koncz *et al.*, 1992; Azpiroz-Leehan and Feldmann, 1997), and tagging with transposons such as Ac/Ds, En/Spm, or Mu (see review of Ramachandran and Sundaresan, 2001).

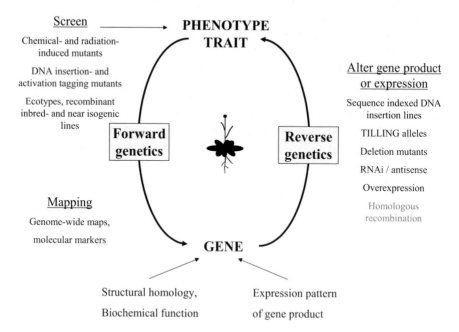

Figure 10.1   Scheme outlining the approaches of forward and reverse genetics. Important steps, resources and keywords used in this chapter are shown.

Both kinds of mutagenic DNA elements insert more-or-less randomly into gene-rich regions of chromosomes and can therefore theoretically be used to create 'loss-of-function' mutants, or mutants with reduced expression, for any gene. Certain tailor-made transposons and T-DNAs that carry strong enhancers or a promoter directing transcription into the region flanking the insertion can lead to 'gain-of-function' or 'activation tagged' mutants (Walden *et al.*, 1994; Martienssen, 1998; Weigel *et al.*, 2000; Marsch-Martinez *et al.*, 2002). These mutations, in which ectopic activation of a flanking gene promotes a mutant phenotype, are usually dominant or semi-dominant. Thus they are visible in the T1-generation, although they are somewhat rare, as they require the activating DNA elements to insert in the vicinity of a gene without disrupting the transcription unit. Insertion mutants can also be designed to contain so-called reporter genes, such as those encoding beta-glucuronidase or green fluorescent protein, that reveal the expression patterns of the chromosomal gene at the insertion site ('gene- and enhancer traps', as described in Topping and Lindsey, 1995; Martienssen, 1998 and references therein; Campisi *et al.*, 1999; http: //genetrap.cshl.org/). Lines with very specific expression patterns (cell type, developmental stage, environmental condition) and lines that might have been missed in conventional insertion mutant screens (e.g. homozygous-lethal insertions in heterozygous mutants) can be identified and the corresponding genes and promoters isolated (Sundaresan *et al.*, 1995).

T-DNA tagging involves *Agrobacterium*-mediated plant transformation using vectors based on native Ti-plasmids. DNA present within two imperfect repeats, the left and right borders, is transferred into a plant by *Agrobacterium* during the infection process, and integrates into the plant genome (Zupan and Zambryski, 1995; Sheng and Citovsky, 1996; and references therein). A key feature for the production of T-DNA insertion mutants is the availability of efficient transformation systems that are rapid and easy to use, such as the *Agrobacterium*-based vacuum-infiltration (Bechtold *et al.*, 1993), 'floral-dip' (Clough and Bent, 1998) and 'floral-spray' protocols. Such methods make it easy to produce thousands of independent T1-transformants, and bypass the tissue culture and regeneration phase and the related problem of somaclonal variations or reduced fertility common to previous transformation methods (Valvekens *et al.*, 1988; van den Bulk *et al.*, 1990). It seems possible to adapt such high-throughput methods to other plants, such as rice and other cereals (Komari *et al.*, 1998; Bent, 2000; Jeon *et al.*, 2000), permitting the creation of large populations of T-DNA insertion mutants in crop species. The basis of the *in planta* methods is that the oocyte of the female gametophyte is transformed after meiosis and presumably during fertilization (see Bent, 2000 and references therein). Transformed T1-plants are selected at a frequency of ~1–3% using selectable markers (e.g. herbicide or antibiotic resistance) encoded within the T-DNA, and each transformed plant carries on average ~1.5 T-DNA copies (Aspiroz-Leehan and Feldmann, 1997). Based on an *Arabidopsis* genome size of 125–130 Mb, and an average gene size of 2.8 kb (The Arabidopsis Genome Initiative, 2000), it can be predicted that approximately 70,000 lines should be screened to detect an insertion in a given gene of interest at 95% probability (Bouchez and Höfte, 1998). However, finding an insertion in a smaller gene with relatively high probability requires screening of a considerably greater number of lines (Krysan *et al.*, 1999).

Transposon tagging has some advantages over T-DNA tagging. For plant species that are difficult to transform, transposons are more efficient because a single transformation event can be used to generate thousands of mutations. Transposons can move around in the genome, and many independent insertions (up to 200 with 'mutator') can be generated in one line (Chandler and Hardeman, 1992; Das and Martienssen, 1995). However, T-DNA tagging results in stable insertions, while transposon tagging requires the removal of transposase activity for insertion stability. Hence, the smaller number of transposon-tagged lines that need to be screened to find a mutant of interest is offset by the difficulty of the subsequent analysis to determine the nature of the insertion event. Another advantage/disadvantage of transposons is their tendency to jump into locations near their original integration site (Jones *et al.*, 1990; Bancroft and Dean, 1993; Ito *et al.*, 2002). While this limits the effectiveness of whole-genome mutagenesis from a single transposon donor site, it raises the possibility of knocking out tandem-repeat genes or clustered gene families, both of which are common in the *Arabidopsis* genome (The Arabidopsis Genome Initiative, 2000), and probably most other plant genomes. Finally, transposon insertions in a gene undergo somatic and germinal reversion, which can provide evidence of a gene's function without the need for functional complementation (Liu and Crawford, 1998 and references therein).

Large DNA-insertion mutant collections have been generated in *Arabidopsis* by various laboratories, altogether comprising several hundred thousand lines. Many of these lines are available through the Arabidopsis Biological Resource Center (ABRC; http://www.biosci.ohio-state.edu/~plantbio/Facilities/abrc/abrchome.htm) and the Nottingham Arabidopsis Stock Center (NASC; http://nasc.nott.ac.uk). In collections of this size it should be possible to find a mutant for any given gene. Large collections of insertion mutants (T-DNA or transposon) also exist for rice, maize, tomato, petunia and snapdragon and additional collections will eventually be created in other species.

A major advantage of DNA-insertion mutagenesis is that the identification and retrieval of the disrupted gene, which is tagged by a DNA tag of known sequence, can be very straightforward. Using methods such as plasmid-rescue (Koncz *et al.*, 1992; Mandal *et al.*, 1993), adapter-mediated polymerase chain reaction (PCR) (Rosenthal and Jones, 1990; Siebert *et al.*, 1995), or thermal-asymmetric-interlaced PCR (Liu *et al.*, 1995) the plant genomic DNA flanking a DNA tag is cloned and sequenced. With the entire *Arabidopsis* genome sequence available, this information is sufficient to immediately localize an insertion by a simple database search.

However, insertion mutagenesis also has major disadvantages. For example, DNA-insertions can be complex (e.g. tandem or inverted repeats, truncated T-DNA or ragged borders) or unpredictable, making the identification of a flanking sequence tag extremely difficult (Gheysen *et al.*, 1990; Nacry *et al.*, 1998). Since several insertions or chromosomal rearrangements can occur in insertion mutant genomes, it is also necessary to confirm that a tagged gene is responsible for an observed phenotype. This is usually accomplished by backcrossing and subsequent co-segregation analysis, and/or by functional complementation of the mutant with a wild-type copy of the gene. Furthermore, in *Arabidopsis* it has been found, and particularly in a remarkable recent large-scale study (Budziszewski *et al.*, 2001), that up to two thirds of insertion mutations do not co-segregate with their mutant phenotypes. In such cases there is no known way to isolate the corresponding mutations other than with a map-based approach (see section 10.4). Additional disadvantages include the inability to generate conditional mutations required for the identification and characterization of essential genes, or the inability to generate specific point mutations in a gene. This last objective is a prerequisite for detailed functional analysis and for the generation of an allelic series of mutations within a specific gene that allow a full range of possible phenotypes to be explored. For example, while homozygous T-DNA insertion lines resulting in full loss-of-function alleles cannot be isolated for *AtCesA3* (W. Scheible, unpublished result), *AtCesA3* antisense lines have been achieved (Burn *et al.*, 2002) and shown to exhibit a severe dwarfing phenotype. Moreover, specific point mutations in the gene confer resistance to the cellulose-synthesis inhibitor isoxaben (Scheible *et al.*, 2001), providing a link between cellulose synthesis, hormone signaling and stress responses (Ellis *et al.*, 2002), or lignification (Caño-Delgado *et al.*, 2003).

The use of T-DNA and transposon insertion mutagenesis coupled with PCR-based screening of multi-dimensionally pooled genomic DNA of insertion mutants is a well-established tool for reverse genetics in *Arabidopsis* (Krysan *et al.*, 1996, 1999; Speulman *et al.*, 1999; Tissier *et al.*, 1999). Furthermore, systematic high-throughput cloning, sequencing and cataloguing of insertion sites by several academic and commercial facilities in Europe and the United States (http://www.arabidopsis.org/links/insertion.html) now dramatically facilitates the identification of *Arabidopsis* insertion mutants in a gene of interest. A simple database search is now often sufficient to quickly identify and request seeds from insertion lines that collectively contain disruptions in a large majority of *Arabidopsis* genes. For example, the Salk Institute Genome Analysis Laboratory (SIGNAL) has received NSF funding to create a sequence-indexed library of ~150,000 T-DNA insertion mutations in the *Arabidopsis* genome. The data are made available via a web-accessible graphical interface, named T-DNA Express (http://signal.salk.edu/cgi-bin/tdnaexpress), that provides gene code-, text- and DNA sequence-based searches of the database. All isolated T-DNA flanking sequences are also deposited into GenBank and seed requests can be directly filed through the webpage of The Arabidopsis Information Resource (TAIR, www.arabidopsis.org). Seeds from the T-DNA insertion lines are deposited with the ABRC and the NASC and the stock centres subsequently propagate and distribute seeds to individual investigators. This resource provides researchers with very convenient access to mutants in their genes of interest, allowing the testing of hypotheses regarding gene function at an unprecedented rate.

10.3.3 Additional reverse genetics resources for mutant alleles

An additional new type of resource for loss-of-function alleles in any gene has been generated in the form of the *Arabidopsis* and rice DELETEAGENE deletion mutant populations from Maxygen-Davis (Li *et al.*, 2001), that were created using fast neutron bombardment. Because fast neutron mutagenesis is applicable to all plant genetic systems, the method has the potential to enable reverse genetics for a wide range of plant species. The *Arabidopsis* collection already comprises over 50,000 lines with an estimated ten random gene deletions per line. The size of a deletion is usually 1–4 kb, but can also be >10 kb. Li *et al.* (2001) predicted that 130,000 deletion lines will be sufficient to yield a 99% probability of success in isolating a deletion in any target locus. This number is considerably lower than that required for insertion mutagenesis (Krysan *et al.*, 1999). While the generation of deletion lines is easy and highly efficient, the PCR-screening for a deletion line in a given gene of interest is not as straightforward as screening insertion libraries, requiring both primers to be specific to the target locus. Since the size of the deletions varies, it is not possible to know beforehand whether the primer annealing sites exist in the deletion lines. Hence it will usually be necessary to test several nested primer pairs to obtain an amplifiable PCR product that differs in size between a wild-type and a deletion mutant. It is also important to note that the estimated number of ten deletions per line will require several time-consuming rounds of backcrossing and PCR-selection to purify the genetic background of any interesting line. In addition, the size of the deletions will often lead to the loss-of-function of neighbouring genes. While this is undesirable when single genes are to be analysed, it represents a major opportunity for understanding and analysing the ~4,000 tandem-arrayed twin genes with potentially redundant or overlapping biological functions (e.g. members of the *CslB* and *CslG* gene families; Richmond and Somerville, 2000) that exist in the *Arabidopsis* genome (The Arabidopsis Genome Initiative, 2000), and this is probably a common feature in plant genomes (Somerville and Somerville, 1999). Hence, collections of deletion mutants represent a very valuable complement to the existing collections of DNA-insertion lines.

In the case that tandemly arrayed gene duplications ('twins') are not close enough to both be inactivated by a single deletion, or if a suitable deletion line cannot be identified, other routes have to be explored to obtain the appropriate double mutant. Producing the double mutant by genetic recombination will usually not be feasible, but a somewhat labour-extensive approach, using transposons or T-DNAs containing a transposon, might help to generate such double mutants. Since transposons tend to jump into locations near their original integration site (Das and Martienssen, 1995; Martienssen, 1998; Ito *et al.*, 2002), a 'launch-pad' line, containing an insertion in, or close to, one of the two gene copies can be used to generate a new transposon tagged population, thereby increasing the probability of finding the double mutant of interest. A set of 103 *Arabidopsis* launchpad lines that contain *Ac/Ds* transposons is already available through the ABRC.

Targeted gene disruption using homologous recombination (HR) could be another solution to eliminate tandem genes simultaneously by gene replacement. HR is the primary tool for gene knockouts and allele substitutions in bacteria, yeast, and mammals. The moss *Physcomitrella patens* is useful in this regard as a model system for plants (Schaefer, 2001, 2002) and plant cell wall synthesis (Lee *et al.*, 2001). In higher plants, the unresolved problem of illegitimate recombination (Puchta and Hohn, 1996; Mengiste and Paszkowski, 1999) still makes targeted gene disruption by HR difficult, though there has been somewhat limited success on a gene-by-gene basis (Miao and Lam, 1995; Kempin *et al.*, 1997), and development of methods for directed mutations and site-specific recombination is on the list of mid-term goals for the 2010 program (Somerville and Dangl, 2000). HR in plants can be stimulated by overexpression of bacterial proteins involved in HR (Shalev *et al.*, 1999) and fine tuning the HR system through genetic modification may eventually allow HR to become also a routine tool for molecular and genetic studies of plants.

Since DNA-insertions and radiation-induced deletions usually cause loss-of-function alleles when they occur within coding sequence, and directed allele replacement is not yet feasible in plants, there is an obvious need for alternative approaches for the reverse genetic investigation of essential genes. TILLING (Targeting Induced Local Lesions IN Genomes), a high-throughput screening method for induced point mutations, is able to yield the required sub-lethal alleles for those genes (McCallum *et al.*, 2000; Colbert *et al.*, 2001) and is available, as an NSF-funded service, to scientists involved in basic research (http://tilling.fhcrc.org:9366/). In brief, DNA extracted from pools of M2 plants from EMS-mutagenized seeds are screened for the presence of point mutations in a gene of interest by PCR amplification of the gene and analysis of the PCR products for single base-pair mismatches by denaturing HPLC, or by using the CEL1 enzyme from celery, which cleaves DNA at mismatches (Oleykowski *et al.*, 1998). Once a mismatch has been confirmed by sequencing, seeds from a mutant *Arabidopsis* plant can be obtained via the ABRC. TILLING has great potential because it can be applied to many organisms and efforts are under way to generate TILLING populations in other species, such as rice and maize.

Strategies to downregulate the expression level of a given gene by transcriptional or post-transcriptional gene silencing (e.g. antisense approach or co-suppression; Meyer and Saedler, 1996; Depicker and van Montagu, 1997; Vaucheret and Fagard, 2001; Baulcombe, 2002) have been widely used, but are somewhat unpredictable and inefficient, since the approaches often result in partial, rather than strong or complete, loss of gene function and usually require the production and screening of a substantial number of transgenics in order to identify an individual which is significantly silenced. More importantly, if the target gene is a member of a multigene family, it may be impossible to specifically affect the activity of only the target gene. A relatively recent method for gene-silencing, termed RNAi (RNA inhibition), based on producing double-stranded, self-complementary (hairpin) RNA from bi-directional transcription of genes in transgenic plants is more useful for reliable, effective gene inactivation (Waterhouse *et al.*, 1998; Chuang and Meyerowitz, 2000; Wesley *et al.*, 2001; Baulcombe, 2002). Hairpin RNA constructs containing self-complementary regions, ranging from ~1 kb to less than 0.1 kb, result in efficient silencing in a wide range of plant species, and inclusion of an intron in these constructs has a consistent enhancing effect. Therefore, short gene-specific regions of a transcribed sequence (e.g. 5′- or 3′ UTR sequences) or highly conserved regions may be used to obtain silencing in only one gene or multiple members of a gene family, respectively. These intron-containing hairpin RNA (ihpRNA) constructs generally give 90% of independent transgenic plants showing silencing, and the degree of silencing with these constructs is much greater than that obtained using either antisense or co-suppression constructs (Wesley *et al.*, 2001). Use of developmentally regulated or inducible promoters, in combination with the ihpRNA approach, could also significantly obviate problems associated with the lethality of some loss-of-function mutations. As with the DELETEAGENE strategy, gene silencing by ihpRNA can be readily applied to a wide range of plant species for gene function analysis. To make RNAi technology compatible with high-throughput functional genomics and reverse genetics approaches, it is necessary to facilitate the production of ihpRNA-constructs. This can be achieved by adapting plant RNAi vectors to GATEWAY™ technology (www.invitrogen.com), where restriction enzyme digestions and ligations are replaced by a single *in vitro* recombination step. GATEWAY™-compatible plant transformation vectors for RNAi have been recently described (Helliwell *et al.*, 2002; Karimi *et al.*, 2002) and are also available for other applications, such as gene over-expression under control of the cauliflower mosaic virus 35S promoter, the production of in-frame N- and C-terminal fluorescent protein and antibody-tag fusions for protein localization studies, or promoter-β-glucuronidase fusions for sensitive gene expression studies at the cellular level (Karimi *et al.*, 2002; Nakagawa, 2002).

For many applications, particularly in species other than *Arabidopsis*, where the routine production of transformants is difficult, virus-induced gene silencing (VIGS) represents another means of suppressing gene function (Baulcombe, 1999). This method exploits the fact that some or all plants have a surveillance system that can specifically recognize viral nucleic acids and mount a sequence-specific suppression of viral RNA accumulation. By inoculating plants with a recombinant virus containing part of a plant gene, it is possible to rapidly silence the endogenous plant gene (Kjemtrup *et al.*, 1998; Ruiz *et al.*, 1998). This approach has already been successfully applied to silence a tobacco cellulose synthase, leading to much shorter internode lengths, small leaves, a dwarf phenotype and reduced cellulose content in cell walls of infected plants (Burton *et al.*, 2000).

Over-expression can be a potentially useful complement to targeted gene disruption and gene silencing. Strong constitutive, inducible (e.g. by hormones, antibiotics, heavy metals or environmental cues) or tissue- and developmentally-specific promoters can be used to this end (Reynolds, 1999), and can lead to strong ectopic gene expression. Over-expression of a given gene can also lead to co-suppression, the homology-dependent gene silencing of both the introduced transgene(s) and the endogenous gene (for reviews see above). Untargeted over-expression of genes is achieved by activation tagging with T-DNA and transposons. However, the information that is obtainable by over-expression may not be easily associated with the normal biological role for a given, yet uncharacterized gene. Over-expression/ectopic expression of gene products in tissues and conditions where they are not properly regulated can alter the biological function of the gene product and lead to additional and new phenotypes. For example, the ectopic expression of the maize transcription factor *Lc*-gene under the control of a 35S promoter leads to anthocyanin pigmentation in various tissues, including roots, which are not normally pigmented (Ludwig *et al.*, 1990). Also, the ectopic expression of the *Arabidopsis LEC1* transcription factor gene is sufficient to induce embryo development in vegetative cells (Lotan *et al.*, 1998). Although such information does not fully reveal the normal biological function of a given gene, it might nevertheless be helpful with uncharacterized genes, genes that act redundantly, or genes that are required during multiple stages of the plant life cycle and whose loss of function results in early embryonic lethality (Weigel *et al.*, 2000).

10.3.4 Finding phenotypes for knockout mutants; running the gauntlet

Since plants are sessile, they are forced to adapt to life in changing biotic and abiotic environments. Hence, many genes might only be required and expressed in the specific cells and under the specific conditions in which its product makes a contribution to the fitness and survival of the organism (Brown and Botstein, 1999). Consequently, many reverse-genetic loss- or reduction-of-function mutants can be expected to be conditional. In accordance with this assumption is the general observation that only a small fraction (1–3%) of the knockout or knock-down mutants in *Arabidopsis* show a visible phenotype when grown in 'normal' greenhouse conditions (Azpiroz-Leehan and Feldmann, 1997; Bouché and Bouchez, 2001; Marsch-Martinez *et al.*, 2002). Finding the conditions under which a lesion in a specific gene yields a phenotype can be very difficult and is often the most time-consuming step in a reverse genetics approach towards the determination of the biological role of a gene. In such cases, an informative phenotype may conceivably only be revealed following a careful study of the 'reverse genetic' mutant by transcript-, proteome-, metabolite-, or, in the case of cell-wall related mutants careful cell wall polymer analysis, and by morphological and microscopic analysis in a wide range of experimental and environmental conditions throughout the life cycle. These conditions might include selective conditions at high plant density and competition between genotypes (e.g. mutant and corresponding wild-type) for resources such as light, water and nutrients. The *Arabidopsis* potassium-channel T-DNA insertion mutant *akt1-1* is a good example of the need for careful evaluation of possible phenotypes. This mutant is phenotypically indistinguishable from wild-type plants when grown on many nutrient media. However, growth of *akt1-1* plants is significantly inhibited compared with wild-type on media containing 100 µM potassium in the presence of ammonium (Hirsch *et al.*, 1998; Spalding *et al.*, 1999). Knockout mutations can also have low penetrance and might be revealed only by multigenerational population studies. For example, adult *Arabidopsis* plants that are homozygous for various mutant actin alleles appear to be morphologically normal and fully fertile. However, when grown as populations descended from a single heterozygous parent, the mutant alleles are found at extremely low frequencies relative to the wild-type in the F2 and following generations, and thus appear to be deleterious (Gilliland *et al.*, 1998).

The search for a phenotype in a mutant can be helped and guided significantly when additional information on gene activity is available before analysis. This includes mining of available transcript databases, such as the Stanford Microarray Database (http://genome-www5.stanford.edu/MicroArray/SMD), and possibly EST libraries, especially if they were obtained from specific tissues, organs or conditions, proteome databases, analysis of promoter-GUS and protein expression patterns.

In addition, gene functions that seem to be non-essential or redundant (see above) might require the generation of double, triple, etc. mutants with lesions in different family members with overlapping expression patterns before phenotypic changes can be found in specific conditions (Leung *et al.*, 1997; Liljegren *et al.*, 2000; Pelaz *et al.*, 2000). Another scenario that merits some consideration is the possibility that a gene, although expressed, has lost its function during the course of evolution. Since contemporary analyses only provide a snapshot of evolution, it is possible that some genes are currently experiencing significant negative selection and have already suffered mutations that render them dysfunctional, or are expressed in cells or at times where they cannot associate with appropriate binding partners or substrates. Simultaneously, genes with new or altered functions that are potentially beneficial to plant fitness slowly emerge, for example, by gene duplication and subsequent divergence.

10.4 Forward genetics in the post-genome era

Forward or classical genetics starts with the screening for mutants showing a particular phenotype, such as a trait likely to be related to changes in the quantity or structure of a cell wall polymer or other cell wall-related factors, followed by their characterization, and works towards the identification of the corresponding, unknown gene(s) (Figure 10.1). Since the approach does not make any assumption about the sequence or function of a gene, it has the potential to assign functions to new, uncharacterized genes (currently at least 40% of the *Arabidopsis* genes still fall in this category) and to provide a link between genes of suspected biochemical function with unexpected phenotypes, or, in short, to teach 'new biology'. The anonymity of the genes under study also represents the major limitation of forward genetics, since double mutants in genes with largely redundant function that can compensate for each other are not accessible.

Many *Arabidopsis* cell wall mutants have been identified using various genetic screens over the last few years; for detailed information consult the overview by Fagard *et al.* (2000b). The screens include those directly aiming for the identification of mutants with altered cell wall polysaccharide composition using chromatographic approaches (Chapple *et al.*, 1992; Reiter *et al.*, 1993, 1997). Several classes of *Arabidopsis* mutants with altered cell walls have also been identified in various screens for developmental and organ- or cell-type specific growth defects, including cytokinesis defects in embryos (Lukowitz *et al.*, 1996; Nickle and Meinke, 1998; Lukowitz *et al.*, 2001; Gillmor *et al.* 2002), isotropic-expansion of root cells at non-permissive temperatures (Baskin *et al.*, 1992), defects in cell-expansion (Desnos *et al.*, 1996; Nicol *et al.*, 1998), the production of collapsed xylem elements (Turner and Somerville, 1997), or the lack of trichome birefringence caused by secondary cellulose deposition (Potikha and Delmer, 1995).

Many of the cell wall mutants isolated in such screens were from ethyl methanesulfonate (EMS) mutagenized collections, for obvious reasons: EMS is easy to use and a highly efficient mutagen, typically causing several hundred single base-pair exchanges in an *Arabidopsis* genome (Colbert *et al.*, 2001). Besides leading to the loss of gene function by introduction of early stop codons, frameshifts, unacceptable substitutions or elimination of splice-sites, EMS can also induce single amino acid substitutions in the gene product, giving rise to functionally altered protein isoforms with special properties, such as conditional phenotypes. Based on the number of mutations per genome, it can be calculated that in order to find a mutation in any given, average-sized *Arabidopsis* gene, at a 95% probability level, approximately 10,000 EMS-lines should be screened, which is approximately seven times lower than the number needed for T-DNA insertion lines. Since no sequence context is available, chemically (EMS-) induced point mutations cannot be rapidly localized and hence require a mapping (positional cloning) approach to identify the corresponding mutant gene. This has been the major difficulty with EMS-induced and radiation-induced deletion mutants, since it is dependent on factors such as the availability of chromosomal maps and sequence contigs, and the possession of an extensive collection of mapped visible or molecular markers. With the availability of the genome sequences for *Arabidopsis* and rice, and the creation of large numbers of mapped molecular markers (e.g. http://www.arabidopsis.org/; http://rgp.dna.affrc.go.jp/Publicdata.html) these bottlenecks have mostly disappeared in these species. Consequently, by systematically exploiting the available sequence information, positional cloning now typically involves just a few basic molecular biology procedures and as little as a few months to isolate almost any mutation that can be mapped (see Lukowitz *et al.*, 2000 and Jander *et al.*, 2002 for a deeper discussion of this topic).

With the tremendous momentum gained by the availability of the entire *Arabidopsis* genome sequence, the forward genetics approach has been very successfully used in the recent past, and has led to the identification of expected cell wall biosynthetic genes (e.g. Bonin *et al.*, 1997; Arioli *et al.*, 1998; Taylor *et al.*, 1999, 2000; Fagard *et al.*, 2000a), and also a number of unexpected genes. Good examples are the recent positional mapping of the *Arabidopsis cyt*1 and *knf* mutations that both lead to defects in cell wall biogenesis and severely reduced cellulose levels at the embryo stage (Lukowitz *et al.*, 2001; Gillmor *et al.*, 2002). The *CYT1* gene encodes a mannose-1-phosphate guanylyltransferase catalysing the formation of GDP-mannose, which is, among other things, required for the synthesis of the core glycan-chain attached to N-linked glycoproteins. The *KNF* gene encodes alpha-glucosidase I, the enzyme that catalyses the first step in N-linked glycan processing. Taken together, these new results point to the requirement of N-glycosylation and N-glycan processing for cellulose biosynthesis. Other examples are the recent identifications of the *Arabidopsis* homeodomain leucine zipper gene *IFL1*, that regulates interfascicular fibre differentiation (Zhong and Ye, 1999), the *ELP1* gene that encodes a chitinase-like gene and causes ectopic lignification, altered cell shape and incomplete walls in piths of inflorescence stems (Zhong *et al.*, 2002), and the *Arabidopsis KOBITO* gene that encodes a putative scaffolding plasma membrane protein necessary for normal synthesis of cellulose during cell expansion (Pagant *et al.*, 2002).

It can be anticipated that in the near future the wealth of information regarding the extent of genetic diversity between *Arabidopsis* accessions will further simplify genetic fine mapping. Using bulked segregant analysis (Michelmore *et al.*, 1991; Lukowitz *et al.*, 2000), together with 'genotyping chips' (Hacia, 1999), the high number and density of SNP and INDEL polymorphisms/molecular markers present in the *Arabidopsis* genome might enable chromosome 'landing' on a genetically defined region that is so small it encompasses only a limited number of open reading frames. Similarly, the availability of full-genome Affymetrix DNA chips (see section 10.6.2) should also make the identification of radiation-induced deletions very straightforward. A mutation in one of the candidate genes can then be spotted by either expression analysis, sequencing or mutation detection techniques, including single-strand conformation polymorphism (SSCP) analysis and chemical or enzymatic cleavage of mismatches (Cotton *et al.*, 1998). Alternatively, small (10–20 kb) overlapping genomic fragments covering the genomic region can be transformed into mutants to determine the gene of interest via functional complementation of the mutant phenotype (Scheible *et al.*, 2001). Considering that more than 15% of the ~27,000 *Arabidopsis* genes might be involved in some aspect of cell wall metabolism (Carpita *et al.*, 2001), that ~40% of *Arabidopsis* genes do not have an attributed biochemical function, and that less than 10% have an attributed biological function, it is likely that the map-based cloning approach will remain of key importance to gain knowledge about the cell wall also in the era of reverse genetics.

Genetic variation among natural populations (ecotypes/accessions) is another highly valuable but yet largely unexploited forward genetics resource. It is likely that natural variation stems from special change-of-function alleles that have been selected for during evolution, helping the population to cope with the habitat, and which are unlikely or impossible to create simultaneously by mutagenesis in the laboratory. *Arabidopsis* is well suited for the study of natural variation because little heterozygosity exists in natural populations, and hundreds of ecotypes/accessions originating from many different climates and latitudes are readily available through the *Arabidopsis* stock centres. Natural variation between *Arabidopsis* accessions has been shown to exist for pathogen disease resistance, and quantitative traits like freezing tolerance, flowering time, seed size, water use efficiency, etc. (see Alonso and Koornneef, 2000). Quantitative ecotype variation has recently also been found for cell wall traits, such as xyloglucan content among *Arabidopsis* accessions

Bayreuth-0 and Shahdara (Markus Pauly, personal communication) and hemicellulose content in maize (Hazen and Walton, 2002).

The mapping of the genes causing quantitative trait variation in an *Arabidopsis* ecotype cross is a somewhat lengthy but nowadays more feasible process, since the genome sequence is complete. However, it still requires the availability of a set of several hundred recombinant inbred lines (RILs), near isogenic lines (NILs), and the establishment of a sufficiently large and robust marker set, that can be used to genotype the RILs and NILs (Figure 10.2). Such resources have been created for some ecotype combinations (see homepages of ABRC and NASC; Loudet *et al.*, 2002) and additional facilities are being produced (e.g. Dr Alan Lloyd, University of Texas, USA; Dr Thomas Altmann; Max-Planck Institute of Molecular Plant Physiology, Golm, Germany). Once these resources are available, a quantitative trait that is most extreme in a small subset of the RILs or NILs can be associated with one, or very few, small genomic regions. A suspect gene in that region can then be directly investigated, or can be further confined by map-based cloning (see above) using a segregating F2-population from a cross between the NIL and the respective wild-type.

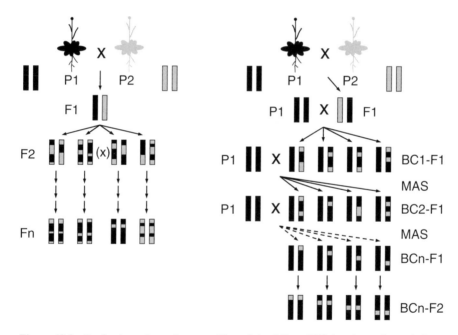

Figure 10.2 Production scheme for recombinant inbred lines (RILs) and near isogenic lines (NILs) required for the mapping of genes underlying natural variation between ecotypes (i.e. P1 and P2). (a) Recombinant inbred lines, RILs, are derived by successively selfing single plants from the progeny of individual F2 plants (single seed descend) until homozygosity is achieved at the Fn generation (n = 7–9). To increase the number of recombination events along chromosomes F2 plants can also be crossed randomly, depicted by (X), before selfing is started. (b) Near isogenic lines, NILs, with known single segment substitutions are obtained from an F1 ecotype cross by recurrent rounds of backcross (BC) breeding (BC-F1 × P) and marker-assisted selection (MAS).

10.5 Technologies for transcript profiling and their use to study cell wall formation and differentiation

Various technologies have been developed in recent years that allow the large-scale detection, quantification and differential expression analysis of transcripts. These include high-throughput sequencing of cDNA libraries, also referred to as EST sequencing (Newman *et al.*, 1994), serial analysis of gene expression (Velculescu *et al.*, 1995; Adams, 1996; www.sagenet.com), cDNA-AFLP (Bachem *et al.*, 1996), and 'reverse Northern' solid-support approaches like high-density cDNA arrays on nylon membranes, cDNA microarrays on glass slides and photolithographic oligonucleotide arrays (gene chips). The DNA array-based methods have made great advances in recent years, are academically and commercially available for several plant species including *Arabidopsis* and rice, and are currently the most widely used.

10.5.1 EST sequencing

A collection of expressed sequence tags (ESTs) typically reflects the level and complexity of gene expression in the sampled organism or tissue at the level of transcript accumulation. However, genes that are expressed at very low levels, in a small number of cells or under a limited number of conditions might not be represented by ESTs and therefore remain undetected in EST-sequencing projects. When EST sequences are generated from tissue-specific cDNA libraries, it is also possible to extract information about the localization and transcript abundance of given genes (Ewing *et al.*, 1999). In this regard the EST information from some plant species, such as *Medicago truncatula*, tomato (http://www.sgn.cornell.edu/), or sugarcane (http://sucest.lbi.dcc.unicamp.br) is better than others, based on the number of different tissue-specific libraries that have been sampled, and whether or not those libraries have been normalized in some way.

EST sequencing has been used to investigate gene expression in wood forming tissues of loblolly pine (Allona *et al.*, 1998) and poplar (Sterky *et al.*, 1998). In the latter report, approximately 5,700 ESTs were generated that represented more than 3,700 unique transcripts. About 4% of the ESTs were involved in various aspects of cell wall formation, such as lignin and cellulose synthesis, another 5% were similar to developmental regulators and members of known signal transduction pathways, and 2% were involved in hormone biosynthesis. An additional 12% of the ESTs showed no significant similarity to any other DNA or protein sequences in existing databases, thus the absence of these sequences from public databases might indicate a specific role for these proteins in wood formation.

10.5.2 DNA-array based approaches

Nylon arrays (Desprez *et al.*, 1998) represent the simplest array-based technology. They are constructed by spotting hundreds or thousands of target DNA fragments, usually EST clones or PCR products, in a grid-format onto commercial nylon membranes. The utility and quality of this methodology has been improved by advances in spotting and the use of two radioactive probe labels (e.g. ^{32}P and ^{33}P), but inherent disadvantages, such as the variability in the amount of spotted target DNA, the dimensional instability of the nylon matrix, the variable degree of accessibility and ease of hybridization of the spotted DNA, and the variable background across the nylon surface, can greatly increase errors in measurement and the quantitative quality of the data. This might not be of great relevance for qualitative screening, and particularly if the resulting interesting clones are subsequently confirmed by other means, but it renders the approach less suitable for whole genome scans, including weakly expressed genes and genes expressed in low-abundance cell-types, or usage as a platform for expression databases.

Glass microarrays have several advantages over nylon arrays. The glass support is stable and the spot density of target clones is sufficiently high to represent all genes of an organism on a single small glass slide. Most importantly, they give superior data compared to nylon filters. The technique uses two fluorescent dyes, such as Cy3 and Cy5, to label cDNA populations derived from polyA-mRNA or total RNA of two different biological samples to be compared. Emission signals of the fluorochromes are linear over a broad range, which makes it possible to evaluate abundant and rare probe species in the same experiment. The two labeled cDNA probes are simultaneously hybridized in a small volume to excess amounts of the target fragments, permitting direct comparison of the two samples. For RNA samples derived from small amounts of tissue, or even single cells, PCR methods can be used to produce sufficient material for fluorescent probes. These features, along with the lower background hybridization on glass slides, improve the detection of rare mRNA species. The fluorescence hybridization signal for each target is quantified with a laser scanner and used to assess mRNA abundance in a given tissue.

Due to the limitations of hybridization in very small volumes and with target DNA molecules that have several bonds to their support matrix, it is likely that DNA microarrays are not always quantitative and thus not an accurate means for measuring the absolute abundance of a message. As with the nylon arrays, the amount of spotted target DNA at each spot can vary on and between microarrays. Therefore, it is possible to hybridize to saturation for one sample at one spot, but not for the other sample. Such saturation can give misleading quantitation and it is difficult to control for such mistakes. In such a case it would not be possible to reliably measure the relative change in expression between two samples. This problem of variability, and also the problem of differential incorporation of the two dyes, is in part compensated for by the use of two-colour detection and reverse labeling of the cDNA preparations, i.e. labeling twice, once with each dye. The hybridizations are then conducted simultaneously and the ratios between the two fluorescence signals are simultaneously determined. By using the ratios of the two dyes, the variability and the amount of spotted DNA have less of an influence on the quantification. Testing the reproducibility of microarray data with technical and biological replicates is crucial before results can be viewed with confidence.

Microarray projects have many of the same problems as other large-scale genome projects, particularly the challenges of implementation and investment costs, and organization and handling of a huge inventory of EST clones or subcloned genomic PCR-fragments. A tremendous amount of labour and expense has to be invested in selecting, growing, sequencing, maintaining and arraying the large number of clones, and simultaneously avoiding cross contamination, degradation or mislabeling. Obviously, high-quality microarrays demand maintenance of infrastructure; therefore glass microarray technology is valuable only if large numbers of arrays are generated. For *Arabidopsis*, this task has been taken over by both academic (http: //afgc.stanford.edu) and commercial (Agilent Technologies, Incyte Genomics) services and array providers.

Another serious problem of many microarrays is the potential occurrence of cross-hybridization between similar genes. Since target clones are usually either partial cDNA sequences (ESTs) or PCR products amplified from genomic DNA, it is unlikely that the fragments are gene-specific and yet also represent all genes of an organism with a minimum of redundancy. This is especially true for the considerable number of genes that exist in multigene families. It was estimated that sequences with >70% identity over >200 nucleotides in length are likely to exhibit some degree of cross-hybridization under standard conditions (Richmond *et al.*, 1999).

The range of applications for DNA-based microarrays is quite diverse (Eisen and Brown, 1999): microarrays are applicable to all organisms for which cDNA can be prepared, and even anonymous fragments of genomic DNA can be used as targets. In principle, there is no need for sequencing of any of the target sequences unless interesting results are seen during experimentation. Microarrays also allow for simultaneous gene expression studies in biochemically interacting species, such as plants and their pathogens. Furthermore, the microarray technique can also be adapted for use in 'reverse Southerns', in which complex DNA probes (e.g. bacterial artificial chromosomes [BACs] or total genomic DNA) are used in place of cDNA (Shalon *et al.*, 1996). This technique can be useful for defining large insertions or deletions in the genome. In addition, insertion sites of transposons or T-DNAs can be verified using microarrays, by hybridizing with DNA flanking the insertion elements produced using inverse PCR (Lemieux *et al.*, 1998). Lastly, microarrays constructed from genomic fragments might be used in combination with labeled DNA binding proteins to determine candidate target genes (Bulyk *et al.*, 1999; Kehoe *et al.*, 1999).

Microarrays have been widely used in recent years, primarily in *Arabidopsis* but also in other plants, to study changes in transcript abundance in mutants with defined lesions, and metabolic and developmental processes (e.g. Wang *et al.*, 2000; Ruuska *et al.*, 2002; http://afgc.stanford.edu). Some of these studies can be related to some aspect of cell wall biology. For example, Hertzberg *et al.* (2001) studied the transcript profiles of ~3,000 genes during different developmental stages of xylogenesis in poplar. This analysis revealed that genes encoding lignin and cellulose biosynthetic enzymes, as well as a number of transcription factors and other potential regulators of xylogenesis, are under strict developmental stage-specific transcriptional regulation. Similarly, Ebskamp *et al.* (2001) used a cDNA microarray with over 1,600 flax cDNAs to analyse gene expression in different parts of the flax stem in the fast growing stage, when the transition between primary and secondary cell wall synthesis takes place. Ebskamp *et al.* (2001) also developed a cDNA microarray with ~3,000 hemp clones to monitor genes involved in cell wall composition.

Surprisingly, no *Arabidopsis* cell wall related microarray studies have been published to date, although the species nevertheless seems to be well suited to a study of processes involving wall metabolism, such as secondary xylem formation, which resembles that of an angiosperm tree stem (Chaffey *et al.*, 2002). However, the Arabidopsis Functional Genomics Consortium (afgc.stanford.edu) makes all their microarray transcript data publicly available through the Stanford Microarray Database, and cell wall researchers can extract meaningful expression information for many of their genes of interest from over 400 microarray slides. For example, Richmond and Somerville (2001) used the AFGC data to survey the expression histories of 18 cellulose synthase and cellulose synthase-like genes with respect to tissue specificity and differential expression under salt, hormone or light treatments. Similarly, Schultz *et al.* (2002) more recently systematically mined and summarized the AFGC expression data of 35 members of the arabinogalactan protein (AGP) gene family. Besides microarrays, cDNA-AFLP technology has also been used to look for differential expression of cell wall related genes during the formation of tracheary elements in the *Zinnia* mesophyll cell system (Milioni *et al.*, 2001).

The most recent and most complex technology for large-scale transcript analysis is the DNA chip technology from Affymetrix (www.affymetrix.com) (Lockhart *et al.*, 1996, 2000). This technology is based on high-density oligonucleotide arrays synthesized directly on derivatized glass slides using a combination of photolithography and oligonucleotide chemistry (Lipshutz *et al.*, 1999). The infrastructure required to produce oligonucleotide arrays exceeds what may be reasonably maintained by an academic centre and the technology is dependent on prior gene sequence information, thus limiting the approach to organisms for which genomic or high quality EST sequence collections exist. However, the commercial and unrestricted availability of 'whole-genome' *Arabidopsis* oligonucleotide chips (representing ~24,000 genes) from Affymetrix, the obvious advantages (see below) and reasonable costs make this technology attractive for many plant scientists. The general approach for monitoring gene expression with this technology is to synthesize approximately 10–20 oligonucleotides for each gene (Wodicka *et al.*, 1997) with a long linker arm, ensuring that each base of the oligonucleotide is equally accessible to hybridization in solution. The oligonucleotides are usually 25-mers and are designed in pairs: one perfectly complementary to the probe, and a companion that is identical apart from a single base difference in the central position. Because the oligonucleotides are only 25 base pairs long, a single central mismatch will result in a considerably lower level of hybridization. The mismatches serve as sensitive internal controls for hybridization specificity and comparison of the perfect match signals with the mismatch signals allows low-intensity hybridization patterns from rare transcripts to be obtained. Given the availability of the entire *Arabidopsis* genome sequence, gene-specific oligonucleotides can readily be designed, even for members of large and highly conserved families. The surface of DNA chips can be entirely covered with several hundreds of thousands of different oligonucleotides of defined sequence, thus it is possible to represent an entire plant genome on one chip. Furthermore, the background fluorescence is usually very low and uniform due to the high oligonucleotide density. Since considerable standardization is used in the manufacturing of the chip, its reproducibility is also very high, reducing costs and labour for technical replicates, and enabling researchers to produce directly comparable high-quality data. Since chips are extremely reproducible and the amount of oligonucleotide at each location is the same, there is no need for two-colour fluorescence detection, allowing the use of a single probe with very high fluorescent efficiency. The currently used labeling method utilizes biotinylation followed by streptavidin coupled to phycoerythrin. This greatly increases the sensitivity of detection, which is important for the analysis of low abundance transcripts. It is foreseeable that chip technology will be the preferred and dominant platform for the generation of expression databases in the future. Besides their use in gene expression studies, gene chips can also be used for genotypic analysis (e.g. mutation analysis), re-sequencing, using variant detector arrays (Hacia, 1999), and large-scale genetic mapping of single nucleotide polymorphisms (SNPs) (Wang *et al.*, 1998). This latter use may revolutionize genetic mapping and marker-assisted breeding in plants (Somerville and Somerville, 1999).

Oligonucleotide microarrays representing ~8,000 genes have been already used to investigate the coordinated circadian regulation of gene expression in *Arabidopsis* (Harmer *et al.*, 2000). In this study, several genes thought to be involved in cell wall relaxation and loosening (e.g. an expansin, a polygalacturonase and auxin efflux carriers) all peaked toward the end of the subjective day, while transcripts encoding enzymes thought to be implicated in the synthesis of cell wall sugars and polymers (two cellulose synthase-like proteins and a dTDP-D-glucose 4,6-dehydratase) peaked later, towards the end of the subjective night, suggesting that after cell wall relaxation and expansion, new cell wall material is laid down to reinforce the enlarged cell.

10.5.3 Real-time RT-PCR

Gene expression analysis by quantitative real-time RT-PCR is a useful complementary approach to the microarray and gene chip technologies. For example, a comparison of the three technologies by Holland (2002) has recently shown that a high percentage of yeast transcription factors are expressed at levels too low (<<1 transcript per cell) for quantification with the microarrays and gene chips, whereas real-time RT-PCR is still capable of generating reproducible expression levels of these genes in unicellular yeast. It is reasonable to assume that a comparable situation also exists for plants, where various cell-types, including those that are highly specialized and present in low abundance, may coexist in the same tissue during a range of physiologically or developmentally different stages, and in which a ~3-times larger number of genes are present. Besides the particularly high sensitivity, the RT-PCR approach also has a very flexible setup, enabling researchers to add or remove genes from their analysis at any time. However, if many genes or large gene families are to be analysed by RT-PCR, the labour and cost required for the design and synthesis of gene-specific primer pairs and the analysis, including biological replicates, can quickly exceed what is affordable for an individual laboratory.

Using RT-PCR, Yokohama and Nishitani (2001) recently analysed the expression of the *Arabidopsis* family of genes encoding the 33 xyloglucan endotransglucosylases-hydrolases (XTHs) involved in the construction and modification of the cellulose/xyloglucan framework (Rose *et al.*, 2002). Since some members of the *XTH* gene family share 90% identity at the nucleotide level, conventional RNA blot analysis was not possible. The RT-PCR analysis revealed that the members vary largely in their absolute expression levels and that many members show organ-specific expression in adult plants and are regulated by different phytohormones. In an attempt to more comprehensively monitor the expression of cell-wall related genes in these conditions, Yokohama *et al.* (2002) have recently assembled a gene-specific oligo DNA-microarray consisting of 762 clones encoding, besides the members of the *XTH* family, 40 cellulose synthases and cellulose synthase-like proteins, 12 callose/glucan synthases, 38 glucosyltransferases, 111 pectinesterases, 69 peroxidases, 77 β-1,3 glucanases, 24 β-1,4 glucanases, 67 polygalacturonases, 20 xylosidases, 26 pectate lyases, 18 -fucosyl and -xylosyltransferases, 20 xylosidases, 25 galactosidases, 23 chitinases, 38 extensins, 35 expansins, 30 glycine-rich proteins, and others. Selected genes of interest were also subjected to a quantitative real-time RT-PCR analysis to quantify their expression levels more precisely. From these studies the authors concluded that a combination of the microarray and the real-time RT-PCR procedures afford excellent insight into the precisely coordinated regulatory system in the transcriptions of the cell wall-related genes during both organ development and responses to hormonal stimuli.

10.6 Proteomic analysis of plant cell walls

Global gene expression and regulation can be characterized by monitoring mRNA abundance, using gene chip or microarray techniques (Schena *et al.*, 1995; Brown and Botstein, 1999), as described in section 10.5.2. However, reliance on this technique as the sole tool for profiling gene expression has a number of limitations. In addition to the technical hurdles outlined above, challenges at the biological level include the following:

1. Gene expression is regulated not only transcriptionally, but also translationally and post-translationally. Several studies have described poor correlations between changes in the levels of specific mRNAs and their corresponding proteins, and large scale quantitative analyses have reported correlation coefficient values of only 0.48–0.61 (Anderson and Seilhammer, 1997; Anderson and Anderson, 1998; Gygi *et al.*, 1999; Ideker *et al.*, 2001).

2. Substantial regulation of cellular events occurs at the protein level with no
    apparent changes in mRNA abundance. Post-translational modification
    of proteins can result in a dramatic increase in protein complexity without
    a concomitant increase in gene expression. Indeed, on average, one yeast
    gene encodes between 1 and 3 distinct modified proteins while in humans
    this number typically escalates to between 3 and 6, and in some cases, up to
    20 derivatives from a single gene (Wilkins *et al.*, 1996).
3. DNA sequence and mRNA expression profiling provide little information
    about subcellular localization beyond the identification of putative target-
    ing signal sequences and structural domains.
4. Proteins are often more stable than mRNAs (Anderson and Anderson,
    1998), and rare transcripts from highly regulated genes are often not repre-
    sented in EST databases (Bouchez and Höfte, 1998).

These limitations indicate that the full value of large-scale profiling of mRNA
populations, or transcriptomics, will not be realized if considered in isolation, but
rather, the benefits will be seen when used in conjunction with other techniques for
analysing global gene expression. In this regard, there is an obvious need to develop
equivalent large-scale initiatives to characterize protein populations.

Proteomics, or the study of the protein complement of the genome, represents one
of the most rapidly developing and innovative fields in genome-scale research (Rossignol, 2001). By characterizing quantitative and qualitative characteristics of global
protein expression, including polypeptide synthesis, degradation, post-translational
modification and interactions with other cellular components, proteomics promises
to span the gap between genomic DNA sequence and biological state. Proteomics
also provides a means to validate predicted gene sequence, since the presence of an
open reading frame (ORF) does not guarantee the existence of a functional gene
product. For example, one study of predicted genes from *Mycoplasma genitalium*
had an error rate of at least 8% (Brenner, 1999). It is certain that sequencing errors
and inaccurate predictions of ORFs also lead to misannotation of plant genome EST
sequences. The sequencing of the corresponding proteome thus provides a means to
validate the putative ORFs and annotations.

10.6.1 Developments in proteomics technologies

The systematic characterization of complex protein populations can be pursued
using many approaches and numerous reviews have recently been published describing developments in diverse areas of proteomics, including protein structure,
function and protein-protein, or protein-ligand interactions (e.g. Mann *et al.*, 2001;
Schmid, 2002; Yanagida, 2002; Yarmush and Jayaraman, 2002). A number of reviews have also been specifically devoted to plant proteomics (Rossignol, 2001; van
Wijk, 2001; Guo *et al.*, 2002; Kersten *et al.*, 2002; Roberts, 2002). Consequently, the
multiplicity of proteomics-related experimentation and emerging technologies are

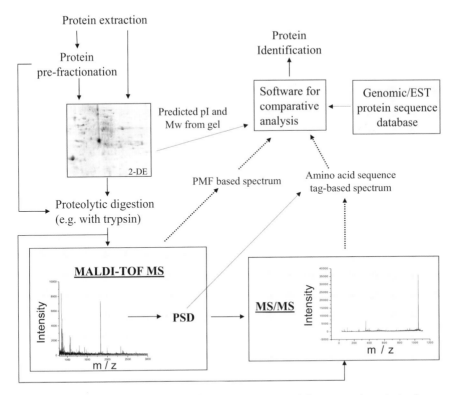

Figure 10.3   Schematic diagram outlining a common approach for proteomic analysis of complex protein mixtures. Following extraction from the biological sample, protein mixtures may be separated by two dimensional gel electrophoresis, or first subjected to a prefractionation step, such as some type of liquid chromatography (e.g. ion exchange or size exclusion-based chromatography). Following separation, a protein of interest is hydrolysed by a site-specific protease (e.g. trypsin) and the daughter peptides are analysed by mass spectrometry, resulting in a peptide mass fingerprint (PMF), generated using MALDI-TOF MS, or an amino acid sequence, obtained by tandem MS or using a post-source decay (PSD) approach with the MALDI-TOF instrument. This information can be combined with the predicted pIs or molecular weight (Mw) of the targeted protein, and used to screen the database of known genes and proteins.

not reiterated here in detail. Instead, a brief overview is provided of the most widely used approach to cataloging complex protein populations (Figure 10.3): high-resolution two-dimensional gel electrophoresis (Görg *et al.*, 2000) followed by protein sequencing and identification using mass spectrometry (Dongre *et al.*, 1997; Andersen and Mann, 2000; Pandey and Mann, 2000). This should provide the reader with a basic vocabulary and list of acronyms with which the detailed literature can more easily be tackled. In addition, the URLs of some useful proteomics related websites are provided in Table 10.1.

Table 10.1 Examples of internet-accessible proteomics related resources.

Resource	URL	Notes
Protein and DNA sequence databases	http://www.ncbi.nlm.nih.gov	National Center for Biotechnology Information (NCBI)
	http://www.ebi.ac.uk	European Bioinformatics Institute (EBI)
	http://www.ddbj.nig.ac.jp	DNA Data Bank of Japan (DDBJ)
2D-Gel databases	http://www.expasy.ch	Expert Protein Analysis System (ExPASy) [WorldWide 2D Gel portal]
	http://www.expasy.org/ch2d/2DHunt/	2-D electrophoresis Finder at ExPASy
	http://semele.anu.edu.au/2d/2d.html	2D-gel database for rice and *Medicago truncatula* at Australian National University, Canberra
	http://aestivum.moulon.inra.fr/imgd/	Maize Genome Database at INRA, Gif-sur-Yvette, France
	http://sphinx.rug.ac.be:8080/ppmdb/index.html	Plant Plasma Membrane Database (PPMdb) at Gent University, Belgium
	http://www.gartenbau.uni-hannover.de/genetic/AMPP	The Arabidopsis mitochondrial proteome project at Abteilung Angewandte Genetik, Universität Hannover, Germany
	http://us.expasy.org/ch2d/	Various (human, mouse, Arabidopsis, *Dictyostelium discoideum*, *Escherichia coli* and *Saccharomyces cerevisiae*) at ExPASy
	http://www.pierroton.inra.fr/genetics/2D/index.html	Maritime pine at Forestry Research Station of Cestas (INRA Bordeaux)
Search engines (from peptide fragments after digestion)	http://www.matrixscience.com	Mascot
	http://prowl.rockefeller.edu/	Profound (and pepfrag.)
	http://prospector.ucsf.edu/	Protein prospector and various tools
Other useful search engines	http://www.expasy.org/tools/tagident.html	TagIdent tool (formerly GuessProt): Retrieves SWISS-PROT entries closest to given pI and mol. wt.
	http://www.expasy.org/tools/glycomod/	Glycomod: Calculate oligosaccharide structure based on mass
	http://www.ncbi.nlm.nih.gov	BLAST (sequence relationship search tool for proteins and nucleotides) at NCBI

10.6.2 Two-dimensional gel electrophoresis-based protein separation and quantitation

Over the last two decades, two-dimensional gel electrophoresis (2DE) has been the method of choice for separation of complex protein mixtures (Jungblut *et al.*, 1996; Jungblut and Thiede, 1997). This approach entails fractionation of proteins by a combination of isoelectric focusing in the first dimension, separating the proteins based on their charge using an immobilized pH gradient (IPG) gel over a broad pI range (typically pI 3–10) or narrow range (e.g. 1 pI unit), and sodium dodecyl sulphate (SDS) polyacrylamide gel electrophoresis (PAGE) in the second dimension, separating proteins based on molecular weight (Görg *et al.*, 2000). The increase in popularity of 2D-PAGE in the last ten years has been due mainly to improvements in gel resolution and reproducibility (Rabilloud, 2002). This has resulted from technical developments such as commercial IPG systems that use overlapping narrow range pI IPG separations (Görg *et al.*, 2000), the availability of fluorescent dyes such as SYPRO RUBY that have high sensitivity and a broad dynamic range enabling low abundance (~ 3 ng) proteins to be visualized (Berggren, *et al.*, 2000), and other incremental advances in electrophoretic technology (Patton *et al.*, 2002). Consequently, the spatial topography of 2D gels is more reproducible, with higher protein loading and faster spot identification by commercially available software (Wilkins *et al.*, 1996). Thus, 2DE has become a core component in many proteomics based analyses (Jungblut *et al.*, 1996; Herbert *et al.*, 2001).

However, 2DE has a number of limitations and is notoriously poor at separating and resolving heavily glycosylated or hydrophobic proteins. Although there are reports of successful separation of membrane proteins (Mikami *et al.*, 2002) and glycoproteins (Packer and Harrison, 1998) using 2DE, it is generally accepted that complete proteomes are not accessible using this strategy for front-end protein separation. Instead, there is an increasing interest in using multi-dimensional liquid chromatography interfaced directly with mass spectrometers, as outlined in some of the reviews listed above.

10.6.3 Mass spectrometry as a proteomics tool

Technological developments in mass spectrometry (MS) over the last few years have ensured that this is now the method of choice for both protein identification and characterization of post-translational modifications (Jungblut and Thiede, 1997; Packer and Harrison, 1998; Costello, 1999; Wilkins *et al.*, 1999; Mann *et al.*, 2001). New ionization methods and mass analysers have heralded major improvements in mass accuracy, resolution, sensitivity and ease of use, and have extended the applicability of MS to characterize large intact macromolecules (Costello, 1999; Loo *et al.*, 1999; Whitelegge *et al.*, 1999). The major technological advance in this regard has been the development of desorption techniques that enable large volatile molecules, such as proteins, to be analysed without extensive fragmentation. These desorption techniques have been described as 'soft desorption' ionization due to the limited fragmentation of the parent molecules and have made mass spectrometers increasingly attractive analytical instruments for biologists. Two types of MS in particular represent powerful tools for proteomic studies (detailed in an excellent review by Mann *et al.*, 2001): matrix-assisted laser desorption/ionization time-of-flight (MALDI-TOF) MS (Liang *et al.*, 1996; Cohen and Chait, 1997; Gevaert and Vanderkerckhove, 2000) and electrospray ionization (ESI) MS, or MS/MS (Mann and Wilm, 1995; Wilm *et al.*, 1996). These may be used individually or together for high resolution/throughput characterization of complex protein mixtures (Figure 10.3).

Once a protein of interest has been separated by 2DE, or some alternative approach, it is subjected to proteolytic degradation using a site-specific protease (commonly trypsin) and the digest is subjected to MS analysis. MALDI-TOF MS is used to measure the masses of the daughter peptides to a high degree of mass accuracy, generating a 'peptide mass fingerprint' (PMF) which is virtually unique to each polypeptide due to the variability in amino acid sequences and the relative distribution of protease cleavage sites between proteins. The PMF, together with the estimated pI and molecular mass values of the parent protein, can then be used to query non-redundant sequence databases containing lists of theoretical peptides predicted from the amino acid sequences of all known protein sequences and open reading frames of ESTs. Thus, 'experimentally observed' and 'theoretical' peptide digests are directly compared. Proteins in the database query are scored and ranked based on criteria such as the number of matching peptide masses and the mass accuracy (Figure 10.3). Peptide mass fingerprinting can be automated and thus has the advantage of providing a high-throughput analytical platform.

In many cases, however, peptide mass fingerprinting gives inconclusive results, such as more than one candidate protein/gene from the database being assigned a high probability score for a match to the experimentally determined PMF. In this case, the sample can be 'de-novo' sequenced using tandem mass spectrometry (MS/MS). The first step of tandem MS involves ionization of a sample and separation based upon the mass-to-charge ratio (m/z) of the primary ions. An ion with a specific m/z value is then selected, fragmented, and the fragment ions detected after passing through the second mass spectrometer. This process produces a series of fragment ions that differ by single amino acids, allowing a portion of the peptide sequence, termed an 'amino acid sequence tag', to be determined and used for database searching. When a peptide has been identified in the database, the theoretical fragmentation pattern can be predicted and compared to the observed MS/MS spectrum for assignment of other peaks that can validate the identification. This procedure theoretically can be repeated for every fragmented peptide in the sample, leading to additional verification or identification of other proteins in the sample. MALDI instruments equipped with an ion reflector can also be used to generate peptide sequencing by post-source decay (PSD) (Spengler *et al.*, 1992). The combination of data obtained using MALDI-TOF MS and MS/MS allows protein characterization with a very high mass accuracy (Blackstock and Weir, 1999), in addition to providing partial amino acid sequence and information about post-translational modifications (Hochstrasser, 1998; Bardor *et al.*,

1999; Harvey, 2001). Furthermore, typically only femtomolar amounts of a protein are needed for protein identification (Blackstock and Weir, 1999).

This field is developing extremely rapidly and many new separation and analytical techniques are emerging, together with new instrumentation. The ability to miniaturize and compartmentalize allows smaller sample quantities to be used at a greater speed with a higher tolerance for contaminants. Future developments may include 'lab on-a-chip' platforms involving microfluidics and nanoseparation devices coupled with high-resolution mass spectrometers, such as Fourier transform mass spectrometers (FTMS). It is likely that within a few years the current approaches for proteomic analysis will have evolved dramatically to increase throughput, emphasize the characterization of post-translational modification and protein complexes and enhance the ease and speed of comparative proteomics.

10.6.4 Subcellular proteomics

Another major advantage of proteomics as a platform for functional genomic studies, and one that is of particular interest to scientists who study subcellular compartments such as the cell wall/apoplast, is that by separating subcellular protein fractions from the total complement of cellular proteins, the localization of gene products may be addressed. Such information may prove invaluable with respect to predicted function for both uncharacterized 'new' proteins and proteins that are present in unexpected locations, as has been reported in animals (Fialka et al., 1997; Scianimanico et al., 1997). Such studies also reveal the dynamic nature of the protein populations in specific subcellular compartments in response to developmental and environmental signals (Masson and Rossignol, 1995). Moreover, in addition to providing important information about location, the fractionation of specific subcellular compartments or organelles results in the enrichment of proteins that are relatively rare in total protein extracts, thus increasing the proportion of the proteome that can be characterized. Thus, the separation of subcellular protein fractions from the total complement of cellular proteins, and the association of a protein species with a particular organelle, provides an important layer of information that cannot be readily obtained by studying DNA sequence.

To date, several examples of subcellular proteomic analyses in plants have been described. These have included studies of organelles such as chloroplasts and their constituent membranes (Peltier et al., 2000, 2002; Vener et al., 2001), mitochondria (Kruft et al., 2001; Millar et al., 2001; Bardel et al., 2002; Werhahn and Braun, 2002), endoplasmic reticulum (Maltman et al., 2002), peroxisomes (Fukao et al., 2002), and other subcellular fractions, including membrane transport proteins (Barbier-Brygoo et al., 2001), GPI-anchored cell surface proteins (Sherrier et al., 1999; Borner et al., 2002), the cytoskeleton (Davies et al., 2001), and the plasma membrane (Santoni et al., 1998, 2000; Prime et al., 2000). Such analyses have underscored the value of pursuing a subcellular approach; for example only a few years ago it was reported that approximately 80% of the plasma membrane-specific proteins correspond to gene products with no known function (Santoni et al., 1998). Efforts to systematically study the cell wall subproteome have only recently been reported, as described in more detail in the next section.

10.6.5 Plant cell walls as targets for proteomic studies

As outlined above, the analysis of subcellular proteomes by pre-fractionation of organelles (Righetti *et al.*, 2001; Kaiser *et al.*, 2002), or isolation of protein populations from discrete subcellular compartments prior to protein separation has the dual benefits of providing insight into the localization of the proteins and also enhancing the detection of low abundance proteins that might otherwise be undetectable in total protein extracts (Mann *et al.*, 2001). This strategy may be applied to study the dynamics of protein populations in the cell wall compartment and apoplast milieu, and also processes underlying wall biosynthesis and metabolism.

10.6.5.1 Cell wall synthesis

As described in detail in Chapter 6, cell wall polysaccharides are synthesized both at the plasma membrane (e.g. cellulose and callose) and in the Golgi (e.g. matrix glycans). Some proteoglycan synthesis or polysaccharide priming may also occur in the ER. Three basic approaches have been used to study wall polysaccharide biosynthetic enzymes and to identify the corresponding genes. The first is to purify the biosynthetic enzymes in order to obtain amino acid sequence information for use in identifying cDNA clones or genes. This approach has recently led to the successful identification of two cDNAs encoding cell wall biosynthetic glycosyltransferases (Edwards *et al.*, 1999; Perrin *et al.*, 1999). A second approach has been to screen mutagenized plant populations for wall biosynthesis mutants, resulting in the identification of genes involved in the synthesis of the nucleotide-sugar substrates (Reiter *et al.*, 1997; Reiter and Vanzin, 2001; Reiter, 2002), and genes with homology to putative cellulose synthases (Turner and Somerville, 1997; Arioli *et al.*, 1998; Delmer, 1999). A third approach is to use apparent conserved motifs for glycosyltransferases to screen sequence databases (Richmond and Somerville, 2001; Bonetta *et al.*, 2002; Reiter, 2002).

Despite recent progress in this area (see Chapter 6), much remains to learnt about the enzymes and corresponding genes responsible for cell wall biosynthesis and, in particular, for the formation of the primary wall matrix polysaccharides. Thus, a proteomic study of the proteins in the Golgi, ER and plasma membranes would provide a repertoire of candidate cell wall biosynthetic enzymes.

10.6.5.2 The cell wall/apoplast: a dynamic subcellular compartment

Relatively few studies to date have focused on the apoplast and cell wall subproteome. Given the profound influence of the wall on most aspects of plant growth and development, the proteomic analysis of cell walls holds enormous potential for both identifying new wall proteins and for characterising the expression and regulation of known proteins.

When defined as a subcellular compartment that includes the apoplastic space, the many classes of cell wall proteins may be divided into two basic categories: (a) structural proteins (reviewed in Chapter 4) that are typically immobilized within the wall (Showalter, 1993) and which comprise approximately 5–10% of the wall dry weight (Cassab and Varner, 1988), or (b) soluble apoplastic proteins, including many enzymes, that are more readily extracted from walls. However, the distinction between 'wall proteins' and 'apoplastic proteins' is somewhat artificial since the interactions *in muro* between many extracellular proteins and the wall are likely to be highly complex, transient and certainly disrupted during experimental extraction. For example, arabinogalactan proteins (AGPs) are often described as structural proteins (Showalter, 1993); however, they are typically highly soluble and appear to associate only loosely with the wall (Nothnagel, 1997). In contrast, α-expansins are a class of proteins with no proposed structural role and which have several characteristics of enzymes, and yet are relatively insoluble and somewhat resistant to extraction (Cosgrove, 1998). For the purpose of this chapter, the terms 'extracellular', 'apoplastic' and 'wall-associated' proteins are used synonymously, implying localization in the apoplast, and therefore a physical proximity to the wall, rather than a *de facto* direct or biologically significant interaction.

Over the last few decades, an appreciation has developed for the complexity and multi-functional nature of cell walls. The primary wall is now regarded as a complex, highly dynamic structure that exhibits substantial spatial and temporal variation in architecture and composition. This is achieved through the action of enzymes that coordinate wall synthesis, deposition, reorganization and selective disassembly. The tightly regulated expression of specific wall proteins has been observed in association with a range of developmental events and in response to external stimuli, including:

1. Cell Expansion: As outlined in Chapter 8, the biochemical mechanism of cell expansion likely reflects the activities of a complex battery of enzymes including polysaccharide hydrolases, transglycosylases and other proteins acting, perhaps synergistically (Fry, 1995; Cosgrove, 1999; Rose and Bennett, 1999). Additional enzymes are believed to play a role in cross-linking polysaccharides at the cessation of expansion to rigidify the wall (Cassab and Varner, 1988). Cell expansion is regulated by environmental stimuli such as light, gravity, anoxia, water stress and hormones (Hoson, 1998) and numerous reports describe correlations between these regulatory factors and changes in the structure of primary wall polysaccharides and expression of wall-modifying enzymes/genes/activities.

2. Cell Differentiation/Morphogenesis: Directional cell elongation and the consequent asymmetric growth that leads to organogenesis are likely to require both regulated wall synthesis and loosening (see Chapter 8) and several reports have associated wall modifying proteins/genes with organ formation (Fleming *et al.*, 1997; Cho and Kende, 1998; Reinhardt *et al.*, 1998; Yung *et al.*, 1999). Reorganization and modification of primary wall

architecture is also a prominent feature of terminal cellular differentiation, such as is seen in fruit ripening and organ abscission (see Chapter 9). A range of wall-modifying enzymes have also been associated with these processes and, interestingly, many of the proteins associated with expansion-related wall loosening also appear to be ripening-related (Chapter 9 and Rose and Bennett, 1999).

3. Defence: A number of extracellular proteins are believed to play a crucial role in plant defence (Dietz, 1996; Sakurai, 1998), including many 'pathogenesis-related' (PR) proteins (Stinzi et al., 1993), and several studies have reported a change in the population of wall proteins in response to wounding (Li and McLure, 1990), insect infestation (van der Westhuizen and Pretorius, 1996) and fungal infection (Olivieri et al., 1998; Hiilovaara-Teijo et al., 1999).

4. Molecular Transport: The apoplast is an important metabolic compartment for the transport and delivery of ions, assimilates and other metabolites (Leigh and Tomos, 1993), and wall-localized enzymes involved in the generation and translocation of assimilates have been described (Brown et al., 1997; N'tchobo et al., 1999).

5. Responses to Environmental Stresses: Quantitative or qualitative changes in cell wall-associated protein populations have been detected in response to environmental variables, including exposure to ozone or sulfur dioxide (Pfanz et al., 1990), heavy metals (Blinda et al., 1997), osmotic stress and water deficit (Marshall et al., 1999), and cold stress (Marentes et al., 1993). These examples represent only a subset of the developmental and environmental factors that are known to influence the protein composition of the apoplast. Any perturbation of plant stasis appears likely to result in secretion or turnover of cell wall/extracellular proteins (Blinda et al. 1997; Hoson, 1998).

These factors hint at the complexity and dynamic nature of the cell wall/apoplast subproteome.

Compared with some other plant subcellular proteomes, such as the chloroplast and plasma membrane, the cell wall has received relatively little attention. In part this probably reflects the technical challenges of isolating cell wall protein fractions that have not been contaminated by cytosolic proteins and proteins from organelles within the protoplast. However, several approaches have been described to isolate cell wall protein populations.

10.6.6 Proteomic analysis of secreted proteins

One approach to obtaining proteins that are localized in the apoplast/cell wall that minimizes contamination with proteins from within the protoplast involves the use of suspension-cultured plant cells. Robertson et al. (1997) used this system in a pioneering study of secreted proteins from five plant species that represented the first attempt to systematically separate and identify large numbers of cell wall-related proteins. Protein populations were isolated from the suspension cell culture medium and from cell wall fractions of intact cells that were sequentially washed with a series of solvents that left the plasma membrane intact. The proteins were separated by 1-D SDS-PAGE and more than 200 protein bands subjected to N-terminal amino acid sequencing, which yielded sequence information for approximately two thirds. Numerous families of known cell wall-localized proteins were identified although, interestingly, a large proportion of the proteins could not be assigned to a protein class based on sequence homology. In many cases, this may reflect the scarcity of sequence information for some of the species that were studied (e.g. carrot). Approximately 30% of the identified proteins from *Arabidopsis* could not be classified and, since publication of this paper, the *Arabidopsis* genome sequence has become available. A re-evaluation of the published N-terminal sequences with the substantially expanded sequence databases now allows a considerably greater proportion of the proteins to be assigned a putative function, although a significant number of proteins still cannot be classified. (S.J. Lee and J. Rose, unpublished data).

A similar strategy has been used to study secondary cell wall synthesis using tobacco cells expressing high cytokinin levels (Blee *et al.*, 2001), which consequently have highly thickened cell walls and exhibit many of the expected characteristics of cells that are actively synthesizing secondary walls. The complement of proteins that were extracted from the cell walls of this tobacco line appeared substantially different from that seen in the equivalent study of tobacco primary wall proteins reported by Robertson *et al.* (1997). While many novel proteins were identified, other sequences indicated the presence of proteins that are related to secondary wall formation, such as peroxidase and polyphenol oxidase/laccase, a lysine-rich protein and extensin.

Cell cultures have also been used to study cell wall construction and reorganization in yeast by identifying the proteins that are secreted into the culture medium of *Saccharomyces cerevisiae* protoplasts that are actively regenerating a cell wall (Pardo *et al.*, 2000). The authors reported the identification of several known proteins involved in wall construction.

10.6.7 Isolation of cell wall-bound proteins

Many reports describe the isolation of a 'cell wall protein fraction' by tissue homogenization in a buffer containing a low salt concentration, followed by consecutive washing of the cell wall pellet with a low salt buffer to remove cytosolic protein contaminants, and then solutions containing high concentrations of salt (e.g. 1.5 M NaCl) to release proteins that are ionically bound to the wall. However, since the polygalacturonate component of cell wall pectin can essentially act as a polyanionic matrix, positively charged proteins from within the protoplast have the potential to bind to the wall once the plasma membrane has been ruptured. In some cases a cytosolic protein can associate so strongly with the wall that a high salt buffer does not disrupt the interaction and a detergent such as SDS is subsequently required to re-solubilize the protein from the wall-enriched pellet (R.S. Saravanan and J. Rose, unpublished data). Therefore, considerable caution should be used when classifying proteins that are isolated using this disruptive technique as 'cell wall proteins', and ideally other approaches should be used to verify their subcellular localization.

This type of approach was used in a recent report describing a proteomic analysis of the cell walls extracted from *Arabidopsis* suspension cells (Chivasa *et al.*, 2002), where a cell wall fraction was removed from disrupted cells and sequentially extracted with calcium chloride and urea. The authors used 2-DE followed by MALDI-TOF MS analysis to identify 69 different proteins, which included numerous known wall proteins with well-established biochemical functions, a number of unclassified proteins and several polypeptides whose location in the wall is unexpected. These latter two classes of proteins demonstrate the potential value of this approach in identifying new cell wall/apoplastic proteins and wall-localized biochemical pathways. Such experiments will provide a platform for subsequent functional studies. The authors also acknowledged the possibility of contamination of the cell wall protein fraction and the need for additional confirmatory analyses using techniques such as immunolocalization.

While suspension cells provide a convenient source of homogenous plant material that can be rapidly regenerated, and from which cell walls and wall proteins may easily be obtained, they represent an artificial biological system. The complement of wall proteins in complex plant tissues is likely to be significantly different, given the associated cellular heterogeneity, and exhibit substantial spatial and temporal variability. Different tissues and cell types will have distinct subsets of wall proteins with diverse functions, and this variability will emerge through careful proteomic analysis of subtypes of plant material. Similarly, little is currently known about the dynamic aspects of wall protein populations during cell growth and differentiation, although some preliminary results have recently started to emerge. For example, an examination of cell wall-associated proteins from the developing xylem of compression and non-compression wood of Sitka spruce (McDougall, 2000) resulted in the identification of several differentially expressed proteins, including oxidases that may contribute to secondary wall formation. Preliminary studies of cell wall-associated proteins from different stages of ripening tomato fruit also suggest that comparative proteomics will be a valuable tool to help elucidate complex processes that involve the coordinated action of multiple enzymes, such as cell wall disassembly during fruit softening (R.S. Saravanan, S. Bashir, and J. Rose, unpublished data; see also Chapter 9).

An alternative experimental approach to isolate cell wall/apoplastic proteins from complex tissues, and one that can be adapted to minimize contamination with cytosolic proteins, is to use pressure-rehydration and vacuum infiltration protocols to extract the apoplastic fluid from the target sample. This approach has been used extensively to extract extracellular proteins from several tissue types, including roots, leaves, stems, fruit and tubers (Ruan *et al.*, 1995, 1996; Blinda *et al.*, 1997; Olivieri *et al.*, 1998; Hiilovaara-Teijo *et al.*, 1999; Yu *et al.*, 1999). In addition, a range of solutions can be infiltrated into the tissues to release different subsets of proteins, such as buffers containing high salt concentrations to release proteins that are ionically bound to the wall. The disadvantages of this technique are that the protein yield is typically low and great care has to be taken to avoid cell lysis, so throughput is generally slow, and many wall-localized proteins cannot be recovered without rupturing the plasma membrane.

To conclude, cell wall proteomics is a rapidly emerging field that should provide unprecedented qualitative and quantitative information regarding the complexity of protein populations that contribute to cell wall synthesis and that are resident in the plant wall.

10.7 Glycomics

In conjunction with the emerging '-omics' fields that are described above, 'glycomics' has been coined as a phrase to describe the analysis of the glycan complement of an organism (Hirabayashi and Kasai, 2000, 2002). The 'glycome', as proposed by Hirabayashi *et al.* (2001) is envisaged as the entire spectrum of glycoconjugates, including glycoproteins, glycoplipids and proteoglycans. The concept of the glycome has developed around non-plant systems and so the spectrum of plant cell wall polysaccharides and oligosaccharides has not yet been proposed to fall within the scope of glycomics. However, it seems only a matter of time before extensive initiatives are instigated to characterize 'plant glycomes', and that these will encompass not only the glycans that are targeted in animals, but also the plant cell wall glycoproteins and proteoglycans, and the complex carbohydrates that comprise plant wall oligo- and polysaccharides.

10.8 New and emerging technologies to detect and screen for changes in cell wall polymers

A considerable number of genes in the *Arabidopsis* genome are predicted to be involved in the backbone synthesis and the decoration and developmental restructuring of cell wall polymers other than cellulose. For example, at least eight enzymes are probably involved in the synthesis of fucogalactoxyloglucans, and the synthesis of rhamnogalacturonan II probably requires at least 21 glycosyltransferases. To date, a very limited number of these genes have been identified with biochemical or classical genetics approaches (Bonin *et al.*, 1997; Edwards *et al.*, 1999; Perrin *et al.*, 1999; Vanzin *et al.*, 2002) and the detection of cell wall changes in loss-of-function mutants corresponding to suspect genes remains a major challenge. Besides reasons such as redundant gene functions, the lethality or scarcity of some mutations, and overly subtle visible/biochemical phenotypes, a likely explanation for this bottleneck between genetic change and defined cell wall change is the lack of suitable detection methods. Traditional cell wall analysis, involving sequential solvent extraction and glycosidic linkage determination, is time consuming and requires the homogenization of complex multicellular plant tissues (e.g. roots, stems or fruit), thereby masking tissue- or cell type specific phenotypes. Hence, established and new technological platforms and tools, such as Fourier transform infrared (FTIR) microspectroscopy (see Chapter 2), enzymatic oligosaccharide fingerprinting methods (Lerouxel *et al.*, 2002) and antibodies to detect specific polysaccharide epitopes (see Chapter 3), are needed that are able to quickly identify an altered cell wall at the cellular level or at least in small tissue samples.

FTIR microspectroscopy is a method that can rapidly and quantitatively probe cell wall components and cross-links from tissue samples as small as $10 \times 10 \ \mu m^2$, by identifying polymers and functional groups *'in muro'* without derivatization of the sample (McCann *et al.*, 1992; Séné *et al.*, 1994; McCann *et al.*, 1997). FTIR spectroscopy, in combination with data-compression methods, such as linear discriminant analysis and principal component analysis (Chen *et al.*, 1998), is a suitable tool to screen large numbers of plants from different species for a broad range of cell wall related phenotypes. This includes mutant populations, ecotypes and plants grown under different experimental conditions. Similarly, studies of changes in cell wall composition and structure during plant cell division, growth and differentiation are all possible with FTIR. An NSF-funded project is currently using discriminant analysis of FTIR spectra as a high-throughput genetic end product screen to identify *Arabidopsis* and maize T-DNA and transposon tagged insertion mutants in wall biogenesis-related genes, for which no prior function was known (http:// plantgenome.sdsc.edu/AwardeesMeeting/poster_Carpita.pdf). The team also uses the reverse genetics approach to examine the effects of DNA insertions into genes that are suspected to be involved in some aspect of cell wall biosynthesis. This latter approach provides a broad matrix of cell wall mutants and conditions that serve the establishment of FTIR spectral libraries, which should be useful to diagnose FTIR-selected mutants of unknown genetic background, or to confirm the molecular mode of action of cytokinesis- and cell-wall inhibiting herbicides.

A similar approach, using FTIR spectroscopy in combination with hierarchical clustering analysis, is being taken to group genotypes based on the similarities between spectra (http://www-biocel.versailles.inra.fr/herman/fig1herman.html). The aim is to obtain information about the chemical changes in unknown samples from the clustering with samples with known changes, requiring the establishment of a reference collection of a large number of spectra obtained in highly standardized conditions (e.g. from dark-grown *Arabidopsis* hypocotyls of the same age). Validation of the method was obtained by showing that alleles for the same locus (e.g. *prc*1, *kob*1 or *mur*1) were clustered in terms of the spectra they generated. Moreover, by clustering with known mutants and wild-types treated with specific inhibitors, novel mutants with defects in cellulose or pectin synthesis, as well as alterations in the cortical cytoskeleton have been identified.

Enzymatic oligosaccharide fingerprinting is a more recent, rapid and efficient approach to study plant cell wall composition and, *nota bene*, polymer structure (Lerouxel *et al.*, 2002). The basis of the approach is to treat cell wall material, obtained from less than 50 mg plant fresh weight, with wall degrading enzymes that possess very specific cleavage properties, and to subject the enzymatically generated wall oligosaccharides to analytical procedures including high-performance anion-exchange-pulsed-amperometric detection liquid chromatography (HPAE-PAD), fluorophore-assisted carbohydrate electrophoresis (FACE; see also Goubet *et al.*, 2002) and MALDI-TOF mass spectrometry (MS). The presence/absence of specific fragment peaks (in case of MALDI-TOF MS or HPAE-PAD) or FACE-bands or a strong change in peak area or band intensity of the test sample (e.g. mutant, treated, tissue A), compared to the reference spectrum (e.g. wild-type, untreated, tissue B), are indicators for altered sugar-linkages and composition. For example, degradation of cell wall material with a xyloglucan-specific endoglucanase from *Aspergillus* (Pauly *et al.*, 1999), which cleaves the xyloglucan backbone after non-substituted glucose residues, and analysis of the fragments by MALDI-TOF MS has been shown to unambiguously and reproducibly identify the *Arabidopsis* cell wall mutants *mur1*, *mur2* and *mur4* (see Figure 10.4; Lerouxel *et al.*, 2002). It is likely that the MALDI-TOF MS approach will soon be extended to other classes of wall polysaccharides, such as pectins and arabinoxylans, and will provide a throughput high enough to be useful to characterize and screen for mutants that have very subtle alterations in their cell wall polymer structure.

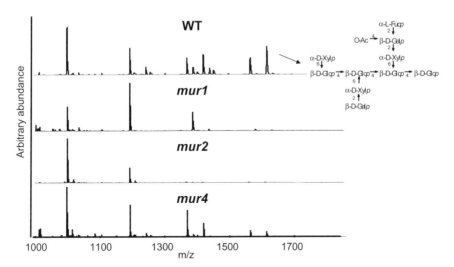

Figure 10.4 MALDI-TOF mass spectra of xyloglucan oligosaccharides released from wild-type (WT) and leaf cell wall material from the *Arabidopsis mur* mutants. The *mur1* mutant is affected in the *de novo* synthesis of GDP-fucose (Zablackis *et al.*, 1996), the *mur2* mutant in the transfer of the fucosyl moiety to xyloglucan (Vanzin *et al.*, 2002), and *mur4* in the epimerization of UDP-xylose to UDP-arabinose (Burget *et al.*, 1999). The structures of the oligosaccharides corresponding to the major ion signals are known (Pauly *et al.*, 2001). One of the structures corresponding to *m/z* of 1597 Da is exemplified in the upper right corner. Figure kindly provided by Dr Markus Pauly (Max Planck Institute of Molecular Plant Physiology, Golm, Germany).

10.9 Outlook

The 'post-genome era' is an expression that is being used with increasing frequency and one that can be interpreted in many different ways, and that has a variety of implications depending on context. In this chapter, an overview is provided of some of the new approaches that are currently being employed, and that are still emerging, for characterizing plant cell wall structures and functions at the genome scale. Some of these techniques are broadly applicable to many areas of biology, while others are more specifically useful for cell wall studies, such as FTIR screening of cell wall mutants. These approaches have a common theme in that they take a high-throughput genome-scale approach to elucidating a biological system but, as with any new paradigm, these technologies are merely additional tools for understanding complex systems and certainly do not represent an 'end point', or an approach that will replace existing disciplines, such as biochemistry, physiology and cell biology, that are described in some of the other chapters. On the contrary, the post-genome world has the exciting potential to unify many such fields in a multidisciplinary platform, thereby allowing the intricacy of the cell wall to be tackled simultaneously by researchers with expertise in diverse areas. A major challenge for the future will be to accommodate, amalgamate and interpret the vast number of diverse datasets that are being generated at an ever-increasing rate, and in the face of new emerging technologies it will become ever more important that cell wall studies be viewed through the lens of integrated scientific disciplines.

Acknowledgements

We thank Dr Markus Pauly for providing Figure 10.4.

References

Adams, M.D. (1996) Serial analysis of gene expression: ESTs get smaller. *Bioessays*, 18, 261–262.

Allona, I., Quinn, M., Shoop, E. *et al.* (1998) Analysis of xylem formation in pine by cDNA sequencing. *Proc. Natl. Acad. Sci. USA*, 95, 9693–9698.

Andersen, J.S. and Mann, M. (2000) Functional genomics by mass spectrometry. *FEBS Lett.*, 480, 25–31.

Anderson, L. and Seilhammer, J. (1997) A comparison of selected mRNA and protein abundances in human liver. *Electrophoresis*, 18, 533–537.

Anderson, N.L. and Anderson, N.G. (1998) Proteome and proteomics: new technologies, new concepts, and new words. *Electrophoresis*, 19, 1853–1861.

The *Arabidopsis* Genome Initiative (2000) Analysis of the genome sequence of the flowering plant *Arabidopsis thaliana*. *Nature*, 408, 796–815.

Arioli, T., Peng, L., Betzner, A.S. *et al.* (1998) Molecular analysis of cellulose synthesis in *Arabidopsis*. *Science*, 279, 717–720.

Aubourg, S. and Rouzé P. (2001) Genome annotation. *Plant Physiol. Biochem.*, 39, 181–193.

Azpiroz-Leehan, R. and Feldmann, K.A. (1997) T-DNA insertion mutagenesis in *Arabidopsis*: going back and forth. *Trends Genet.*, 13, 152–156.

Bachem, C.W.B., Oomen, R.J. and Visser R.G.F. (1996) Transcript Imaging with cDNA-AFLP: a step-by-step protocol. *Plant Mol. Biol. Rep.*, 16, 157–173.

Bancroft, I. and Dean, C. (1993) Transposition pattern of the maize element Ds in *Arabidopsis thaliana. Genetics*, 134, 1221–1229.

Barbier-Brygoo, H., Gaymard, F., Rolland, N. and Joyard, J. (2001) Strategies to identify transport systems in plants. *Trends Plant Sci.*, 10, 420–425.

Bardel, J., Louwagie, M., Jaquinod, M. *et al.* (2002) A survey of the plant mitochondrial proteome in relation to development. *Proteomics*, 2, 880–896.

Bardor, M., Loutelier-Bourhis, C., Marvin, L. *et al.* (1999) Analysis of plant glycoproteins by matrix-assisted laser desorption ionisation mass spectrometry: application to the N-glycosylation analysis of bean phytohemagglutinin. *Plant Phys. Biochem.*, 37, 319–325.

Baskin, T., Betzner, A., Hoggart, R., Cork, A. and Williamson, R. (1992) Root morphology mutants in *Arabidopsis thaliana. Aust. J. Plant Physiol.*, 19, 427–437.

Baulcombe, D.C. (1999) Fast forward genetics based on virus-induced gene silencing. *Curr. Opin. Plant Biol.*, 2, 109–113.

Baulcombe, D. (2002) RNA silencing. *Current Biol.*, 12, R82-R84.

Bechtold, N., Ellis, J. and Pelletier, G. (1993) *In planta Agrobacterium*-mediated gene transfer by infiltration of adult *Arabidopsis thaliana* plants. *C. R. Acad. Sci., Paris*, 316, 1194–1199.

Bent, A.F. (2000) *Arabidopsis in planta* transformation: uses, mechanisms, and prospects for transformation of other species. *Plant Physiol.*, 124, 1540–1547.

Berggren, K., Chernokalskaya, E., Stenberg, T.H. *et al.* (2000) Background-free high sensitivity staining of proteins in one- and two-dimensional sodium dodecyl sulfate-polyacrylamide gels using a luminescent ruthenium complex. *Electrophoresis*, 21, 2509–2521.

Blackstock, W.P. and Weir, M.P. (1999) Proteomics: quantitative and physical mapping of cellular proteins. *Trends Biotechnol.*, 17, 121–127.

Blee, K.A., Wheatley, E.R., Bonham, V.A. *et al.* (2001) Proteomic analysis reveals a novel set of cell wall proteins in a transformed tobacco cell culture that synthesizes secondary walls as determined by biochemical and morphological parameters. *Planta*, 212, 404–415.

Blinda, A., Koch, B., Ramanjulu, S. and Dietz, K.-J. (1997) *De novo* synthesis and accumulation of apoplastic proteins in leaves of heavy metal-exposed barley seedlings. *Plant Cell Environ.*, 20, 969–981.

Bonin, C., Potter, I., Vanzin, G., and Reiter, W.D. (1997) The *MUR1* gene of *Arabidopsis thaliana* encodes an isoform of GDP-D-mannose-4,6-dehydratase, catalysing the first step in the de novo synthesis of GDP-L-fucose. *Proc. Natl. Acad. Sci. USA*, 94, 2085–2090.

Bonnetta, D.T., Facette, M., Raab, T.K. and Somerville, C.R. (2002) Genetic dissection of plant cell-wall biosynthesis. *Biochem. Soc. Trans.*, 30, 298–301.

Borner, G.H.H., Sherrier, D.J., Stevens, T.J., Arkin, I.T. and Dupree, P. (2002) Prediction of glycosylphosphatidylinositol-anchored proteins in Arabidopsis. A genomic analysis. *Plant Physiol.*, 129, 486–499.

Bouché, N. and Bouchez, D. (2001) *Arabidopsis* gene knockout: phenotypes wanted. *Curr. Opin. Plant Biol.*, 4, 111–117.

Bouchez, D. and Höfte, H. (1998) Functional genomics in plants. *Plant Physiol.*, 118, 725–732.

Brenner, S.E. (1999) Errors in genome annotation. *Trends Genet.*, 15, 132–133.

Brown, P.O. and Botstein, D. (1999) Exploring the new world of the genome with DNA microarrays. *Nature Genet.*, 21, 33–37.

Brown, M.M., Hall, J.L. and Ho, L.C. (1997) Sugar uptake by protoplasts isolated from tomato fruit tissues during various stages of fruit growth. *Physiol. Plant.*, 101, 533–539.

Budziszewski, G.J., Lewis, S.P., Glover, L.W. *et al.* (2001) *Arabidopsis* genes essential for seedling viability: isolation of insertional mutants and molecular cloning. *Genetics*, 159, 1765–1778.

van den Bulk, R.W., Löffler, H.J.M., Lindhaut, W.H. and Koornneef, M. (1990) Somaclonal varia-
    tion in tomato: effect of explant source and a comparison with chemical mutagenesis. *Theor.
    Appl. Genet.*, 80, 817–825.
Bulyk, M.L., Gentalen, E., Lockhart, D.J. and Church, G.M. (1999) Quantifying DNA-protein
    interactions by double-stranded DNA arrays. *Nature Biotechnol.*, 17, 573–578.
Burget, E.G. and Reiter, W.-D. (1999) The mur4 mutant of arabidopsis is partially defective in the
    de novo synthesis of uridine diphospho-L- arabinose. *Plant Physiol.*, 121, 383–389.
Burn, J.E., Hocart, C.H., Birch, R.J., Cork, A.C. and Williamson, R.E. (2002) Functional analysis
    of the cellulose synthase genes *CesA1, CesA2*, and *CesA3* from *Arabidopsis. Plant Physiol.*,
    129, 797–807.
Burton, R.A., Gibeaut, D.M., Bacic, A. *et al.* (2000) Virus-induced silencing of a plant cellulose
    synthase gene. *Plant Cell*, 12, 691–705.
Campisi, L., Yang, Y., Yi, Y. *et al.* (1999) Generation of enhancer trap lines in *Arabidopsis* and
    characterization of expression patterns in the inflorescence. *Plant J.,* 17, 699–707.
Caño-Delgado, A., Penfield, S., Smith, C., Catley, M. and Bevan, M. (2003) Reduced cellulose
    synthesis invokes lignification and defense responses in *Arabidopsis thaliana. Plant J.*, 34,
    351–362
Carpita, N.C. (1996) Structure and biogenesis of the cell walls of grasses. *Annu. Rev. Plant Physiol.
    Plant Mol. Biol.*, 47, 445–476.
Carpita, N.C. and Gibeaut, D.M. (1993) Structural models of the primary cell walls in flowering
    plants: consistency of molecular structure with the physical properties of the walls during
    growth. *Plant J.*, 3, 1–30.
Carpita, N.C. and McCann, M. (2000) The cell wall, in *Biochemistry and Molecular Biology of
    Plants* (eds B.B. Buchanan, W. Gruissem and R.J. Jones), American Society of Plant Biolo-
    gists, Rockville, MA, pp. 52–109.
Carpita, N., Tierney, M. and Campbell, M. (2001) Molecular biology of the plant cell wall: search-
    ing for the genes that define structure, architecture and dynamics. *Plant Mol. Biol.,* 47, 1–5.
Cassab, G.I. and Varner, J.E. (1988) Cell wall proteins. *Annu. Rev. Plant Physiol. Plant Mol. Biol.,*
    39, 321–353.
Chaffey, N., Cholewa, E., Regan, S. and Sundberg, B. (2002) Secondary xylem development in
    Arabidopsis: a model for wood formation. *Physiol. Plant.,* 114, 594–600.
Chandler, V.L. and Hardeman, K.J. (1992) The Mu elements of *Zea mays. J. Adv. Genet.*, 30,
    77–122.
Chapple, C., Vogt, T., Ellis, B. and Somerville, C. (1992) An *Arabidopsis* mutant defective in the
    general phenyl propanoid pathway. *Plant Cell*, 4, 1413–1424.
Chen, L., Carpita, N.C., Reiter, W.-D., Wilson, R.H., Jeffries, C. and McCann, M. (1998) A rapid
    method to screen for cell-wall mutants using discriminant analysis of Fourier transform
    infrared spectra. *Plant J.*, 16, 385–392.
Chivasa S., Ndimba, B.K., Simon, W.J. *et al.* (2002) Proteomic analysis of the *Arabidopsis thali-
    ana* cell wall. *Electrophoresis*, 23, 1754–1765.
Chuang, C-F., and Meyerowitz, E.M. (2000) Specific and heritable genetic interference by double-
    stranded RNA in *Arabidopsis thaliana. Proc. Natl. Acad. Sci. USA,* 97, 4985–4990.
Clough, S.J. and Bent, A. (1998) Floral-dip: a simplified method for *Agrobacterium*-mediated
    transformation of *Arabidopsis thaliana. Plant J.*, 16, 735–743.
Cohen, S.L. and Chait, B.T. (1997) Mass spectrometry of whole proteins eluted from sodium do-
    decyl sulfate-polyacrylamide gel electrophoresis gels. *Anal. Biochem.* 247, 257–267.
Colbert, T.G., Till, B.J., Tompa, R. *et al.* (2001) High-throughput screening for induced point
    mutations. *Plant Physiol.*, 126, 480–484.
Cosgrove, D.J. (1998) Cell wall loosening by expansins. *Plant Physiol.,* 118, 333–339.
Cosgrove, D.J. (1999) Enzymes and other agents that enhance cell wall extensibility. *Annu. Rev.
    Plant Mol. Biol.*, 50, 391–417.

Costello, C.E. (1999) Bioanalytical applications of mass spectrometry. *Curr. Opin. Biotechnol.*, 10, 22–28.

Cotton, R.G.H., Edkins, E. and Forrest, S. (eds) (1998) *Mutation Detection: A Practical Approach,* IRL Press, Oxford.

Cutler, S. and Somerville, C.R. (1997) Cellulose synthesis: cloning in silico. *Curr. Biol.,* 7, R108-R111.

Das, L. and Martienssen, R. (1995) Site-selected transposon mutagenesis at the hcf106 locus in maize. *Plant Cell*, 7, 287–294.

Davies, E., Stankovic, B., Azama, K., Shibata, K. and Abe, S. (2001) Novel components of the plant cytoskeleton: a beginning to plant 'cytomics'. *Plant Sci.*, 160, 185–196.

Delmer, D.P. (1999) Cellulose biosynthesis: exciting times for a difficult field of study. *Annu. Rev. Plant Physiol. Plant Mol. Biol.*, 50, 245–276.

Depicker, A. and van Montagu, M. (1997) Post-transcriptional gene silencing in plants. *Curr. Opin. Cell Biol.*, 9, 373–382.

Desnos, T., Orbovic, V., Bellini, C. *et al.* (1996) *Procuste*1 mutants identify two distinct genetic pathways controlling hypocotyl cell elongation, respectively, in dark- and light-grown *Arabidopsis* seedlings. *Development*, 122, 683–693.

Desprez, T., Amselem, J., Caboche, M. and Höfte, H. (1998). Differential gene expression in *Arabidopsis* monitored using cDNA arrays. *Plant J.*, 14, 643–652.

Dietz, K.J. (1996) Functions and responses of the leaf apoplast under stress. *Prog. Bot.,* 58, 221–254.

Dongre, A.R., Eng, J.K. and Yates, J.R. III (1997) Emerging tandem-mass-spectrometry techniques for the rapid identification of proteins. *Trends Biotechnol.*, 15, 418–425.

Ebskamp, M., Busink, H., Pogodina, N. *et al.* (2001) Monitoring of genes involved in cell wall composition in flax and hemp using cDNA microarrays. In *Abstract book of the 9th International Cell Wall Meeting*, p. 137.

Edwards, M.E., Dickson, C.A., Chengappa, S., Sidebottom, C., Gidley, M.J. and Reid, J.S.G. (1999) Molecular characterisation of a membrane-bound galactosyltransferase of plant cell wall matrix polysaccharide biosynthesis. *Plant J.*, 19, 691–697.

Eisen, M.B. and Brown, P.O. (1999) DNA arrays for analysis of gene expression, *Methods in Enzymology, vol. 303* (ed. S. Weissman). Academic Press, San Diego, pp. 179–205.

Ellis, C., Karafyllidis, I., Wasternack, C. and Turner, J.G. (2002) The *Arabidopsis* mutant *cev*1 links cell wall signaling to jasmonate and ethylene responses. *Plant Cell*, 14, 1557–1566.

Ewing, R.M., Kahla, A.B., Poirot, O., Lopez, F., Audic, S. and Claverie, J.M. (1999) Large-scale statistical analyses of rice ESTs reveal correlated patterns of gene expression. *Gen. Res.*, 9, 950–959.

Fagard, M., Desnos, T., Desprez, T. *et al.* (2000a) *PROCUSTE1* encodes a cellulose synthase required for normal cell elongation specifically in roots and dark-grown hypocotyls of *Arabidopsis*. *Plant Cell*, 12, 2409–2423.

Fagard, M., Höfte, H. and Vernhettes, S. (2000b) Cell wall mutants. *Plant Physiol. Biochem.*, 38, 15–25.

Faik, A., Price, N.J., Raikhel, N.V. and Keegstra, K. (2000) An *Arabidopsis* gene encoding a xylosyltransferase involved in xyloglucan biosynthesis. *Proc. Natl. Acad. Sci. USA,* 99, 7797–7802.

Favery, B., Ryan, E., Foreman, J. *et al.* (2001) *KOJAK* encodes a cellulose synthase-like protein required for root hair cell morphogenesis in *Arabidopsis. Genes Dev.*, 15, 79–89.

Fialka, I., Pasquali, C., Lottspeich, F., Ahorn, H. and Huber, L.A. (1997) Subcellular fractionation of polarized epithelial cells and identification of organelle-specific proteins by two-dimensional gel electrophoresis. *Electrophoresis*, 198, 2582–2590.

Fleming, A.J., McQueen-Mason, S., Mandel, T. and Kuhlemeier, C. (1997) Induction of leaf primordia by the cell-wall protein expansin. *Science*, 276, 1415–1418.

Fry, S.C. (1995) Polysaccharide-modifying enzymes in the plant cell wall. *Annu. Rev. Plant Physiol. Plant Mol. Biol.*, 46, 497–520.

Fukao, Y., Hayashi, M. and Nishimura, M. (2002) Proteomic analysis of leaf peroxisomal protein in greening cotyledons of *Arabidopsis thaliana*. *Plant Cell Physiol.*, 43, 689–696.

Gevaert, K. and Vanderkerckhove, K. (2000) Protein identification methods in proteomics. *Electrophoresis*, 21, 1145–1154.

Gheysen, G., Herman, L., Breyne, P., Gielen, J., van Montagu, M. and Depicker, A. (1990) Cloning and sequence analysis of truncated T-DNA insertions from *Nicotiana tabacum*. *Gene*, 94, 155–163.

Gilliland, L.U., McKinney, E.C., Asmussen, M.A. and Meagher, R.B. (1998) Detection of deleterious genotypes in multigenerational studies. I. Disruptions in individual *Arabidopsis* actin genes. *Genet.*, 149, 717–725.

Gillmor, C.S., Poindexter, P., Lorieau, J., Palcic, M.M. and Somerville, C. (2002) Alpha-glucosidase I is required for cellulose biosynthesis and morphogenesis in *Arabidopsis. J. Cell Biol.*, 156, 1003–1013.

Goff, S.A., Ricke, D. *et al.* (2002) A draft sequence of the rice genome (*Oryza sativa* L. ssp. Japonica). *Science*, 296, 92–100.

Görg, A., Obermaier, C., Boguth, G. *et al.* (2000) The current state of two-dimensional electrophoresis with immobilized pH gradients. *Electrophoresis*, 21, 1037–1053.

Goubet, F., Jackson, P., Deery, M.J. and Dupree, P. (2002) Polysaccharide analysis using carbohydrate gel electrophoresis: a method to study plant cell wall polysaccharides and polysaccharide hydrolases. *Anal. Biochem.*, 300, 53–68.

Guo, Y.M., Shen, S.H., Jing, Y.X. and Kuang, T.Y. (2002) Plant proteomics in the post-genomic era. *Act. Bot. Sin.*, 44, 631–641.

Gygi, S.P., Rist, B., Gerber, S.A., Turecek, F., Gelb, M.H. and Aebersold, R. (1999) Quantitative analysis of complex protein mixtures using isotope-coded affinity tags. *Nat. Biotechnol.*, 17, 994–999.

Hacia, J.G. (1999) Resequencing and mutational analysis using oligonucleotide microarrays. *Nature Genet.*, 21, 42–47.

Harmer, S.L. and Kay, S.A. (2000) Microarrays: determining the balance of cellular transcription. *Plant Cell*, 12, 613–615.

Harvey, D.J. (2001) Identification of protein-bound carbohydrates by mass spectrometry. *Proteomics*, 1, 311–328.

Hazen, S.P. and Walton, J.D. (2002) Identification of QTL associated with cereal cell wall hemicellulose content, in *Abstract book of the Plant Cell Wall Biosynthesis Meeting*, Lake Arrowhead, May 2002, p. 21.

Hazen, S.P., Scott-Craig, J.S. and Walker, J.D. (2002) Cellulose synthase-like genes of rice. *Plant Physiol.*, 128, 336–340.

Helliwell, C., Wesley, V., Wielopolska, A., Wu, R.-M., Bagnall, D. and Waterhouse, P. (2002) High throughput gene silencing in *Arabidopsis*. In *Abstract book of the 13th International Conference on Arabidopsis Research*, pp. 1–17.

Henrissat, B., Coutinho, P.M. and Davies, G.J. (2001) A census of carbohydrate-active enzymes in the *Arabidopsis* genome. *Plant Mol. Biol.*, 47, 55–72.

Herbert, B.R., Harry, J.L., Packer, N.H., Gooley, A.A., Pedersen, S.K. and Williams, K.L. (2001) What place for polyacrylamide in proteomics? *Trends Biotechnol.*, 19(S), S3–S9.

Hertzberg, M., Aspeborg, H., Schrader, J. *et al.* (2001) A transcriptional roadmap to wood formation. *Proc. Natl. Acad. Sci. USA*, 98, 14732–14737.

Hiilovaara-Teijo, M., Hannukkala, A., Griffith, M., Yu, X.-M. and Pihakaski-Maunsbach, K. (1999) Snow-mold-induced apoplastic proteins in winter rye leaves lack antifreeze activity. *Plant Physiol.*, 121, 665–673.

Hirabayashi, J. and Kasai, K. (2000) Glycomics, coming of age! *Trends Glycosci. Glycotechnol.*, 12, 1–5.

Hirabayashi, J. and Kasai, K. (2002) Separation technologies for glycomics. *J. Chromatogr. B*, 771, 67–87.

Hirabayashi, J., Arata, Y. and Kasai, K. (2001) Glycome project: concept, strategy and preliminary application to *Caenorhabditis elegans*. *Proteomics*, 1, 295–303.

Hirsch, R.E., Lewis, B.D., Spalding, E.P. and Sussman, M.R. (1998) A role for the AKT1 potassium channel in plant nutrition. *Science*, 280, 918–921.

Hochstrasser, D.F. (1998) Proteome in perspective. *Clin. Chem. Lab. Med.*, 36, 825–836.

Holland, M.J. (2002) Transcript abundance in yeast varies over six orders of magnitude. *J. Biol. Chem.*, 277, 14363–14366.

Hong, Z., Delauney, A.J. and Verma, D.P.S. (2001) A cell plate-specific callose synthase and its interaction with phragmoplastin. *Plant Cell*, 13, 755–768.

Hoson, T. (1998) Apoplast as the site of response to environmental signals. *J. Plant Res.*, 111, 167–177.

Ideker, T., Thorsson, V., Ranish, J.A. *et al.* (2001) Integrated genomic and proteomic analyses of a systematically perturbed metabolic network. *Science*, 292, 929–934.

Ito, T., Motohashi, R., Kuromori, T. *et al.* (2002) A new resource of locally transposed dissociation elements for screening gene-knockout lines in silico on the *Arabidopsis* genome. *Plant Physiol.*, 129, 1544–1556.

Jander, G., Norris, S.R., Rounsley, S.D., Bush, D.F., Levin, I.M. and Last, R.L. (2002) *Arabidopsis* map-based cloning in the post-genome era. *Plant Physiol.*, 129, 440–450.

Jeon, J.S., Lee, S., Jung, K.H. *et al.* (2000) Technical Advance: T-DNA insertional mutagenesis for functional genomics in rice. *Plant J.*, 22, 561–570.

Jones, J.D.G., Carland, F.C., Lim, E., Ralston, E. and Dooner, H.K. (1990) Preferential transposition of the maize element Activator to linked chromosomal locations in tobacco. *Plant Cell*, 2, 701–707.

Jungblut, P. and Thiede, B. (1997) Protein identification from 2-DE gels by MALDI mass spectrometry. *Mass Spectrom. Rev.*, 16, 145–162.

Jungblut, P., Thiede, B., ZimnyArndt, U. *et al.* (1996) Resolution power of two-dimensional electrophoresis and identification of proteins from gels. *Electrophoresis*, 17, 839–847.

Kaiser, C.A., Chen, E.S. and Losko, S. (2002) Subcellular fractionation of secretory organelles. *Method Enzymol.*, 351(C), 325–388.

Karimi, M., Inze, D. and Depicker, A. (2002) GATEWAY™ vectors for *Agrobacterium*-mediated plant transformation. *Trends Plant Sci.*, 7, 193–195.

Kawagoe, Y. and Delmer, D.P. (1997) Pathways and genes involved in cellulose biosynthesis. *Genet. Eng.*, 19, 63–87.

Kehoe, D.M., Villand, P. and Somerville, S. (1999) Technical focus: DNA microarrays for studies of higher plants and other photosynthetic organisms. *Trends Plant Sci.*, 4, 38–41.

Kempin, S.A., Liljegren, S.J., Block, L.M., Rounsley, S.D. and Yanofsky, M.F. (1997) Targeted disruption in *Arabidopsis*. *Nature*, 389, 802–803.

Kersten, B., Burkle, Kuhn, E.J. *et al.* (2002) Large scale plant proteomics. *Plant Mol. Biol.*, 48, 133–141.

Kjemtrup, S., Sampson, K.S., Peele, C.G. *et al.* (1998) Gene silencing from plant DNA carried by a geminivirus. *Plant J.*, 14, 91–100.

Komari, T., Hiei, Y., Ishida, Y., Kumashiro, T. and Kubo, T. (1998) Advances in cereal gene transfer. *Curr. Opin. Plant Biol.*, 1, 161–165.

Koncz, C., Nemeth, K., Redei, G. and Schell, J. (1992) T-DNA insertional mutagenesis in *Arabidopsis*. *Plant Mol.Biol.*, 20, 963–976.

Kruft, V., Eubel, H., Jänsch, L., Werhahn, W. and Braun, H.-P. (2001) Proteomic approach to identify novel mitochondrial proteins in Arabidopsis. *Plant Physiol.*, 127, 1694–1710.

Krysan, P.J., Young, J.C. and Sussman, M.R. (1999) T-DNA as an insertional mutagen in *Arabidopsis*. *Plant Cell*, 11, 2283–2290.

Krysan, P.J., Young, J.C., Tax, F. and Sussman, M.R. (1996) Identification of transferred DNA insertions within *Arabidopsis* genes involved in signal transduction and ion transport. *Proc. Natl. Acad. Sci. USA*, 93, 8145–8150.

Lee, K.J.D., Knight, C.D. and Knox, J.P. (2001) *Physcomitrella*, a model system for the study of plant cell walls, in *Abstract book of the 9th International Cell Wall Meeting*, p. 240.

Leigh, R.A. and Tomos, A.D. (1993) Ion distribution in cereal leaves: pathways and mechanisms. *Phil. Trans. Royal Soc. Lond. B.*, 341, 75–86.

Lemieux, B., Aharoni, A. and Schena, M. (1998) DNA chip technology. *Mol. Breeding*, 4, 277–289.

Lerouxel, O., Choo, T.-S., Seveno, M. *et al.* (2002) Rapid structural phenotyping of plant cell wall mutants by enzymatic oligosaccharide fingerprinting. *Plant Physiol.*, 130, 1754–1763.

Leung, J., Merlot, S. and Giraudat, J. (1997) The *Arabidopsis ABSCISIC ACID-INSENSITIVE2* (*ABI2*) and *ABI*1 genes encode homologous protein phosphatases 2C involved in abscisic acid signal transduction. *Plant Cell*, 9, 759–771.

Li, Z.-C. and McClure, J.W. (1990) Soluble and bound apoplastic proteins and isozymes of peroxidase, esterase and malate dehydrogenase in oat primary leaves. *J. Plant Physiol.*, 136, 398–403.

Li, X., Song, Y., Century, K. *et al.* (2001) A fast neutron deletion mutagenesis-based reverse genetics system for plants. *Plant J.*, 27, 235–242.

Li, Y., Darley, C.P., Ongaro, V. *et al.* (2002) Plant expansins are a complex multigene family with an ancient evolutionary origin. *Plant Physiol.*, 128, 854–864.

Liang, X., Bai, J., Liu, Y.H. and Lubman, D.M. (1996) Characterisation of SDS-PAGE separated proteins by matrix-assisted desorption/ionization peptide mass spectrometry. *Anal. Chem.*, 68, 1012–1018.

Liljegren, S.J., Ditta, G.S., Eshed, Y., Savidge, B., Bowman, J.L. and Yanofsky, M.F. (2000) SHATTERPROOF MADS-box genes control seed dispersal in *Arabidopsis. Nature*, 404, 766–770.

Lipshutz, R.J., Fodor, S.P.A., Gingeras, T.R. and Lockhart, D.J. (1999) High-density synthetic oligonucleotide arrays. *Nature Genet.*, 21, 20–24.

Liu, D. and Crawford, N.M. (1998) Characterization of the germinal and somatic activity of the *Arabidopsis* transposable element *Tag*1. *Genetics*, 148, 445–456.

Liu, Y.G., Mitsukawa N., Oosumi T. and Whittier, R.F. (1995) Efficient isolation and mapping of *Arabidopsis thaliana* T-DNA insert junctions by thermal asymmetric interlaced PCR. *Plant J.*, 8, 457–463.

Lockhart, D.J. and Winzeler, E.A. (2000) Genomics, gene expression and DNA arrays. *Nature*, 405, 827–836.

Lockhart, D.J., Dong, H., Byrne, M.C. *et al.* (1996) Expression monitoring by hybridization to high-density oligonucleotide arrays. *Nature Biotechnol.*, 14, 1675–1680.

Loo, J.A., DeJohn, D.E., Du, P., Stevenson, T.I. and Loo, R.R.O. (1999) Application of mass spectrometry for target identification and characterization. *Med. Res. Rev.,* 19, 307–319.

Lotan, T., Ohto, M.-A., Yee, K.M. *et al.* (1998) *Arabidopsis LEAFY COTYLEDON*1 is sufficient to induce embryo development in vegetative cells. *Cell*, 93, 1195–1205.

Loudet, O., Chaillou, S., Camilleri, C., Bouchez, D. and Daniel-Vedele, F. (2002) Bay-0 × Shahdara recombinant inbred lines population: a powerful tool for the genetic dissection of complex traits in *Arabidopsis. Theor. Applied Genet.*, 104, 1173–1184.

Ludwig, S.R., Bowen, B., Beach, L. and Wessler, S.R. (1990) A regulatory gene as a novel visible marker for maize transformation. *Science*, 247, 449–450.

Lukowitz, W., Mayer, U. and Jürgens, G. (1996) Cytokinesis in the *Arabidopsis* embryo involves the syntaxin-related KNOLLE gene product. *Cell*, 84, 61–71.

Lukowitz, W., Gillmor, C.S. and Scheible, W.-R. (2000) Positional cloning in *Arabidopsis*: why it feels good to have a genome initiative working for you. *Plant Physiol.*, 123, 795–805.

Lukowitz, W., Nickle, T.C., Meinke, D.W., Last, R.L., Conklin, P.L. and Somerville, C.R. (2001) *Arabidopsis cyt1* mutants are deficient in a mannose-1-phosphate guanylyltransferase and point to a requirement of N-linked glycosylation for cellulose biosynthesis. *Proc. Natl. Acad. Sci. USA,* 98, 2262–2267.

Maltman, D.J., Simon, W.J., Wheeler, C.H., Dunn, M.J., Wait, R. and Slabas, A.R. (2002) Proteomic analysis of the endoplasmic reticulum from developing and germinating seeds of castor (*Ricinus communis*). *Electrophoresis,* 23, 626–639.

Mandal, A., Lang, V., Orczyk, W. and Palva, E.T. (1993) Improved efficiency for T-DNA-mediated transformation and plasmid rescue in *Arabidopsis thaliana. Theor. Appl. Genet.,* 86, 621–628.

Mann, M., Hendrickson, R.C. and Pandey, A. (2001) Analysis of proteins and proteomes by mass spectrometry. *Annu. Rev. Biochem.,* 70, 437–473.

Mann, M. and Wilm, M. (1995) Electrospray mass spectrometry for protein characterisation. *Trends Biochem. Sci.,* 20, 291–224.

Marentes, E., Griffith, M., Mlynarz, A. and Brush, R.A. (1993) Proteins accumulate in the apoplast of winter rye leaves during cold acclimation. *Physiol. Plant.,* 87, 499–507.

Marsch-Martinez, N., Greco, R., van Arkel, G., Herrera-Estrella, L. and Pereira, A. (2002) Activation tagging using the En-1 maize transposon system in *Arabidopsis. Plant Physiol.,* 129, 1544–1556.

Marshall, J.G., Dumbroff, E.B., Thatcher, B.J., Martin, B., Rutledge, R.G. and Blumwald, E. (1999) Synthesis and oxidative insolubilization of cell wall proteins during osmotic stress. *Planta,* 208, 401–408.

Martienssen, R.A. (1998) Functional genomics: probing plant gene function and expression with transposons. *Proc. Natl. Acad. Sci. USA,* 95, 2021–2026.

Masson, F. and Rossignol, M. (1995) Basic plasticity of protein expression in tobacco leaf plasma membrane. *Plant J.,* 8, 77–85.

McCallum, C.M., Comai, L., Greene, E.A. and Henikoff, S. (2000) Targeting Induced Local Lesions IN Genomes (TILLING) for Plant Functional Genomics. *Plant Physiol.,* 123, 439–442.

McCann, M.C., Hammouri, M.K., Wilson, R.H., Belton, P.S. and Roberts, K. (1992) Fourier transform infrared microspectroscopy is a new way to look at plant cell walls. *Plant Physiol.,* 100, 1940–1947.

McCann, M.C., Chen, L., Roberts, K. *et al.* (1997) Infrared microspectroscopy: sampling heterogeneity in plant cell composition and architecture. *Physiol. Plant.,* 100, 729–738.

McDougall, G.J. (2000) A comparison of proteins from the developing xylem of compression and non-compression wood of branches of Sitka spruce (*Picea sitchensis*) reveals a differentially expressed laccase. *J. Exp Bot.,* 5, 1395–1401.

Meinke, D.W., Cherry, J.M., Dean, C., Rounsley, S.D. and Koornneef, M. (1998) *Arabidopsis thaliana*: a model plant for genome analysis. *Science,* 282, 662–682.

Mengiste, T. and Paszkowski, J. (1999) Prospects for the precise engineering of plant genomes by homologous recombination. *Biol. Chem.,* 380, 749–758.

Meyer, P. and Saedler, H. (1996) Homology-dependent gene silencing in plants. *Annu. Rev. Plant Physiol. Plant Mol. Biol.,* 47, 23–48.

Miao, Z.H. and Lam, E. (1995) Targeted disruption of the *TGA*3 locus in *Arabidopsis thaliana. Plant J.,* 7, 359–365.

Michelmore, R.W., Paran, I. and Kesseli, R.V. (1991) Identification of markers linked to disease-resistance genes by bulked segregant analysis: a rapid method to detect markers in specific genomic regions by using segregating populations. *Proc. Natl. Acad. Sci. USA,* 88, 9828–9832.

Mikami, S., Kishimoto, T., Hori, H. and Mitsui, T. (2002) Technical improvement to 2D-PAGE of rice organelle membrane proteins. *Biosci. Biotech. Bioch.,* 66, 1170–1173.

Milioni, D., Sado, P.-E., Stacey, N.J., Domingo, C., Roberts, K. and McCann, M.C. (2001) Differential expression of cell-wall related genes during the formation of tracheary elements in the *Zinnia* mesophyll cell system. *Plant Mol. Biol.,* 47, 221–238.

Millar, A.H., Sweetlove, L.J., Giegé, O. and Leaver, C.J. (2001) Analysis of the *Arabidopsis* mitochondrial proteome. *Plant Physiol.*, 127, 1711–1727.

Nacry, P., Camilleri, C., Courtial, B., Caboche, M. and Bouchez, D. (1998) Major chromosomal rearrangements induced by T-DNA transformation in *Arabidopsis. Genetics*, 149, 641–650.

Nakagawa, T. (2002) Gateway binary vectors (pGWBs) for efficient transformation of plants, in *Abstract book of the 13th International Conference on Arabidopsis Research*, 1–39.

Nam, J., Mysore, K.S., Zheng, C., Knue, M.K., Matthysse, A.G. and Gelvin, S.B. (1999) Identification of T-DNA tagged *Arabidopsis* mutants that are resistant to transformation by *Agrobacterium. Mol. Gen. Genet.*, 261, 429–438.

Newman, T., De Bruijn, F.J., Green, P. *et al.* (1994) Genes galore: a summary of methods for accessing results from large-scale partial sequencing of anonymous *Arabidopsis* cDNA clones. *Plant Physiol.*, 106, 1241–1255.

Nickle, T.C. and Meinke, D.W. (1998) A cytokinesis-defective mutant of *Arabidopsis* (cyt1) characterized by embryonic lethality, incomplete cell walls, and excessive callose accumulation. *Plant J.*, 15, 321–332.

Nicol, F., His, I., Jauneau, A., Vernhettes, S., Canut, H. and Höfte, H. (1998) A plasma membrane-bound putative endo-1,4-ß-D-glucanase is required for normal wall assembly and cell elongation in *Arabidopsis. EMBO J.*, 17, 5563–5576.

Nothnagel, E.A. (1997) Proteoglycans and related components in plant cells. *Int. Rev. Cyt.*, 174, 195–291.

N'tchobo, H., Dali, N., Nguyen-Quoc, B., Foyer, C. and Yelle, S. (1999) Starch synthesis in tomato remains constant throughout fruit development and is dependent on sucrose supply and sucrose synthase activity. *J. Exp. Bot.*, 338, 1457–1463.

Oleykowski, C.A., Bronson Mullins, C.R., Godwin, A.K. and Yeung, A.T. (1998) Mutation detection using a novel plant endonuclease. *Nuc. Acids Res.*, 26, 4597–4602.

Olivieri, F., Godoy, A.V., Escande, A. and Casalongué, C.A. (1998) Analysis of intercellular washing fluids of potato tubers and detection of increased proteolytic activity upon fungal infection. *Physiol. Plant.*, 104, 232–238.

Packer, N.H. and Harrison, M.J. (1998) Glycobiology and proteomics: Is mass spectrometry the Holy Grail? *Electrophoresis*, 19, 1872–1882.

Pagant, S., Bichet, A., Sugimoto, K. *et al.* (2002) *KOBITO*1 encodes a novel plasma membrane protein necessary for normal synthesis of cellulose during cell expansion in *Arabidopsis. Plant Cell*, 14, 2001–2013.

Pandey, A. and Mann, M. (2000) Proteomics to study genes and genomes. *Nature*, 405, 837–846.

Pardo, M., Ward, M., Bains, S. *et al.* (2000) A proteomic approach for the study of *Saccharomyces cerevisiae* cell wall biogenesis. *Electrophoresis*, 21, 3396–3410.

Patton, W.F., Schulenberg, B. and Steinberg, T.H. (2002) Two-dimensional gel electrophoresis; better than a poke in the ICAT? *Curr. Opin. Biotech.* 13, 321–328.

Pauly, M., Andersen, L.N., Kaupinen, S. *et al.* (1999). A xyloglucan-specific endo-β-1,4-glucanase from *Aspergillus aculeatus*: expression cloning in yeast, purification, and characterization of the recombinant enzyme. *Glycobiol.*, 9, 93–100.

Pauly, M., Qin, Q., Greene, H., Albersheim, P., Darvill, A.G. and York, W.S. (2001a) Changes in the structure of xyloglucan during cell elongation. *Planta*, 212, 842–850.

Pauly, M., Eberhard, S., Albersheim, P., Darvill, A. and York, W.S. (2001b) Effects of the *mur1* mutation on xyloglucans produced by suspension-cultured *Arabidopsis thaliana* cells. *Planta*, 214, 67–74.

Pear, J.R., Kawagoe, Y., Schreckengost, W.E., Delmer, D.P. and Stalker, D.M. (1996) Higher plants contain homologs of the bacterial *celA* genes encoding the catalytic subunit of cellulose synthase. *Proc. Natl. Acad. Sci. USA*, 93, 12637–12642.

Pelaz, S., Ditta, G.S., Baumann, E., Wisman, E. and Yanofsky, M.F. (2000) B and C floral organ identity functions require SEPALLATA MADS-box genes. *Nature*, 405, 200–203.

Peltier, J.B., Friso, G., Kalume, D.E. *et al.* (2000) Proteomics of the chloroplast: systematic identification and targeting analysis of lumenal and peripheral thylakoid proteins. *Plant Cell,* 12, 319–341.

Peltier, J.B., Emanuealksson, O., Kalume, D.E. *et al.* (2002) Central functions of the lumenal and peripheral thylakoid proteome of Arabidopsis determined by experimentation and genome-wide prediction. *Plant Cell,* 14, 211–236.

Perrin, R.M., DeRocher, A.E., Bar-Peled, M. *et al.* (1999) Xyloglucan fucosyltransferase, an enzyme involved in plant cell wall biosynthesis. *Science,* 284, 1976–1979.

Pfanz, H., Dietz, K.-J., Weinerth, I. and Oppmann, B. (1990) Detoxification of sulfur dioxide by apoplastic peroxidases. In *Sulfur Nutrition and Sulfur Assimilation in Higher Plants* (eds H. Rennenberg, C.H. Brunold, L.J. De Kok and I. Stulen), SPB Academic Publishing, The Hague.

Potikha, T. and Delmer, D.P. (1995) A mutant of *Arabidopsis thaliana* displaying altered patterns of cellulose deposition. *Plant J.,* 7, 453–460.

Prime, T.A., Sherrier, D.J., Mahon, P., Packman, L.C. and Dupree, P. (2000) A proteomic analysis of organelles from *Arabidopsis thaliana. Electrophoresis,* 21, 3488–3499.

Puchta, H. and Hohn, B. (1996) From centimorgans to base pairs – homologous recombination in plants. *Trends Plant Sci.,* 1, 340–348.

Rabilloud, T. (2002) Two-dimensional gel electrophoresis in proteomics: Old, old fashioned, but it still climbs up the mountains. *Proteomics,* 2, 3–10.

Ramachandran, S. and Sundaresan, V. (2001) Transposons as tools for functional genomics. *Plant Physiol. Biochem.,* 39, 243–252.

Reinhardt, D., Wittwer, F., Mandel, T. and Kuhlemeier, C. (1998) Localized upregulation of a new expansin gene predicts the site of leaf formation in the tomato meristem. *Plant Cell,* 10, 1427–1437.

Reiter, W.D. (2002) Biosynthesis and properties of the plant cell wall. *Curr. Opin. Plant Biol.,* 5, 536–542.

Reiter, W.D. and Vanzin, G.F. (2001) Molecular genetics of nucleotide sugar interconversion pathways in plants. *Plant Mol. Biol.,* 47, 95–113.

Reiter, W.D., Chapple, C.C.S. and Somerville, C. (1993) Altered growth and cell wall in a fucose deficient mutant of *Arabidopsis. Science,* 261, 1032–1035.

Reiter, W.D., Chapple, C.C.S. and Somerville, C. (1997) Mutants of *Arabidopsis thaliana* with altered cell wall polysaccharide composition. *Plant J.,* 12, 335–345.

Reynolds, P.H.S. (ed.) (1999) *Inducible Gene Expression in Plants.* CABI Publishing, New York.

Richmond, C.S., Glasner, J.D., Mau, R., Min, H. and Blattner, F.R. (1999) Genome-wide expression profiling in *Escherichia coli* K-12. *Nuc. Acids Res.,* 27, 3821–3835.

Richmond, T. and Somerville, C. (2000) The cellulose synthase superfamily. *Plant Physiol.,* 124, 495–498.

Richmond, T., and Somerville, C. (2001) Integrative approaches to determining *Csl* function. *Plant Mol. Biol.,* 47, 131–143.

Righetti, P.G., Castagna, A. and Herbert, B. (2001) Prefractionation techniques in proteome analysis: a new approach identifies more low-abundance proteins. *Anal. Chem.,* 73, 320A–326A.

Roberts, J. (2002) Proteomics and a future generation of plant molecular biologists. *Plant Mol. Biol.,* 48, 143–154.

Robertson, D., Mitchell, G.P., Gilroy, J.S., Gerrish, C., Bolwell, G.P. and Slabas, A.R. (1997) Differential extraction and protein sequencing reveals major differences in patterns of primary cell wall proteins from plants. *J. Biol. Chem.,* 1272, 15841–15848.

Rose, J.K.C. and Bennett, A.B. (1999) Cooperative disassembly of the cellulose-xyloglucan network of plant cell walls: parallels between cell expansion and fruit ripening. *Trends Plant Sci.,* 4, 176–183.

Rose, J.K.C., Braam, J., Fry, S.C. and Nishitani, K. (2002) The XTH family of enzymes involved in xyloglucan endotransglucosylation and endohydrolysis: current perspectives and a new unifying nomenclature. *Plant Cell Physiol.*, 43, 1421–1453.

Rosenthal, A. and Jones, D.S.C. (1990) Genomic walking and sequencing by oligo-cassette mediated polymerase chain reaction. *Nuc. Acids Res.*, 18, 3095–3096.

Rossignol, M. (2001) Analysis of the plant proteome. *Curr. Opin. Biotech.*, 12, 131–134.

Ruan, Y.-L., Mate, C., Patrick, J.W. and Brady, C.J. (1995) Non-destructive collection of apoplastic fluid from developing tomato fruit using a pressure dehydration procedure. *Aust. J. Plant Physiol.*, 22, 761–769.

Ruan, Y.-L., Patrick, J.W. and Brady, C.J. (1996) The composition of apoplastic fluid recovered from intact developing tomato fruit. *Aust. J. Plant Physiol.*, 23, 9–13.

Ruiz, M.T., Voinnet, O. and Baulcombe, D.C. (1998) Initiation and maintenance of virus-induced gene silencing. *Plant Cell*, 10, 937–946.

Ruuska, S.A., Girke, T., Benning, C. and Ohlrogge, J.B. (2002) Contrapuntal networks of gene expression during *Arabidopsis* seed filling. *Plant Cell*, 14, 1191–1206.

Santoni, V., Doumas, P., Rouquié, D. *et al.* (1998) Use of a proteome strategy for tagging proteins present at the plasma membrane. *Plant J.*, 16, 633–641.

Sakurai, N. (1998) Dynamic function and regulation of apoplast in the plant body. *J. Plant Res.*, 111, 133–148.

Santoni, V., Kieffer, S., Desclaux, D., Masson, F. and Rabilloud, T. (2000) Membrane proteomics: use of additive main effects with multiplicative interaction model to classify plasma membrane proteins according to their solubility and electrophoretic properties. *Electrophoresis*, 21, 3329–3344.

Sarria, R., Wagner, T.A., O'Neill, M.A. *et al.* (2001) Characterization of a family of *Arabidopsis* genes related to xyloglucan fucosyltransferase 1. *Plant Physiol.*, 127, 1595–1606.

Saxena, I.M., Brown, R.M., Fevre, M., Geremia, R.A. and Henrissat, B. (1995) Multidomain architecture of beta-glycosyl transferases: implications for mechanism of action. *J. Bacteriol.*, 177, 1419–1424.

Schaefer, D.G. (2001) Gene targeting in *Physcomitrella patens. Curr. Opin. Plant Biol.*, 4, 143–150.

Schaefer, D.G. (2002) A new moss genetics: targeted mutagenesis in *Physcomitrella patens. Annu. Rev. Plant Biol.*, 53, 477–501.

Scheible, W.-R., Eshet, R., Richmond, T., Delmer, D.P. and Somerville, C.R. (2001) Modifications of cellulose synthase confer to isoxaben and thiazolidinone herbicides in *Arabidopsis Ixr1* mutants. *Proc. Natl. Acad. Sci. USA,* 98, 10079–10084.

Schena, M., Shalon, D., Heller, R., Chai, A., Brown, P.O. and Davis, R.W. (1995) Quantitative monitoring of gene expression patterns with a complementary DNA microarray. *Science*, 270, 467–470.

Schmid, M.B. (2002) Structural proteomics: the potential of high-throughput structure determination. *Trends Microbiol.*, 10, S27–S31 Suppl. S.

Schultz, C.J., Rumsewicz, M.P., Johnson, K.L., Jones, B.J., Gaspar, Y.M. and Bacic, A. (2002) Using genomic resources to guide research directions. The arabinogalactan gene family as a test case. *Plant Physiol.*, 129, 1448–1463.

Scianimanico, S., Pasquali, C., Lavoie, J., Huber, L.A., Gorvel, J.-P. and Desjardins, M. (1997) Two-dimensional gel electrophoresis of endovacuolar organelles. *Electrophoresis*, 18, 2566–2572.

Seki, M., Narusaka, M., Yamaguchi-Shinozaki, K. *et al.* (2001) *Arabidopsis* encyclopedia using full-length cDNAs and its application. *Plant Physiol. Biochem.*, 39, 211–220.

Seki, M., Narusaka, M., Kamiya, A. *et al.* (2002) Functional annotation of a full-length *Arabidopsis* cDNA collection. *Science*, 296, 141–145.

Séné, C.F.B., McCann, M.C., Wilson, R.H. and Grinter, R. (1994) FT-Raman and FT-infrared spectroscopy: an investigation of five higher plant cell walls and their components. *Plant Physiol.*, 106, 1623–1633.

Shalev, G., Sitrit, Y., Avivi Ragolski, N., Lichtenstein, C. and Levy, A.A. (1999) Stimulation of homologous recombination in plants by expression of the bacterial resolvase RuvC. *Proc. Natl. Acad. Sci. USA*, 96, 7398–7402.

Shalon, D., Smith, S.J. and Brown, P.O. (1996) A DNA-microarray system for analyzing complex DNA samples using two-color fluorescent probe hybridization. *Genome Res.*, 6, 639–645.

Sheng, J. and Citovsky, V. (1996) *Agrobacterium*-plant cell DNA transport: Have virulence proteins, will travel. *Plant Cell*, 8, 1699–1710.

Sherrier, D.J., Primer, T.A. and Dupree, P. (1999) Glycosylphosphatidylinositol-anchored cell-surface proteins from *Arabidopsis*. *Electrophoresis*, 20, 2027–2035.

Showalter, A.M. (1993) Structure and function of plant cell wall proteins. *Plant Cell*, 5, 9–23.

Siebert, P.D., Chenchik, A., Kellogg, D.E., Lukyanov, K.A. and Lukyanov, S.A. (1995) An improved PCR method for walking in uncloned genomic DNA. *Nuc. Acids Res.*, 23, 1087–1088.

Somerville, C. and Dangl, J. (2000) Plant biology in 2010. *Science*, 290, 2077–2078.

Somerville, C. and Somerville, S. (1999) Plant functional genomics. *Science*, 285, 380–383.

Spalding, E.P., Hirsch, R.E., Lewis, D.R., Qi, Z., Sussman, M.R. and Lewis, B.D. (1999) Potassium uptake supporting plant growth in the absence of AKT1 channel activity: inhibition by ammonium and stimulation by sodium. *J. Gen. Physiol.*, 113, 909–918.

Spengler, B., Kirsch, D., Kaufman, R. and Jaeger, E. (1992) Peptide sequencing by matrix-assisted laser desorption mass spectrometry. *Rapid Commun. Mass Spectrom.*, 6, 105–108.

Speulman, E., Metz, P.L.J., van Arkel, G., Bas te Lintel Hekkert, B., Stiekema, W.J. and Pereira, A. (1999) A two-component Enhancer-Inhibitor transposon mutagenesis system for functional analysis of the *Arabidopsis* genome. *Plant Cell*, 11, 1853–1866.

Sterky, F., Regan, S., Karlsson, J. *et al.* (1998) Gene discovery in the wood-forming tissues of poplar: analysis of 5,692 expressed sequence tags. *Proc. Natl. Acad. Sci. USA*, 95, 13330–13335.

Stinzi, A., Heitz, T., Prasad, V. *et al.* (1993) Plant 'pathogenesis-related' proteins and their role in defense against pathogens. *Biochimie.*, 75, 687–706.

Sundaresan, V., Springer, P., Volpe, T., Haward, S. and Jones, J.D.G. (1995) Patterns of gene action in plant development revealed by enhancer trap and gene trap transposable elements. *Genes Dev.*, 9, 1797–1810.

Taylor, N.G., Laurie, S. and Turner, S.R. (2000) Multiple cellulose synthase catalytic subunits are required for cellulose synthesis in *Arabidopsis*. *Plant Cell*, 12, 2529–2539.

Taylor, N.G., Scheible, W.-R., Cutler, S., Somerville, C.R. and Turner S.R. (1999) The *irregular xylem*3 locus of *Arabidopsis* encodes a cellulose synthase required for secondary cell wall synthesis. *Plant Cell*, 11, 769–779.

Tissier, A.F., Marillonnet, S., Klimyuk, V. *et al.* (1999) Multiple independent defective suppressor-mutator transposon insertions in *Arabidopsis*: a tool for functional genomics. *Plant Cell*, 11, 1841–1852.

Topping, J.F. and Lindsey, K. (1995) Insertional mutagenesis and promoter trapping in plants for the isolation of genes and the study of development. *Transgen. Res.*, 4, 291–305.

Turner, S.R., and Somerville, C.R. (1997) Collapsed xylem phenotype of *Arabidopsis* identifies mutants deficient in cellulose deposition in the secondary cell wall. *Plant Cell*, 9, 689–701.

Valvekens, D., van Montagu, M. and Lijsebettens, M.V. (1988) *Agrobacterium tumefaciens*-mediated transformation of *Arabidopsis thaliana* root explants by using kanamycin selection. *Proc. Natl. Acad. Sci. USA*, 85, 5536–5540.

Vanzin, G.F., Madson, M., Carpita, N.C., Raikhel, N.V., Keegstra, K. and Reiter, W.-D. (2002) The *mur2* mutant of *Arabidopsis thaliana* lacks fucosylated xyloglucan because of a lesion in fucosyltransferase AtFUT1. *Proc. Natl. Acad. Sci. USA*, 99, 3340–3345.

Vaucheret, H. and Fagard, M. (2001). Transcriptional gene silencing in plants: targets, inducers and regulators. *Trends Genet.*, 17, 29–35.

Velculescu, V.E., Zhang, L., Vogelstein, B. and Kinzler, K.W. (1995) Serial analysis of gene expression. *Science*, 270, 484–487.

Vener, A.V., Harms, A., Sussman, M. and Vierstra, R. (2001) Mass spectrometric resolution of reversible protein phosphorylation in photosynthetic membranes of *Arabidopsis thaliana*. *J. Biol. Chem.*, 276, 6959–6966.

Walden, R., Fritze, K., Hayashi, H., Miklashevichs, E., Harling, H. and Schell, J. (1994) Activation tagging: a means of isolating genes implicated as playing a role in plant growth and development. *Plant Mol. Biol.*, 26, 1521–1528.

Wang, D.G., Fan, J.-B., Siao, C.-J. *et al.* (1998) Large-scale identification, mapping, and genotyping of single-nucleotide polymorphisms in the human genome. *Science*, 280, 1077–1082.

Wang, R., Guegler, K., LaBrie, S.T. and Crawford, N.M. (2000) Genomic analysis of a nutrient response in *Arabidopsis* reveals diverse expression patterns and novel metabolic and potential regulatory genes induced by nitrate. *Plant Cell*, 12, 1491–1509.

Wang, X., Cnops, G., Vanderhaeghen, R., De Block, S., van Montagu, M. and van Lisjebettens, M. (2001) *AtCslD3*, a cellulose synthase-like gene important for root hair growth in *Arabidopsis*. *Plant Physiol.*, 126, 575–586.

Waterhouse, P.M., Graham, M.W. and Wang, M.B. (1998) Virus resistance and gene silencing in plants can be induced by simultaneous expression of sense and antisense RNA. *Proc. Natl. Acad. Sci. USA*, 95, 13959–13964.

Weigel, D., Ahn, J.H., Blázquez, M.A. *et al.* (2000) Activation tagging in *Arabidopsis*. *Plant Physiol.*, 122, 1003–1013.

Werhahn, W. and Braun, H.-P. (2002) Biochemical dissection of the mitochondrial proteome from *Arabidopsis thaliana* by three-dimensional gel electrophoresis. *Electrophoresis*, 23, 640–646.

Wesley, V., Helliwell, C.A., Smith, N.A. *et al.* (2001) Construct design for efficient, effective and high-throughput gene silencing in plants. *Plant J.*, 27, 581–590.

van der Westhuizen, A.J. and Pretorius, Z. (1996) Protein composition of wheat apoplastic fluid and resistance to russian wheat aphid. *Aust. J. Plant Physiol.*, 23, 645–648.

Whitelegge, J.P., Le Coutre, J., Lee, J.C. *et al.* (1999) Toward the bilayer proteome, electrospray ionization-mass spectrometry of large, intact transmembrane proteins. *Proc. Natl. Acad. Sci. USA*, 96, 10695–10698.

van Wijk, K.J. (2001) Challenges and prospects of plant proteomics. *Plant Physiol.*, 126, 501–508.

Wilkins, M.R., Sanchez, J.C, Golley, A.A. *et al.* (1996) Progress with proteome projects: Why all proteins expressed by a genome should be identified and how to do it. *Genet. Eng. Rev.*, 13, 19–50.

Wilkins, M.R., Gasteiger, E., Gooley, A.A. *et al.* (1999) High-throughput mass-spectrometric discovery of protein post-translational modifications. *J. Mol. Biol.*, 289, 645–657.

Wilm, M., Neubauer, G. and Mann, M. (1996) Parent ion scans of unseparated peptide mixtures. *Anal. Chem.*, 68, 527–533.

Wodicka, L., Dong, H., Mittmann, M., Ho, M.-H. and Lockhart, D.J. (1997) Genome-wide expression monitoring in *Saccharomyces cerevisiae*. *Nature Biotechnol.*, 15, 1359–1367.

Yanagida, M. (2002) Functional proteomics; current achievements. *J. Chromatog. B*, 771, 89–106.

Yarmush, M.L. and Jayaraman, A. (2002) Advances in proteomic technologies. *Annu. Rev. Biomed. Eng.*, 4, 349–373.

Yokohama, R. and Nishitani, K. (2001) A comprehensive expression analysis of all members of a gene family encoding cell-wall enzymes allowed us to predict cis-regulatory regions involved in cell-wall construction in specific organs of *Arabidopsis*. *Plant Cell Physiol.*, 42, 1025–1033.

Yokohama, R., Imoto, K. and Nishitani, K. (2002) Comprehensive analyses of gene families involved in cell-wall construction in *Arabidopsis*, in *Abstract book of the 13th International Conference on Arabidopsis Research*, pp. 1–31.

Yu, J., Hu, S., Wang, J. *et al.* (2002) A draft sequence of the rice genome (*Oryza sativa* L. ssp. *indica*). *Science*, 296, 79–92.

Yu, Q., Tang, C., Chen, Z. and Kuo, J. (1999) Extraction of apoplastic sap from plant roots by centrifugation. *New Phytol.*, 143, 299–304.

Yung, M.-H., Schaffer, R. and Putterill, J. (1999) Identification of genes expressed during early Arabidopsis carpel development by mRNA differential display: characterization of *ATCEL2*, a novel endo-1,4-β-D-glucanase. *Plant J.*, 17, 203–208.

Zablackis, E., Huang, J., Muller, B., Darvill, A.G. and Albersheim, P. (1995) Characterization of the cell-wall polysaccharides of *Arabidopsis thaliana* leaves. *Plant Physiol.*, 107, 1129–1138.

Zablackis, E., York, W.S., Pauly, M., Hantus, S., Reiter, W.-D., Chapple, C.C.S., Albersheim, P. and Darvill, A.G. (1996) Substitution of L-fructose by L-galactose in cell walls of Arabidopsis mur1. *Science*, 272, 1808–1810.

Zhong, R. and Ye, Z.-H. (1999) *IFL1*, a gene regulating interfascicular fiber differentiation in *Arabidopsis*, encodes a homeodomain-leucine zipper protein. *Plant Cell*, 11, 2139–2152.

Zhong, R., Kays, S.J., Schroeder, B.P. and Ye, Z.-H. (2002) Mutation of a chitinase-like gene causes ectopic deposition of lignin, aberrant cell shapes, and overproduction of ethylene. *Plant Cell*, 14, 165–179.

Zupan, J.R. and Zambryski, P. (1995) Transfer of T-DNA from *Agrobacterium* to the plant cell. *Plant Physiol.*, 107, 1041–1047.

Index